# Medicinal Chemistry
## Principles and Practice

Second Edition

# Medicinal Chemistry

## Principles and Practice

## Second Edition

**Edited by**

**Frank D. King**
*GlaxoSmithKline, Harlow, UK*

ISBN 0-85404-631-3

A catalogue record for this book is available from the British Library



Published by The Royal Society of Chemistry,
Thomas Graham House, Science Park, Milton Road,
Cambridge CB4 0WF, UK

Registered Charity Number 207890

For further information see our web site at www.rsc.org

Typeset by Keytec Typesetting Ltd.
Printed in Great Britain by TJ International Ltd, Padstow, Cornwall

# Foreword

The excitement and challenge described in my foreword to the first edition in 1994 are in even starker juxtaposition today. The human genome has been unravelled and, if anything, scientific advance is even faster, but these advances have increasingly revealed the complexity of disease processes. While great strides are being made in understanding disease, improvements in treatment are proving to be incremental rather than huge quantum leaps.

This in no way should discourage us from the search for better medicines as these are sorely needed. Two of the biggest issues facing society concern healthcare.

For poor countries to develop economically they must improve their healthcare status. Public health measures such as clean water and better nutrition are first priority to be followed by immunisation programmes and better use of antibiotics to tackle infectious diseases. This can only be a start, however, and an increasing research effort (from a pitiful, low level) is now being committed to the diseases most commonly encountered in less developed countries.

In the developed economies, the demographics show a dramatically aging population. There is a huge need for better medical treatments of the chronic, degenerative diseases associated with age, otherwise we shall have neither the financial nor the human resources available to care for these large numbers of older people. Society faces a choice. Either we maintain incentives and regulatory practices which support innovation and the approval of drugs providing good quality of life until people 'die healthy' from some relatively acute episode, or we ration limited resources to the large numbers of elderly living relatively poor quality lives and who are becoming an increasing burden on society.

To scientists in our industry, the choice should be clear. It is our role to argue for and provide the innovation; so, good luck in your research.

Sir Tom McKillop
Chief Executive Officer
AstraZeneca

# Preface

Since the publication of the first edition of this book in 1994, much has changed for the medicinal chemist within the pharmaceutical industry. The advances in biotechnology now mean that nearly all human protein targets can be cloned and expressed in sufficient quantities for high throughput screening and for structural studies. With the advances in computing, these proteins can be visualised in 3-D, and rapid docking and *in silico* experiments, such as virtual screening and *de novo* design, can be performed. Increasing computational power has also been used to support the more traditional QSAR approaches with ever increasing sophistication. The advent of LC-MS-MS (SCIEX) now means that pharmacokinetics (PK) plays a key role in drug discovery within the lead optimisation (LO) process. Thus, it is now feasible to optimise compounds for programmes for which there is no reliable animal model either of disease or known pharmacological effect, simply by optimising their PK properties. The increasing accumulation of PK data has led to the development of algorithms for predicting many of the factors affecting drug absorption, distribution and elimination which give structural insights into factors affecting these properties. In addition, *in vitro* measurements of membrane permeability, cytochromes P450 inhibition and transporter affinity have been increasingly incorporated as standard assays. Within chemistry itself, combinatorial chemistry has matured as an approach and has largely evolved into the synthesis of singles in multidimensional arrays, both for lead seeking and lead optimisation. Thus, high throughput chemistry is now yet another weapon in the armament of the medicinal chemist.

With the advent and maturity of all these technologies, driving the changes, it was felt that an update of this book was required. However, the basic principles of medicinal chemistry have not really changed – the optimisation of a molecule that interacts with the desired biological target, to meet the criteria set for entry into development. Therefore, this edition has attempted to retain the basics, but to express these in the light of the new technologies. Thus, a chapter on combinatorial chemistry has been included and the chapter on molecular biology has been expanded to cover the methods and terms that the medicinal chemist is likely to meet in programme discussions. Other chapters have been updated, whilst still retaining the fundamental principles and a new chapter on toxicology has been included, which will be the next traditionally post-candidate function to be brought into the lead optimisation phase. In addition, the chapter on post-candidate development has focused on chemical development, reflecting the main interaction between medicinal chemistry and the development functions. In every case, the authors have been encouraged to illustrate their topics with real examples, and three new case histories have been included that give insights into successful projects.

Finally I should like to thank all of the people who have helped with the second edition of this book. In particular, I thank the authors of the chapters, without whom this book would not have been completed. I would particularly like to take this opportunity of specially mentioning Gary Price, who died very recently. He was an excellent scientist, a good friend and colleague, who will be sorely missed by many at GSK. I would also like to thank everyone at Beecham, then SB, now GSK who have helped me to formulate my ideas, many of which are included within this book, in particular Charlie Fake, Mike Hadley and Peter Machin. I would also like to thank Joe Martin and Sally Redshaw for assistance in writing the chapter on Saquinavir, and Jenny King and Judith Quinlan for their help with the statistics. Last, but not least, I would like to thank my wife for her support and help in proof reading the manuscript.

# Contents

**Contributors** xxi

**Abbreviations** xxiii

**General References** xxvii

**Chapter 1** **Drug–Receptor Interactions** 1
*Rob Leurs*

1 Introduction 1
2 Receptor Proteins 2
3 Mechanism of Drug Action: Agonism and Antagonism 5
4 Receptor Subtypes 6
5 Drug Selectivity 9
6 Quantification of Drug–Receptor Interactions 10
The Receptor Binding of Ligands 11
From Receptor Binding to Effect 15
The Occupation Theory of Clark 15
Partial Agonists and Intrinsic Activity 15
Partial Agonism and Drug Efficacy 16
Receptor Antagonism 18
7 From Antagonists to Inverse Agonists 21
8 Final Considerations 23
9 Selected Reading 23

**Chapter 2** **An Introduction to Ion Channels** 25
*Brian Cox*

1 Introduction 25
2 Structure, Function and Classification 26
3 Representative Examples 29
Ligand-gated Channels 29
Voltage-gated Channels 34
Inward Rectifier and Miscellaneous Channels 38
4 Conclusion 40
5 References 40

**Chapter 3** **Intracellular Targets** **42**
*Eric Hunt, Neil Pearson, Timothy M. Willson and Andrew Takle*
1 Introduction 42
2 Antibacterial Inhibitors of Protein Synthesis 42
3 Bacterial Topoisomerase Inhibitors: Fluoroquinolones 47
Mode of Action 47
Structure–Activity Relationships 47
Conclusions 52
4 Nuclear Receptors 52
Nature's Gene Switches 53
The Orphan Nuclear Receptors 55
Nuclear Receptors as Targets for Drug Discovery 57
5 Protein Kinases 57
Inhibition of Protein Kinases 59
Inhibitors of Epidermal Growth Factor Receptor Kinase (EGFR) 60
Inhibitors of p38 MAP Kinase 61
Conclusion 62
6 References 62

**Chapter 4** **Enzyme Inhibitors** **64**
*David A. Roberts and Walter H.J. Ward*

1 Introduction 64
2 Enzyme Inhibitor Categories 67
Kinetics of Substrate Utilisation 67
$IC_{50}$ Values Reflect Affinity and Assay Conditions 70
Mechanisms of Reversible Inhibition, Affinity and Selectivity 72
Tight Binding Inhibitors 77
Slow Binding and Irreversible Inhibitors 77
3 Opportunities in Drug Design 79
Exploitation of Enzyme Kinetics in Drug Discovery 79
Isothermal Titration Calorimetry 80
Structure-based Design 81
4 Classes of Enzymes and Examples of Enzyme Inhibitors 82
Synthetases: Inhibition of Thymidylate Synthase 82

Reductases: Inhibition of Hydroxymethylglutaryl-CoA Reductase (HMG-CoA Reductase) 83
Kinases: Anilinoquinazolines as Inhibitors of Protein Kinases 84
Proteases: Inhibition of Angiotensin Converting Enzyme and Renin 85
5 Concluding Remarks 89
Acknowledgements 89
6 References 90
Additional Reading 90

**Chapter 5 Biological Evaluation of Novel Compounds 91**
*Gary W. Price, Graham J. Riley and Derek N. Middlemiss*

1 Introduction 91
2 Primary Screens 93
Broad Spectrum Evaluation of Compounds 99
SB-236057 100
3 Secondary Assays 101
*In Vitro* Evaluation of Compounds 101
Binding Assays in Native Tissue *versus* Recombinant Systems 101
Evaluation of Functional Activity of Compounds 102
Agonists/Antagonists/Inverse Agonists 102
Assays for Evaluating Functional Activity 103
GTP$\gamma$S Binding 104
Reporter Gene Assays 107
Measuring Function in Native Tissues 108
[$^3$H]5HT Release from Brain Slices 109
4 Drug Metabolism and Pharmacokinetics (DMPK) 111
5 Pharmacodynamic Assays 111
Requirements of a P.D. assay 111
Examples of P.D. assays 112
SB-236057 112
6 Animal Models – Pre-clinical Proof of Concept 113
Proof of Mechanism Models 114
Disease Models 115
7 Acknowledgements 116
8 References 116

**Chapter 6** **Pharmacokinetics** **118**
*Phillip Jeffrey*

1 Introduction 118
2 The Process and Terminology of Drug Delivery 119
3 The Blood (or Plasma) Concentration–Time Curve 120
4 Bioavailability 124
5 Clearance 125
6 Volume of Distribution 129
7 Half-life 133
8 'Advanced' Pharmacokientics 136
9 Conclusions 137
10 References 137
11 Bibliography 137

**Chapter 7** **Drug Metabolism** **138**
*Stephen E. Clarke*

1 Introduction 138
2 Distribution of Drug Metabolism Enzymes 138
3 The Drug Metabolising Enzymes 139
Phase I Metabolising Enzymes 141
Cytochrome P450 141
CYP3A4 142
CYP2D6 145
CYP2C9 145
CYP1A2 147
CYP2C19 147
CYP2C8 148
Other Oxidative Enzymes 148
Flavin Monooxygenase 148
Monoamine Oxidase 149
Aldehyde Oxidase 149
Xanthine Oxidase 150
Phase II Conjugation 150
Glucuronidation 150
Sulfation 151
Acetylation 151
Glutathione 152
4 Conclusions 153
5 References and Bibliography 153

**Chapter 8 Toxicology in the Drug Discovery Process** **155**
*Susan M. Evans, Elisabeth George and C. Westmoreland*

1 Introduction 155
2 Overview of Toxicity Assessment in Drug Discovery 157
Disease Selection, Target Identification and Lead Series Identification 158
Lead Compound Optimisation/Candidate Compound Selection 159
Pre-clinical Safety Assessment and Clinical Development 159
3 *In Silico* Systems for Assessing Toxicity 159
4 Introduction to *In Vitro* Systems 161
High Throughput Toxicity Screening 166
Cytotoxicity Screens Predictive of *In Vivo* Toxicity 166
Cytotoxicity Screens Used to Assess *In Vitro* Therapeutic Indices 168
Cytotoxicity Screens Used to Validate Potential Drug Candidates in Cellular Pharmacology Screens 168
*In vitro* Screens for Specific Toxicities 169
Genetic Toxicology 169
Target Organ Toxicity 170
Hepatotoxicity 171
Nephrotoxicity 172
Haematotoxicity 172
5 *In Vivo* Toxicology in Candidate Selection 173
6 The Use of New Technologies in Safety Assessment 176
Toxicogenomics 176
Proteomics 177
Nuclear Magnetic Resonance 178
7 Summary 179
8 References 179

**Chapter 9 Chemical Development** **182**
*Paul Smith*

1 Introduction 182
2 Illustrative Examples 190

The Importance of Quality 190
3,3-Dimethylindoline. An Example of Route Discovery and Development 190
3 Future Trends 192
4 Acknowledgement 193
5 References 193
Further General Reading 194

**Chapter 10 Physicochemical Properties 195**
*Han van de Waterbeemd*

1 Introduction 195
2 Solubility 196
Dissolution and Solubility 196
Measurement of Solubility 198
Calculation of Solubility 199
3 Lipophilicity 200
Definitions and Lipophilicity Scales 200
The Information Content in log*P* 202
The Major Contributions 202
Diff ($\log P^{N-I}$) 202
$\Delta\log P$ 202
Measurement of log*D*/log*P* 202
From Shake-flask to High Throughput 202
Difficulties with Alternative Solvent Systems 202
Estimation of log*P* and log*D* 203
4 Hydrogen Bonding 203
5 Molecular Size 204
6 Ionisation Constants 205
Ionisation/Protonation State 205
Estimation of $pK_a$ 206
7 Electronic Properties 207
8 Estimation of Other Molecular Properties 208
Computational Properties 208
Ligand–Receptor Interactions 208
9 Relationships to Drug Disposition 210
Estimation of Gastrointestinal Absorption 210
Estimation of Brain Penetration 211
Estimation of Pharmacokinetic Properties 212
10 References 213

**Chapter 11 Quantitative Structure–Activity Relationships** **215**
*David J. Livingstone*

1 Introduction 215
2 Background to QSAR 217
3 Compound Selection 220
4 Describing Chemical Structure 223
5 Building QSAR Models 225
Multiple Linear Regression 226
Principle Component Methods 228
Data Display 230
Classified Data 232
3D QSAR 236
6 Artificial Intelligence 238
7 Summary 240
8 References 241

**Chapter 12 Computational Chemistry and Target Structure** **243**
*Colin Edge*

1 Introduction – The Basic Toolkit 243
Graphics Programs 243
Protein Modelling Programs 245
Programs That Combine Receptor and Ligand 246
Abstract Site Models 246
2 Structural Information for Computational Chemistry 246
X-ray Crystallography 246
Nuclear Magnetic Resonance 247
Structural Databases 247
3 The Use of Structure in Drug Design 248
Docking and Virtual Screening 248
Prediction of Binding Energies 249
*De Novo* Design 250
4 Related (Homologous) Structures 250
Protein Sequence Alignment 251
Homology Modelling 251
Membrane Proteins – Difficult Cases 252
5 The Absence of Target Structural Information 253
Quantitative Structure–Activity Relationships 253
Molecular Descriptors 254

6 When the Target Structure Is Almost Irrelevant 254
Drug *Versus* Leads 254
Drug-like Qualities 255
Blood–Brain Barrier 256
7 Conclusion 257
8 References 257

**Chapter 13 Patent Medicine 260**
*Bill Tyrrell*

1 Introduction 260
2 What Are Patents? 262
3 What Is Patentable? 265
Novelty 265
Inventive Step 266
Sufficiency 267
Utility or 'Industrial Application' 268
4 'Patentese' 269
5 Applying for Patents 271
6 Prosecution and Litigation 276
7 Trips and US Practice 278
8 The Biotech Revolution 280
9 Advance Module 283
10 Sites for Sore Eyes 285
11 Conclusion 286
12 References and Notes 287

**Chapter 14 An Introduction to Molecular Biology 291**
*Ralph Rapley and Robert J. Slater*

1 Introduction 291
2 Nucleic Acids 291
3 Proteins 293
4 The Flow of Genetic Information 294
5 DNA Replication 294
6 Transcription 296
7 RNA Processing 297
8 Protein Synthesis 297
Activation of Amino Acids 298
Translation 299
9 The Genetic Code 300
10 Post Translational Modification 302

11 The Control of Transcription and Translation 302
12 Genomics 302
13 Nucleic Acid Analysis and Recombinant DNA Technology 304
Nucleic Acid Extraction Techniques 304
Electrophoresis of Nucleic Acids 305
Restriction Mapping of DNA Fragments 306
Nucleic Acid Blotting and Hybridisation 307
Production of Gene Probes 308
DNA Gene Probe Labelling 309
The Polymerase Chain Reaction 309
Elements Involved in the PCR 310
Primer Design in the PCR 311
PCR Amplification Templates 312
Applications of the PCR 313
Recombinant DNA Technology and Gene Libraries 315
Digesting Genomic DNA Molecules 315
Ligating DNA Molecules 315
Considerations in Gene Library Preparations 316
Screening Gene Libraries 317
Screening Expression cDNA Libraries 317
Nucleotide Sequencing of DNA 319
PCR Cycle Sequencing 321
Automated Fluorescent DNA Sequencing 321
Maxam and Gilbert Sequencing 322
14 Bioinformatics and the Internet 322
15 Human Genome Mapping Project 324
16 References 325

**Chapter 15 Strategy and Tactics in Drug Discovery 327**
*Frank D. King*

1 Introduction 327
2 Target Identification and Validation 328
3 Lead Identification 330
4 Lead Optimisation 332
5 Development Candidate 334
6 Back-up/Follow-up 335
7 Optimising the Chances for Success 335

8 Decision Making in Medicinal Chemistry 338
Pharmacophore 338
Bioisosteres 341
Pharmacophoric Bioisosterism 341
Template Bioisosterism 341
Conformational Restriction 342
Improved Affinity 344
Improved Selectivity 345
Improved Chemical/Metabolic Stability 346
Pro-drugs 347
Soft Drugs 350
Data Interpretation 351
Chemistry 355
9 Patents 357
10 Conclusion 358
11 References 358

**Chapter 16 Combinatorial Chemistry: Tools for the Medicinal Chemist 359**
*Morag A.M. Easson and David C. Rees*

1 Introduction 359
2 Concepts in Combinatorial Chemistry 360
Compound Libraries and Arrays 360
Mix and Split Synthesis 361
3 Impact of CC on the Drug Discovery Process 363
4 Solid-phase Synthesis of 'Drug-like' Molecules 364
Linkers and the Solid Support 364
On-bead Monitoring 366
Encoded Libraries 366
Scope of Reactions and Structures on Solid Phase 367
Singles *Versus* Mixtures 368
5 Solution-phase Library Synthesis 368
Strategies for Solution-phase Synthesis 368
Parallel Solution-phase Libraries 369
Support-bound Reagents and Scavengers 370
Multi-component Condensations 371
6 Solid Phase *Versus* Solution Phase 372
7 Laboratory Automation and Equipment 373
Revolution at the Bench 373
Synthesis 374
Purification 376

Analysis 376
8 Biological Activity from Compound Libraries 377
Lead Generation Compound Libraries 377
Lead Optimisation Compound Libraries 378
Examples of Library Structures Demonstrating Biological Activity 379
9 Conclusions 379
10 References 381

**Chapter 17 The Identification of Selective 5-$HT_{2C}$ Receptor Antagonists: A New Approach to the Treatment of Depression and Anxiety 382**
*Steven M. Bromidge*

1 Introduction to Depression 382
2 Rationale for 5-$HT_{2C}$ Antagonists in Depression 383
3 Initial Lead: Identification of SB-200646 383
4 Conformational Restriction: Identification of SB-206553 384
5 Molecular and Receptor Modelling Studies 385
6 Bioisosteric Replacement of the *N*-Methylindole 387
7 Identification of Biarylcarbamoylindolines 391
8 Bispyridyl Ethers: Identification of SB-243213 393
9 Synthesis of SB-243213 394
10 Summary 395
11 References 395

**Chapter 18 The Identification of the HIV Protease Inhibitor Saquinavir 397**
*Frank D. King*

1 Introduction 397
2 Primary Assay 399
3 Inhibitor Design 399
Identification of the Minimum Inhibitor Sequence 399
Optimisation of the N-terminus 400
Optimisation of the Proline 401
X-ray Structures 404
4 Synthesis of Saquinavir 405
5 Clinical Data 405
6 Conclusion 406
7 References 406

**Chapter 19 Discovery of Vioxx (Rofecoxib)** **407**
*Frank D. King*

1 Introduction 407
2 Lead Molecules 409
3 Identification of Rofecoxib (MK-966) 410
Assays 410
Medicinal Chemistry 411
4 *In Vivo* Activity of Rofecoxib 412
5 Clinical Results 413
6 Conclusion 414
7 References 414

**Chapter 20 NK1 Receptor Antagonists** **415**
*Chris Swain*

1 Introduction 415
2 Medicinal Chemistry Programme 416
Reducing Calcium Channel Activity 417
Improving the Duration of Action 419
Non-CNS Penetrant Compounds 421
3 Profile of MK-869 Clinical Candidate 423
4 Clinical Results 424
Emesis 424
Pain 425
Psychiatric Indications 426
Acknowledgements 427
5 References 427

**Appendices**

1 Ranking of Key Ethical Drug Products in 2000 (US$ Sales Value) 428
2 Summary of Receptor Properties 430
3 Plot of Molar Concentration *vs.* $g\,ml^{-1}$ for Different Molecular Weights 440
4 Table of Molar Concentration *vs.* $g\,ml^{-1}$ for Different Molecular Weights 441
5 Conversion Table for $IC_{50}$ ($K_i$) to $pIC_{50}$ ($pK_i$) 441

**Subject Index** **442**

# Contributors

**S.M. Bromidge,** *GlaxoSmithKline, Via Alessandro Fleming 2, Verona, Italy*
**S.E. Clarke,** GlaxoSmithKline, The Frythe, Welwyn, Herts. AL6 9AR, UK
**B. Cox,** *Novartis Horsham Research Centre, Wimblehurst Road, Horsham, West Sussex RH12 5AB, UK*
**M.A.M. Easson,** *Organon, Motherwell, Lanark ML1 5SH, UK*
**C. Edge,** *GlaxoSmithKline New Frontiers Science Park, Third Avenue, Harlow, Essex CM19 5AW, UK*
**S.M. Evans,** *GlaxoSmithKline, Park Road, Ware, Herts. SG12 0DP, UK*
**E. George,** *GlaxoSmithKline, Park Road, Ware, Herts. SG12 0DP, UK*
**E. Hunt,** *GlaxoSmithKline, New Frontiers Science Park, Third Avenue, Harlow, Essex CM19 5AW, UK*
**P. Jeffrey,** *GlaxoSmithKline, The Frythe, Welwyn, Herts. AL6 9AR, UK*
**F.D. King,** *GlaxoSmithKline, New Frontiers Science Park, Third Avenue, Harlow, Essex CM19 5AW, UK*
**R. Leurs,** *Department of Pharmacochemistry, Vrije Universiteit, De Boelelaan 1083, 1081 HV Amsterdam, The Netherlands*
**D.J. Livingstone,** *ChemQuest, Delamere House, 1 Royal Crescent, Sandown, Isle of Wight PO36 8LZ, UK*
**D.N. Middlemiss,** *GlaxoSmithKline, New Frontiers Science Park, Third Avenue, Harlow, Essex CM19 5AW, UK*
**N. Pearson,** *GlaxoSmithKline, New Frontiers Science Park, Third Avenue, Harlow, Essex CM19 5AW, UK*
**G.W. Price,** *GlaxoSmithKline, New Frontiers Science Park, Third Avenue, Harlow, Essex CM19 5AW, UK*
**R. Rapley,** *Department of Biosciences, University of Hertfordshire, Hatfield, Herts. AL10 9AB, UK*
**D.C. Rees,** *Organon, Motherwell, Lanark ML1 5SH, UK*
**G.J. Riley,** *GlaxoSmithKline, New Frontiers Science Park, Third Avenue, Harlow, Essex CM19 5AW, UK*
**D.A. Roberts,** *AstraZeneca, Mereside, Alderley Park, Macclesfield, Cheshire SK10 4TG, UK*
**R.J. Slater,** *Department of Biosciences, University of Hertfordshire, Hatfield, Herts. AL10 9AB, UK*
**P. Smith,** *GlaxoSmithKline, New Frontiers Science Park, Third Avenue, Harlow, Essex CM19 5AW, UK*
**C. Swain,** *Merck Sharp & Dohme, Neurosciences Research Centre, Terlings Park, Harlow, Essex CM20 1QR, UK*

**A. Takle,** *GlaxoSmithKline, New Frontiers Science Park, Third Avenue, Harlow, Essex CM19 5AW, UK*

**A.W. Tyrrell,** *GlaxoSmithKline, New Horizons Court, Great West Road, Brentford TW8 9ET, UK*

**H. van de Waterbeemd,** *Pfizer Central Resesrch, Sandwich, Kent CT13 9NJ, UK*

**W.H.J. Ward,** *AstraZeneca, Mereside, Alderley Park, Macclesfield, Cheshire SK10 4TG, UK*

**C. Westmoreland,** *GlaxoSmithKline, Park Road, Ware, Herts. SG12 0DP, UK*

**T.M. Willson,** *GlaxoSmithKline, Research Triangle Park, North Carolina 27709, USA*

# Abbreviations

A – adenine
ABPI – Association of the British Pharmaceutical Industry
ACE – angiotensin converting enzyme
ACh – acetylcholine
Acyl-CoA – cholesterol acyltransferase
AD – Alzheimer's disease
ADME – absorption, distribution, metabolism and elimination
ADP – adenosine diphosphate
AE – adverse experience (event)
AI – artificial intelligence
AMP – adenosine monophosphate
AMPA – $\alpha$-amino-3-hydroxy-5-methyl-4-isoxazole propionate
ANN – artificial neural networks
API – active pharmaceutical ingredient
ATP – adenosine triphosphate
AUC – area under the curve
BBB – blood brain barrier
$B_{max}$ – maximum number of binding sites
BK – high conductance channels
C – cytosine
$C_{max}$ – maximum concentration
cAMP – adenosine 3′,5′-cyclic monophosphate
CC – combinatorial chemistry
cDNA – DNA fragments complementary to mRNA
cGMP – current good manufacturing practice
CHO – Chinese hamster ovary
CIE – chemotherapy-induced emesis
CL – clearance
$CL_i$ – intrinsic clearance
CNS – central nervous system
CoMFA – comparative molecular field analysis
COX – cyclooxygenase
CPA – carboxypeptidase A
CTX – clinical trial exemption certificate
CV – cross validation
CYP – cytochrome P450
D – dopamine
D – distribution coefficient
ddNTP – dideoxyribonucleoside triphosphate
DEREK – deductive estimation of risk from existing knowledge
DMPK – drug metabolism and pharmacokinetics
DNA – deoxyribonucleic acid
dNTP – deoxyribonucleoside triphosphate
dUMP – deoxyuridine monophosphate
EACPR – enoyl (ACP) reductase
EBI – european bioinformatics institute
$EC_{50}$ – concentration giving 50% of the maximal response

$ED_{50}$ – dose giving 50% of the maximal response
EDTA – ethylene diamine tetraacetic acid
EGFR-TK – epidermal growth factor – tyrosine kinase
ELS – evaporative light scattering
EPC – European patent convention
EPO – European Patent Office
EST – expressed sequence tag
FDA – Food and Drug Administration
FLIPR – fluorometric imaging plate reader
FMN – flavin mononucleotide
FMO – flavin monooxygenase
FRET – fluorescence resonance energy transfer
FTIM – first time in man
G – Gibbs free energy
G – guanine
GA – genetic algorithm
GABA – $\gamma$-aminobutyric acid
GC – gas chromatography
GDP – guanidine diphosphate
GI – gastrointestinal
GLP – good laboratory practice
Glu – glutamine/glutamate
GPCR – G-protein coupled receptor
GST – glutathione-S-transferase
GTP – guanidine triphosphate
H – enthalpy
H – histamine
HA – hydrogen bond acceptor
HD – hydrogen bond donor
HEK – human embryonic kidney
hERG – human ether-a-go-go related gene
HIV – human immunodeficiency virus
HMGCoA – hydroxymethyl-glutaryl-CoA
HPLC – high pressure (performance) liquid chromatography
HRP – horse radish peroxidase
HSV – herpes simplex virus
HTS – high throughput screening
5-HT – 5-hydroxytryptamine, serotonin
$IC_{50}$ – concentration giving 50% inhibition
ICV – intra-cerebro vascular
ID – intra-duodenal
IDS – information disclosure statement
IK – intermediate conductance channels
IL – interleukin
ILC – immobilised liposome chromatography
IM – intra-muscular
IND – investigational new drug application
IP – intra-peritoneal
IP – intellectual property
IPV – intra-hepatic portal vein
ITC – isothermal titration calorimetry
IV – intravenous
KA – kainic acid/kainate
KNN – $k$-nearest neighbour
$K_D$ – dissociation constant
$K_i$ – inhibition constant
$K_i'$ – apparent inhibition constant
$K_m$ – Michaelis constant
LC – liquid chromatography
$LD_{50}$ – (lethal) dose causing 50% mortality
LDL – low density lipoprotein
LO – lead optimisation
LSD – least significant difference
m – muscarinic

MAO – monoamine oxidase
MAD – maximum absorbable dose
MAPK – mitogen-activated protein kinase
MAS – minimum acceptable solubility
MCCs – multi-component condensations
mCPP – *meta*-clorophenyl-piperazine
MEC – minimum effective concentration
MED – male erectile dysfunction
MEP – molecular electrostatic potentials
MLP – molecular lipophilic potentials
MLR – multiple linear regression analysis
mM – milimolar
$\mu$M – micromolar
MO – molecular orbital
mRNA – messenger ribonucleic acid
MPTP – 1-methyl-4-phenyl-1,2,5,6-tetrahydropyridine
MTC – maximum tolerated concentration
MTD – maximum tolerated dose
MTT – 3- (4,5-dimethylthiazol-2-yl)-2,5-diphenyltetrazolium bromide
MR – molar refractivity
MS – mass spectrometry
MW – molecular weight
nACh – nicotinic acetylcholine
NADH – nicotinamide adenine dinucleotide
NAPDH – NADH phosphate
NAT – *N*-acetyl transferase
NCE – novel chemical entity
NIH – national institute of health
NK – neurokinin
NLM – non-linear mapping
nM – nanomolar
NMDA – *N*-methyl-D-aspartate
NMR – nuclear magnetic resonance
NSAIDS – non-steroidal anti-inflammatory drugs
P – partition coefficient
PAH – *para*–amino hippurate
PAPS – 3′-phosphoadenosine–5′-phosphosulfate
PCA – principle component analysis
PCR – polymerase chain reaction
PCT – patent cooperation treaty
PD – pharamcodynamic
PD – Parkinson's disease
PDB – protein data bank
PFGE – pulsed field gel electrophoresis
PG – prostaglandin
PK – pharmacokinetic
p$K_i$ – negative log of the inhibition (affinity) constant
PL – phospholipase
PLS – partial least squares
PO – *per os* (by mouth)
PPAR – peroxisomal proliferator-activated receptor
PSA – polar surface area
PTK – protein tyrosine kinase
QSAR – quantitative structure–activity relationship
R&D – research and development
RE – regulatory exclusivity
RF – radiofrequency
RFLP – restriction fragment polymorphism
RNA – ribonucleic acid
rRNA – ribosomal ribonucleic acid
RT – reverse transcriptase

RTECS – registry of toxic effects of chemical substance
S – entropy
S9 – supernatant following 9000 gav. centrifugation
SAR – structure–activity relationship
SC – subcutaneous
SD – standard deviation
SED – standard error of the difference
SEM – standard error of the mean
SK – small conductance channels
SNS – sensory neurone specific
SP – substance P
SPE – solid phase extraction
SPS – solid phase synthesis
SSRI – selective serotonin reuptake inhibitor
STK – serine–threonine kinase
T – thymine
$t_{1/2}$ – half life
$T_{max}$ – time to reach maximum concentration
TM – transmembrane helix
7-TMR – 7 transmembrane receptor
TNF – tumour necrosis factor
tRNA – transfer ribonucleic acid
TS – thymidylate synthase
TS – transition state
TRIPS – trade-related aspects of intellectual property rights
TX – thromboxane
U – uracil
UDP – uracil diphosphate
UTP – uracil triphosphate
UGP – UDP-glucoronosyl transferase
UK – United Kingdom
US – United States of America
USPTO – US Patent and Trademark office
V – volume of distribution
$V_{ss}$ – volume of distribution at steady state
WDI – world drug index

# General References

## Reference Books

'Comprehensive Medicinal Chemistry', C. Hansch, J.C. Emmett, P.D. Kennewell, C.A. Ramsden, P.G. Sammes and J.B. Taylor (eds.), 1990, Vols 1–6, Pergammon, Oxford.

'Burger's Medicinal Chemistry', M.E. Wolff (ed.), 5th Edn., 1995, John Wiley & Sons, New York.

'The Merck Index', M.J. O'Neil, A. Smith and P.E. Heckleman (eds.), 13th Edn. 2001, Merck & Co., Whitehouse Station, NJ.

'Medicines Compendium 2002', Datapharm Communications Ltd., London.

'Goodman & Gilman's: The Pharmacological Basis of Therapeutics' J.G. Hardman, L.E. Limbird and A. Goodman Gilman (eds.), 10th Edn., 2001, McGraw-Hill, New York.

'Medicinal Chemistry: An Introduction', G. Thomas, 2000, John Wiley & Sons, Chichester.

'An Introduction to Medicinal Chemistry', G.L. Patrick, 1995, Oxford University Press, Oxford.

'The Organic Chemistry of Drug Design and Action', R.B. Silverman, 1992, Academic Press, San Diego.

## Journals and Review Periodicals

Advances in Drug Research; Annual Reports in Medicinal Chemistry; Antimicrobial Agents and Chemotherapy; Bioorganic and Medicinal Chemistry; Bioorganic and Medicinal Chemistry Letters; Chemical and Pharmaceutical Bulletins; Current Medicinal Chemistry; Current Opinion in Drug Discovery & Development; Drug Design and Delivery; Drug Discovery Today; Drug News and Perspectives; Drugs of the Future; European Journal of Medicinal Chemistry; Journal of Combinatorial Chemistry; Journal of Computational Chemistry; Journal of Computer-aided Molecular Design; Journal of Medicinal Chemistry; Medicinal Chemistry Research; Medicinal Research Reviews; Pharmacochemistry Library; Progress in Drug Research; Progress in Medicinal Chemistry; QSAR; Trends in Pharmacological Sciences

CHAPTER 1

# Drug–Receptor Interactions

ROB LEURS

## 1 INTRODUCTION

In our current society medicines play an extremely important role in the management of health care problems and are considered to become even more important in the coming years. With our knowledge of the entire human genome the elucidation of the complete human proteome (the complete set of expressed proteins) is foreseen in the next decade. Why are these developments so exciting for the area of medicinal chemistry? To grasp the importance of these developments one needs to understand that proteins are often the molecular targets of drugs. As such, the elucidation of the human proteome is expected to lead to the discovery of many new drug targets and to bring new therapies for important human diseases.

What kind of proteins should we think of if we want to develop new drugs? It is not easy to give clear answers to this questions, as often we need first to define the underlying patho-physiological mechanism of our disease of interest. Depending on the proteins involved in these processes one might select a few of them as potential target. So, if it is not easy to tell what targets will be the 'golden' choice for new diseases, what are the targets of our currently used medicines? What can we learn from those drugs that have proven to be successful in the past? In general, it is considered that drugs act *via* the interaction with four kind of regulatory proteins:

- *Enzymes* (*e.g.* Aspirin inhibits the enzyme cyclooxygenase).
- *Carriers* (*e.g.* Prozac inhibits serotonin transporter).
- *Ion channels* (*e.g.* local anaesthetics inhibit $Na^+$ channels).
- *Receptor proteins* (*e.g.* the anti-allergic Zyrtec® blocks the histamine $H_1$ receptor).

The concept of proteins as drug targets is not novel and was already suggested at the end of the 19th and the beginning of the 20th centuries. Ehrlich and Langley both contributed to the idea that compounds displayed biological activity by binding to cellular constituents (Ehrlich: '*corpora non agunt, nisi fixata*', which tells us that 'agents do not work, unless bound') that were soon named '*receptors*' (Langley: '*receptive substances*'). One could consider that every protein that acts as the molecular target for a certain drug should be called a receptor. However, this term has been reserved mainly for those proteins that play an important role in the intercellular communication *via* chemical messengers. As such, enzymes, ion channels and carriers are usually not classified as receptors. The term receptors is mostly reserved for those protein structures that serve as (intra)cellular antennas for chemical messengers. Upon recognition of the appropriate chemical signal (also referred to as ligands) the receptor proteins transmit the signal into a biochemical change in the target cell *via* a wide variety of possible pathways.

For many years receptors, the cellular antennas, remained hypothetical structures, until in the 1970s the development of radioactive ligands led to the visualization and quantification of binding sites for drugs in tissues or isolated cells. Nowadays, detailed structural information (X-ray, NMR) of a variety of receptor proteins is known and this has led to the development of detailed insights in the molecular processes involved in drug–receptor interactions. Combined with the pivotal role of receptors in physiology, these proteins have become a favourite class of proteins for the development of drugs.

## 2 RECEPTOR PROTEINS

Our current knowledge of drug action is, for a large part, due to our insights into (patho)-physiological processes. It is now clear that many chemical mediators, ranging from small molecules to large proteins, play essential roles in the intercellular communication involved in *e.g.* hormonal regulation or neurotransmission *via* the interaction with highly specific receptor proteins. To grasp the importance of this bimolecular interaction one should consider that even important functions like vision, smell and taste rely on the interaction of physical (photons) or chemical messengers with highly specific receptor proteins in the eye, nose or tongue. Similarly, various hormones and neurotransmitters all have their specific receptor proteins in a variety of target tissues.

Do all these 'receptors' have a common structure? In light of the very diverse chemical structures of the so-called 'first-messengers' (see Figure 1 for a small selection) and their remarkable specificity of action one would expect a wide diversity in receptor protein structure. However, the overall structure of receptor proteins is often not so divergent, suggesting that signal transmission *via* receptor proteins is governed by a limited number of basic mechanisms that are utilised in an extremely efficient way. One distinguishes four super-families of receptor proteins, which cover most of the relevant receptor proteins. These four receptor families are:

- *Ligand-gated ion channels*, which are membrane-bound receptors, directly linked to an ion channel. Examples include the nicotine acetylcholine receptor and the $GABA_A$ receptor.
- *G-protein coupled receptors,* which are membrane-bound receptors coupled to G-proteins. After activation of the G-proteins a variety of biochemical signal transduction pathways can be activated. Many chemical messengers, like hormones and various neurotransmitters, act through G-protein coupled receptors.
- *Tyrosine kinase-linked receptors,* which are membrane bound receptors and contain an intrinsic enzymatic function (tyrosine kinase activity) in their intracellular domain. Upon activation *e.g.* by ligands like insulin, the receptor is activated and is able to phosphorylate tyrosine residues of other intracellular proteins. Protein phosphorylation is one of the underlying mechanisms of the regulation of protein function.

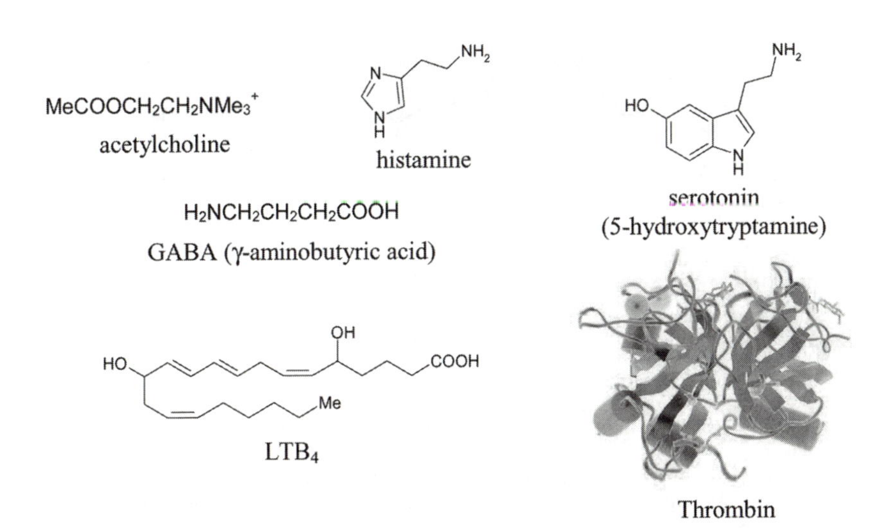


**Figure 1** *Structures of selected ligands for G-protein coupled receptors*

- *Intracellular receptors regulating gene transcription,* which are located in the cytosol. Upon binding of the appropriate chemical signal, *e.g.* steroid hormones, the activated receptors translocate to the nucleus and initiate gene transcription.

Although the individual members of each receptor family show considerable variation in amino acid composition, each receptor family shows a typical molecular architecture (Figure 2). The *ligand-gated ion channels* are an assembly of 4–5 subunits, each predicted to span the plasma membrane several times. The actual channel for specific ions is formed by the interface of the various subunits. Ligand binding resides at the N-terminus of specific subunits and binding of the endogenous ligand results in the opening of the channels. The *G-protein-coupled receptors* are single polypeptide chains of 300–600 amino acids, which are predicted to span the plasma membrane seven times. The seven $\alpha$-helical transmembrane domains are thought to form a hydrophilic pocket within the membrane. Most small ligands acting as drugs at these targets are thought to bind within this pocket. However, endogenous activators can, depending

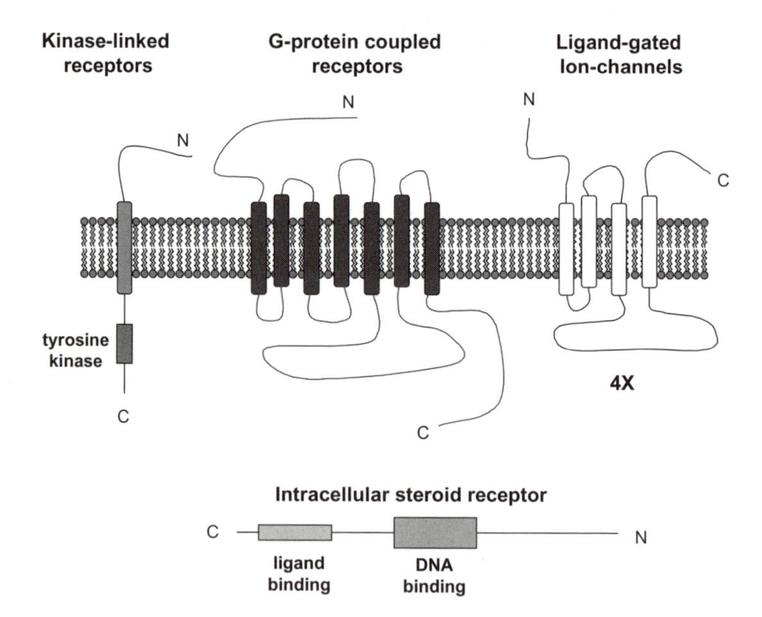


**Figure 2** *Schematic representation of the four major classes of receptor proteins. Except for the steroid receptors, the receptor proteins are localised in the cell membrane. The ligand-gated ion channels are made up of an assembly of 4–5 subunits, which each contain four transmembrane domains*

on the ligand–receptor pair, bind within the pocket to the extracellular loops and/or the N-terminus. Upon activation by the appropriate ligand, G-protein coupled receptors transfer the signal *via* conformational alterations to a member of the family of G-proteins, which, in turn, can activate or inhibit various enzymes or modulate ion channel activity.

The *tyrosine kinase receptor* proteins are also membrane bound and, like the G-protein coupled receptors, single polypeptide chains. The receptors contain a large N-terminus at the outside of the cells, which is involved in ligand recognition. The ligand binding site is connected to the intracellular effector (tyrosine kinase activity) domain *via* a single transmembrane domain. Upon ligand binding tyrosine kinase receptors form dimers, which leads to the stimulation of tyrosine kinase activity. Finally, the *intracellular receptors* all have a conserved DNA-binding domain (zinc-fingers) attached to a C-terminal, variable ligand binding domain. For activation the ligands have to enter the cell and since the effects are produced as a result of altered gene transcription, they rely on new protein synthesis and are thus inherently slow in onset.

## 3 MECHANISM OF DRUG ACTION: AGONISM AND ANTAGONISM

To appreciate fully the usefulness of receptor proteins as drug targets one needs to realize another important concept in drug action. Receptor proteins are primarily made for the recognition of endogenous ligands, which upon binding to the receptor proteins give rise to a cellular effect. These ligands are called agonists. In many instances one can foresee the usefulness of exogenously administered (synthetic) agonists. For example, in patients with diabetes one can control the energy metabolism by injection of recombinant insulin to supplement the reduced endogenous levels. In asthmatic patients one can relieve airway problems by the inhalation of so-called $\beta_2$-agonists that will relax the smooth muscle of the airways. These drugs will bind to and activate the $\beta_2$-adrenergic receptor (normally activated by noradrenaline) in the smooth muscle cells and thereby relax the smooth muscle that is contracted *via* endogenous ligands (*e.g.* histamine and leucotrienes), which are released as part of the underlying pathological mechanism of asthma. Through the stimulation of the $\beta_2$-receptor one counteracts the action of other signalling molecules. This principle of drug action is called physiological antagonism; the physiological response of muscle contraction is prevented by its counterpart of muscle relaxation.

The concept of physiological antagonism is important in drug action, but not as important as a more direct approach to counteract the action of

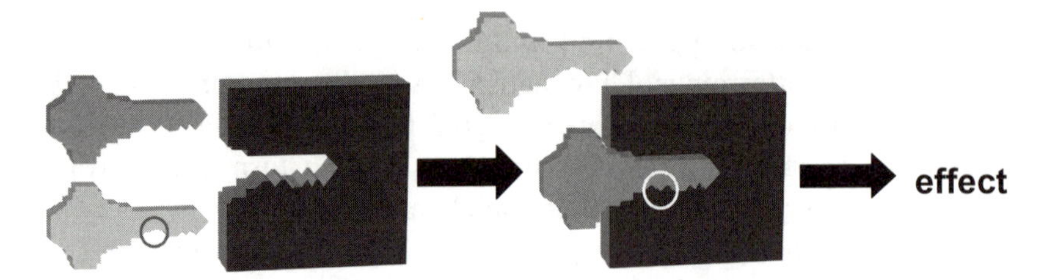


**Figure 3** *Lock-and-key principle for receptor–ligand interactions. Only one of the keys (ligands) fits perfectly into the lock (receptor) and will be able to open the lock (give a response). The small difference between the two keys is indicated by the circle. The 'imperfect' key will fit in the lock, but is not able to open the lock. By sitting in the lock the imperfect key prevents the perfect key getting into the lock. One could regard an antagonist as an imperfect key and a receptor agonist as the perfect key*

endogenous signalling molecules, the use of receptor antagonists. It had already been recognized in the early days of medicinal chemistry and pharmacology that certain ligands do not produce an effect on their own upon interaction with a receptor protein (*i.e.* act as agonist). In contrast, these ligands can inhibit the action of endogenous signalling molecules by the simple binding to the appropriate receptor protein. This concept can be easily understood if one considers the *'lock-and-key' principle* for ligand–receptor interaction (Figure 3). Only keys (ligands) that fulfil all criteria for fitting into the lock (receptor) can open the door (produce an effect). Some keys can fit in the lock, but not perfectly. Consequently, they cannot open the door yet, by fitting into the lock, these keys prevent the original key from fitting into the lock and opening the door. The concept of receptor antagonism is extremely important in medicinal chemistry and is very often the underlying mechanism of drug action. To prevent *e.g.* the constriction of airway smooth muscle in asthmatic conditions one can *e.g.* administer receptor antagonists that prevent the actions of the signalling molecules causing muscle contraction (*e.g.* histamine and leukotriene antagonists).

## 4 RECEPTOR SUBTYPES

It is now well known that endogenous chemical messengers often use more than one receptor protein to elicit their effects. The fact that endogenous signals can act as ligand for more than one receptor protein substantially increases the versatility of signalling molecules. The same signalling molecule can produce different effects in different tissues *via* interaction with different receptor proteins.

Initially, receptor subtypes were discovered on the basis of differential sensitivity of specific physiological responses to drugs. However, in the genomics era the discovery of new genes has revolutionised this area and has brought us a large number of previously unknown receptor proteins. This shift in paradigm is nicely illustrated by the developments in the field of histamine receptors.

The first evidence for the existence of a specific histamine receptor came from the work of Bovet and Staub (1937), who reported the weak competitive inhibitory effects of piperoxan (Figure 4) against the effects of histamine. This discovery led to the development of several potent so-called 'antihistamines' (Figure 4), which were able to inhibit the contraction of histamine of airway smooth muscle and to relieve certain symptoms of allergic disease. However, none of these compounds could block the stimulatory actions of histamine on *e.g.* gastric acid secretion. This led Ash and Schild (1962) to the hypothesis that there are probably two subtypes of histamine receptors referred to as the histamine $H_1$ and $H_2$ receptors. Both subtypes can be stimulated by histamine, but the actions of histamine can be antagonised by a subtype-specific antagonist. This hypothesis was definitively established in 1972 when Nobel-prize winning Sir James Black and his co-workers identified burimamide (Figure 4), a compound that competitively antagonised the actions of histamine on gastric acid secretion. Various burimamide derivatives (*e.g.* cimetidine, ranitidine, Figure 4) have been developed as $H_2$ antagonists for the treatment of gastric ulcers and have been major block-busters for many years. The observation in 1983 that histamine can inhibit its own synthesis

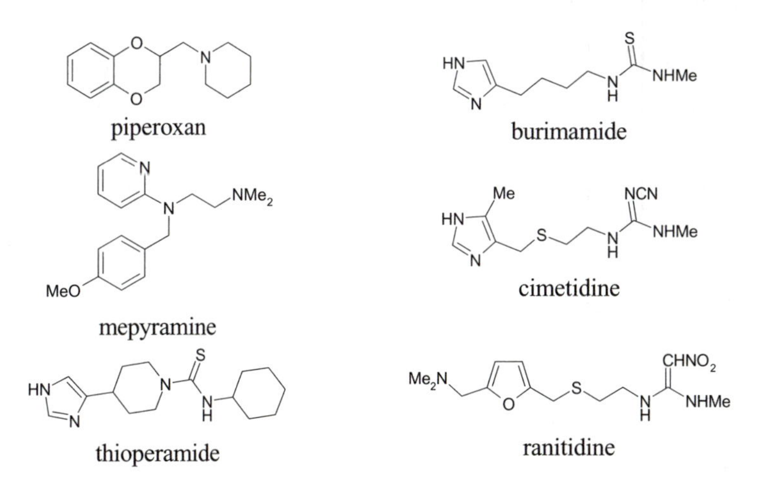


**Figure 4** *Chemical structures of the ligands used to pharmacologically differentiate histamine $H_1$, $H_2$ and $H_3$ receptor*

**Table 1** *The major members of the family of biogenic amines are acetycholine, dopamine, (nor)adrenaline, histamine and serotonin; they all act* via *a diverse set of receptors, which besides G-protein coupled receptors, can also belong to the family of ion channels (*e.g. *for acetylcholine and serotonin)*

| | | *G-protein coupled receptors* | *Ion channels* |
|---|---|---|---|
| Acetylcholine | $MeCOOCH_2CH_2NMe_3^+$ | m1–m5 | Nicotinic receptors<br>Motoric endplate |
| Dopamine | 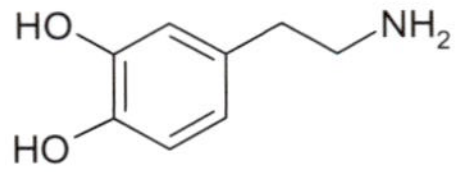 | $D_1$–$D_5$ | |
| Noradrenaline | 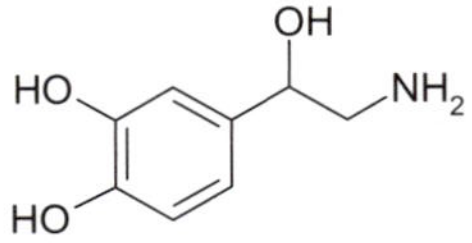 | $\alpha_1$, $\alpha_2$<br>$\beta_1$–$\beta_3$ | |
| Histamine | 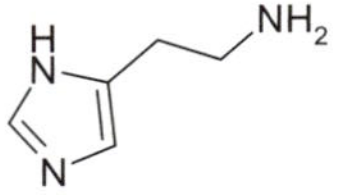 | $H_1$–$H_4$ | |
| Serotonin<br>(5-hydroxy-tryptamine) | 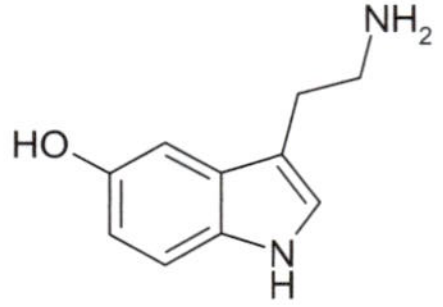 | 5-$HT_1$, 5-$HT_{2A}$–5-$HT_{2C}$,<br>5-$HT_4$–5-$HT_7$ | 5-$HT_3$ |

and release from brain slices suggested the existence of a histamine $H_3$ receptor; this hypothesis was confirmed by the discovery of unique agonists and antagonists (*e.g.* thioperamide, Figure 4) for this receptor as well.

From these developments one can learn that for many years receptor classification has relied on the development of receptor subtype selective agonists and antagonists. However, with the introduction of molecular biology in the area of receptor research, this paradigm has completely changed. In the last 15 years most new receptors have been suggested on the basis of the discovery of new genes. Again, the field of histamine receptors nicely illustrates this development. Until Autumn 2000 three genes encoding the previously known $H_1$, $H_2$ and $H_3$ receptor were cloned. A new gene, encoding an $H_4$ receptor subtype, was recently identified in a genome database (so-called '*in silico* cloning'), as a gene that was very similar to the $H_3$ receptor gene. The encoding receptor is expressed in the intestine and immune cells and binds several histamine receptor ligands with a distinct profile. The *in vivo* function of this new receptor is so far unknown, but potentially offers new options for therapeutic intervention.

Often it is seen that endogenous ligands mediate their actions through one or more receptors of just one superfamily. For example, all of the known histamine receptors ($H_1$–$H_4$) belong to the family of G-protein coupled receptors. However, there are some important exceptions to this rule. Within the family of (simple) biogenic amine neurotransmitters (Table 1) important differences can already be noticed. Whereas histamine, dopamine and noradrenaline act only *via* (various) G-protein coupled receptors, acetylcholine and serotonin act *via* both G-protein coupled receptors and ligand-gated ion channels.

## 5 DRUG SELECTIVITY

To understand drug action one should be aware of the important issue of drug selectivity. In general, a drug is known for its activity at one target. However, some drugs can display activity at a variety of targets. The specificity of the ligand–receptor interaction can be compared with the combination of a lock with specific keys. Some locks can accommodate a variety of keys, whereas the opposite is also true. Some keys can fit more than one lock. Often such drugs show preference for one type of receptor, but at higher concentrations they can also interact with other targets. The three structurally closely related $H_3$ agonists imetit, immepip and ($R$)-$\alpha$-methylhistamine are *e.g.* all equipotent at the $H_3$ receptor (a G-protein coupled receptor), but *in vivo* two of them also show agonistic activity at either the serotonin $5HT_3$ receptor (a ligand-gated ion channel;

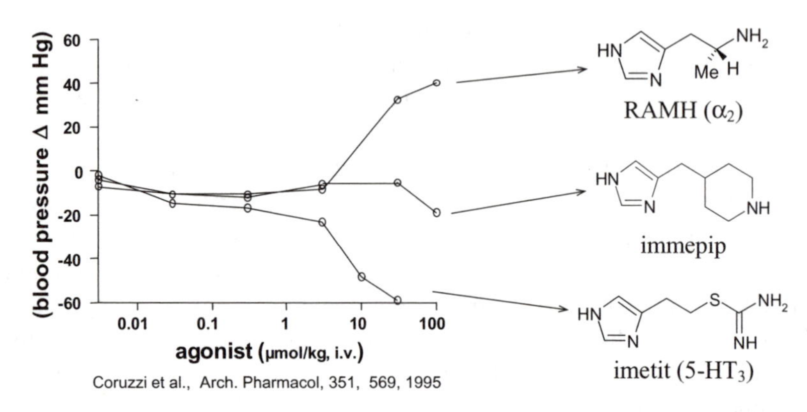


**Figure 5** *Selectivity of three histamine $H_3$ receptor agonists* in vivo. *The three agonists were administered peripherally to rats and the blood pressure was monitored. As can be seen from the graphs the three agonists show a different profile in this experimental model. None of the effects are antagonised by an $H_3$ receptor antagonist. The effects of imetit and (R)-α-methylhistamine (RAMH) are blocked by respectively the $5HT_3$ antagonist ondansetron and the $\alpha_2$ antagonist yohimbine*

imetit) or the $\alpha_2$-adrenergic receptor (a G-protein coupled receptor; (*R*)-$\alpha$-methylhistamine) (Figure 5). This example illustrates that for newly developed drugs one needs to know the full spectrum of its biological activities at a variety of known targets in order to be able to explain its action in complex physiological systems. Of course the *in vivo* application is the most challenging one in this respect. Often drug side-effects can be explained on the basis of a known interaction with another receptor site.

## 6 QUANTIFICATION OF DRUG–RECEPTOR INTERACTIONS

For the development of new drugs detailed quantitative measurements on drug–receptor–ligand interactions and receptor specificity are one of the requirements. The interest in physiology and pharmacology at the end of the 19th century resulted in the availability of a variety of relatively simple experimental models to evaluate the action of (endogenous) chemical substances. These models often represented isolated tissues, which could be used for measurements for several hours once the tissues were transferred to so-called organ bath chambers, containing physiological salt solution supplemented with *e.g.* $O_2$ and glucose. With these experimental models it became feasible to measure drug responses in a reproducible and accurate manner. Nowadays, more simple model systems (isolated cells, semi-purified recombinant receptor proteins) are available to determine quantitatively receptor–ligand interactions.

## The Receptor Binding of Ligands

With early, simple physiological models one noticed already that drug effects are not linearly related to the concentration of added drug, but showed a hyperbolic relationship. At a certain concentration of the ligand no further increase in the response is obtained (Figure 6). If one plots the concentration on a log-scale the dose–response curve becomes sigmoidal. These graphs resemble exactly the well-known relationships for enzyme–substrate interaction, the hyperbolic Michaelis–Menten equation.

In fact, similar mathematical models have been applied to the receptor–ligand interaction. Clark's occupation theory was the first model that could describe the observations of drug action in isolated tissues. In this theory the receptor–ligand interaction is considered to be a bimolecular interaction, in which the receptor–ligand complex is responsible for the generation of the biological effect. Clark assumed that the effects of a drug were proportional to the fraction of receptors occupied by the drug. Consequently, for a maximal effect the drug has to occupy all receptors.

In Clark's theory, the agonist (A) interacts in a reversible way with the receptor (R) and the formed complex (AR) gives rise to the effect:

$$A + R \Leftrightarrow AR \rightarrow \text{effect}$$

In equilibrium, the rate of the forward reaction of an agonist A reversibly bound to its receptor R is proportional to the concentration of A and R, and the proportionality constant is denoted by $k_1$:

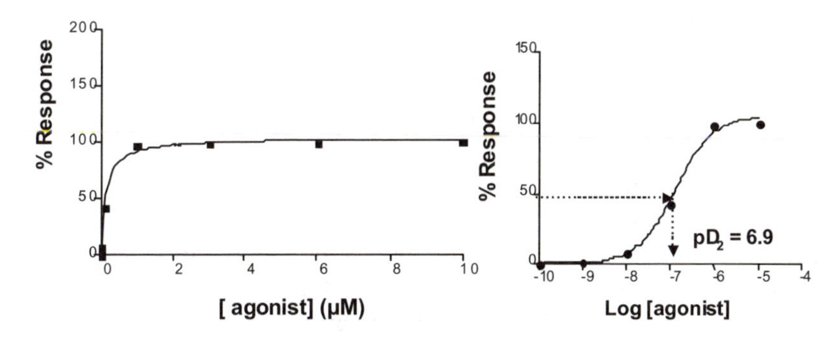


**Figure 6** *Schematic representation of the relationship between the concentration of an agonist and the observed response. The right-hand graph shows the same data-set plotted on a log scale. From the sigmoidal dose–response curve one can obtain a parameter that is often used to compare different ligands: the log[agonist] that causes 50% of the maximal response (also called $pD_2$). The agonist with the highest $pD_2$ value is the most potent one*

$$\text{rate of association} = k_1[\text{A}][\text{R}]$$

Similarly, the rate of the backward reaction, in which the agonist–receptor complex dissociates again, is proportional to the concentration of the AR complex:

$$\text{rate of dissociation} = k_{-1}[\text{AR}]$$

In equilibrium the rate of the forward reaction equals the rate at which existing AR complexes dissociate: $k_1[\text{A}][\text{R}] = k_{-1}[\text{AR}]$. In other words, within a certain period of time the same number of molecules A will bind to and dissociate from the receptor.

In equilibrium the dissociation constant $K_D$ can be described as follows:

$$K_D = \frac{k_{-1}}{k_1} = \frac{[\text{A}][\text{R}]}{[\text{AR}]}$$

The dissociation constant is a measure of the affinity of a ligand for its receptor. The lower the value of $K_D$ of a drug, the higher the affinity for its receptor. As such the $K_D$ value is very often used as a parameter to compare the potency of drugs.

The definition of $K_D$ can also be used to derive mathematical relationships between receptor occupancies and the concentration of the ligand. In the time of Clark it was impossible to directly determine the binding of a ligand to receptor proteins. Measurements were made indirectly by studying agonist responses. Since the 1970s radioactive receptor ligands (also called radioligands) have become available and it is now possible to accurately determine receptor occupancy by labelled drugs. For sake of simplicity we will now first deduce the equations that describe receptor–ligand binding and then bring that back into the historical perspective of Clark and couple the process of ligand binding to the generation of a biological response.

Nowadays the determination of binding of ligands to receptors is relatively easy with the availability of selective radioactive (or fluorescent) ligands (either agonist or antagonist) for a large number of receptors. With very simple techniques we can measure the binding of a radioligand to a receptor. In brief, one incubates the radioligand with appropriate biological material (*e.g.* cell membranes expressing the receptor) in a buffer, waits for equilibrium and then rapidly (within a second) separates the cell membranes (with bound radioligand) from the incubation mixture. Most often the separation is performed by simple filtration over glass fibre filters. The radioactivity remaining on the filter can be determined easily.

In the next equations the receptor–radioligand complex ([RL]) is called Bound ($B$). The ligand that is not bound to receptor is called Free ($F$). As both can be measured experimentally, one can substitute these terms in the equation for the equilibrium dissociation constant:

$$\frac{[\mathrm{L}][\mathrm{R}]}{[\mathrm{RL}]} = \frac{k_{-1}}{k_1} = K_\mathrm{D} = \frac{F[\mathrm{R}]}{B}$$

We also know that the total amount of receptor present, $R_{tot}$ (*i.e.* the maximal number of binding sites, also called $B_{max}$) equals the receptor–ligand complex RL and the free amount of receptors R:

$$R_{tot} = B_{max} = [\mathrm{RL}] + [\mathrm{R}] = B + [\mathrm{R}];$$

this can be rewritten as:

$$[\mathrm{R}] = B_{max} - B$$

Using this equation we can substitute [R] in the previous equation:

$$K_\mathrm{D} = \frac{F(B_{max} - B)}{B}$$

leading to:

$$FB_{max} - FB = K_\mathrm{D}B$$

if we now multiply by $B$ one gets:

$$K_\mathrm{D}B + FB = FB_{max}$$

Rearrangement leads to:

$$B(K_\mathrm{D} + \mathrm{F}) = FB_{max}$$

This results in the final equation:

$$B = \frac{FB_{max}}{K_\mathrm{D} + F}$$

which describes the hyperbolic relationship between the free concentration of ligand L and the amount of ligand that will form a receptor–ligand complex (Figure 7). This equation is completely in accordance with the Michaelis–Menten equation of substrate–enzyme reactions,

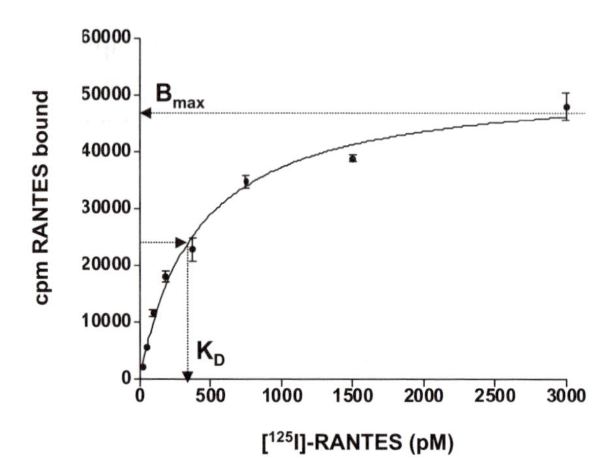


**Figure 7** *Saturation binding analysis of the binding of radio-labelled* $^{125}$*I-RANTES to chemokine CCR1 receptor expressed in COS-7 cells. Increasing concentrations of the radio-labelled chemokine were incubated with membranes containing the CCR1 receptor. After reaching equilibrium membranes were separated by rapid filtration and the radioactivity remaining bound to the receptor was determined. From this saturation binding isotherm one can calculate the maximum number of receptors* ($B_{max}$) *and the equilibrium dissociation constant (affinity)* $K_D$ *at 50% of the* $B_{max}$

$$V = \frac{[S]V_{max}}{K_m + [S]},$$

in which $V$ stands for the rate of catalysis, $V_{max}$ for the maximal rate, [S] for the concentration of substrate, and $K_m$ for the Michaelis constant, which informs us about the affinity of the substrate for the enzyme.

The equation on receptor occupancy can be further rearranged to derive the so-called fractional occupancy $f$:

$$f = \frac{B}{B_{max}} = \frac{F}{K_D + F}$$

From this equation the two biologically meaningful constants, $K_D$ (drug affinity) and $B_{max}$ (maximal receptor number) can be determined with proper experimental design and as can be seen in Figure 7 the equation, $B_{max}$ can be determined at very high concentrations of the radioactive ligand ($f = 1$) whereas $K_D$ equals the free concentration of ligand that results in a fractional occupancy of 0.5, *i.e.* when it occupies exactly half of the total number of receptors. The $K_D$ and $B_{max}$ values are easily deduced from the graph as shown in Figure 7.

## From Receptor Binding to Effect

*The Occupation Theory of Clark.* In the occupation theory of Clark, the generation of a biological effect by agonists is directly related to the formation of the receptor–ligand complex:

$$\mathrm{A} + \mathrm{R} \Leftrightarrow \mathrm{AR} \rightarrow \text{effect}$$

Consequently, the generation of the response is determined by the formation of the receptor–ligand complexes; *i.e.* in the equation for receptor binding one can replace the terms $B$ and $B_{\max}$ with $E$ (effect) and $E_{\max}$, leading to:

$$\text{effect} = E_{\max} \frac{[\mathrm{A}]}{K_{\mathrm{D}} + [\mathrm{A}]}$$

This equation describes nicely the observed hyperbolar relationship of agonist–effect relationships; the affinity parameter is obtained by determining the concentration of A that results in half-maximal effect.

Often these relationships are presented on a log-scale, resulting in the sigmoidal relationships. From these curves the $-\log K_{\mathrm{D}}$ value, or $\mathrm{p}D_2$ value, is often used quantitatively to compare different ligands. In Figure 8, ligand A is more potent than ligand B and this is reflected in its higher $\mathrm{p}D_2$ value.

*Partial Agonists and Intrinsic Activity.* The occupation theory of Clark turned out to be true for just a limited number of cases. Some drugs acting at the same receptor can elicit different maximal effects at maximal receptor occupancy (Figure 8), a feature that can not be explained by the occupation theory. To account for these discrepancies, Ariens (1954) introduced the term intrinsic activity ($\alpha$) to describe the relationship between the effect and the receptor occupancy.

$$E = \alpha[\mathrm{AR}]$$

where $E$ is the effect, $\alpha$ is the intrinsic activity of the drug, and [AR] the concentration of drug–receptor complex.

For an agonist producing the maximal response (full agonist), $\alpha$ was defined as 1, for antagonists (no effect) $\alpha$ was defined as 0. For so-called partial agonists $\alpha$ was defined between 0 and 1 depending on the maximal effect these ligands could elicit; the intrinsic activity is defined as:

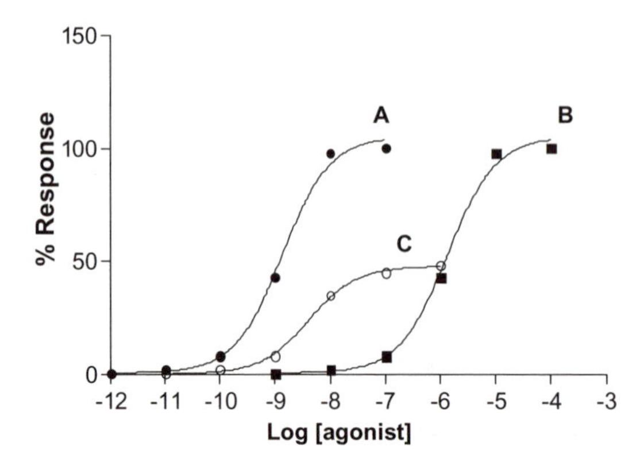


**Figure 8** *Dose–response curves for three different agonists at one receptor. Agonists A and B are both able to fully activate the receptor and to elicit a maximal response. However, ligand A is active at much lower concentrations,* i.e. *its $pD_2$ value is higher. Agonist C is active at lower concentrations compared to agonist B, but acts as a partial agonist*

$$\alpha = \frac{E_{\max}(\text{partial agonist})}{E_{\max}(\text{full agonist})}$$

The introduction of the intrinsic activity was an important hallmark in the classification of drugs. It separated two different properties of molecules which determine the final outcome of the interaction with receptor proteins. The affinity of the ligand ($K_D$) for the receptor determines the receptor occupancy, whereas after binding, ligands need to posses a second property (Ariens tried to define that with intrinsic activity) in order to activate the receptor. This notion can also be deduced from Figure 8. The partial agonist C shows a higher affinity than the full agonist B, but is less effective in the generation of a biological response.

*Partial Agonism and Drug Efficacy.* The introduction of the intrinsic activity can be used to describe the existence of partial agonists, but still considers the final effect to be proportionally related to the number of receptor–ligand complexes. Yet, in several cases it was observed that a partial agonist has different agonistic properties in different tissues, although activating the same receptor. This observation is illustrated by the effects of the $H_2$ receptor ligand burimamide in either the guinea-pig right atrium or Chinese hamster ovary cells, overexpressing the human $H_2$ receptor (Figure 9). At the right atrium burimamide does not lead to any increase in frequency of heart beats (in contrast to the agonist histamine). However, using the isolated cell system one can observe weak residual

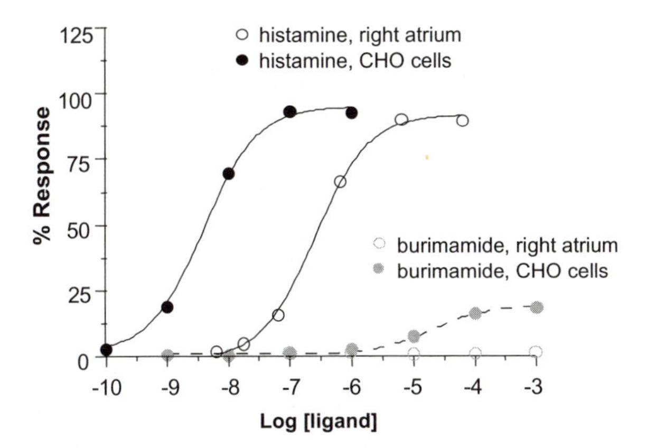


**Figure 9** *Effect of histamine and burimamide at the $H_2$ receptor in the guinea-pig right atrium (causing an increase in frequency of contractions) or transfected Chinese hamster ovary (CHO) cells (causing a rise in the level of cAMP). The effect of different $H_2$ receptor expression levels is apparent. Only at high receptor expression levels (CHO cells) can the weak partial agonism of burimamide be observed*

agonist properties of burimamide ($\alpha = 0.16$). How to explain these discrepancies?

Independent from Ariens, Stephenson (1956) introduced the concept of efficacy '*e*' as a property of the drug to explain the relationships between occupancy and response. According to Stephenson, agonist activation results in the generation of a stimulus, which is translated by the tissue into a final response. The efficacy of an agonist is the parameter that indicates its ability to generate the stimulus (instead of the final response in the theory of Ariens). The response, $R$, of a tissue is an unknown function of the stimulus, $S$, that is specific for the tissue:

$$R = f(S)$$

The stimulus $S$ was defined following the theories of Clark:

$$S = e\frac{[\mathrm{A}]}{K_\mathrm{D} + [\mathrm{A}]}$$

In this theory, agonists induce a certain degree of stimulus, which dependent on the transducer function, will be translated into a certain degree of response. The discrepancies in the effects of ligands in different tissues can be explained by differences in the transducer function, $f$. The function is highly dependent upon *e.g.* receptor density and the coupling of receptor occupancy to the ultimate response.

How to translate this back to Figure 9? In the guinea-pig right atrium the level of $H_2$ receptor expression is approximately 100-fold lower compared to the recombinant cell system. If one assumes that every activated receptor can give rise to the production of a certain amount of intracellular second messenger (cAMP in the case of the $H_2$ receptor), substantially lower amounts of cAMP will be produced by burimamide and histamine in the guinea-pig atrium in comparison to the isolated cells. In the guinea-pig atrium elevated cAMP levels are needed to increase the heart frequency. From the data in Figure 9 one should conclude that the increase in cAMP levels in the guinea-pig atrium induced by burimamide is not sufficient to modulate the frequency. In other words, the stimulus $S$ is too low to result in a final response.

If we examine the responses to the full agonist histamine in the same two experimental systems, we see a leftward shift of the dose–response curve in the CHO cells (high expression) compared to the guinea-pig right atrium (low expression). As histamine has the same affinity for the $H_2$ receptor in the two experimental models, one can deduce from these data that histamine only occupies a small part of the available receptors in the CHO cells to elicit a full response. Such observations have led to the introduction of the term 'receptor reserve'. This term is used to indicate the fraction of receptors that are unoccupied by an agonist when the maximal agonist response is obtained. For G-protein coupled receptors, the presence of receptor reserve can be understood if one realises that one agonist-occupied receptor can activate many G-proteins. At a certain level of receptor occupation all available G-proteins will be activated and a further increase in the occupancy will not lead to a further increase in G-protein activation.

## Receptor Antagonism

So far, we have mainly dealt with agonism. However, as stated before, drugs acting as receptor antagonists are very important in current drug therapy. In general one can distinguish two forms of antagonism, competitive antagonism and non-competitive antagonism.

Competitive antagonism is based on the principle that an agonist or antagonist can bind to the same recognition site(s) on the receptor and when both agonist and antagonist are present concomitantly, they will compete for receptor binding. The ability of a competitive antagonist to influence the receptor occupancy by an agonist (and therefore to elicit a response) is determined by the affinity for the receptor and concentration of both the agonist and antagonist. The important characteristic of competitive antagonism is that the antagonism can always be reversed by increased concentrations of the agonist (Figure 10). For non-competitive antagonism, increasing the agonist concentration does not result in a reversal of the effects of the antagonist (Figure 10). This can occur *e.g.* when the antagonist does not bind reversibly (*e.g.* alkylating agents or slowly dissociating ligands) or binds at a different site on the receptor and thereby modulates the receptor conformation in a way that agonist action cannot take place, so-called allosteric modulation.

Competitive antagonism is most commonly observed and has proven to be a good concept for drug development. As for agonists, receptor affinities can be obtained from binding studies in a straightforward manner, using the same equations. However, in functional studies quantitative measurements can also be made for antagonist. When agonist (A) and the competitive antagonist (B) are both present one has to consider two equilibria at the same time:

$$A + R \Leftrightarrow AR \rightarrow \text{effect} \quad \text{and} \quad B + R \Leftrightarrow BR$$

If an antagonist occupies part or the total number of receptors, less AR will be formed and consequently a lower effect will be observed. The effects of competitive antagonists on agonist-induced responses is studied

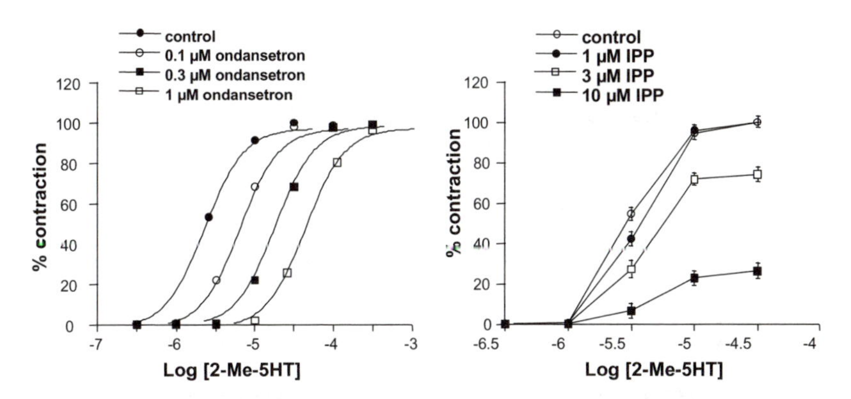


**Figure 10** *Dose–response curves for the $5HT_3$ agonist 2-methyl-serotonin, eliciting the contraction of the guinea-pig ileum. The actions of the agonist are competitively antagonised by the classical $5HT_3$ antagonist ondansetron and non-competitively antagonised by iodophenpropit (IPP), originally developed as an $H_3$ antagonist. These data again illustrate the concept of drug specificity. Specificity is highly dependent on the actual ligand concentration. At 10 μM, iodophenpropit is blocking both the $H_3$ and the $5HT_3$ receptor and specificity for the $H_3$ receptor is lost*

by determining the dose–response relationships for the agonists in the absence or in the presence of increasing concentrations of antagonists. If one assumes that equal agonist occupancies in the absence or in the presence of antagonist will produce equal response one can derive the following relationship:

$$\frac{[A']}{[A]} = 1 + \frac{[B]}{K_B}$$

in which [A] is the concentration of agonist, [A′] is the concentration of agonist A, giving the same response as [A] in the presence of concentration [B] of the antagonist B, and $K_B$ is the equilibrium dissociation constant for the antagonist B.

The ratio between the two agonist concentrations is also referred to as the dose ratio and is used in the equation below, as in the method derived by Schild to determine antagonist potencies.

$$\text{Log(dose ratio} - 1) = \log[B] - \log K_B$$

The $pA_2$ is often used to compare antagonists and is defined as the negative logarithm of the molar concentration of antagonist that produces a dose ratio of 2. This will lead to a rearrangement of the Schild equation to:

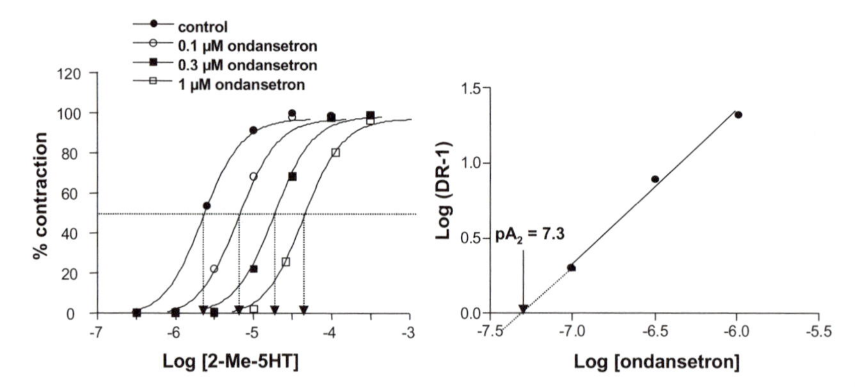


**Figure 11** *Schild-plot analysis of competitive antagonism by ondansetron of the 2-methyl-serotonin-induced contractions of the guinea-pig ileum. From the various sigmoidal curves the $pD_2$ values are determined and used to calculate the dose-ratio. Subsequently the Schild-plot is constructed. The intercept of the straight line with the* x-*axis is the $pA_2$ value, the parameter that is often used to compare different antagonists. The antagonist with the highest $pA_2$ value is the most potent one*

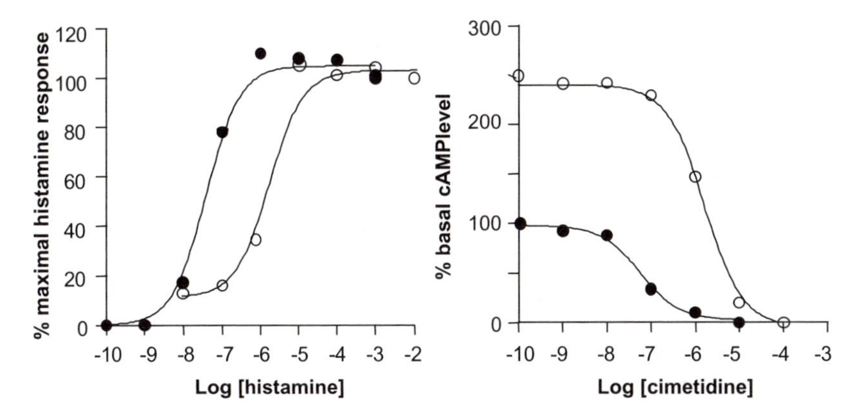


**Figure 12** *Antagonism by burimamide of the actions of the agonist histamine (left-hand panel) and the inverse agonist cimetidine (right-hand panel, see Section 3) in transfected Chinese hamster ovary cells. The effects of the ligands alone (closed circles) or in the presence of 100 μM burimamide (open circles) were determined on the level of cAMP in transfected cells. As shown in Figure 9, burimamide acts as a partial agonist in this system. This explains why in the presence of burimamide (open circles) the curve starts at an elevated level. As burimamide is not as effective as histamine, it acts as an antagonist for the actions of histamine. The same is true for the response of the inverse agonist cimetidine*

$$pA_2 = -\log K_B$$

Using a Schild-plot (Figure 11) the $pA_2$ can be easily determined experimentally.

A final issue to consider is the antagonism observed with partial agonists. In view of their limited ability to cause receptor stimulation, these ligands will act as antagonists when applied together with a full agonist (Figure 12).

## 7 FROM ANTAGONISTS TO INVERSE AGONISTS

So far we have considered antagonists within the classical scheme of Clark, Ariens and Stephenson; *i.e.* being 'silent' ligands. However, in 1982 Braestrup reported on the discovery of a so-called inverse-agonist for the benzodiazepine–GABA receptor complex. Normally benzodiazepines act at the $GABA_A$ receptor, a ligand-gated ion channel, which is very important in the regulation of brain function. Benzodiazepines bind at a different site than the endogenous activator GABA and agonists at the benzodiazepine binding site (*e.g.* diazepam) increase the GABA affinity.

Yet, some ligands (*e.g.* methyl 6,7-dimethoxy-4-ethyl-$\beta$-carboline-$\varepsilon$-carboxylate) were identified that had the opposite action of diazepam. Instead of increasing the affinity for GABA, a reduction in affinity was induced by these so-called inverse agonists. To complicate the situation further, one can also find benzodiazepine ligands (*e.g.* flumazenil) that do not affect the GABA affinity, but can antagonize the action of both the benzodiazepine agonist and inverse agonist. To explain these findings it was suggested that the GABA–benzodiazepine receptor exists in two conformations which are in equilibrium, an open channel form, the active conformation with high affinity for GABA, and a closed form, an inactive conformation with low affinity for GABA. In this scheme diazepam shows high affinity for the active conformation, thereby stabilising it, and stabilising the binding of GABA to the activated conformation. For the inverse agonists the inactive conformation would be favoured.

In the 1990s the concept of a multiple receptor conformation was extended to the large family of G-protein coupled receptors. It is now clear that G-protein coupled receptors can signal to a G-protein to some extent without any agonist. Usually this signal is very small and is not noticed. However, with the development of recombinant cell systems, which express receptors at high densities, the agonist-independent signalling has become quite clear. Many compounds previously identified as antagonist appear to act as (partial) inverse agonists, *i.e.* they diminish the constitutive activity of the receptor (Figure 13).

To explain the existence of negative efficacy one now assumes that agonist-independent activity is secondary to spontaneous isomerisation of the receptors between an inactive (R) and an active ($R^*$) state, which couples to the G-protein. This isomerisation involves conformational changes, which to some extent occur spontaneously in G-protein coupled receptors. Of importance is the notion that the isomerisation to $R^*$ can be induced by specific mutations that probably disrupt intramolecular constraints within the receptor protein. This feature has proven to be the mechanistic basis of a variety of genetic diseases. Agonists preferentially bind to the active state of the receptor and thus shift the equilibrium towards $R^*$. Inverse agonists preferentially bind to the inactive state of the receptor and shift the equilibrium towards R. Neutral antagonists are those drugs that present similar affinities for both R and $R^*$, favouring neither state of the receptor. By definition, neutral antagonists are able to block both the action of an agonist and inverse agonist. An example of such interaction is shown in Figure 12 for the $H_2$ receptor. The weak partial agonist burimamide can competitively antagonise both the action of the agonist histamine and the inverse agonist cimetidine.

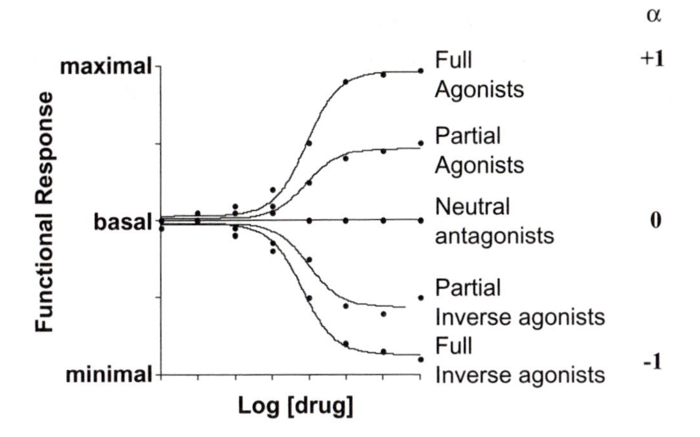


**Figure 13** *Schematic representation of the full spectrum of activities that ligands can display when a receptor shows a considerable level of agonist-independent activity. Besides the known (partial) agonism, the scheme is now completed with (partial) inverse agonists with intrinsic activities between 0 and −1. As can be seen in the right-hand panel of Figure 12, the histamine $H_2$ receptor shows* e.g. *constitutive activity, which can be inhibited by the inverse agonist cimetidine, previously considered to be an $H_2$ antagonist*

## 8 FINAL CONSIDERATIONS

With the elucidation of the human genome a wide variety of new proteins will be discovered and be considered as interesting drug targets. Most likely many of these proteins will again be receptors, which to some extent we have learned to target with selective ligands. With the developments in chemistry and cell biology it will be possible to start efficient drug development programmes for these new targets, despite our limited knowledge of these targets in physiology. The developed ligands will be essential to learn the role of these new targets in physiology and to validate the target in a patho-physiological condition. As such, medicinal chemistry is still instrumental for the development of new drugs!

## 9 SELECTED READING

1. E.A. Barnard, *Trends Biochem. Sci.*, 1992, **17**, 368.
2. G. Carpenter, *Annu. Rev. Biochem.*, 1987, **56**, 881.
3. S.R. Coughlin, *Curr. Opinion Cell Biol.*, 1994, **6**, 191.
4. R.M. Evans, *Science*, 1988, **240**, 889.
5. T. Kenakin, *Pharmacological Analysis of Drug Receptor Interactions*, Raven Press, New York, 1993.

6. G. Milligan *et al.*, *Trends Pharmacol. Sci.*, 1995, **16**, 10.
7. International Human Genome Sequencing Consortium, Initial sequencing and analysis of the human genome, *Nature (London)*, 2001, **409**, 860.

CHAPTER 2

# An Introduction to Ion Channels

BRIAN COX

## 1 INTRODUCTION

Ion channels are a ubiquitous class of proteins that control fundamental physiological events, for example all muscle contractions including heartbeat, the ciliary clearance process in the lungs, the degranulation process of immune cells, the regulation of blood pressure and electrolyte balance within the kidneys, fluid secretion in the salivary glands, the sensory reception of sound, light, odour and touch, and the generation, propagation and integration of all electrical signals in the brain and central nervous system. Consequently, ion channels represent very attractive targets for manipulation across many therapeutic areas. Many of the top 100 selling drugs are modulators of ion channels and the total annual world-wide sales of ion channel modulators exceeds \$6 billion. Major therapeutic

**Table 1** *Currently marketed ion channel modulators*

| *Drug* | *Type* | *Mechanism* | *Revenue* $10^6$ *\$yr*$^{-1}$ |
|---|---|---|---|
| Istin/Norvasc | Antihypertensive | L-type Calcium channel | 2470 |
| Adalat | Antihypertensive | L-type Calcium channel | 972 |
| Cardizem | Antiarrhythmic | L-type Calcium channel | 736 |
| Procardia | Antiarrhythmic | L-type Calcium channel | 720 |
| Zofran | Antiemetic | $5\text{-HT}_3$ | 626 |
| Dormicium | Anxiolytic | GABA-A | 521 |
| Neurontin | Anticonvulsant | Calcium channel? | 465 |
| Lamictal | Anticonvulsant | Sodium channel | 150 |
| | | Total (% of drug sales) | \$6660 × $10^6$ (10%) |

*Others:* Local anaesthetics, neuromuscular blockers, analgesics, oral hypoglycaemics and general anaesthetics.

indications include hypertension, cardiac arrhythmia, anxiety, epilepsy, pain and chemotherapy-induced emesis (Table 1). This chapter attempts to condense what is a massive scientific area into a few pages, but it is hoped that it will ignite the reader's interest in this fascinating area, a comprehensive reference list of carefully selected reviews has been included to aid further study.

## 2 STRUCTURE, FUNCTION AND CLASSIFICATION

Ion channels utilise just four ions as charge-carriers: $Na^+$, $K^+$, $Ca^{2+}$ and $Cl^-$ and accomplish all their functions by forming an ion specific membrane-spanning aqueous ion-conduction pathway. Ion movement down an electrochemical gradient delivers net charge to the cell and thereby changes the membrane voltage or potential, which ultimately controls cellular responses. Figure 1 is a generic representation of an ion channel spanning the cell membrane. In the resting situation, the cell cytoplasm contains low levels of sodium, calcium and chloride and high levels of potassium relative to the extracellular environment.

Channels are often assembled from several sub-units, the ion conducting protein is referred to as the $\alpha$-sub-unit that in itself may be a single protein or aggregated multiple copies of a single protein. The $\alpha$-sub-unit may also

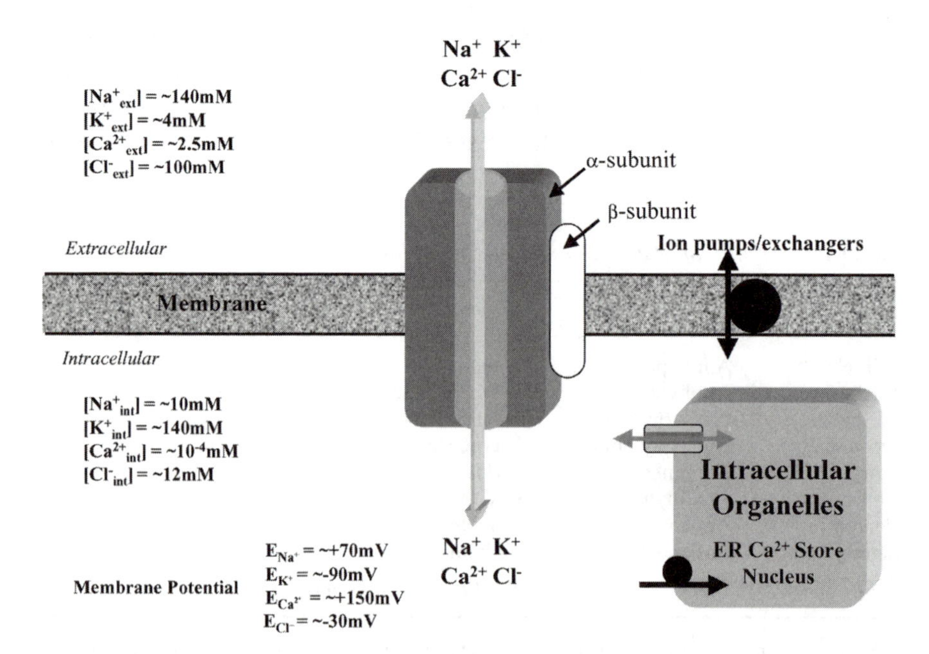


**Figure 1** *Schematic representation of an ion channel*

be accompanied by one or more supporting sub-units referred to as $\beta$, $\gamma$, $\delta$ *etc*. Each protein sub-unit is folded into a complex tertiary structure that spans the cell membrane, Figure 2 is a schematic representation of the structure of the voltage-gated sodium channel. Four homologous segments each spanning the membrane six times are assembled together to produce the ion-conducting cylindrical pore.

Although not the subject of this chapter, it is also worthy of note that other cellular processes have an influence on ion levels within the cell, for example ion pumps and exchangers, which require energy to operate, also span the cell membrane. Intracellular ion stores, in particular calcium stores, also play an important role in certain processes, such as signal transduction.

An intriguing question that has been the subject of many debates and a number of theories over the years is how can ion channels be selective for one ion? Clearly it cannot be due to ionic diameter alone because for example, how does a potassium channel exclude sodium ions considering the larger size of the potassium ion? The latest proposal is that channels possess a *selectivity filter*, a narrow region that acts as a molecular sieve (Figure 3). The diameter of the filter firstly necessitates the shedding of most of the waters of hydration from the ion, which then electrostatically binds to charged and polar amino acid residues. As the loss of water of hydration is not energetically favourable, an ion will only be able to pass through the channel if the binding energy of interaction with the key amino acid residues can compensate for the energy loss on dehydration. This compensation can only occur if the hydrated ion fits and binds optimally to the binding site. Thus, selectivity is a product of specific size and energy interactions very much like substrate–enzyme or ligand–receptor interactions.[1]

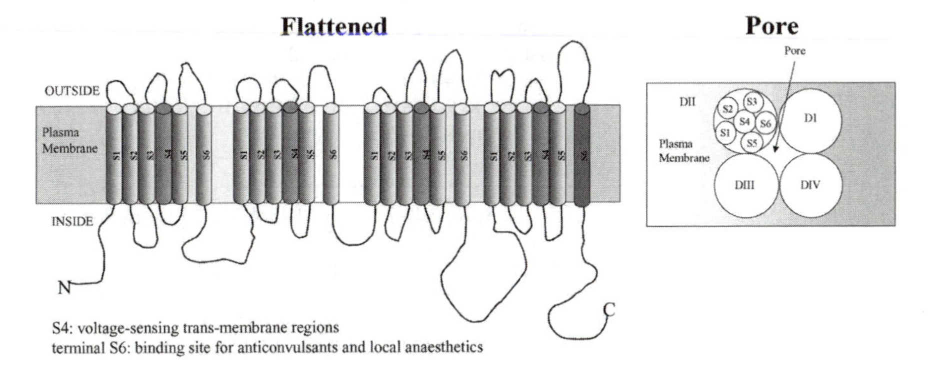


**Figure 2** *Schematic representation of the tertiary structure of a sodium channel*

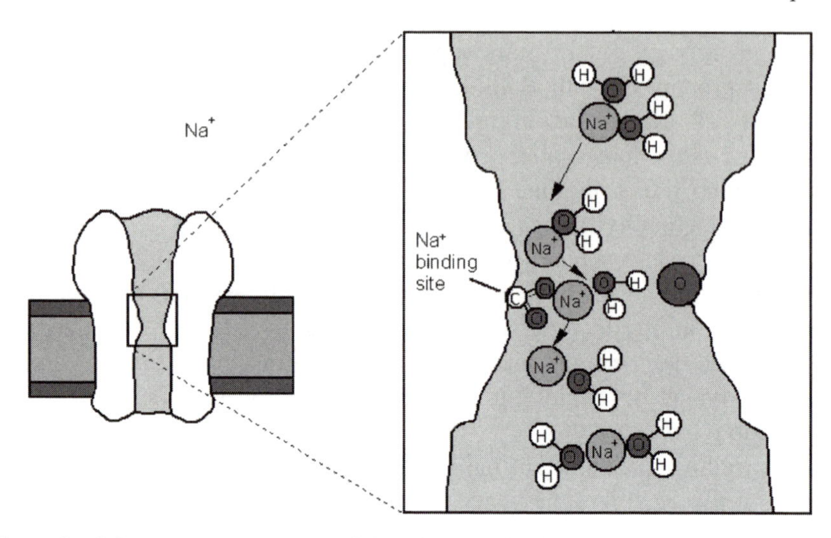


**Figure 3** *Schematic representation of the selectivity filter of a sodium channel*

**Table 2** *Classification of ion channels*

| | | |
|---|---|---|
| *Ligand-gated* | *Extracellular* | Cationic: nACh ($\alpha$1–$\alpha$9),[2] Glutamate (NMDA, AMPA, Kainate (KA)),[3] $5HT_3$,[4] ATP ($P_2X_{1-7}$),[5]<br>Anionic: $GABA_A$, $GABA_C$[6] |
| | *Intracellular* | Cationic: $K_{Ca}$s ($SK_{Ca}$1–3, $IK_{Ca}$1, $BK_{Ca}$ (maxi-K)),[8,9] Ca $IP_3R$ (I–III),[10] Ryanodine (RyR (1–3))[11,12]<br>Anionic: CLCA1–4[14,15] |
| *Voltage-gated* | | Calcium: $Ca_v$1.1–1.4 (L-type), $Ca_v$2.1–2.3 (P,Q,N,R-types), $Ca_v$3.1–3.3 (T-type)[19,20]<br>Sodium: $Na_v$1.1–1.9[16–18]<br>Potassium: $K_v$1.1–1.7,1.10, $K_v$2.1–2.2, $K_v$3.1–3.4, $K_v$4.1–4.3, $K_v$5.1, $K_v$6.1–6.2, $K_v$8.1, $K_v$9.1–9.3, hERG, KCNQ1–5, minK, miRP1–2.[8,9]<br>Chloride: ClC0–5[21] |
| *Inward rectifiers* | | Kir1.1–1.3, Kir2.1–2.4, Kir3.1–3.4, Kir4.1, Kir6.1–6.2, Kir7.1, SUR1–2.[9]<br>Two pore: (TWIK1–2, TREK, TASK, TRAAK)[9] |
| *Intercellular + miscellaneous channels* | | Gap junction: connexins $\alpha_1$(Cx43), $\beta_1$(Cx32), $\beta_2$(Cx26),[30] Aquaporins (AQP1)[31] |

Ion channels can be classified by consideration of their gene source; however, they are normally classified by the ion that they conduct and by the way in which their opening and closing is regulated, referred to as *gating*. This approach results in four groupings: *ligand-gated channels* that are regulated by the binding of a molecule or ion to the channel where the binding site can be *extracellular* (outside the cell) or *intracellular* (inside the cell); *voltage-gated channels* which are regulated by changes in the membrane potential of the cell; *inwardly-rectifying channels* that modulate the resting membrane potential and can be regulated by a variety of intra- and extracellular factors and, finally, *intercellular and miscellaneous channels* which include the gap junction channels (connexins) and the aquaporins (Table 2).

## 3 REPRESENTATIVE EXAMPLES

### Ligand-gated Channels

This group or super-family of channels is regulated by the binding of a molecule or ion to the *extracellular* or *intracellular* face of the channel. Figure 4 depicts the operation of an extracelluar ligand-gated channel, examples of which are: nicotinic acetylcholine (nACh),[2] ionotropic glutamate (iGlu),[3] 5-hydroxytryptamine (serotonin) (5-$HT_3$),[4] ATP[5] and $\gamma$-aminobutyric acid A ($GABA_A$)[6] receptor-channels. Binding of an agonist opens the channel, as is the case for the binding of the endogenous agonist acetylcholine **1** to the neuronal nicotinic acetylcholine channel (Figure 5). The classification of this channel as "nicotinic" is a reference to the ability of the alkaloid nicotine **2** to act as an exogenous agonist of this channel. Conversely, drugs and toxins can act to block the binding of the endogenous agonist, thus biasing the channel towards the closed state. This binding may be either of low energy and reversible or of high energy and not reversible. For the nicotinic acetylcholine channel an example of a reversible blocker is Atracurium (Tracium®) **3** the neuromuscular blocking drug used in surgery designed from d-tubocurarine **4** an alkaloid extracted from the South American arrow poison curare.[7] The polypeptide snake venom $\alpha$-bungarotoxin is an example of a non-reversible blocker of this channel.

The glutamate (Glu) **5** and $GABA_A$ channels mediate the vast majority of excitatory and inhibitory neurotransmission in the brain and can be found paired at the synapse to effectively balance post-synaptic neuronal activity (Figure 6). The three glutamate channels are further divided into three types based on their pharmacology, defined by their response to the reasonably selective agonists; *N*-methyl-D-aspartate (NMDA) **6**,

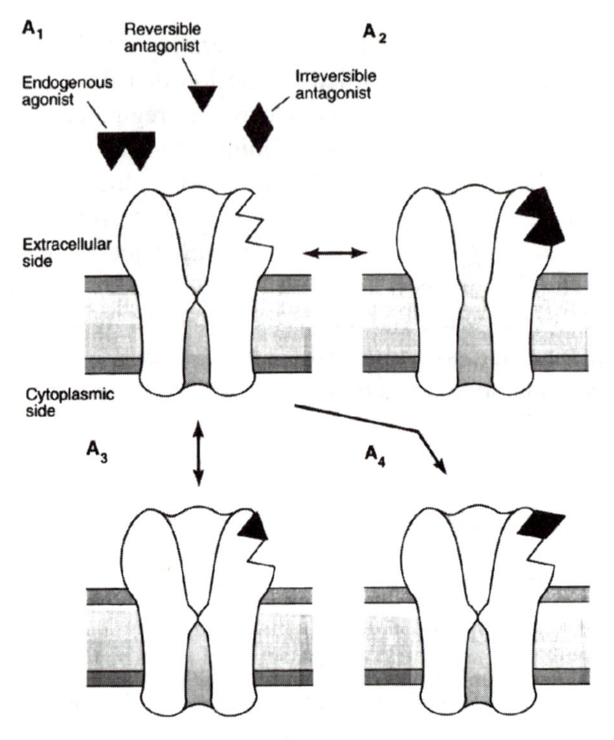


**Figure 4** *Schematic representation of the operation of an extracellular ligand-gated channel*

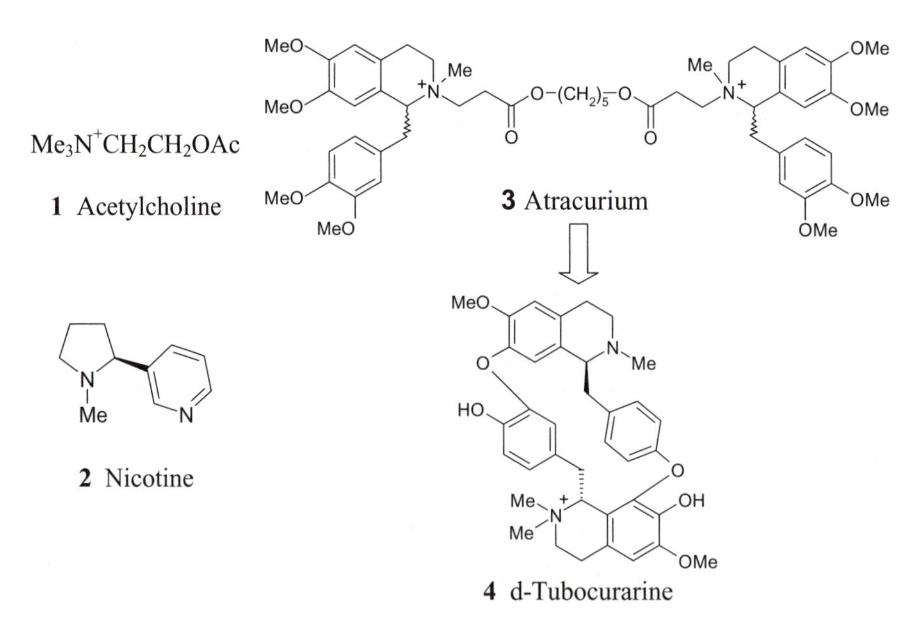


**Figure 5** *Ligands acting at the nicotinic acetylcholine (nACh) extracellular ligand-gated channel*

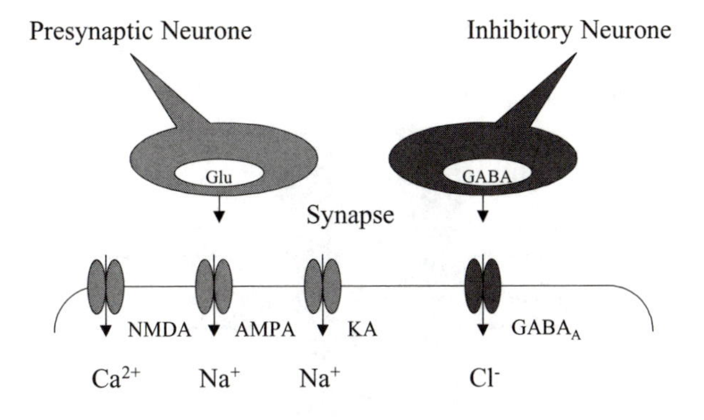


**Figure 6** *Schematic representation of the role of the glutamate/GABA extracellular ligand-gated channel*

$\alpha$-amino-3-hydroxy-5-methyl-4-isoxazole propionate (AMPA) **7** and kainic acid (Kainate, KA) **8** (Figure 7).

For many ion channels, modulation of the channel can be mediated through sites other than the natural ligand binding site. This is illustrated by the multiple binding sites, often referred to as allosteric or regulatory sites, known for modulators of the NMDA receptor-channel, which again have been defined pharmacologically by the ligands that bind to them. The GABA$_A$ receptor channel has an allosteric modulatory site to which the benzodiazepine anxiolytics (*e.g.* diazepam **10**) bind and they enhance the inhibitory action of GABA **9** by increasing the frequency of channel opening in response to GABA binding (Figure 8).

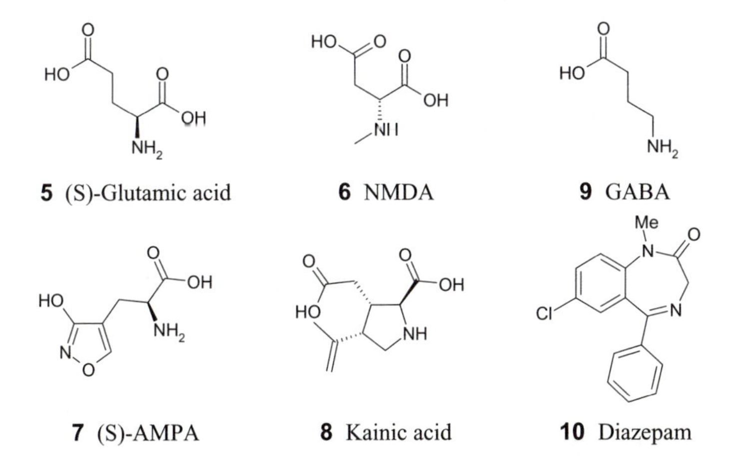


**Figure 7** *Ligands acting at the glutamate/GABA extracellular ligand-gated channel*

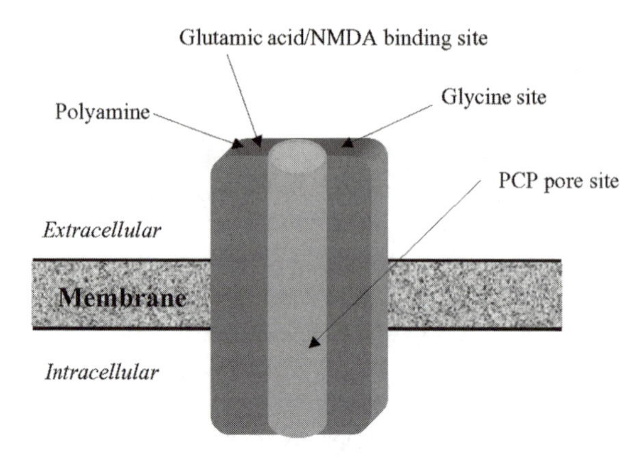


**Figure 8** *Schematic representation of the role of the glutamate NMDA extracellular ligand-gated channel showing different binding sites*

The 5-$HT_3$ channel is unique in being the only 5-HT **11** binding protein that is an ion channel, the rest (5-$HT_1$, 5-$HT_2$, 5-$HT_{4-7}$ and their subtypes) are all 7-trans-membrane receptors (Figure 9). A number of 5-$HT_3$ antagonist have been developed for use in the prevention of chemotherapy, radiation-induced and post-operative nausea and vomiting, these include ondansetron (Zofran®) **12** and granisetron (Kytril®) **13**. Recently, other therapeutic applications have been targeted, *e.g.* the introduction of alosetron (Lotronex®) **14** for irritable bowel syndrome (IBS).

Intracellular ligand-gated channels operate in much the same way as their extracellular gated counterparts except the ligand binds to the

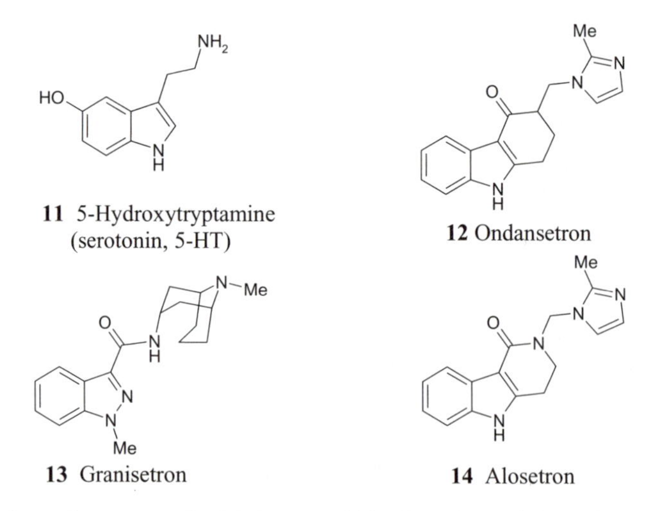


**Figure 9** *Ligands acting at the 5-$HT_3$ extracellular ligand-gated channel*

intracellular surface of the channel or the channel is part of some intracellular structure. Many of the examples of this type of channel are gated by the binding of calcium ions or are involved in intracellular calcium mobilisation. These include the calcium-activated potassium channels,[8,9] the inositol triphosphate family[10] and the ryanodine receptor (sarcoplasmic reticulum calcium channel).[11,12] The calcium-activated potassium channels, unlike the vast majority of potassium channels, are not voltage-gated (see later) but regulated by changes in intracellular calcium levels. They are all constitutively linked to the protein calmodulin, a ubiquitous mediator of calcium sensitivity. These channels were initially sub-divided on the basis of their conductance and their sensitivities to a variety of toxins, but more recently this classification has been complemented by the discovery that three distinct genes encode these subfamilies. The sub-families are the high conductance or Maxi-K ($BK_{ca}$), the intermediate conductance ($IK_{ca}$) and the small conductance ($SK_{ca}$) channels. The opening of these channels is responsible for membrane hyperpolarisation and compounds that modulate the channel towards the open state would have a beneficial effect on hyperactivity situations.

Much attention has been focussed on the BK channel openers for use in conditions such as stroke, asthma (for airway relaxation) and bladder hyperactivity. The oxindole, BMS-204352 **15**, a potent BK opener, is the compound to have progressed the furthest to date (Figure 10).

There are a number of toxin blockers of this family of channels, the scorpion toxin charybdotoxin blocks the BK and the IK channels, whereas members of the SK channel sub-family are blocked by apamin, the eighteen amino acid peptide neurotoxin present in the venom of the European honeybee. Interestingly, apamin is displaced from its binding site by the alkaloid tubocurarine **4**, an example of cross-sensitivity between ion channels towards certain ligands. Consideration of this activity shown by compounds like tubocurarine led to the design of selective SK blockers such as UCL-1848 **16**, the therapeutic application of which remains to be determined.[8,9,13] There are also a number of intracellular ligand gated

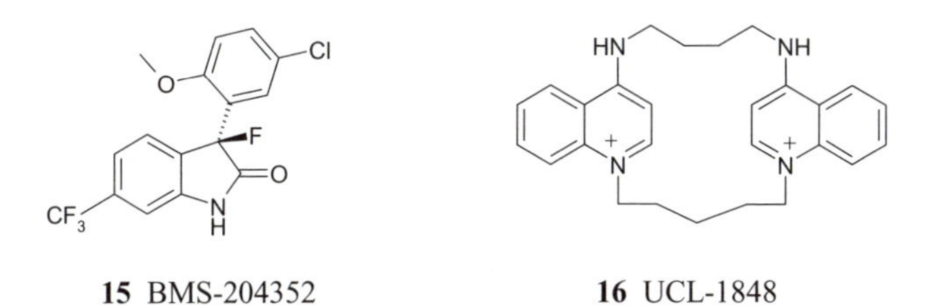


**15** BMS-204352 **16** UCL-1848

**Figure 10** *Ligands acting at the calcium-activated potassium channels family of intracellular ligand-gated channels*

channels conducting anions, *e.g.* the calcium activated chloride channels, which are found mainly in the secretory epithelia of the lungs and gut, and have been implicated in having a role in cystic fibrosis and asthma.[14,15]

## Voltage-gated Channels

The other major sub-division in the classification of ion channels is that of those that are gated (opened or closed) by changes in membrane potential. Change in membrane potential is associated with a number of processes in a whole variety of cell types. However, it is particularly important for the functioning of excitable cells such as those found in the nervous system and muscle. Unlike ligand-gated channels, where binding of an endogenous ligand triggers a conformational change in the channel protein structure leading to channel opening or closing, for voltage-gated channels the conformational change leading to opening and closing of the channel is triggered by movement of charged regions of the channel back and forth with changes in the electrical field of the membrane (Figure 11).[1]

Voltage-gated channels respond to a change in membrane potential by moving from the closed resting state ($A_1$) to a transient open state, whereupon depolarisation of that segment of the membrane then occurs. After this has occurred, the channel then enters a prolonged refractory or inactivated state. In an active neurone, this is the process by which a nerve impulse is transmitted, consecutive groups of channels open and close, spreading this membrane depolarisation or 'action potential' down the neurone to the next synapse or to a target tissue *e.g.* muscle.

The voltage-gated channels are sub-divided into groups depending on which ion they conduct: sodium,[16–18] calcium,[19,20] potassium[8,9] and chloride.[21] Structurally the four families are very similar, generally large proteins with multiple trans-membrane spanning regions; this is depicted for the sodium channel in Figure 2. The $\alpha$-sub-unit is composed of four

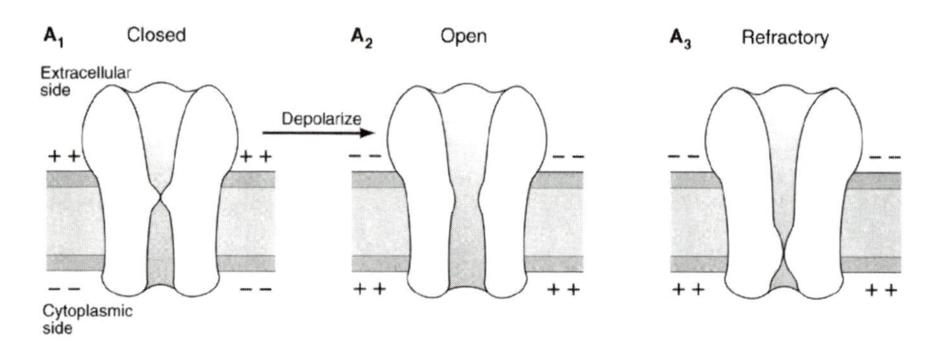


**Figure 11** *Schematic representation of the operation of a voltage-gated channel*

homologous domains (I to IV), each of which contains six trans-membrane segments (1 to 6). The S4 trans-membrane segments serve as voltage sensors, the S5 and S6 segments and the SS1/SS2 segment (P loop) between them together form the walls of the sodium-ion-conducting pore.[22]

Although voltage-gated ion channels are not gated by the binding of an endogenous ligand, modulation can be mediated through exogenous ligands by binding to allosteric sites. Nature has exploited this with a large number of examples, the blocking and opening of these ion channels so important for key neuronal processes has given rise to exquisitely potent and selective toxins used by their 'owners' for hunting or for defence. For example, the conotoxins from the marine cone snail are blockers of the neuronal (N)-type calcium channel and are used to paralyse its prey, tetrodotoxin is the nerve paralysing toxin present in the puffer fish, which blocks neuronal sodium channels, and the plant pyrethrenoid insecticides are neuronal sodium channel openers.

For voltage-gated calcium channels, kinetic, pharmacological and molecular biological studies have identified the existence of six different channel types: L, N, P, Q, R and T. The T-type channel is kinetically distinct from the others in that it is activated by only relatively small changes in membrane potential. The channels are further classified by their sensitivity to a variety of chemical tools and by the molecular identity of the $\alpha_1$ pore-forming unit.[23] The top selling ion channel modulators are L-type calcium channel blockers used in the prophylaxis of angina and as anti-hypertensives. These compounds relax vascular muscle, dilate coronary and peripheral arteries, and to a varying degree depress the contractility of the mycocardium in the heart. The compounds largely fall into two structural classes (except the benzothiazepinone diltiazem **20**), the dihydropyridines and the phenylalkylamines (Figure 12). The recently introduced amlodipine **18** is a long acting version of nifedipine, which was one of the first compounds to be discovered in this area.[24]

Attempts to identify blockers of the neuronal or N-type calcium channel as potential analgesics has been the subject of much recent research.[25,26] The peptide $\omega$-conotoxin ($\omega$-Conopeptide-MVIIA, SNX-111 (Ziconotide®)) **23** has found use in the treatment of severe chronic pain but requires intrathecal administration (Figure 13). The drive within the pharmaceutical industry has been to discover an orally-active, selective, small molecule modulator to overcome the disadvantages of intrathecal administration. Several approaches have been taken to lead identification in the field; by the use of high throughput screening campaigns (*e.g.* **21**), by the modification of known drugs shown to exhibit N-type channel blockade in addition to an existing activity (*e.g.* **22**) and by ligand-based

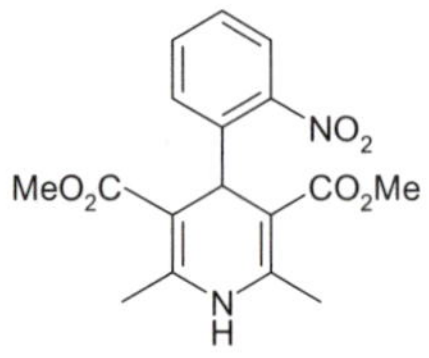

**17** Nifedipine (Adalat/Procardia®)

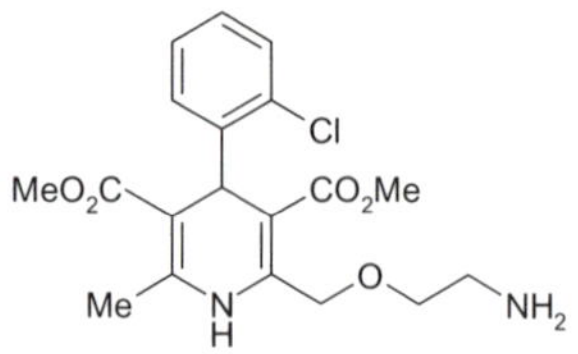

**18** Amlodipine (Istin/Norvasc®)

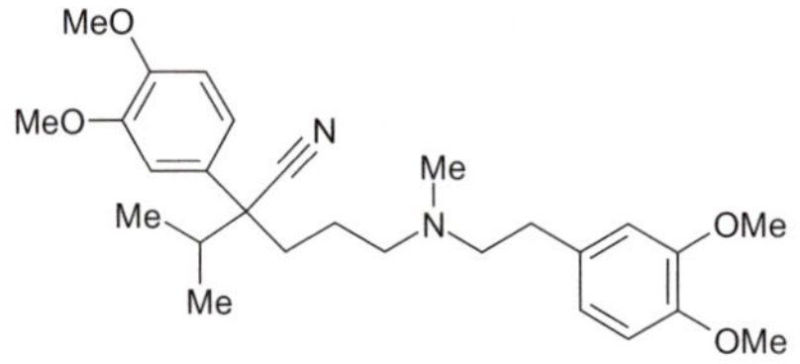

**19** Verapamil (Covera®)

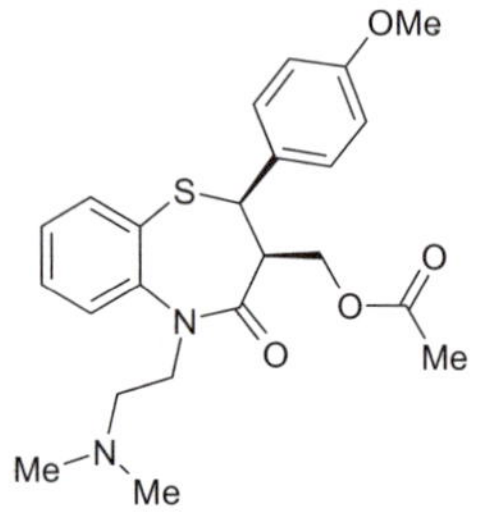

**20** Diltiazem

**Figure 12** *Ligands acting at the voltage-gated L-type calcium channel*

approaches making use of the three-dimensional solution structure of $\omega$-Conopetide-MVIIA **23**. Compound **24** was designed employing an alkyl phenyl ether scaffold for what was found to be the key binding amino acids of **23**, (Arg-10, Leu-11, Tyr-13).

Modulators of voltage-gated sodium channels are also very important; openers find use as insecticides, whereas for blockers, similar to the N-type calcium channel area, the emphasis has been on finding small, selective molecules. Interestingly, the majority of sodium channel blockers used in therapy were discovered well before their mode of action was determined. Examples of which are the local anaesthetics such as procaine **25**, class I antiarrhythmics such as lignocaine **26** and anticonvulsants such as phenytoin **27** and carbamazepine **28** (Figure 14). More recently better tolerated compounds such lamotrigine (Lamictal®) **29** have entered use as anti-epileptics. Only comparatively recently have the various channel subtypes been identified and their tissue distribution determined. These include four brain sub-types, ($Na_v$1.1–1.3 and $Na_v$1.6), skeletal muscle ($Na_v$1.4), cardiac ($Na_v$1.5) PN1 ($Na_v$1.7) and sensory neurone specific (SNS) ($Na_v$1.8–1.9) subtypes. Efforts are now focussed on finding sub-type selective blockers; SNS is of particular interest as an analgesic target, which as the name suggests is highly implicated in pain transmission.[16,17]

The voltage-gated potassium channels are an enormous family of proteins consisting of many diverse structural types, for example some 20 mammalian $K_v$ genes have been cloned and assigned to sub-families ($K_v$1,

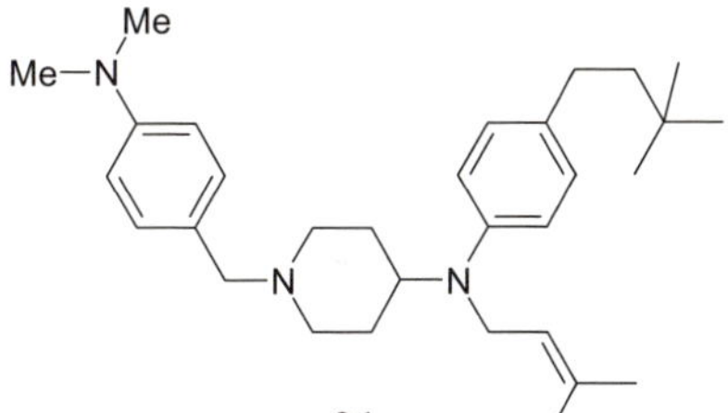

**Figure 13** *Ligands acting at the voltage-gated N-type calcium channel*

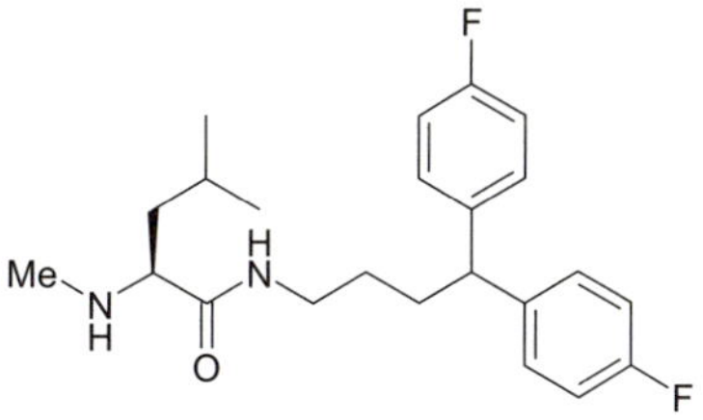

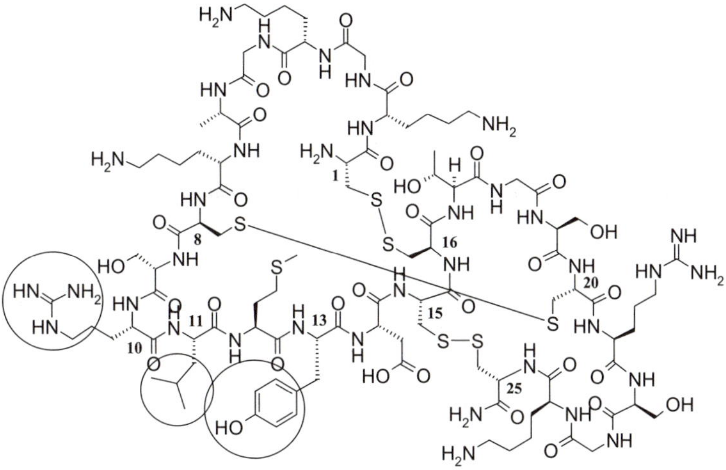

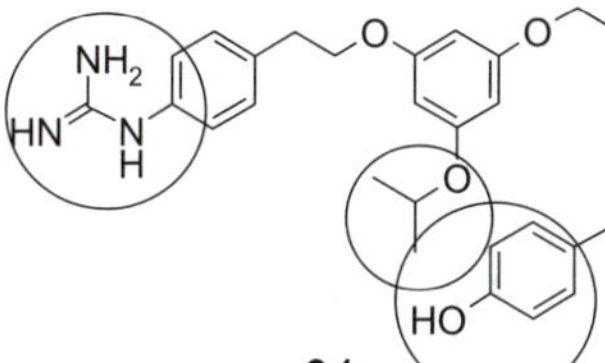

$K_v2$ *etc.*) of which there are individual variants (*e.g.* $K_v1.1$–$K_v1.7$). $K_v1.3$ plays an essential role in the stimulation and maintenance of cellular proliferation of a number of key immune cells, which has made it an attractive target for immuno-suppression.[9] The KCNQ channel and its sub-types have been shown to be responsible for the potassium current known as the M-current which plays a critical role in the regulation of neuronal excitability. The recently discovered KCNQ activator retigabine **30** is being investigated as a potential anti-convulsant (Figure 15).

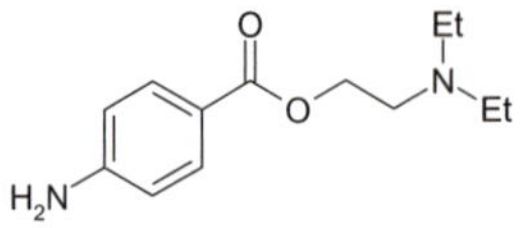

**25** Procaine 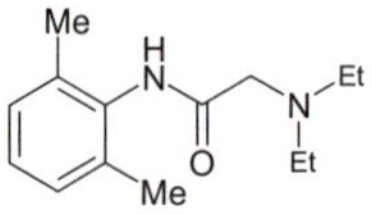 **26** Lignocaine

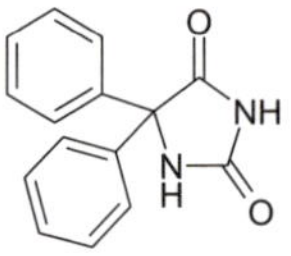

**27** Phenytoin 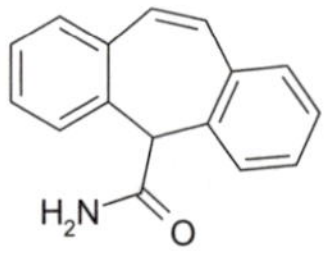 **28** Carbamazepine 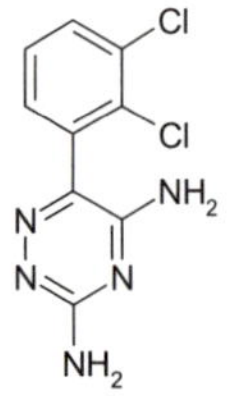 **29** Lamotrigine

**Figure 14** *Ligands acting at the voltage-gated sodium channels*

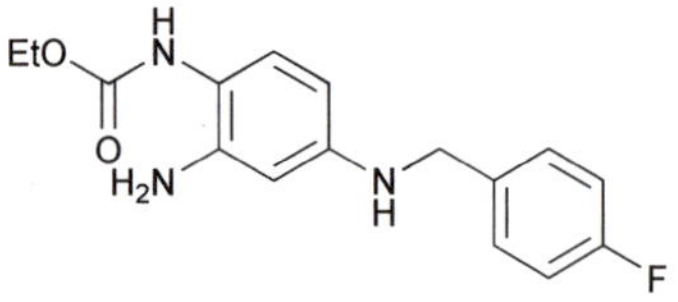

**30** Retigabine

**Figure 15** *Ligand acting at the voltage-gated potassium channels*

## Inward Rectifier and Miscellaneous Channels

The last group of channels to be discussed contains the inward rectifiers and miscellaneous types such as intercellular channels. The inward rectifiers are a distinct family of potassium channels (Kirs) containing two trans-membrane segments, these channels conduct current in the inward direction. These channels are differentially assembled with auxiliary sub-units (SURs) to form the $K_{ATP}$ channels which are inhibited by intracellular ATP and are unique as they act to couple cellular energy metabolism to membrane electrical activity.[9,27,28] Openers of this channel, *e.g.* diazoxide **31**, are antihypertensive, acting by relaxation of the smooth muscle within the walls of the arteries (Figure 16). Glibenclamide **32** is a blocker of the $K_{ATP}$ channels found in the pancreatic $\beta$-cells which secrete insulin, it binds to the SUR (sulfonylurea receptor) sub-unit of the $K_{ATP}$ assembly and is effective as antidiabetic drug for non-insulin dependant (type 2) diabetes.[28]

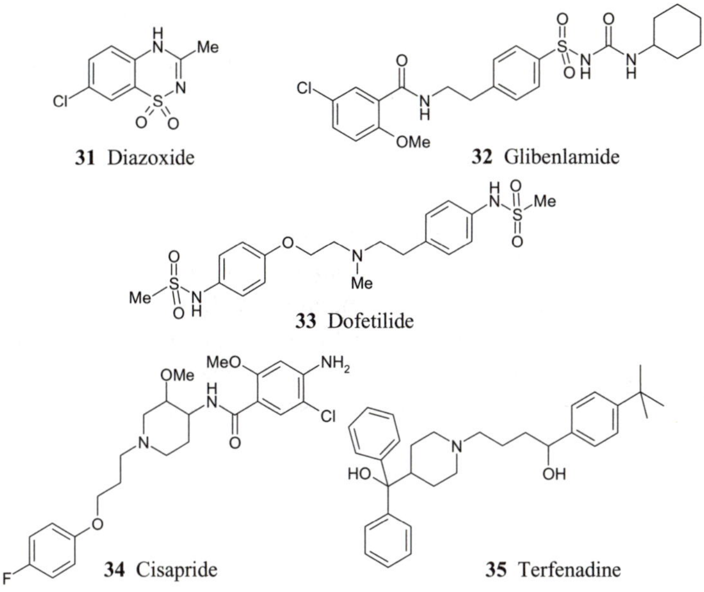

**31** Diazoxide

**32** Glibenlamide

**33** Dofetilide

**34** Cisapride

**35** Terfenadine

**Figure 16** *Ligands acting at the $K_{ATP}$ and hERG/miRP channels*

Another potassium channel co-assembly which has been the subject of much research recently is the hERG (human ether-a-go-go related gene) and miRP (minK-related peptide); this complex contributes the rapid or $I_{kr}$ component to the cardiac repolarisation phase. Blockade of this complex causes prolongation of the cardiac QT interval which has given rise to therapeutically useful antiarrythmic agents such as dofetilide **33**, the methanesulfonilide motif of which appears to be key for binding. In 'normal' patients channel blockade can potentially give rise to polymorphic ventricular tachycardia or torsades de pointes which can, in susceptible individuals, cause sudden death.[9,29] Recently, a number of marketed drugs have been withdrawn due to adverse affects associated with hERG blockade, these include the gastroprokinetic cisapride (Prepulsid®) **34** and the antihistamine terfenadine (Triludan®) **35**. The association of hERG blockade with QT prolongation and adverse affects have led to the recommendation that all new drugs should undergo a rigorous assessment of risk prior to human dosing.

Intercellular channels such a the gap junction family (connexins) connect neighbouring cells and mediate direct cell-to-cell communication by allowing the passage of nutrients, metabolites and small biological molecules such as second messengers.[30] Aquaporins are passive but selective pores for the transport of water and glycerol and are widespread in animals. There are ten different members of the mammalian aquaporin family and a number of disease states have been found to be associated with channel irregularities. Bacteria also contain aquaporins making them novel targets for antibacterial agents.[31]

## 4 CONCLUSION

To the medicinal chemist the sheer range, diversity and multifunctional nature of ion channels represent an amazing wealth of targets for potential therapeutic intervention. Genomic mapping will surely highlight many new families and help to distinguish the role and function of many sub-members of already known families. The complex modes of gating of the channels are indeed a big challenge for screen design and an equally big challenge for the medicinal chemist looking to discover potent and selective drugs with which to modulate them.

## 5 REFERENCES

1. S.A. Siegelbaum and J. Koester, in *Principles of Neural Science*, E.R. Kandel, J.H. Schwartz and T.M. Jessell (eds.), Elsevier, New York, 3rd edition, 1991, Chapter 5, p. 66.
2. J.A. Dani, *Biol. Psychiatry*, 2001, **49**, 166.
3. H. Brauner-Osborne *et al.*, *J. Med. Chem.*, 2000, **43**, 2609.
4. L.M. Gaster and F.D. King, *Med. Res. Rev.* 1997, **17**, 163.
5. S.S. Bhagwat and M. Williams, *Eur. J. Med. Chem.*, 1997, **32**, 183.
6. M. Chebib and G.A.R. Johnston, *J. Med. Chem.*, 2000, **43**, 1427.
7. R.D. Waigh, *Chem. Br.*, 1988, **24**, 1209.
8. G.J. Kaczorowski and M.L. Garcia, *Curr. Opin. Chem. Biol.*, 1999, **3**, 448.
9. M.J. Coghlan *et al.*, *J. Med. Chem.*, 2001, **44**, 1627.
10. S.K. Joseph, *Cell. Signal.*, 1996, **8**, 1.
11. R. Zucchi and S. Ronca-Testoni, *Pharmacol. Rev.*, 1997, **49**, 1.
12. J.L. Sutko *et al.*, *Pharmacol. Rev.*, 1997, **49**, 53.
13. N.A. Castle, *Perspect. Drug Discovery Des.*, 1999, **15/16**, 131.
14. A.D. Gruber *et al.*, *Curr. Genomics*, 2000, **1**, 201.
15. A. Nakanishi *et al.*, *Proc. Natl. Acad. Sci. USA*, 2001, **98**, 5175.
16. J.J. Clare *et al.*, *Drug Discovery Today*, 2000, **5**, 506.
17. T. Anger *et al.*, *J. Med. Chem.*, 2001, **44**, 115.
18. W.A. Catterall, *Neuron*, 2000, **26**, 13.
19. W.A. Catterall, *Annu. Rev. Cell Dev. Biol.*, 2000, **16**, 521.
20. D.J. Triggle *et al.*, *Ann. NY Acad. Sci.*, 1991, **635**, 123.
21. T.J. Jentsch *et al.*, *Ann. NY Acad. Sci.*, 1993, **707**, 285.
22. W.A. Catterall, *Nature (London)*, 2001, **409**, 988.

23. E. McCleskey, *Curr. Opin. Neurobiol.*, 1994, **4**, 304.
24. D. Rampe and D.N. Triggle, *Prog. Drug Res.*, 1993, **40**, 191.
25. B. Cox and J.C. Denyer, *Expert Opin. Ther. Pat.*, 1998, **8**, 1237.
26. B. Cox, *Curr. Rev. Pain*, 2000, **4**, 488.
27. N. Inagaki and S. Seino, *Jpn. J. Physiol.*, 1998, **48**, 397.
28. S.J.H. Ashcroft, *J. Membr. Biol.*, 2000, **176**, 187.
29. J.S. Mitcheson *et al.*, *Proc. Natl. Acad. Sci. USA*, 2000, **97**, 12329.
30. M.M. Falk, *Eur. J. Cell Biol.*, 2000, **79**, 564.
31. M.S.P. Sansom and R.J. Law, *Curr. Biol.*, 2001, **11**, R71.

CHAPTER 3

# Intracellular Targets

ERIC HUNT, NEIL PEARSON, TIMOTHY M. WILLSON AND ANDREW TAKLE

## 1 INTRODUCTION

Many targets that have provided successful drugs, many proteases, G-protein coupled receptors and ion channels are accessible to small molecule modulators directly from the extra-cellular environment. However, many targets are intracellular and they provide an additional challenge to the medicinal chemist, that of cellular penetration. This chapter briefly introduces the reader to some of the more important intracellular targets with references included for further information.

## 2 ANTIBACTERIAL INHIBITORS OF PROTEIN SYNTHESIS

Protein synthesis has both of the key attributes for an antibacterial target: it is essential for bacterial growth, and the macromolecular assemblies carrying out this process in prokarytic and eukaryotic cells are sufficiently different to enable selective inhibition. Although ribosome-mediated protein synthesis follows essentially the same pathway in bacterial and mammalian cells, the components of 70S bacterial ribosome are fewer and contain gross structural differences from those of the more complex 80S ribosome of eukaryotes. The majority of antibacterial protein synthesis inhibitors interfere directly with normal ribosome function and include such diverse structures as the macrolides, streptogramins, lincosamides, aminoglycosides, chloramphenicol, pleuromutilins, fusidic acid, tetracyclines and oxazolidinones. There are also major differences between the bacterial and mammalian enzymes involved in charging transfer-RNAs with amino acids prior to their delivery to the ribosome and use in protein synthesis. Selective inhibitors of these bacterial enzymes include mupirocin (isoleucyl-tRNA synthetase inhibitor) and indolmycin (tryptophanyl-tRNA synthetase inhibitor).

The protein synthesis inhibitors that have found the widest utility are the macrolide antibiotics, of which erythromycin has been in clinical use for almost fifty years.[1] Erythromycin and other macrolides inhibit peptide chain elongation by binding to the large, or 50S, ribosome subunit. Although the exact molecular interactions are unknown, the binding is tight and reversible and occurs at the ribosomal peptidyl transferase centre. Macrolides are quite lipophilic molecules and enter bacterial cells by passive diffusion. Because of this, their antimicrobial spectrum is confined to Gram-positive organisms and the more permeable Gram-negative species. Enterobacteriaceae and pseudomonads are intrinsically resistant; although macrolides are potent inhibitors of cell-free protein synthesis on ribosomes from these organisms, they are excluded from the cell by the relatively impermeable Gram-negative outer membrane. Nevertheless, macrolides are highly active against organisms commonly encountered in community-acquired respiratory tract infections (including atypical pneumonia), sexually transmitted infections, and skin and soft tissue infections, and have been widely used in treating these diseases.

Erythromycin **1** has excellent antimicrobial properties but has a major shortcoming in that its oral bioavailability is variable and erratic. This has been attributed in part to the compound's extreme lability to stomach acid. In dilute mineral acid, erythromycin is rapidly converted into antibacterially inactive dehydration products by way of reactions involving the 6-hydroxy and 9-keto groups. During the macrolide renaissance of the 1980s, many semi-synthetic analogues of erythromycin were prepared in the search for derivatives with improved acid stability and enhanced oral bioavailability and pharmacokinetics. A number of these semi-synthetic macrolides were successfully developed and entered clinical use, including erythromycin 6-methyl ether (**2**, clarithromycin), a derivative of erythromycin 9-oxime (**3**, roxithromycin) and a ring-expanded 15-membered macrolide (**4**, azithromycin). Clarithromycin, which has much improved oral bioavailability over erythromycin, and azithromycin, which has a greatly extended *in vivo* half-life, have proved particularly successful.[2]

Streptogramins also inhibit peptide chain elongation by binding to the 50S ribosome subunit.[3] The best known members of this group are the natural products pristinamycin and virginiamycin. Uniquely, these antibiotics consist of mixtures of two distinct compounds, streptogramins A and B, which individually bind to separate sites on the 50S ribosome and inhibit the early and late stages of protein synthesis respectively. In combination, the binding of type A streptogramins causes increased binding affinity of type B streptogramins, resulting in marked antibacterial

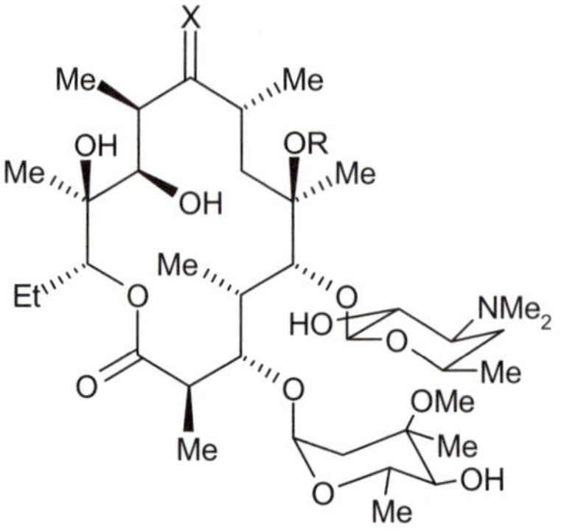

**1** erythromycin: R = H, X = O
**2** clarithromycin: R = Me, X = O
**3** roxithomycin: R = H, X = $NOCH_2(CH_2)_2OMe$

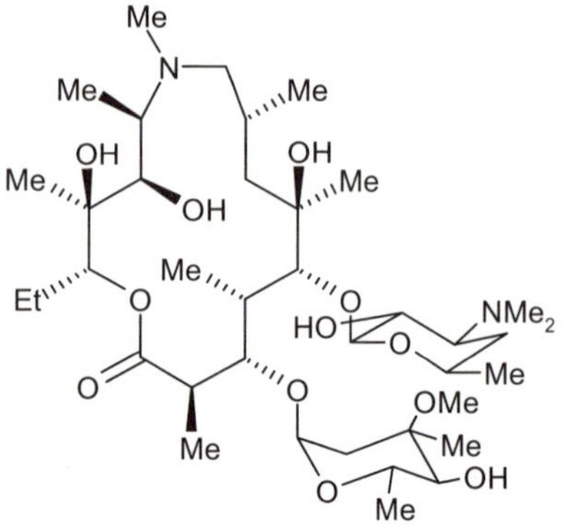

**4** azithromycin

synergy against Gram-positive bacteria and an effect that is bactericidal, in contrast to the bacteriostatic effect of the individual components. Although they are highly active against Gram-positive organisms and are orally absorbed, natural streptogramins have found limited clinical use. Their spectrum of activity, which includes methicillin resistant *Staphylococcus aureus*, is appropriate for treating serious Gram-positive infections in hospitals but this use is precluded by poor solubility in water and lack of injectable formulations. Investigation of semi-synthetic derivatives of pristinamycin during the 1990s identified a number of water soluble derivatives, notably quinupristin **5**, a type B derivative containing the water solubilising quinuclidine group, and dalfopristin **6**, a type A derivative containing the diethylamino-ethylsulfonyl moiety. The quinupristin–dalfopristin combination (Synercid®) has entered clinical use for treating severe infections caused by multiresistant Gram-positive pathogens.

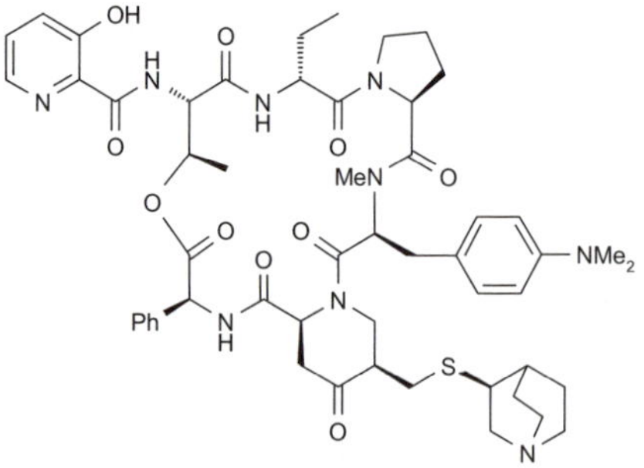

**5** Quinupristin

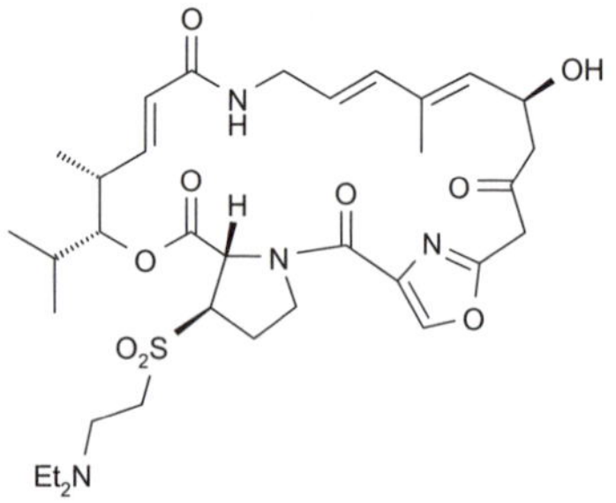

**6** Dalfopristin

The aminoglycosides differ markedly in structure from the macrolides and streptogramins and, not surprisingly, have very different antimicrobial and pharmacokinetic properties. The first member of this group, streptomycin **7**, was discovered during the 1940s and was followed by a large number of natural and semi-synthetic analogues. Rapid emergence of resistance has severely limited the utility of streptomycin, and gentamicin (mixture of **8**–**10**) is now usually regarded as the aminoglycoside of first choice. Aminoglycosides bind to the 30S ribosome subunit and have two major effects: they interfere with initiation of protein synthesis and they cause misreading of mRNA. This latter effect can lead to either premature termination of translation or incorporation of incorrect amino acids in the growing polypeptide chain. These polycationic antibiotics have a spectrum of activity encompassing mainly aerobic Gram-negative bacteria. They diffuse through aqueous porin channels in the Gram-negative outer membrane and cross the cytoplasmic membrane by a mechanism that depends on electron transport and a membrane potential that is negative on the interior. Because this latter process is oxygen-dependent, anaerobic organisms and facultative bacteria growing under anaerobic conditions are resistant. Against susceptible organisms, aminoglycosides are rapidly bactericidal and bacterial killing is concentration dependent. These highly polar, water soluble antibiotics are not adequately absorbed orally and must be administered by injection. Their use is complicated by the fact that all aminoglycosides show ototoxicity and nephrotoxicity and, in comparison with most antibiotics, have relatively narrow therapeutic margins. Despite these limitations, they are important agents and have been widely used in treating serious infections caused by Gram-negative aerobic bacteria.[4]

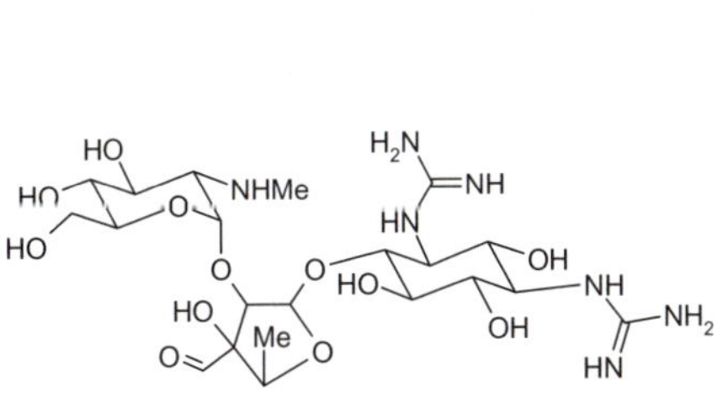

**7** Streptomycin

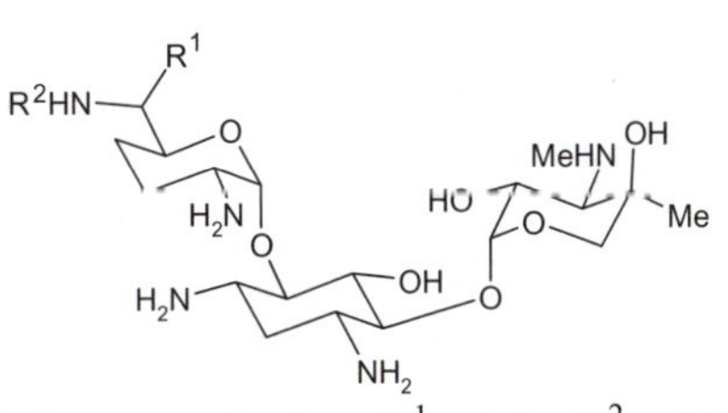

**8** Gentamycin $C_1$: $R^1 = Me, R^2 = Me$
**9** Gentamycin $C_{1a}$: $R^1 = H, R^2 = H$
**10** Gentamycin $C_2$: $R^1 = Me, R^2 = H$

One of the most recent protein synthesis inhibitors to enter clinical use is the synthetic oxazolidinone antibiotic linezolid **11**. Significantly, at the time of its launch, linezolid was the first member of a new class of antibiotic to enter the clinic for more than three decades. Oxazolidinones bind selectively to the 50S ribosome subunit at a site near the interface with the 30S subunit and inhibit translation at the initiation phase of protein synthesis. Linezolid and related oxazolidinones have a spectrum of activity covering mainly Gram-positive organisms, including *S. aureus*, *S. epidermidis*, enterococci and streptococci. Because the oxazolidinone binding site is unique, there is no cross-resistance with other classes of antibiotic. Linezolid has a good pharmacokinetic profile, by both oral and parenteral administration, and is highly effective in treating community- and hospital-acquired infections due to multiresistant Gram-positive cocci.[5]

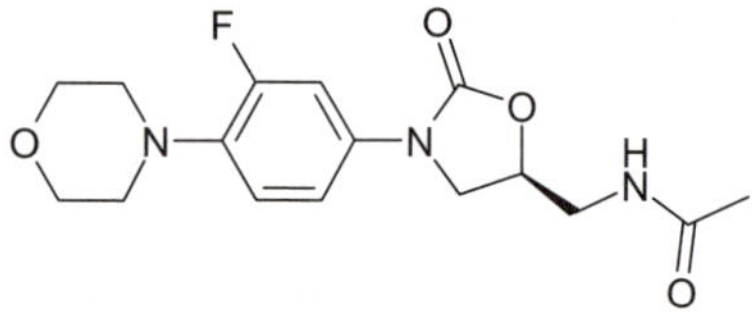

**11** Linezolid

At the start of the 21st century the greatest challenge facing antibacterial chemotherapy is combating the spread of bacterial resistance. The successful introduction of linezolid has prompted wide interest in the oxazolidines, which show no cross-resistance with older, established antibiotic classes. A clear objective would be to expand the utility of the oxazolidines to treating community-acquired respiratory infections. The latest generation of semi-synthetic macrolides, the ketolide antibiotics, address some of the resistance issues of erythromycin and its derivatives, and the most advanced members of this class are undergoing clinical evaluation. Interest has re-emerged in some previously known classes of protein sythesis inhibitors, such as pleuromutilins, which show no cross-resistance with established antibiotics but have not yet been exploited for human medicine. Alongside the work on new and improved classes of protein synthesis inhibitors, it is important to recognise that considerable progess has also been made in elucidating the detailed molecular structure of the 70S ribosome. X-ray crystallographic studies on the 30S ribosome have elucidated the binding interactions of the aminoglycosides.[6] A recent X-ray structure of the 50S ribosome undoutedly heralds similar studies on the binding of antibiotics that interact with the large ribosome subunit.[7]

## 3 BACTERIAL TOPOISOMERASE INHIBITORS: FLUOROQUINOLONES

The discovery that a compound isolated from a commercial preparation of chloroquine was antibacterially active has proven to be the catalyst for forty years of world-wide research into the fluoroquinolone class of antibacterial agents.[8] During this time, the antibacterial spectrum and potency has been improved enormously from the early, modestly potent Gram-negative agents used for treating urinary tract infections, to potent broad spectrum agents suitable for indications such as respiratory tract infections.

In addition to a narrow spectrum of activity, early quinolones such as nalidixic acid **12** had low oral absorption with peak serum levels of less than 0.5 $\mu g\,ml^{-1}$. Hence, intense international research did not begin in earnest until it was shown that in derivatives such as norfloxacin the combination of a fluorine atom at C-6 and a piperazine ring at C-7 afforded both improved anti-Gram-positive activity and pharmacokinetics.

### Mode of Action

The highly condensed state of DNA introduces problems of entanglement, strand unwinding and supercoiling which are addressed by the topoisomerases. The fluoroquinolones inhibit the function of DNA gyrase (bacterial topoisomerase II) and topoisomerase IV.[9] These are type II topoisomerases which transiently cleave both strands of the double helix by forming phosphotyrosyl bonds with active site tyrosines, pass another double helix through this break, then reseal the break. In a given bacterium, the relative sensitivity of these targets will differ with a general trend that DNA gyrase is more sensitive in Gram-negative bacteria and topoisomerase IV is more sensitive in Gram-positive bacteria. It is important to note that no X-ray structure of the inhibitory ternary complex of a topoisomerase, DNA and fluoroquinolone has been determined to date.

### Structure–Activity Relationships

Fluorine as the C-6 substituent is superior to all others and dramatically enhances *in vitro* potency against the target enzymes.[10] The even more dramatic improvement in antibacterial activity suggests an additional enhancement of penetration into the bacterial cell. Hence, research has focused on the 6-fluoroquinolones and 6-fluoronaphthyridones, known generically as the fluoroquinolones. For this class of antibiotics the pharmacophoric unit consists of the 4-oxo moiety, its associated C-3 carboxylic acid and the pyridone nitrogen (N-1). Modifications at C-2, the C-3 carboxyl and C-4 are minimally tolerated whereas all other positions can be varied widely. A basic substituent at C-7 is optimal and is very important for absorption from the gastrointestinal tract. Furthermore, the C-7 substituent has proven to be the most tolerant to modification and can afford major changes in potency and spectrum, as well the ability to adjust absorption, distribution, metabolism and excretion.

Discussion of the SAR will concentrate on the N-1, C-5 , C-7 and C-8 substituents which have afforded marketed antibiotics. Subsequent to the patenting of norfloxacin **13** (Figure 1) in 1978, knowledge of SAR for the fluoroquinolones grew rapidly.[11] Early studies indicated that a small saturated or unsaturated alkyl N-1 substituent afforded good antibacterial

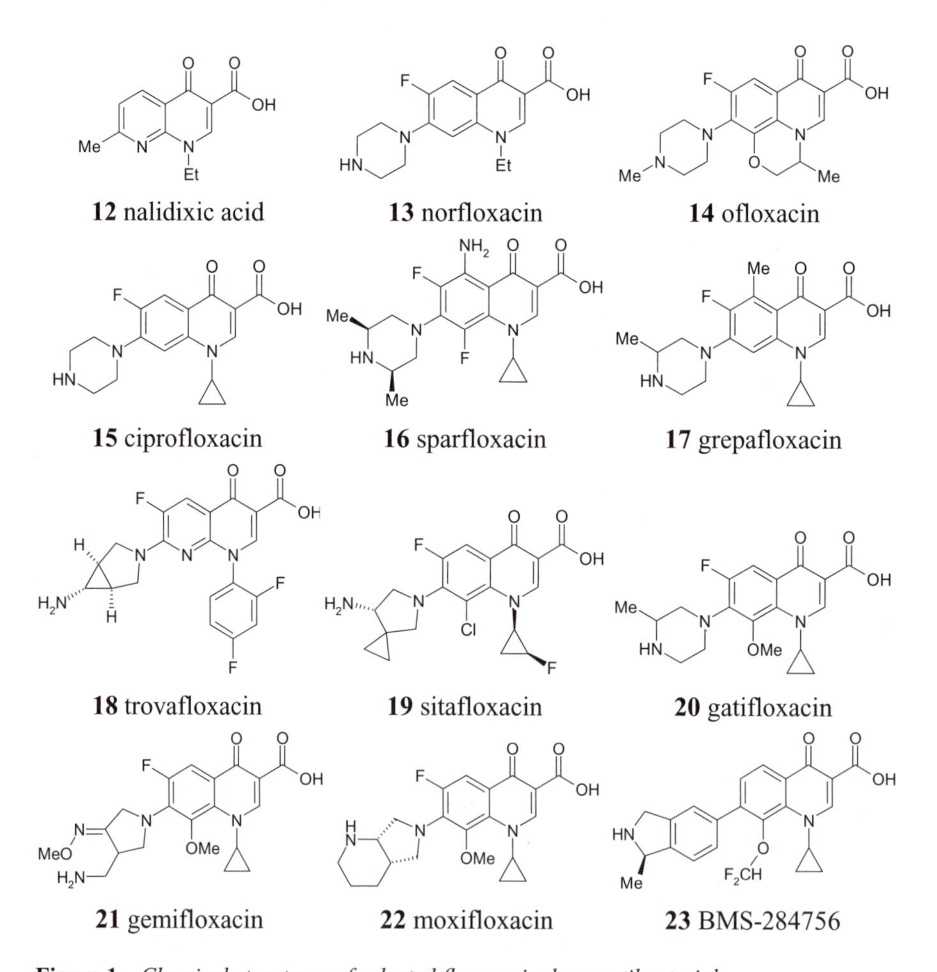


**Figure 1** *Chemical structures of selected fluoroquinolone antibacterials*

activity, especially against the Gram-negative organisms, with ethyl being optimal. In addition, fused N-1 substituents were identified with good activity. These are exemplified by the benzoxazine ofloxacin **14** (marketed in 1985), where this may be considered as a derivative resulting from the fusion of an N-1 ethyl group and a C-8-methoxy group. The more potent *S*-enantiomer has also been marketed more recently. The discovery of ciprofloxacin **15** (marketed in 1986), with an N-1 cyclopropyl substituent, and the even more surprising potent activity seen for N-t-butyl derivatives upset this view. Subsequently, a substituted phenyl moiety was shown to afford potent activity, exemplified by the 2,4-difluorophenyl substituent present in tosufloxacin (marketed in 1990) and the more recent trovafloxacin **18** (1998) (Table 1). These examples may serve as a reminder that a too scholarly or rational interpretation of SAR at early stages of research efforts can be detrimental to progress.

Overall, ciprofloxacin has become the 'gold standard' of the first generation of fluoroquinolones, showing optimal antibacterial activity particularly against the Gram-negative organisms such as Enterobacteriaceae and *Pseudomonas aeruginosa*. However, this agent has a number of limitations, which provided impetus for developing new compounds. Firstly, with hindsight the activity against Gram-positive cocci could be considered to be inadequate for treating respiratory tract infections and resistance rapidly arose, especially in *Staphylococcus aureus*. Secondly, ciprofloxacin is not very potent against the anaerobic organisms commonly encountered in abdominal infections. Thirdly, there are other issues unrelated to enzyme and antibacterial potency issues, such as sub-optimal clearance rates requiring dosing three times a day and significant interactions with theophyline metabolism.

To this day, cyclopropyl is still the most common N-1 substituent found in fluoroquinolones, with the appropriate combination of C-7, C-5 and C-8 substituent or naphthyridinone being required to produce drug candidates. There is ample and increasing evidence that the optimum quinolone will not result from a cut-and-paste assembly of the best individual substituents at each position. This is particularly the case when toxicological effects, physiochemical properties such as solubility and pharmacokinetic factors are considered, rather than just antibacterial potency. Moreover, an acceptable safety profile is becoming increasingly important and difficult to achieve, with high profile clinical issues having highlighted toxicological effects, which are regarded as class effect issues by regulatory authorities. The first marketed agent with Gram-positive activity suitable for treating respiratory tract infections was sparfloxacin **16**, which clearly illustrates this general scenario.

Sparfloxacin contains a 5-amino group which improves Gram-positive

**Table 1** *Comparative* in vitro *minimum inhibitory concentration* ($MIC_{90}$ μg ml$^{-1}$)

| *Organism* | *Ciprofloxacin* (**15**) | *Sparfloxacin* (**16**) | *Trovafloxacin* (**18**) | *Grepafloxacin* (**17**) | *Gatifloxacin* (**20**) | *Moxifloxacin* (**22**) |
|---|---|---|---|---|---|---|
| *S. aureus* MSSA | 0.25–>2 | 0.01–1 | 0.06 | 0.1–0.25 | 0.1–0.13 | 0.06 |
| *S. aureus* MRSA | — | 0.03–16 | 0.05–4 | 4 | 0.2–16 | 4 |
| *S. aureus* CRSA | >64 | >4 | 1–8 | 8–16 | 4–16 | 2 |
| *S. pneumoniae* | 1–8 | 0.5 | 0.12–0.5 | 0.25–0.5 | 0.5 | 0.12–0.25 |
| *E. faecalis* | 0.5–4 | 0.12–2 | 0.12–2 | 0.4–4 | 0.8–2 | 1 |
| *H. influenzae* | >0.008–0.06 | 0.025 | 0.016 | 0.008–0.06 | 0.013–0.016 | 0.06 |
| *B. fragilis* | 4–128 | 1–2 | 0.25–2 | 2–32 | 0.25–1 | 0.12 |
| *E. coli* | >0.01–0.25 | 0.03–0.5 | 0.03–0.5 | 0.06–0.12 | 0.016–0.1 | 0.008 |
| *P. aeruginosa* | 0.25–8 | 6.25–25 | 1–4 | 1–8 | 3.2–32 | 1–4 |

activity when combined with an N-1 cyclopropyl group but not with ethyl or 2,4-difluorophenyl. The extent of this improvement is dependent on the C-8 substituent with fluorine being optimal. A study on the effect of piperazine methylation in sparfloxacin and its analogues showed that *cis*-dimethylation decreased the *in vitro* activity slightly but, importantly, improved both the oral efficacy in animal models and selectivity for bacterial *versus* mammalian topoisomerase II.[12] However, significant issues with phototoxicity and QT interval prolongation have seriously limited usage. The 5-methyl derivative grepafloxacin **17** also exhibits more potent Gram-positive activity than ciprofloxacin, with improved pharmacokinetics resulting in higher levels in the lung and a more than doubled half-life in man. Again, QT-prolongation issues have been seen in the clinical situation leading to the withdrawal of this compound.

The paediatric use of fluoroquinolones has been prevented by the observation that lesions are induced in the cartilage of immature animals. This puts fluoroquinolones at a disadvantage when compared to macrolides or beta-lactams and it would be highly advantageous to remove this effect. In this regard, it is notable that the 6-H derivative BMS-284756 **23** shows no arthropathy in rat studies and is currently in Phase III trials.[13]

The widest variety of structural modifications have been introduced at the C-7 position, with cyclic diamines dominating due to a combination of excellent biological activity and their synthetic accessibility *via* nucleophilic aromatic substitution on a halogenated quinolone precursor. Primarily driven by improvements in Gram-positive activity, intensive investigation of piperazines was followed by an emphasis on amino- and aminomethyl-pyrrolidines. Sitafloxacin **19** and gemifloxacin **21** are two examples of the fruits of these efforts. In the late 1980s more structurally complex C-7 substituents have come to the fore due to their ability to ameliorate adverse effects and improve pharmacokinetic properties. Several groups have studied bicyclic systems with trovafloxacin **18** and moxifloxacin **22** having been marketed and several other examples in late stage clinical trials.

The increasing prevalence of fluoroquinolone resistant bacteria, due to both target mutations and efflux mechanisms, has made potent activity *versus* these pathogens essential for new agents.[14] This has been addressed by increasing the potency against the parent enzymes such that the reduced activity against the mutant enzymes still results in an MIC level within the therapeutically acceptable range. The 8-methoxy substituent present in compounds such as moxifloxacin has been demonstrated *in vitro* to afford this property, attributed to the increased level of bactericidal activity this substitution confers. It remains to be seen whether the expectations for these agents are realised in the clinical situation.

### Conclusions

Since the marketing of the first quinolone antiobiotic, nalidixic acid, these agents have attained increasingly widespread clinical use, with the growth in the market seemingly poised to increase with the recent launch of a new generation of broad spectrum fluoroquinolones such as gatifloxacin, moxifloxacin. However, the recent post-marketing withdrawal of trovafloxacin (liver toxicity issues) and grepafloxacin illustrates how unexpected issues can arise at any time and the need for further research in fluoroquinolones.

## 4 NUCLEAR RECEPTORS

The lipophilic steroid and retinoid hormones control a diverse range of biological activities, including development, reproduction, differentiation and cellular homeostasis, by binding to intracellular nuclear receptors.[15] Acting as transcriptional switches, the nuclear receptors can turn on and off the complex circuitry of gene expression within cells. Several small lipophilic hormones control these switches. Well-known examples are the sex steroids estradiol, testosterone and progesterone, the adrenal stress hormones cortisol and aldosterone, the secosteroid vitamin D, and the non-steroid hormones thyroxine and retinoic acid. These hormones play important roles in human physiology, and several widely prescribed pharmaceutical drugs bind to the nuclear receptors (Table 2). Most of

**Table 2** *Examples of nuclear receptor drugs*

| *Drug®* | *Nuclear receptor* | *Indication* |
|---|---|---|
| Nolvadex | Estrogen receptor | Breast cancer |
| Evista | Estrogen receptor | Osteoporosis |
| Premarin/Prempro | Estrogen and progesterone receptors | Hormone replacement therapy |
| Tri-Cyclen | Estrogen and progesterone receptors | Contraception |
| Casodex | Androgen receptor | Prostate cancer |
| Aldactone | Mineralocorticoid receptor | Congestive heart failiure |
| Flovent/Flonase | Glucocorticoid receptor | Asthma/hay fever |
| Dovonex | Vitamin D receptor | Psoriasis |
| Synthroid | Thyroid hormone receptor | Hypothyroidism |
| Accutane | Retinoid acid receptor | Acne |
| Lopid | Peroxisome proliferator-activated receptor $\alpha$ | Cardiovascular disease |
| Avandia | Peroxisome proliferator-activated receptor $\gamma$ | Adult onset (Type 2) diabetes |

these drugs, however, were developed using traditional pharmacology prior to the cloning of their receptors. Recent advances in the molecular biology of the nuclear receptors, coupled with the solution of their high resolution X-ray crystal structures, has uncovered new opportunities for drug discovery.

## Nature's Gene Switches

The nuclear receptors are related to each other by their amino acid sequence and their function within cells (Figure 2(A)). The receptors use a DNA-binding domain to interact with the regulatory regions of genes, whose expression they control. This domain contains eight cysteines, which form a pair of tetra-coordinate binding sites for zinc atoms. The three-dimensional structure of the DNA-binding domain from the gluco-

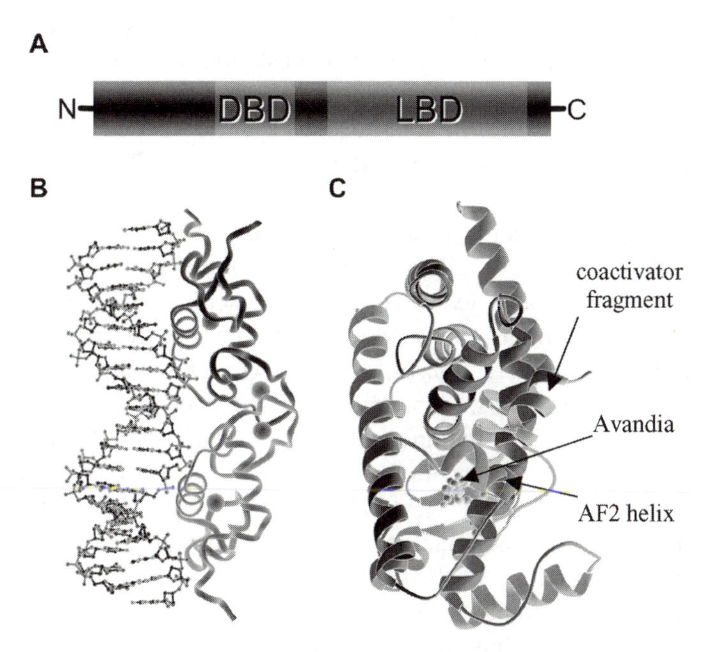


**Figure 2** *The structure of nuclear receptors:* (A) *The nuclear receptors contain a DNA-binding domain (DBD) and a ligand binding domain (LBD), whose amino acid sequences are conserved within the receptor family.* (B) *The structure of the DBD of the glucocorticoid receptor. Two glucocorticoid receptor DBDs bind to DNA as a homodimer. An α-helix from each DBD fits into the major groove of the DNA double helix.* (C) *The LBD of the orphan nuclear receptor PPARγ is composed of three layers of α-helices.*[16] *A molecule of the diabetes drug Avandia is seen inside the receptor*

corticoid receptor reveals that the two zinc atoms allow folding of the protein, such that an $\alpha$-helix is placed into the major groove of the DNA double helix (Figure 2(B)). The amino acids on this $\alpha$-helix are responsible for the sequence specific recognition of the DNA by the glucocorticoid receptor. Nuclear receptors can bind to DNA as monomers, homodimers or heterodimers. The glucocorticoid receptor binds its DNA sequence as a homodimer with a three nucleotide spacing between the two binding sites. Variations in the orientation of the binding sites and their nucleotide spacing allow a large variety of DNA sequences to be recognised by different nuclear receptors.

Adjacent to the DNA-binding domain is a separate domain where lipophilic hormones interact with the receptor. Known as the ligand-binding domain, this portion of the receptor is larger and proportionally more complex than the DNA-binding domain. X-ray crystal structures of several nuclear receptor ligand-binding domains have been solved.[17] These structures show that the domain is constructed of three layers of $\alpha$-helices (Figure 2(C)). The lower half of the domain contains a pocket, into which the lipophilic hormones can bind. Comparison of structures determined in the presence and absence of hormone suggests that the hormone triggers a conformational change in the C-terminal $\alpha$-helix of the domain, known as the AF2 helix. This conformational change can be likened to a molecular switch. In the absence of hormone, the receptor is "off", and the binding of the hormone places the receptor in the "on" position. Several families of accessory proteins have been identified that bind to the AF2 helix. These proteins have the ability to modify the folding of DNA within the nucleus of cells and also to act as bridging factors between the nuclear receptor and a large multiprotein complex that initiates the process of gene transcription. Collectively, these accessory proteins are known as co-activators or co-repressors, depending on whether they increase or decrease gene expression within cells. Conformational changes in the receptor, following hormone binding, lead to release of co-repressors and recruitment of co-activators. Remodelling of the adjacent DNA structure and assembly of the cellular transcriptional machinery results in an increase in the expression of the gene.

Analysis of the available X-ray structures reveals that most of the amino acids lining the ligand binding pocket are hydrophobic. Not surprisingly, the natural ligands that bind to the nuclear receptors are lipophilic molecules with similar molecular volumes. High affinity binding results from the complementarity of the shape of the hormone with the hydrophobic surface of the ligand binding pocket. A few polar amino acids may also contribute to the ability of a receptor to recognise its specific hormone. Thus, nuclear receptor ligands are usually small lipophilic molecules containing a limited number of hydrogen bond donor and acceptor groups. Molecules with these physical properties generally show good cell penetration, which aids the design of synthetic drug molecules targeting these intracellular receptors.

## The Orphan Nuclear Receptors

To date, forty eight human nuclear receptors have been cloned that contain either the signature DNA-binding or ligand-binding domain. Only twelve of these receptors recognise the eight classical steroid and retinoid hormones. The remaining thirty six are termed orphan nuclear receptors, because at the time of their isolation the corresponding hormones were unknown.[18] The presence of so many orphan receptors predicts that there must be many additional lipophilic hormones which are waiting to be identified. An example is the discovery of dietary fatty acids as ligands for the peroxisome proliferator-activated receptors.

A large percentage of the calories that we ingest are in the form of fat. Understanding how the body regulates the storage or disposal of this energy is an important medical issue, since the fat content and composition of the diet can affect human health. For example, diets high in saturated fat are associated with an increase in the incidence of heart disease, diabetes, obesity and cancer. In 1990, Stephen Green at the ICI Central Toxicology Laboratory reported the cloning of a new orphan nuclear receptor. A class of chemicals that cause the hyperproliferation of organelles in the liver, known as peroxisomes, could turn on (activate) the receptor. For this reason, the orphan receptor was called the peroxisome proliferator-activated receptor (PPAR). When two closely related receptors were cloned, the three receptors were designated PPAR$\alpha$, PPAR$\gamma$ and PPAR$\delta$,[19] even though only the original receptor (PPAR$\alpha$) was associated with peroxisome proliferation. All three PPARs could be activated by high concentrations of fatty acids, which led to the intriguing hypothesis that the PPARs might be hormonal receptors for dietary fats and lipids. The identification of two classes of known drug molecules that bind to the PPARs has further strengthened the connection between the PPARs and fat metabolism. The lipid lowering fibrate drugs and the insulin sensitising glitazone drugs bind to PPAR$\alpha$ and PPAR$\gamma$, respectively.[20] The molecular mechanism of action of these drugs shows that they promote the disposal or storage of dietary fat by changing the expression levels of fat metabolising enzymes and lipid binding proteins.

The fibrate drugs lower triglycerides and raise HDL-cholesterol (the "good" cholesterol), two risk factors for the development of cardiovascular disease. Activation of PPAR$\alpha$ by fibrate drugs leads to changes in the expression of genes involved in fat metabolism in the liver. Importantly, the phenomenon of peroxisome proliferation appears to be limited to rodents and does not extend to humans. Fibrate drugs have not been as widely prescribed as the statin class drugs, which lower LDL-cholesterol (the "bad" cholesterol) by inhibition of the enzyme HMG-CoA reductase. This view may be changing, however, as clinicians recognise that it is important to control additional risk factors that affect the progression of coronary heart disease. For example, in 2500 patients with normal levels of LDL-cholesterol, the Veterans Affairs HDL cholesterol Intervention Trial (VA-HIT) showed that a fibrate drug lowered the incidence of heart attacks by 22%.[21] It is possible that super-fibrates with improved activity on human PPAR$\alpha$ will be powerful drugs for management of heart disease, alone or in combination with statin drugs.

Adult onset (Type 2) diabetes is an increasingly common disease in Western society. Epidemiological studies suggest that poor diet, obesity and a sedentary lifestyle are associated with insulin resistance leading to the development of diabetes. Peripheral tissues, such as muscle, use glucose as an energy source when signalled by the peptide hormone insulin. If muscle becomes resistant to the insulin signal, the pancreas must produce greater quantities of insulin to compensate. As patients become increasingly insulin resistant, the pancreas can no longer compensate, resulting in poor glucose control and diabetes. The glitazones were developed as insulin sensitising drugs to correct the underlying cause of this disease. Although their molecular mechanism of action was unknown, several glitazone drugs were recently approved for the treatment of adult onset diabetes. PPAR$\gamma$ was cloned as a transcription factor that controlled the formation of fat cells. Since obesity is a common co-morbidity with adult onset diabetes, it was a major surprise when PPAR$\gamma$ was identified as the receptor for the anti-diabetic glitazone drugs.[22] We still do not fully understand why the regulation of gene expression by PPAR$\gamma$ in fat leads to insulin sensitisation of the muscle in diabetic patients. However, there is accumulating evidence that fat tissue plays an important metabolic role in humans. For example, patients who lack adequate levels of fat tissue (a disease known as lipodistrophy) are highly insulin resistant and often develop diabetes at an early age.

Most diabetics die prematurely from cardiovascular disease. A new generation of anti-diabetic drugs that combines the beneficial effects of the glitazones and fibrates is currently under development. These drugs, which activate both PPAR$\alpha$ and PPAR$\gamma$, should treat the insulin resistance, high triglycerides and low HDL cholesterol, which are commonly seen in adult onset diabetes.

**Nuclear Receptors as Targets for Drug Discovery**

The sequencing of the human genome will reveal the bulk of the new targets for drug discovery in the 21st century. Experience has taught pharmaceutical scientists that there are only a few classes of molecular targets that can be readily manipulated by orally active drugs. The nuclear receptors are one of these classes of tractable drug discovery targets. To date, there has been little evidence of redundancy within this receptor family. Every receptor has been shown to regulate an important physiological process. Thus, each of the remaining orphan receptors has the potential to regulate a new hormone signalling pathway involved in human disease.

Our understanding of the biological role of a nuclear receptor is often built around the knowledge of a single gene, whose expression is up or down regulated. In reality, each receptor controls the expression of hundreds of genes within a cell. Differential gene expression technology, which employs gene chips or glass arrays to identify all the genes whose expression is changed, will add a new level of detail to our understanding of nuclear hormone action. These advances in our understanding of how nuclear receptors regulate gene expression, coupled with new insights from the three-dimensional structures of their ligand-binding domains, augurs well for continued success in manipulating Nature's gene switches to treat human diseases.

## 5 PROTEIN KINASES

Protein kinases are a family of intracellular targets which have caught the attention of the medicinal chemistry community in recent years. These enzymes catalyse the phosphorylation of protein substrates by transferring the terminal phosphate from ATP onto the side-chains of specific amino acid residues in the substrate protein. The best characterised groups of protein kinases are the serine-threonine kinases (STKs), which phosphorylate the side-chain hydroxyl groups of serine and/or threonine residues, and the protein tyrosine kinases (PTKs), which phosphorylate the phenolic hydroxyl groups of tyrosine residues. The latter PTK group can be further subdivided into receptor tyrosine kinases (rPTKs), where the kinase is linked to an extra-cellular receptor, and non-receptor tyrosine kinases.

*In vivo*, the action of protein kinases is countered by that of protein phosphatases, which catalyse the dephosphorylation of phospho-proteins. As the activity of many proteins is dependent on their phosphorylation state, the opposing action of these two groups of enzymes effectively acts as a collection of molecular switches which allow the activity of the substrate proteins to be switched between one state and another (*e.g.* inactive to active or *visa versa*). If two or more protein kinases are arranged in series (*i.e.* the substrate protein(s) is itself a protein kinase which requires phosphorylation for activation) a protein kinase cascade is formed. Such cascades form an essential component of many intracellular signal transduction processes, where transmission of a signal is accomplished by a series of sequential phosphorylation reactions. Furthermore, as each individual protein kinase can phosphorylate a number of substrate molecules, amplification of the signal can also be achieved.[23]

One of the most studied group of protein kinase cascades are the so-called mitogen-activated protein kinase (MAPK) cascades. These cascades enable the transmission of intracellular signals following binding of various extra-cellular ligands (*e.g.* growth factors, cytokines) to their respective cell surface receptors.[24] In the archetypal MAPK cascade (Raf-MEK-ERK, Figure 3), binding of a growth factor to its receptor leads to the activation of a protein known as Ras. Once activated, Ras itself then activates Raf, the first protein kinase in the Raf-MEK-ERK signalling module. The activated Raf then phosphorylates, and hence activates MEK which in turn phosphorylates ERK1 and 2. The phosphorylated ERKs then translocate into the nucleus and activate a number of transcription factors which result in gene expression. Thus the initial ligand–receptor binding event is converted into an intracellular response. However, it should be noted that many parallel signalling pathways exist and that cross-talk between such pathways can also occur. Consequently, any response is

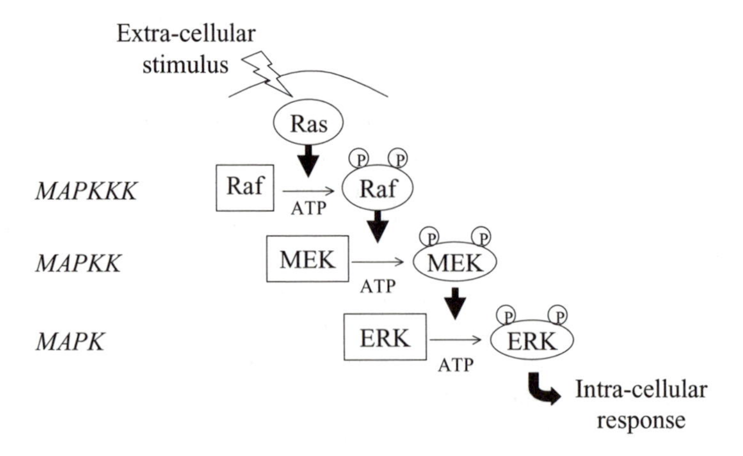


**Figure 3** *The Raf-MEK-ERK protein kinase cascade: MAPK = MAP kinase; MAPKK = MAP kinase kinase; MAPKKK = MAP kinase kinase kinase*

actually an integrated, balanced response to a number of intra-cellular signals rather than due to any single pathway.

### Inhibition of Protein Kinases

Protein kinases are involved in the regulation of many aspects of cellular function, such as cellular metabolism, cell growth and differentiation, apoptosis and immune responses. Inhibitors of these enzymes could, therefore, have utility in the treatment of many disparate diseases and, as such, are being actively sought within the pharmaceutical industry.[25]

The majority of small molecule protein kinase inhibitors reported to date are ATP-competitive inhibitors. If such an inhibitor is to have any potential therapeutic utility, it will need to overcome two hurdles which are specific to protein kinase inhibitors of this type (in addition to that of cellular penetration which is common to all inhibitors of intracellular targets). The first is the issue of selective inhibition of the desired kinase. This was initially thought to be a major problem for ATP-competitive inhibitors, especially when one considers that there are an estimated 500 protein kinases in the human genome. However, in 1994 the Parke-Davies group reported details of a selective inhibitor of the epidermal growth factor receptor tyrosine kinase (EGFR), PD 153035 **24**, which indicated that inhibitors with an acceptable selectivity profile could indeed be obtained.[26] The second is to overcome the high concentration (millimolar) of ATP within the cell. This latter problem is not insurmountable if inhibitors of sufficient potency are identified, but it goes some way to explain the discrepancy often seen between *in vitro* $IC_{50}$s and those from cell-based assays. Caution should also be exercised when comparing *in vitro* $IC_{50}$ data from different sources since the $IC_{50}$s are critically dependent on the concentration of ATP used in the assay.

Protein kinases have proven to be amenable to structural elucidation and, as such, there is a wealth of information available with which to aid structure-based drug design.[27] The available structures include examples of STKs and PTKs, both in their apo-form and in complex with ATP (or its analogues) or small molecule inhibitors. These indicate that the protein kinase family share a structurally conserved catalytic unit, consisting of two major domains linked by a single peptide strand. ATP binds to the interface between these two domains with the adenine base forming two hydrogen bonds, from N-1 and the C-6 amino group, to the linker region of the protein (Figure 4). The structures have also revealed that there are regions in the vicinity of the ATP binding site which are not occupied by ATP and which show structural diversity between different protein kinases. Furthermore, structures of selective protein kinase inhibitors bound to their

**24** PD 153035 **25** ZD 1839 **26** SB-203580

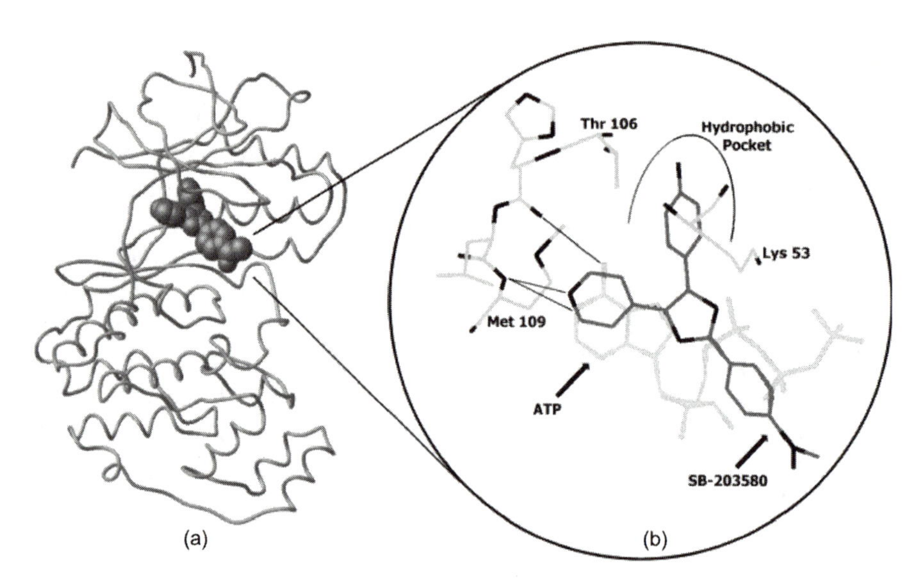


**Figure 4** (a) *An X-ray crystal structure of SB-203580 bound to the ATP binding site of p38α MAP kinase, and* (b) *an active site view indicating the key interactions made between SB-203580 and the protein; hydrogen bonds are depicted by solid lines. Key protein residues in the vicinity of the linker region and the aromatic binding pocket are shown, as is an overlaid structure of ATP (shadow) docked into p38α MAP kinase for comparison*

respective target kinase indicate that, in many cases, the inhibitors access one or more of these regions which may account for their selectivity.

## Inhibitors of Epidermal Growth Factor Receptor Kinase (EGFR)

Epidermal Growth Factor Receptor kinase (EGFR) is a receptor tyrosine kinase, which induces cellular proliferation and differentiation following binding of its ligand. A great deal of evidence has been gathered to implicate aberrant activity of EGFR (either as a result of mutations or over-expression of the enzyme) with the development and progression of a number of human tumours. Indeed, EGFR is often used as prognostic marker since its over-expression correlates with increased death rates in several tumour types. Consequently, inhibitors of EGFR have been much sought after for cancer chemotherapy.[28]

One of the most promising EGFR inhibitors is the quinazoline ZD 1839 **25** from Astra-Zeneca. The quinazoline series of PTK inhibitors is postulated as binding to ATP binding site of the EGFR kinase with the quinazoline N-1, forming a hydrogen bond with a backbone N–H of the linker region. This binding mode then places the anilino group into a hydrophobic pocket not occupied by ATP. ZD 1839 is a potent inhibitor of the EGFR kinase ($IC_{50} \approx 23$ nM) with almost 100-fold selectivity over the closely related c-erbB-2 kinase ($IC_{50} \approx 2$ μM) and excellent oral bioavailability. In cell based assays, it inhibits EGF stimulated cell growth but shows little effect on cells in the absence of EGF stimulation. In *in vivo* models, a daily 10 $mg\,kg^{-1}$ dose caused 50% inhibition of A431 tumour growth, whereas regression to an undetectable size was achieved following daily doses of 200 $mg\,kg^{-1}$ over two weeks. ZD 1839 (Iressa®) is currently in the late stages of clinical development.

## Inhibitors of p38 MAP Kinase

p38 MAP kinase is a serine–threonine protein kinase which lies at the MAPK level of the p38 MAP kinase cascade. The cascade is activated in response to a variety of extra-cellular stimuli, such as environmental and chemical stress signals, and results in a number of down stream events which are both stimulus- and cell type-dependent. In particular, activation of the p38 cascade has been linked to the production of the pro-inflammatory cytokines interleukin-1 (IL-1) and tumour necrosis factor $\alpha$ (TNF-$\alpha$). Inhibitors of p38 have thus been sought for the treatment of a variety of inflammatory diseases such as rheumatoid arthritis.[29]

One of the most widely documented p38 MAP kinase inhibitors is the prototypical pyridinyl-imidazole, SB-203580 **26**. This compound is a potent inhibitor of p38$\alpha$ MAP kinase ($IC_{50}$ 48nM) and is effective at reducing lipopolysaccharide induced TNF-$\alpha$ production in the mouse ($ED_{50}$ 15 $mg\,kg^{-1}$) following oral dosing. SB-203580 also showed an intriguing selectivity profile when screened against the closely related p38$\beta$, $\gamma$ and $\delta$ enzymes. Whilst p38$\beta$ was inhibited with equal potency to p38$\alpha$ ($IC_{50} \sim 50$ nM), both p38$\gamma$ and $\delta$ were insensitive to the activity of SB-203580 ($IC_{50}$s $>$ 10 000 nM). This selectivity profile was rationalised following a series of X-ray crystallography and site-directed mutagenesis (SDM) studies.

The X-ray studies indicated that SB-203580 bound to the ATP binding site of p38$\alpha$, with the pyridine nitrogen forming a hydrogen bond to the backbone NH of Met109 in the linker region of the protein and the 4-fluorophenyl group binding in a hydrophobic pocket, which lies behind the ATP binding site (Figure 4). In p38$\alpha$ and $\beta$ the hydrophobic pocket is bordered by residues Thr106 and Lys53, whereas both p38$\gamma$ and $\delta$ contain a larger methionine residue at position 106. This larger residue is postulated to block access to the hydrophobic pocket, and thus prevent binding of the inhibitor to the enzyme. This theory was supported by the results of a series of SDM studies where p38$\alpha$ was rendered insensitive to SB-203580 following mutation of Thr106 to Met and, conversely, p38$\gamma$ and $\delta$ were made sensitive to SB-203580 following the reverse Met106 to Thr mutation.[29]

### Conclusion

Protein kinases constitute a class of intracellular molecular targets with many potential therapeutic indications. Although the large number of kinases likely to be present in the human genome suggests that the identification of truly specific inhibitors is unlikely, inhibitors with a therapeutically useful profile of activity do appear attainable – indeed a number with very encouraging selectivity profiles have already been identified. A large amount of structural information is also available with which to aid inhibitor design and which has enabled the rationalisation of SAR trends and selectivity profiles. Therefore it appears just a matter of time before a kinase inhibitor is identified which fulfils the promise of this target class.[30]

## 6 REFERENCES

1. H.A. Kirst, in *Prog. Med. Chem.*, G.P. Ellis and D.K. Luscombe (eds.), 1993, **30**, p. 57.
2. D.T.W. Chu, *Curr. Opin. Microbiol.,* 1999, **2**, 467.
3. J.C. Barriere, *et al.*, *Curr. Pharm. Design*, 1998, **4**, 155.
4. I. Phillips and K.P. Shannon, in *Antibiotic and Chemotherapy*, F. O'Grady *et al.* (eds.), Churchill Livingstone, New York, 7th edition, 1997, p. 164.
5. D. Clemett and A. Markham, *Drugs*, 2000, **59**, 815.
6. A.P. Carter *et al.*, *Nature (London)*, 2000, **407**, 340.
7. N. Ban *et al. Science*, 2000, **289**, 905.
8. G.Y. Lesher *et al.*, *J. Med. Pharm. Chem.*, 1962, **5**, 1063.
9. D.C. Hooper, *Drugs*, **58**, 1999, Suppl. 2, 6.
10. J.M. Domagala, *J. Med Chem.*, 1986, **29**, 394.
11. L.A. Mitscher *et al.*, in *Quinolone Antimicrobial Agents*, D.C. Hooper

and J.S. Wolfson (eds.), American Society for Microbiology, Washington DC, 2nd edition.
12. T.D. Gootz and K.E. Brighty, *Med. Res. Rev.*, 1996, **16**, 433.
13. A. Graul *et al., Drugs Future* 1999, **24**, 1324.
14. D.C. Hooper, *Drug Resistance Updates*, 1999, **2**, 38.
15. D.J. Mangelsdorf *et al.*, *Cell*, 1995, **83**, 835.
16. R.T. Nolte *et al.*, *Nature (London)*, 1998, **395**, 137.
17. R.V. Weatherman *et al.*, *Annu. Rev. Biochem.*, 1999, **68**, 559.
18. S.A. Kliewer *et al.*, *Science*, 1999, **284**, 757
19. S.A. Kliewer *et al.*, *Proc. Natl. Acad. Sci. USA*, 1994, **91**, 7355.
20. T.M. Willson *et al.*, *J. Med. Chem.*, 2000, **43**, 527.
21. H.B. Rubins *et al.*, *New Engl. J. Med.*, 1999, **341**, 410.
22. J.M. Lehmann *et al.*, *J. Biol. Chem.*, 1995, **270**, 12953.
23. J.E. Ferrell, *Trends Biochem. Sci.*, 1996, **21**, 460.
24. M.J. Robinson and M.H. Cobb, *Curr. Opin. Cell Biol.*, 1997, **9**, 180.
25. P. Cohen, *Curr. Opin. Chem. Biol.*, 1999, **3**, 459.
26. D.W. Fry *et al.*, *Science*, 1994, **265**, 1093.
27. L.M. Toledo *et al.*, *Curr. Med. Chem.*, 1999, **6**, 775.
28. P. Traxler, *Exp. Opin. Ther. Patents*, 1998, **8**, 1599.
29. J.L. Adams *et al.*, in *Prog. Med. Chem.*, F.D. King and A.W. Oxford (eds.), Elsevier, Amsterdam, 2001, **38**, p. 1.
30. Gleevec®, a bcr-abl tyrosine kinase inhibitor, has recently been approved for the treatment of chronic myeloyenous leukemia (see B.J. Druker, *Trends in Mol. Med.*, 2002, **8**, 514).

CHAPTER 4

# Enzyme Inhibitors

DAVID A. ROBERTS AND WALTER H.J. WARD

## 1 INTRODUCTION

Enzyme inhibition represents a major strategy in drug design and almost one-third of the current top fifty drugs by sales are enzyme inhibitors. The inhibition of an enzyme-catalysed reaction can enable the selective modulation of a variety of biochemical processes such as making cell growth, division and viability untenable, or interrupting major metabolic pathways by blocking the formation of an essential or undesirable metabolite. Enzyme inhibition is complementary to receptor modulation *via* antagonists and in some cases can be used to potentiate the activity of a desirable species by inhibiting its degradation. Thus, the biological activity of species 'B' can be attenuated *via* inhibition of the enzyme involved in its biosynthesis (Figure 1). The same overall effect can be achieved *via* antagonism of the receptor(s) for 'B'. A good example of this is the attenuation of the action of the vasoconstrictor peptide angiotensin II (AII), which can be achieved *via* inhibition of its biosynthesis by the angiotensin converting enzyme (ACE) or *via* AII receptor antagonism.

Given these choices for the attenuation of biochemical mechanisms, the medicinal chemist must firstly decide on whether enzyme inhibition is the most suitable strategy to achieve a particular goal. Various questions can be considered to guide the decision making process: what is known about the structure and mechanism of the enzyme? Has the structure been characterised by high resolution X-ray crystal or NMR analysis or is there an opportunity to derive, computationally, a model structure based on homology with a related, characterised protein? Does the substrate of the enzyme possess biological activity and will its concentration increase and cause undesirable side-effects? Does the target enzyme have only one significant substrate or will other biological processes be affected if the

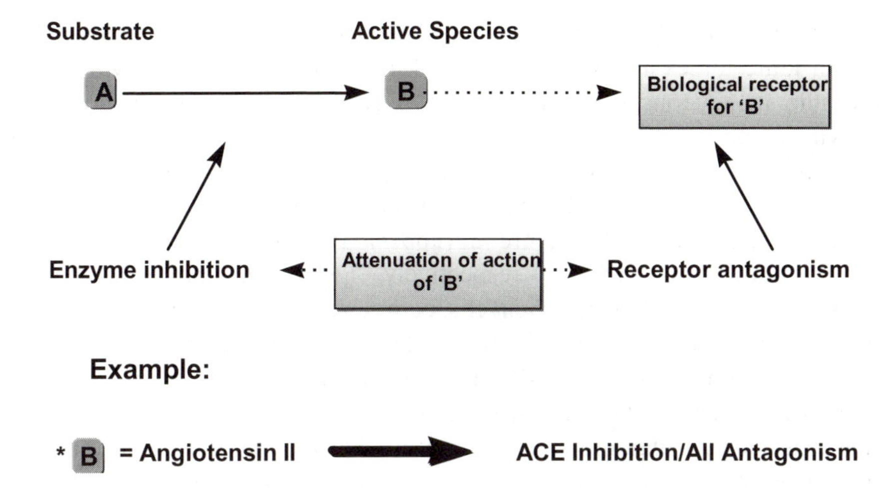


**Figure 1** *Complementarity between enzyme inhibition and receptor antagonism*

enzyme is inhibited? Are there isoforms or related enzymes present which might also be inhibited? If the enzyme is an essential component of an infecting organism, is it unique to that organism, or present in the host also?

Unfortunately, many of these questions will remain unanswered at the outset of a typical drug hunting programme. Nevertheless, if enzyme inhibition is judged to constitute an effective strategy, the question arises as to how the chosen enzyme can be inhibited. An essential first step is to understand how enzymes in general, and the target enzyme in particular, work.

Enzymes are proteins which contain chiral recognition sites for specific substrates. They catalyse chemical reactions at these sites, often causing huge rate enhancements ($10^{10}$–$10^{14}$-fold). In some cases their activity depends on the presence of other organic molecules or ions called cofactors. There are over 3700 functionally distinct enzymes, catalogued according to the Enzyme Commission Classification (see http://expasy.nhri.org.tw/enzyme/).

These enzymes share certain key properties which enable them to achieve catalysis and also provide the medicinal chemist with opportunities to derive general strategies for their inhibition. For example, enzymes contain catalytic functional groups, such as general acids and bases, nucleophiles and metal ions (which can function as Lewis acids or redox systems). Proteases, for example, illustrate the range of functional groups employed by enzymes to facilitate the hydrolysis of peptide bonds, exploiting both general acid/base catalysis (aspartyl and metalloproteases) and nucleophilic catalysis (serine and cysteine proteases). Enzymes also provide a relatively water-free, low dielectric constant environment, in which ionic forces are strengthened and transient intermediates, which are unstable in aqueous environments, have enhanced survival. In addition, enzymes stabilise the transition states of reactions, thereby reducing the activation energy for, and hence increasing the rate of, the reaction. In its simplest schematic form, the course of an enzyme-catalysed reaction can be considered to follow Figure 2 (upper), in which E and S are free enzyme and free substrate respectively, ES is the physically-bound complex of E and S (the Michaelis complex, which has an equilibrium dissociation constant of $K_s$), P the product(s) and EP a physically bound enzyme–product complex. All of these species lie in energy minima, while the transition state for chemical transformation ($ES^{\ddagger}$) of ES into EP lies at an energy maximum. This equation can be expressed as a reaction profile (Figure 2 lower) in which the free energy $\Delta G^{\circ}$ is plotted *versus* the reaction coordinate showing (qualitatively) the progress of the reaction. Some enzymes follow mechanisms where the highest energy maximum is not at the chemical transformation of ES to EP, but at a different step, such as dissociation of product.

Until relatively recently, enzyme inhibitors were designed predominantly on the basis of the substrate ground state structure (so-called substrate analogue inhibitors). Similarly, the binding interactions available to the product structure can also be used in the design of putative inhibitors.

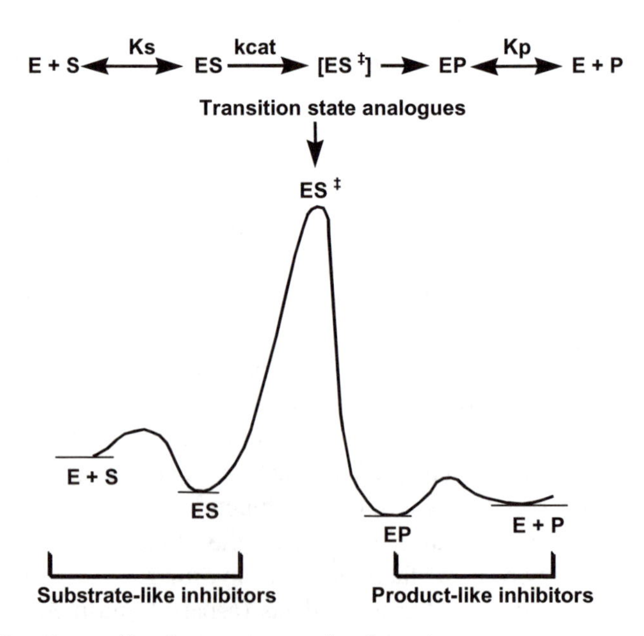


**Figure 2** *Reaction profile of an enzyme-catalysed reaction*

There is little conceptual difficulty with these classes of agent since stable molecular structures (substrates or products) are used to design inhibitors. The strength of the enzyme–substrate interaction in the ground state is far outstripped by that in the fleeting transition state. Thus, enzymes catalyse chemical reactions by virtue of their ability to sequester specifically the relevant transition state, thereby stabilising it and lowering the energy required by E and S to reach it. This suggests that inhibitors with very high enzyme affinity might be obtained if the geometric, steric and electronic nature of the transition state moiety is used as the basis for inhibitor design. A true transition state analogue, with fleeting partial bonds, cannot be synthesised. However, derivatives which incorporate a number of the features likely to be involved in transition state binding can sometimes be prepared in stable form. The improvement in binding affinities of such 'transition state analogues' are often striking and will be illustrated by examples later in this chapter.

## 2 ENZYME INHIBITOR CATEGORIES

### Kinetics of Substrate Utilisation

Enzyme assays are usually designed to ensure that accumulation of product is insufficient to influence rate, so that the kinetics can be described by the Michaelis–Menten equation.[1]

$$v = \frac{[\mathrm{S}]k_{\mathrm{cat}}[\mathrm{E}]_{\mathrm{t}}}{K_{\mathrm{m}} + [\mathrm{S}]} \tag{1}$$

If one assumes that dissociation of substrate is rapid relative to catalysis, then the physical significance of the parameters is as follows:

$K_m$, the Michaelis constant (substrate concentration giving half maximal rate, which approximates to $K_s$ in Figure 2),

$k_{cat}$, the rate constant for generation of product (mol product per mol active site per s, also known as the catalytic constant) and

$[E]_t$, the total concentration of enzyme active sites.

If substrate dissociation is not rapid relative to catalysis, then the magnitudes of $K_m$ and $k_{cat}$ are influenced by several steps in the reaction mechanism. The value of $[E]_t$ is often not known, because the enzyme may not be 100% pure and active. Accordingly, $k_{cat}[E]_t$ is frequently replaced by the term 'maximal velocity', $V_{max}$, which is a more qualitative parameter that describes the rate at saturating substrate. Equation (1) gives a plot of $v$ against [S] (Figure 3), which is linear at low substrate

($[\mathrm{S}] \ll K_\mathrm{m}$, $v \approx [\mathrm{S}]k_\mathrm{cat}[\mathrm{E}]_\mathrm{t}/K_\mathrm{m} \approx [\mathrm{S}]V_\mathrm{max}/K_\mathrm{m}$), curves near the mid-point ($[\mathrm{S}] = K_\mathrm{m}$, $v = k_\mathrm{cat}[\mathrm{E}]_\mathrm{t}/2 = V_\mathrm{max}/2$) and tends towards an asymptote at saturating substrate ($[\mathrm{S}] \gg K_\mathrm{m}$, $v \approx k_\mathrm{cat}[\mathrm{E}]_\mathrm{t} \approx V_\mathrm{max}$). The specificity constant of an enzyme/substrate pair is governed by the ratio $k_\mathrm{cat}/K_\mathrm{m}$. Specificity is high if a substrate is turned over rapidly and bound at high affinity. There is a maximum value of $k_\mathrm{cat}/K_\mathrm{m}$ around $10^9\ \mathrm{s}^{-1}\ \mathrm{M}^{-1}$, reflecting the collision frequency, which is governed by the rate of diffusion in water. Measurements of $k_\mathrm{cat}/K_\mathrm{m}$ can be useful, because values below around $10^5\ \mathrm{s}^{-1}\ \mathrm{M}^{-1}$ suggest that the substrate is poor (may follow an artefactual mechanism), or the catalytic activity may be due to a contaminant in the enzyme preparation. The magnitudes of $K_\mathrm{m}$ and $V_\mathrm{max}$ have been traditionally determined from Lineweaver–Burk plots ($1/v$ against $1/[\mathrm{S}]$, Figure 4), which utilise the following rearrangement of the Michaelis–Menten equation:

$$1/v = 1/V_\mathrm{max} + (K_\mathrm{m}/V_\mathrm{max})(1/[\mathrm{S}]) \tag{2}$$

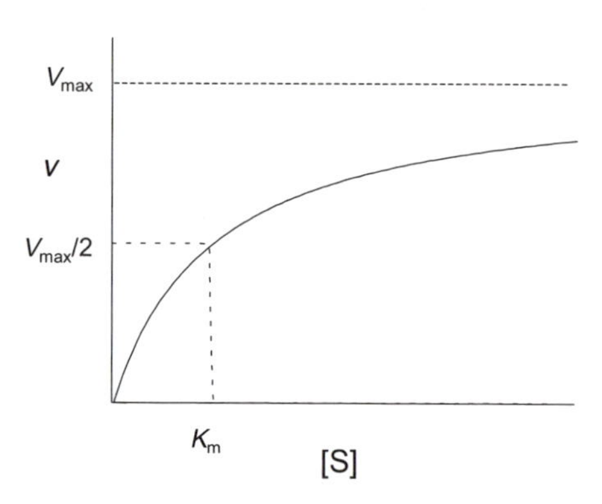


**Figure 3** *Plot of* v versus *[S] for a reaction obeying Michaelis–Menten kinetics*

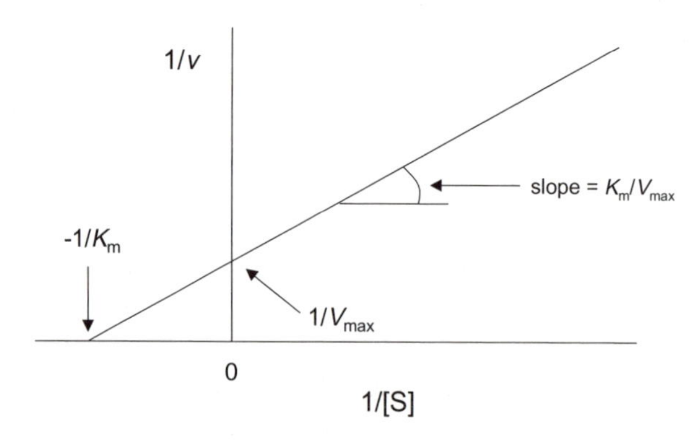


**Figure 4** *Lineweaver–Burk plot of 1/*v versus *1/[S]*

Thus, the intercept on the ordinate is $1/V_{max}$ and the gradient is $K_m/V_{max}$. This approach is often inaccurate, because transformation to $1/v$ introduces statistical bias and usually assigns excessive weighting to the least accurate data points (low rates).[2] Although Lineweaver–Burk plots remain a clear way to present kinetic data, they are not the preferred method to calculate parameter values. For many years, software has been available which allows more accurate non-linear regression directly using the Michaelis–Menten equation.

During the catalytic cycle, the enzyme exists as a free unimolecular species and in intermolecular complexes with substrate, intermediates and product. The majority of enzyme assays aim to follow accumulation of product (or depletion of substrate) when there is an approximation to steady state. This means that, following a rapid approach to equilbrium (the pre-steady state, usually complete in under 1 s), the concentration of all species, apart from product, does not change significantly during the course of the assay. It follows that measured parameter values often reflect several chemical steps.[1] Steady state kinetics, therefore, is an indirect probe of multiple steps, which occur during the assay. Different reaction schemes can give similar, or identical, kinetic profiles. The magnitudes of $K_m$, $V_{max}$ and inhibition constants may vary considerably according to the conditions of the assay.

It is important to establish the validity of any enzyme assay. If using a fragment of the enzyme derived by protein engineering, it should be confirmed that it is a reliable model for the authentic protein in the physiological state. All of the enzyme molecules in the preparation should function properly. Thus, the preparation should be 100% pure with (for monomeric enzymes) the concentration of active sites ($[E]_t$ from tight binding inhibition or isothermal titration calorimetry, see below) being equal to the concentration of protein divided by its $M_r$. Assays may follow the catalytic activity of a tiny fraction of active enzyme, in the presence of a large excess of inactive protein, which may perturb the results. This situation is not detected in common quality control tests, which use denatured enzyme (*e.g.* SDS-PAGE, mass spectrometry, amino acid analysis) and only qualitative activity criteria (*e.g.* $V_{max}$, which compares one batch with another, but gives no indication of how much protein is active). Having established how much protein is active, it is important to check that the active fraction has the correct characteristics (*i.e.* those of the disease state). Key criteria include $k_{cat}$, $K_m$ and inhibitor SAR. These parameters may vary according to activation status (*e.g.* phosphorylation, limited proteolysis, cofactor or regulator binding). It can be important to use the physiological substrate, or check that the kinetic mechanism (rate limiting step, sequence of association of substrates, order of dissociation of products) is the same for model and physiological substrates.

## $IC_{50}$ Values Reflect Affinity and Assay Conditions

Enzyme inhibitors are usually identified in assays performed at a single, fixed concentration of test compound. The next step is to determine the concentration required for 50% inhibition ($IC_{50}$). This value is often used to rank inhibitors in terms of potency and selectivity. Below, we explain how interpretation of $IC_{50}$ values, however, can be difficult, because they may not correlate with affinity, nor with activity under physiological conditions.

It is usually assumed that active compounds equilibrate rapidly with the target enzyme (during a small fraction of the assay time) and lead to reversible inhibition. Irreversible inhibition is discussed later. The methods used to calculate the $IC_{50}$ value should reflect the physical processes that occur during the assay. If this is not the case (*e.g.* due to tight or slow binding inhibition, see below), then the calculated values may be misleading. Most dose–response studies can be considered as an equilibrium between active and inhibited enzyme:

$$E' + I \underset{K_i'}{\rightleftharpoons} E'I \qquad \text{(Scheme 1)}$$

where E′ is all of the catalytically active species with no inhibitor bound, E′I is all species with inhibitor bound and $K_i'$ is the apparent inhibition constant:

$$K_i' = [E'][I]_{free}/[E'I] \qquad (3)$$

$K_i'$ is greater than, or approximately equal to, the dissociation constant of the enzyme–inhibitor complex ($K_d$), which is the absolute measure of binding affinity. It can be shown (see ref. 3) that:

$$IC_{50} = K_i' + [E]_t/2 \qquad (4)$$

This is a very important equation, because it illustrates how the observed value of $IC_{50}$ may be dependent not only upon the apparent affinity ($K_i'$) but also on the concentration of enzyme in the assay ($[E]_t$). In many cases (importantly, not all), the concentration of compound added to give 50% inhibition is much greater than the enzyme concentration. If so, then $K_i'$ is $\gg [E]_t$, so that Equation (4) becomes $IC_{50} \approx K_i'$. Thus, the aim of dose–response studies is to determine the magnitude of the apparent inhibition constant, $K_i'$. The next subsection explains how the value of $K_i'$ is often specific to the conditions of the assay, because it is influenced by the substrate concentration, $K_m$ and mechanism of inhibition. $K_i'$ is calculated from the enzyme activity ([E′] in Scheme 1 above), which is observed at different concentrations of inhibitor. It is essential that the measured parameter (which should be rate, but which is usually amount of product) is a constant linear function of enzyme concentration. This requires validity of the steady state assumption (see above). Curved relationships for rate *versus* enzyme concentration (or product *versus* time) give misleading estimates of $IC_{50}$.

If the total inhibitor concentration approximates to $[I]_{free}$, then Scheme 1 can be used to derive the relationship:

$$v = v_0/(1 + [I]/K_i') \tag{5}$$

where $v$ is the measured rate and $v_0$ is the uninhibited rate. This equation should be used for non-linear regression to calculate the value of $IC_{50}$ ($K_i'$) (and $v_0$) if $K_i' \gg [E]_t$.[1–3] This approach can also be used to estimate the confidence intervals of $IC_{50}$, but only if the equation is modified to a logarithmic form (such as replacing $IC_{50}$ by $10^{-pIC50}$, where $pIC_{50} = -\log IC_{50}$). This is because the $IC_{50}$ value is correct to within a certain multiple (*e.g.* a factor of 2), not a certain number (*e.g.* 0.5 μM). Non-linear regression allows identification of data sets where the equation gives a poor quality of fit. This is often clearly seen in residual plots, which show the difference between the best fit line and each of the data points.[4] A systematic distribution of residuals (several consecutive points all above the line or all below the line) indicates that the equation does not give an adequate description of the data (Figure 5).

$IC_{50}$ values are sometimes estimated by other approaches, such as direct graphical methods (*e.g.* calculation of percent inhibition, plotting against log of inhibitor concentration and then joining the data points). Such approaches are less accurate than non-linear regression, because they require calculation of percent inhibition, which gives increased error and bias in the data points. Direct graphical methods are also less able to identify dubious $IC_{50}$ values (because there are no residual plots). A 'four parameter logistic' equation is sometimes fitted to dose–response data. This includes additional variables (slope factor and background), which are not usually justified and so increase uncertainty in the calculated $IC_{50}$ value. More importantly, they allow additional flexibility in the fit, which often masks complex, but common, behaviour that invalidates the assumption that $IC_{50} = K_i'$ (such as tight or slow binding inhibition, see below).

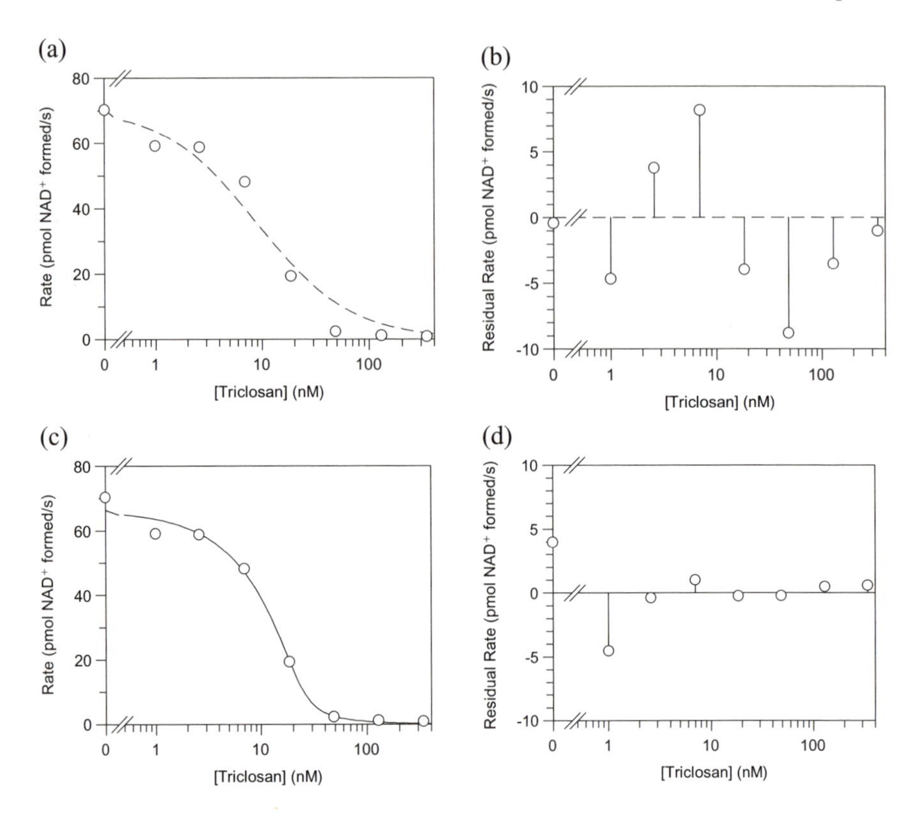


**Figure 5** *Residual plots to assess quality of fit. The activity of enoyl (ACP) reductase (EACPR) was measured at various concentrations of triclosan.*[5] *Two equations were fitted to the measured rates and then the residual differences between the best fit lines and the measured rates were plotted against inhibitor concentration.* (a) *Dashed line indicates the best fit for the standard dose–response relationship (Equation (5)), with* $K_i'$ *incorrectly estimated as 9 nM.* (b) *Poor fit revealed by systematic residuals.* (c) *Solid line indicates the best fit for tight binding inhibition, with* $[E]_t$ *estimated as approximately 23 nM and* $K_i' \ll [E]_t$. (d) *Good fit shown by random residuals*

## Mechanisms of Reversible Inhibition, Affinity and Selectivity

Knowledge of the mechanism of inhibition helps to understand the complex relationships between $IC_{50}$, binding affinity, biological activity and selectivity under physiological conditions. Thus, it aids in the evaluation of active hits from screening campaigns and in the elucidation of SAR. The mechanism of inhibition determines whether the association of the test compound with the target enzyme requires prior binding of another ligand (*e.g.* coenzyme). So, its elucidation helps to ensure that structure-based design is based on relevant intermolecular complexes. The kinetic mechanism of inhibition can be identified from measured rates at different concentrations of substrate and inhibitor. This can be different to the physical mechanism, because steady state kinetic assays allow only indirect monitoring of intermediates (see above). For a single substrate system, the different mechanisms can be described using Scheme 2.

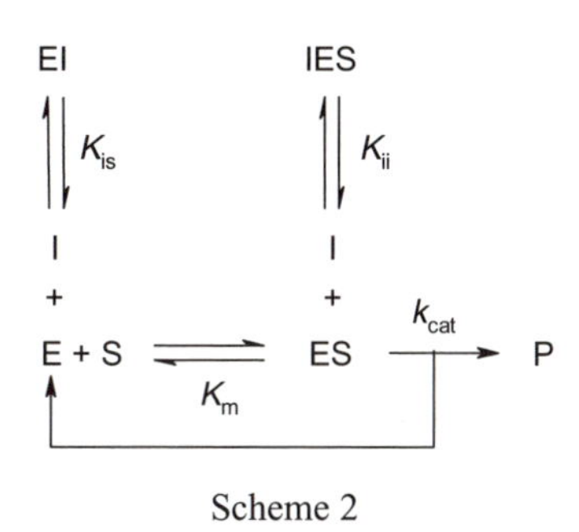


Scheme 2

The terms $K_{is}$ and $K_{ii}$ derive from effects on the slope and intercept of Lineweaver–Burk plots.[2] For single substrate systems, $K_{is}$ is the equilibrium dissociation constant for inhibitor ($K_d$) after extrapolation to zero substrate and $K_{ii}$ is that after extrapolation to saturating substrate. Thus, the mechanism of inhibition gives insight into the true binding affinity. Inhibition is competitive ($K_{is} \ll K_{ii}$) when saturating substrate prevents the effect of the test compound (in Scheme 2, IES is not formed). Competitive kinetics indicate that the inhibitor binds only before the substrate, although they may associate at different sites. For example, inhibition of enoyl (ACP) reductase (EACPR) by triclosan follows competitive kinetics with respect to NADH.[5] This is because NADH and $NAD^+$ use the same site, and the compound binds only to a complex between the enzyme and $NAD^+$.

Uncompetitive inhibition occurs when the compound has no effect at an extrapolation to zero substrate ($K_{is} \gg K_{ii}$, EI is not formed in Scheme 1). Binding of the substrate is required before the inhibitor can associate with the enzyme. For example, a pyridazine (2) and mycophenolic acid (1) both display uncompetitive inhibition of inosine monophosphate (IMP) dehydrogenase when IMP is varied, apparently due to association with an enzyme–intermediate complex at a site vacated by NADH (3) only after IMP has been bound and oxidised (Figure 6).[6] This mechanism requires that there is a slow step after dissociation of one of the products, so that the enzyme–intermediate complex accumulates to allow binding of the inhibitor.

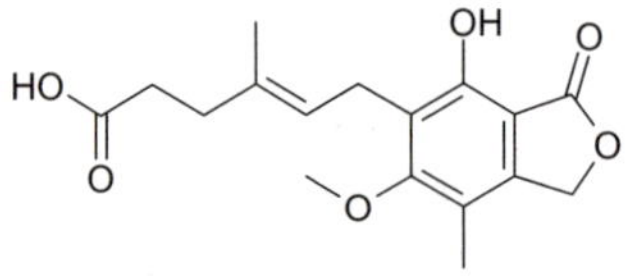

1

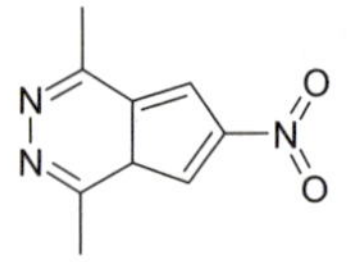

2

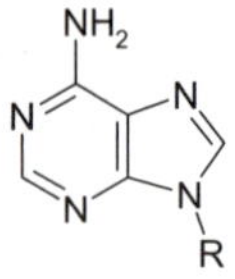

3

R = ribose-diphosphate-ribose-nicotinamide

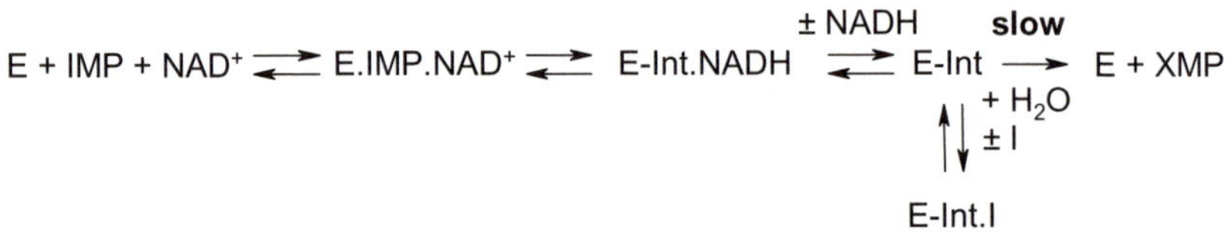

E + IMP + NAD$^+$ ⇌ E.IMP.NAD$^+$ ⇌ E-Int.NADH ⇌ (± NADH) E-Int → (slow, + $H_2O$) E + XMP

E-Int ⇅ (± I) E-Int.I

**Figure 6** *Proposed kinetic mechanism for inhibition of IMP dehydrogenase. E-Int is a catalytic intermediate where there is a covalent bond between the purine C2 of IMP and the S in Cys331. I is the inhibitor, mycophenolic acid* **1** *or a pyridazine*

Inhibition is non-competitive when the compound may be bound at all substrate concentrations (both EI and IES are formed in Scheme 2). Pure non-competitive describes compounds whose activity does not vary with substrate concentration ($K_{is} = K_{ii}$), whereas mixed non-competitive is a general case where there is a tendency towards either competitive or un-competitive, kinetics ($K_{is} \neq K_{ii}$). Non-competitive kinetics may be due to the inhibitor binding away from the catalytic site (at an allosteric site), but this is not always true. For example, like many aldose reductase inhibitors, the sulfonylnitromethane, ZD5522, follows mixed non-competitive kinetics with respect to glucose, apparently because it can bind to enzyme–NADPH complex at the site for glucose substrate, and at the same site on the slowly dissociating enzyme–NADP$^+$ complex after dissociation of the product sorbitol.[7]

The examples above illustrate how the kinetic mechanism of inhibition may give misleading suggestions about the identity of the binding site. One approach to this problem is to determine the kinetic mechanism (see above) of the enzyme in the absence of inhibitors. This provides a basis for interpretation of the behaviour of test compounds. The kinetic mechanism is often not known, so that a less rigorous solution is required. It is frequently useful to compare the test compound with a better characterised inhibitor. For example, inhibition of epidermal growth factor receptor–tyrosine kinase (EGFR-TK) by two anilinoquinazolines [(4-(3-chloroanilino)quinazoline and Iressa®, see later] follows competitive kinetics with respect to ATP and pure non-competitive kinetics when a peptide substrate is varied (Figure 7).[8] A hydrolysis-resistant ATP analogue, 5′-adenylyl $\beta,\gamma$-imidodiphosphate, shows the same pattern, whereas a Tyr to Phe peptide substrate analogue is pure non-competitive with respect to ATP and competitive when the peptide substrate is varied. These results strongly suggest that the anilinoquinazolines interact with an ATP-binding site on EGFR-TK. Another useful approach is to test for competition between different structural classes of inhibitor. Thus, acetic acids, spirohydantoins and sulfonylnitromethanes exhibit kinetic competition for aldose reductase, suggesting that they use overlapping sites.[7] Inhibitors are useful tools to investigate mechanism because, unlike substrates, they are not converted into intermediates and products, which complicate the interpretation of kinetics.

The magnitude of $K_i'$ can be derived for each mechanism of inhibition:[9]

Mixed non-competitive, $$K_i' = K_{is}K_{ii}([S] + K_m)/\{K_{is}[S] + K_{ii}K_m\} \quad (6)$$
Competitive, $$K_i' = K_{is}([S] + K_m)/K_m \quad (7)$$
Uncompetitive, $$K_i' = K_{ii}([S] + K_m)/[S] \quad (8)$$
Pure non-competitive, $$K_i' = K_{is} = K_{ii} \quad (9)$$

These equations have consequences for drug discovery. The magnitude of $IC_{50}$ ($K_i'$) depends not only upon the affinity of the inhibitor ($K_{is}$ or $K_{ii}$), but also on the substrate concentration and $K_m$. $IC_{50}$ values are rarely accurate estimates of binding affinity ($K_d$), although they may reflect ranking and relative shifts, because $IC_{50}$ is a constant multiple of $K_d$. This fortunate situation tends to occur only when the different compounds

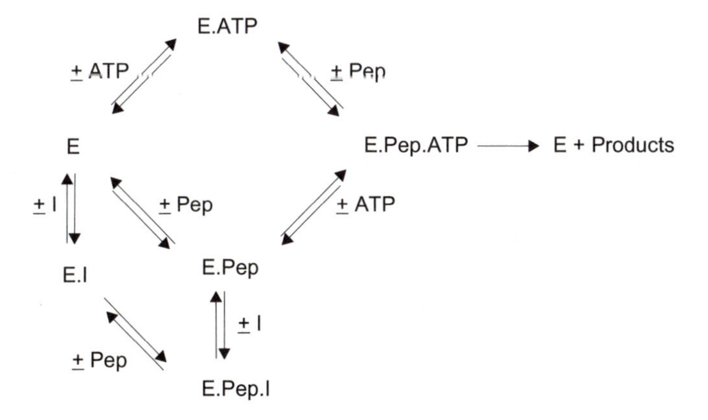


**Figure 7** *Kinetic mechanism for inhibition of EGFR-TK by certain anilinoquinazolines*

follow the same mechanism of inhibition and the conditions for the assays are identical. Failure to satisfy these conditions sometimes explains difficulties in reproducing $IC_{50}$ values, or shifts in potency when moving between enzyme assays, or from enzyme to cell assays. $IC_{50}$ values are often used to evaluate selectivity, but their magnitudes are specific to the assay conditions, so that selectivity may be quite different under physiological conditions. For example, many kinase inhibitors are competitive with respect to ATP and are evaluated in assays where it is at a concentration of up to 10 μM. A compound which is unselective in such assays may be highly selective (over 100-fold) under physiological conditions (approximately 2 mM ATP and 1 mM ADP), because ATP and ADP compete more effectively for some kinases.

The mechanism of inhibition was traditionally determined by examining Lineweaver–Burk plots (Equation (2), Figure 4) at different, fixed concentrations of inhibitor. In theory, competitive kinetics increase the slope and give lines that all intersect at the intercept on the ordinate, which is unchanged. Uncompetitive kinetics give parallel lines at each inhibitor concentration, due to increasing the intercept, without affecting the slope. Non-competitive kinetics give lines which intersect at a single point, which is on the abscissa for pure non-competitive inhibition. In practice, Lineweaver–Burk analysis is usually an unsatisfactory approach to determine mechanism of inhibition, because small errors in measured rates are sufficient to generate multiple points of intersection or make it difficult to determine whether lines are parallel. Multivariate non-linear regression provides a much more rapid, rigorous and reliable method to determine mechanism of inhibition.[2,4] Here, rather than generating a (monovariate) fit of substrate dependence at each of several different concentrations of inhibitor (each fit having different parameter values), the whole data set (multivariate, with [S] and [I] changing) is analysed to give a single set of best fit parameter values and residual sum of squares. Conventional statistical methods can be used to compare the fits obtained using rate equations for different mechanisms of inhibition. For example, the residual sum of squares can be used to estimate the probability that $K_{ii}$ tends to infinity (competitive inhibition), $K_{is}$ tends to infinity (uncompetitive inhibition), or $K_{is} = K_{ii}$ (pure non-competitive inhibition).

An appropriate experimental design is essential in order to distinguish between alternative mechanisms of inhibition.[4] For example, the $IC_{50}$ of a competitive inhibitor increases by a factor of only two in going from $[S] = K_m/100$, to $[S] = K_m$. Conversely, increasing from $[S] = K_m$ to $[S] = 100\ K_m$ gives a 50-fold increase in $IC_{50}$. Thus, competitive and pure non-competitive inhibitors ($IC_{50}$ independent of [S]) are clearly distinguished only in the second experimental design. As a rule, $IC_{50}$ needs to be measured at low and high substrate concentrations ($[S] \ll K_m$ and $[S] \gg K_m$) in order to determine the mechanism of inhibition.

## Tight Binding Inhibitors

Equation (4) shows how, if $K_i'$ is not $> [E]_t$, then $IC_{50} \approx [E]_t/2$. Such a characteristic is known as tight binding inhibition. Dose–response equations for tight binding inhibitors are given by Copeland[3] and Morrison and Walsh.[10] A change in affinity may not alter the measured $IC_{50}$ value. This prevents monitoring of the relationship between the structure of the inhibitor and affinity for the target enzyme. It also becomes impossible to follow how changes in affinity influence the biological activity, for example in cells. Tight binding inhibition is commonly observed in drug discovery, because medicinal chemistry often pursues *in vitro* potency as a key objective. Examples include inhibition of 25 nM aldose reductase by ZD5522,[7] 40 nM IMP dehydrogenase by mycophenolic acid[6] and 80 nM EACPR by triclosan.[5] Tight binding should be suspected if the $IC_{50}$ is low (*e.g.* below 10 nM) and does not change according to the structure of the compound. It is readily detectable because the magnitude of $IC_{50}$ increases linearly with $[E]_t$ ($IC_{50}$ values should be independent of $[E]_t$). Tight binding also gives a steep slope on dose–response plots (readily seen in residual plots, Figure 5). In order to obtain more informative $IC_{50}$ values, tight binding can be avoided by decreasing $[E]_t$ (if the assay is sufficiently sensitive), or by increasing $K_i'$. For example, increasing the concentration of a competing substrate may allow discrimination between different tight binding inhibitors, although it also increases the magnitude of $K_i'$, which complicates comparisons with $IC_{50}$ values collected at lower substrate concentrations.

## Slow Binding and Irreversible Inhibitors

All of the kinetic derivations shown above assume that there is a steady state, because free and bound inhibitor come to equilibrium within a small fraction of the assay time and then remain at the same equilibrium position. This assumption is sometimes not valid, causing the degree of inhibition to change during an assay.[10] A slow onset of inhibition tends to be linked with a tendency towards irreversible inhibition, otherwise inhibition would be weak. Accordingly, assays may overestimate potency if a slow binding inhibitor is preincubated with the enzyme before adding substrate. Potency may be underestimated if there is no preincubation (Figure 8). Slow binding is common in drug discovery,[10] often because potent compounds are assayed at low concentrations, which can give a low

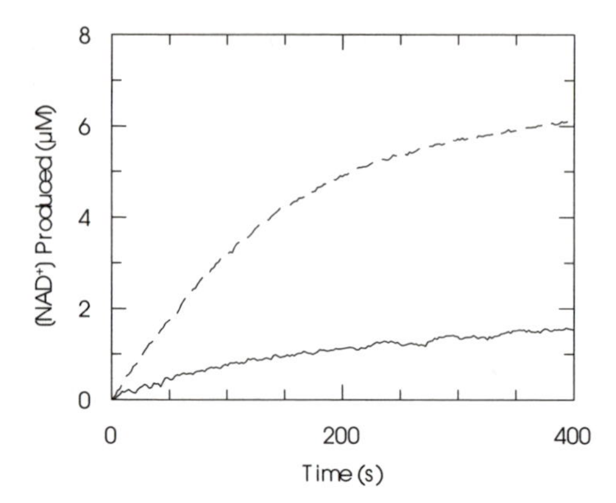


**Figure 8** *Slow binding inhibition of EACPR by triclosan. The graph shows the rate with (solid line) and without (dashed line) a 30 min preincubation of enzyme, 100 µM $NAD^+$ and 150 nM triclosan*[5]

collision frequency between enzyme and inhibitor. Alternative causes include: only a small fraction of the enzyme being in a form which is bound by the inhibitor (*e.g.* triclosan associates with a complex between EACPR and $NAD^+$, which is a weak binding product of the assay),[5] or a slow conformation change, or slow formation of a covalent bond between enzyme and inhibitor. Slow binding may be detected as a curved time-course of product accumulation in the presence of inhibitor, when uninhibited controls are linear. Alternatively, the $IC_{50}$ value may be lower if the enzyme is preincubated with the inhibitor or there may be a slow regain of catalytic activity after dilution of an enzyme/inhibitor mixture. Correct estimation of potency for slow binding inhibitors requires either a suitable preincubation of enzyme and inhibitor, followed by starting the assay with minimal perturbation of the established equilibria (*e.g.* small increase in volume and $[S] < K_m$), or the assay can be started by simultaneous addition of substrate and inhibitor and then an integrated rate equation must be fitted to the curved time-course of product accumulation.[5,10]

The terms slow binding and irreversible reflect the duration of the assay. Thus, inhibition may tend towards irreversible in a 5 minute assay following dilution, although most of the activity may be regained after dialysis for several hours. Irreversible inhibition often involves formation of a covalent bond with the enzyme, but this is not a requirement because reversibility is kinetically defined. A classical example of an irreversible inhibitor that forms a covalent bond with its enzyme target is clavulanic acid which is co-administered with the antibacterial agent amoxycillin in order to inhibit $\beta$-lactamases expressed by resistant bacteria (Figure 9).

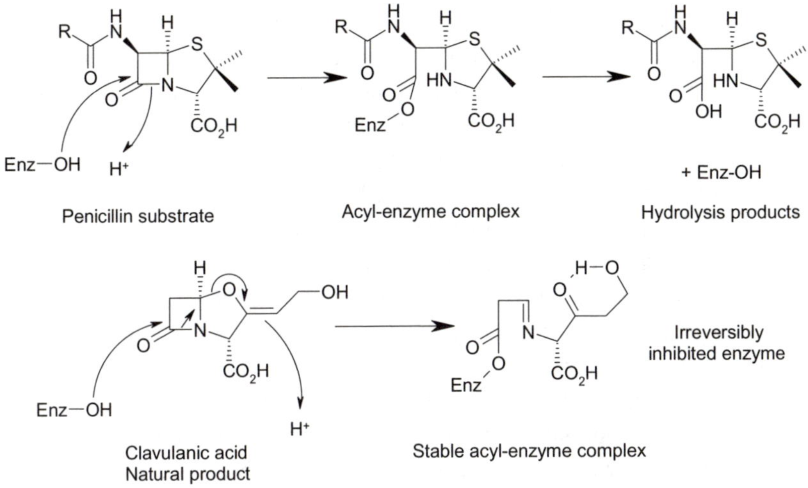

**Figure 9** *Cleavage of the β-lactam nucleus of penicillin by β-lactamase and the irreversible inhibition demonstrated by clavulanic acid*

Likewise, reversible inhibition may involve rapid, reversible formation of a covalent bond with the enzyme. Slow on and off rates for inhibitors may influence pharmacokinetics and pharmacodynamics. Slow off rates may extend duration of action and reduce side-effects.

## 3 OPPORTUNITIES IN DRUG DESIGN

### Exploitation of Enzyme Kinetics in Drug Discovery

The above discussion outlines various types of inhibitor and the question arises as to which is the most suitable for a particular enzyme target. Potent, reversible, competitive inhibitors achieve high affinity for the target enzyme by fitting the active site well, and so are likely to be selective. Moreover, the design of such agents can be based rationally upon knowledge of the enzyme mechanism/structure and the enzyme substrate. However, in many cases, detailed mechanistic/substrate information may not be available and the generation of active leads may be reliant upon screening of compound collections or libraries. This approach may identify compounds which inhibit by binding allosterically, away from the substrate binding site. However, such regulatory sites appear to be rare, and are often employed by a known physiological regulator. A potential problem with competitive inhibitors is that their activity can, in principle, be overcome by substrate accumulation. Reversible, non-competitive inhibitors do not suffer from the latter problem but historically there often has been no rational link with the enzyme mechanism and substrate, and so inhibitor design tended to be more empirical. However, knowledge of the kinetic mechanism often reveals that non-competitive and uncompetitive inhibitors act as analogues of catalytic intermediates or products (*e.g.* see above: inhibition of IMP dehydrogenase by mycophenolic acid, EACPR by triclosan, and aldose reductase by ZD4522).

Irreversible inhibition can produce complete inactivation of the enzyme, which cannot be overcome. However, initial recognition by the enzyme often involves only substrate-like affinity and so high drug concentrations may be required. Also, the enzyme usually processes the inhibitor quite inefficiently as a substrate and in only a fraction of the enzyme–inhibitor complexes does the final nucleophilic attack and trapping occur.

The kinetic mechanism of inhibition gives an indirect indication of the intermolecular complex that is responsible for the biological activity of the test compound. This may be important for structure-based design, because a bound ligand (*e.g.* substrate or cofactor) may form part of the binding site for the inhibitor (*e.g.* triclosan binds to EACPR–$NAD^+$ complex,[5] and Tomudex® binds to a complex of thymidylate synthase and dUMP). In such cases, structures of bimolecular enzyme–inhibitor complexes may not only fail to detect these important interactions but also may contain the inhibitor bound in a different orientation, or even at a different location. Use of relevant structures is required in order to ensure that the medicinal chemist is not misled.

Pharmaceutical companies are building databases of $IC_{50}$ values against different enzymes. Informatics approaches include the use of these values in multivariate QSAR in order to predict potency (known as “affinity fingerprinting”).[11] It has been shown above how many factors influence the magnitude of $IC_{50}$, which complicates comparisons. In order to extract information from databases, there must be careful consideration of the question being asked (*e.g.* How does structure affect affinity? How does structure influence activity in cells?), followed by appropriate data processing.

## Isothermal Titration Calorimetry

Isothermal titration calorimetry (ITC) is increasingly being employed to characterise molecular interactions with target proteins in support of drug discovery.[12] The technique monitors the heat change when a test compound binds to a target protein, enabling precise measurement of $K_d$ values. As it is a direct binding method, the interpretation of results is easier than for indirect approaches, such as enzyme kinetics. ITC in the presence of substrates, products or inhibitors rapidly provides information on the mechanism of action of the test compound. Like tight binding inhibition, ITC allows measurement of stoichiometry, and so evaluation of the proportion of the sample that is functional. ITC can characterise binding to protein fragments and catalytically inactive mutant enzymes, which can provide a useful assessment of the validity of other assays (*e.g.* using tagged proteins, protein fragments where the $M_r$ is restricted to allow NMR, or proteases engineered to decrease autolysis). ITC is the only method which directly measures the standard enthalpy of binding ($\Delta H°$). Enthalpy values in different buffers allow characterisation of proton movement linked to the association of protein and ligand, giving information on the ionisation of groups involved in binding. Biochemical systems exhibit enthalpy–entropy compensation, where increased bonding is offset by an entropic penalty, reducing the magnitude of change in affinity. This means that there is little or no correlation between affinity and $\Delta H°$. When characterising SAR, therefore, most groups involved in binding can be detected as contributing to $\Delta H°$, but not to affinity. Large shifts in $\Delta H°$ may reflect a modified binding mode (several new interactions or use of a different site), or protein conformation changes. Thus, ITC may highlight a discontinuity in SAR, and so identify compounds where experimental 3-D structural data are likely to be particularly valuable in molecular design.[12]

## Structure-based Design

Medicinal chemists are increasingly exploiting data on the 3-D structures of protein–ligand complexes. This tendency has been facilitated by recent improvements in recombinant protein engineering and expression, protein purification, NMR, X-ray crystallography and computational chemistry. For example, lead identification and lead optimisation using SAR by NMR is a powerful approach, which identifies small molecules that bind to proximal sites on the target protein. Molecular linkers between the compounds are then incorporated in order to generate high affinity ligands.[13]

The importance of using relevant protein constructs and intermolecular complexes is discussed above. Structures from X-ray diffraction, where compound and protein are co-crystallised, are more reliable than those where the compound is soaked into a crystal of the protein. This is because compounds may bind at artefactual locations when soaked into crystals.

One of the major contributions of experimental 3-D structure determination arises from its ability to identify groups, which are located at the interface between the test compound and the target protein. This point may appear obvious, but many interfacial groups actually make a barely measurable contribution to affinity, because there is only a small change in free energy as they move from solvent into the binding site.[14] In the absence of structural data, the medicinal chemist can detect only those groups that interact more strongly with solvent (to disfavour binding) or target protein (to promote binding). Thus, analysis of inhibitor SAR tends to be misleading if it is assumed that lack of contribution to affinity indicates a location away from the molecular interface.

## 4 CLASSES OF ENZYMES AND EXAMPLES OF ENZYME INHIBITORS

There are many examples of successful drug hunting based on inhibition of a very wide range of enzyme classes, including proteases, oxygenases, reductases, phosphodiesterases, topoisomerases, transferases, isomerases and kinases. It would be impossible to cover even a fraction of these various inhibitor types in a single chapter. Nevertheless, it is instructive to consider at least a few examples of inhibitor design and discovery which have led either to successful marketed products or interesting and challenging medicinal chemistry.

### Synthetases: Inhibition of Thymidylate Synthase

Thymidylate synthase (TS) catalyses the conversion of dUMP and $N^5,N^{10}$-methylenetetrahydrofolate (5,10-$CH_2$-$FH_4$, **4**) into TMP and dihydrofolate. The enzyme is required for biosynthesis of TMP, in order to allow replication of DNA. TS, therefore, is a target enzyme for cancer therapy. The quinazoline, Tomudex (raltitrexed) **5**, is a structural analogue of 5,10-$CH_2$-$FH_4$, which inhibits TS. It is a cytotoxic agent, which is used mainly in the treatment of colorectal cancer. Tomudex uptake into cells is mediated by a protein, the reduced folate carrier. It is then used as a substrate by folylpolyglutamyl synthase, which catalyses the addition of further glutamate residues. The principal form of Tomudex within murine leukemia (L1210) cells is the tetraglutamate, which has 60-fold higher affinity for TS. The $IC_{50}$ (= 7 nM) for inhibition of L1210 cell growth is below the $K_{is}$ of Tomudex (59 nM), but above that of the tetraglutamate ($K_{is}$ = 0.94 nM). Isolated enzyme studies, therefore, are consistent with cellular studies, which suggest that formation of polyglutamates and concentration within cells are responsible for biological activity.[15]

The catalytic mechanism of TS involves formation of a covalent bond between an active site Cys thiolate and C-6 on the pyrimidine ring of

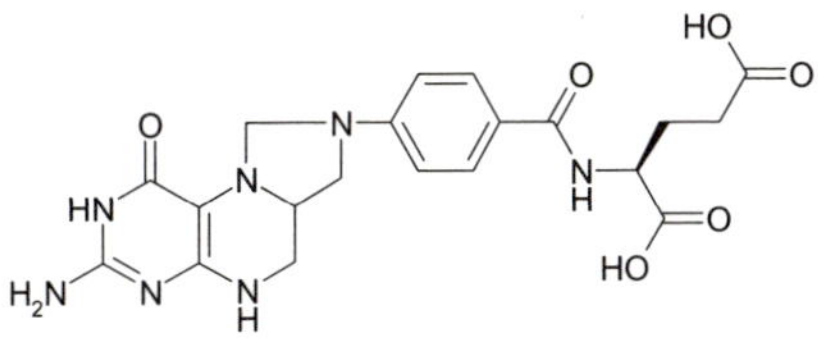

**4**

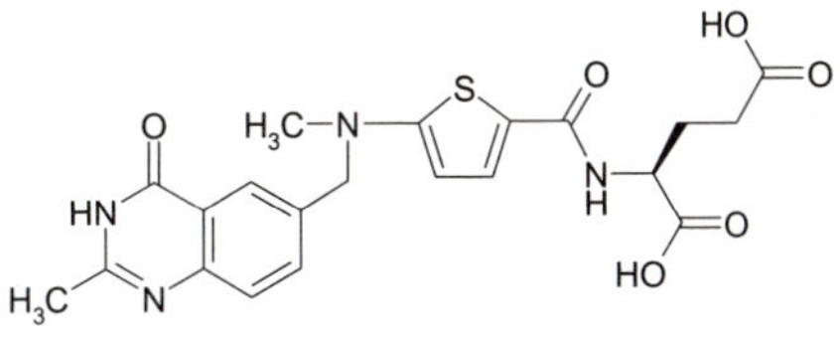

**5**

dUMP. Also, a methylene bridge is formed between N-5 of 5,10-$CH_2$-$FH_4$ and C-5 of the pyrimidine, generating covalent links between the three species in the E-dUMP-$CH_2FH_4$ complex. There is then deprotonation of the pyrimidine C-5 and bonds are broken to yield active enzyme, TMP and dihydrofolate. Tomudex subverts this mechanism by replacing 5,10-$CH_2$-$FH_4$ in the covalent complex between E and dUMP. There are multiple interactions between Tomudex and dUMP within their binding sites on TS.

## Reductases: Inhibition of Hydroxymethylglutaryl-CoA Reductase (HMG-CoA Reductase)

The causal relationship between blood cholesterol and coronary heart disease is well established. It is also known that reduction of blood cholesterol levels by diet or drug therapy helps prevent the sequelae of this disease. Although cholesterol biosynthesis comprises some twenty-six steps it can readily be suppressed by regulation of the enzyme HMG-CoA reductase which catalyses the rate-limiting step, the reduction of HMG-CoA to mevalonate. Inhibition of this enzyme, therefore, provides an attractive opportunity to inhibit cholesterol biosynthesis at an early stage in the cascade. The build-up of unprocessed substrate does not lead to toxicity problems, whereas earlier attempts to inhibit later stage enzymes in the cascade lead to the accumulation of desmosterol in toxic levels. Very potent, competitive inhibitors of HMG-CoA reductase (known as statins) have become available from natural product isolation programmes. These compounds typically follow slow, tight binding kinetics, which makes characterisation difficult. Cultures of *Aspergillus terreus* yielded mevinolin (Figure 10) which is a very potent inhibitor ($K_{is}$ $6.4 \times 10^{-10}$ M), in which the secondary alcohol at the 5-position of the acidic side-chain is postulated to act as a mimetic of the tetrahedral transition state of the reduction reaction. As the prodrug lactone, the compound is absorbed orally into the liver and efficiently inhibits cholesterol biosynthesis in man, thereby lowering circulating low density lipoprotein (LDL) levels and

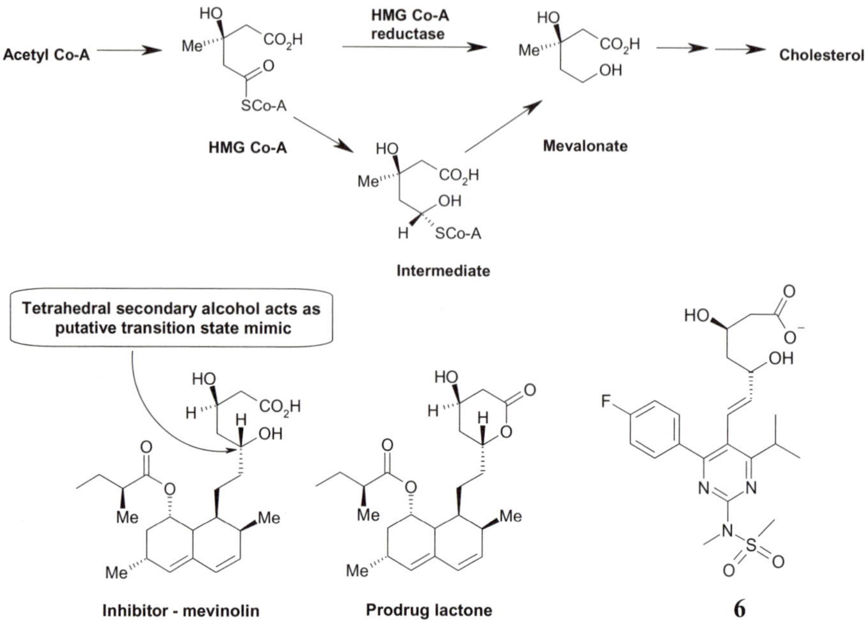

**Figure 10** *HMG-CoA reductase: mechanism and inhibition*

elevating LDL receptor numbers. This profile has, therefore, led to the compound being widely used to control inappropriately high cholesterol levels in large numbers of patients, thereby reducing coronary risk factors such as atherosclerosis.

Crestor® (ZD4522, rosuvastatin) **6** is a new statin, which is a single enantiomer (3*R*, 5*S*), having the same stereochemistry as the proposed transition state for HMG-CoA reductase (Figure 10). It is a potent, slow binding inhibitor ($K_{is}$ around $10^{-10}$ M), which, like other statins, follows competitive kinetics with respect to HMG-CoA and non-competitive kinetics with respect to NADPH.

## Kinases: Anilinoquinazolines as Inhibitors of Protein Kinases

Inhibition of tyrosine kinases is a potential approach for the treatment of cancer. The catalytic mechanism of EGFR-TK was investigated in order to allow structure-based searching for inhibitors.[8] Initial rate studies disproved formation of a phosphoenzyme intermediate, showing catalysis occurs in a ternary complex between enzyme, ATP and peptide. The pH and temperature dependence of catalysis were consistent with a mechanism in which an enzyme carboxylate facilitates deprotonation of the substrate tyrosyl hydroxyl, activating it as a nucleophile to attack the $\gamma$-phosphorus of ATP. This mechanism was used to define a query when searching for inhibitors in a database of predicted 3-D structures. Hits were defined as resembling the ATP $\gamma$-phosphate, tyrosyl hydroxyl and tyrosine aromatic ring, all of which were proposed to interact strongly with the enzyme during catalysis. This search identified several inhibitors which were used to define queries for 2-D searching of a larger database, leading to the discovery of 4-(3-chloroanilino)quinazoline **7**, which is a potent inhibitor of EGFR-TK ($K_{is}$ = 16 nM). The compound functions as an analogue of ATP **8** (see above). Several pharmaceutical companies have subsequently published patents on anilinoquinazolines as inhibitors of a range of protein kinases. Examples include the selective EGFR-TK inhibitor Iressa® (ZD1839) **9**, which is currently in phase III clinical trials for advanced non-small cell lung cancer.

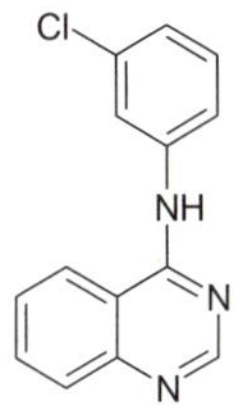

**7** R = ribose triphosphate **8** **9**

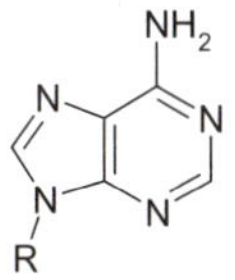

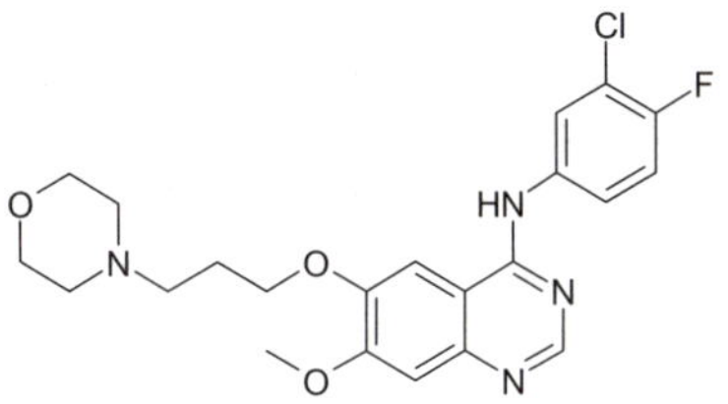

## Proteases: Inhibition of Angiotensin Converting Enzyme and Renin

The renin–angiotensin–aldosterone system is a multi-regulated proteolytic cascade of enzyme-mediated events that converts angiotensinogen into angiotensin I (AI), angiotensin II (AII), and angiotensin III (AIII), and so provides a major regulatory mechanism for the control of blood pressure in mammals (Figure 11). This cascade of events provides the medicinal chemist with the opportunity of blocking the formation of inappropriately high amounts of AII *via* inhibition of either the enzyme renin (aspartyl protease), or the angiotensin converting enzyme (zinc-dependent metallo-protease; ACE).

Inhibition of ACE, which cleaves AI to AII has demonstrated that obstruction of the system, prior to the formation of AII, can effectively lower blood pressure in a large majority of hypertensive patients. The key hypothesis in the development of ACE inhibitors, proposed by Ondetti and Cushman of the Squibb group, was that since ACE is mechanistically related to the well-characterised metallopeptidase carboxypeptidase A (CPA), CPA inhibitors could be modified to produce inhibitors of ACE.

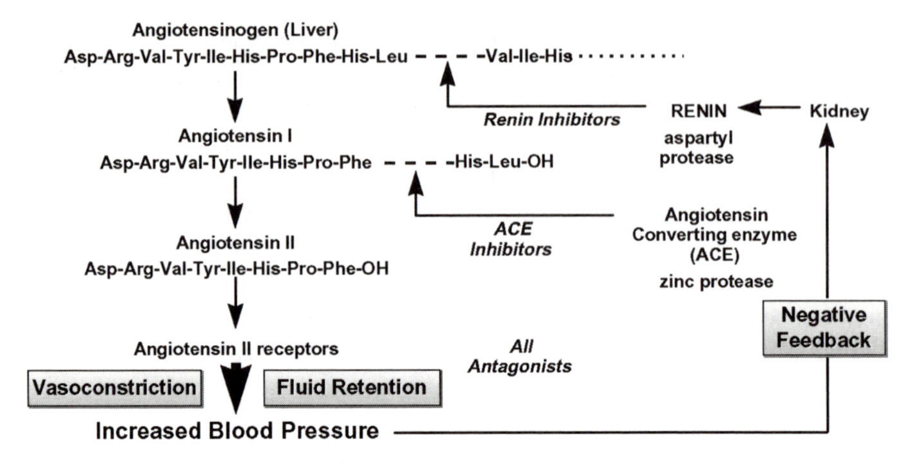


**Figure 11** *Schematic representation of the renin–angiotensin–aldosterone system*

ACE is a metallopeptidase which recognises and cleaves dipeptides from the C-terminus of AI and other related peptides. The enzyme contains a zinc ion at its catalytic site. Because there were no X-ray data available for the ACE active site, it was proposed that the zinc ion could coordinate the scissile amide bond of the substrate, activating it to nucleophilic attack by a conveniently placed water molecule, by analogy with the CPA mechanism (Figure 12). This line of thought eventually led to the discovery and development of several commercially successful agents, the first three of which were captopril **10**, enalapril (prodrug is structure **11**), and lisinopril **12**. These orally active, low molecular weight inhibitors incorporate ligands (thiolate or carboxylate) to bind the zinc ion at the active site of the protease, thereby attaining high affinity. A similar mechanism of

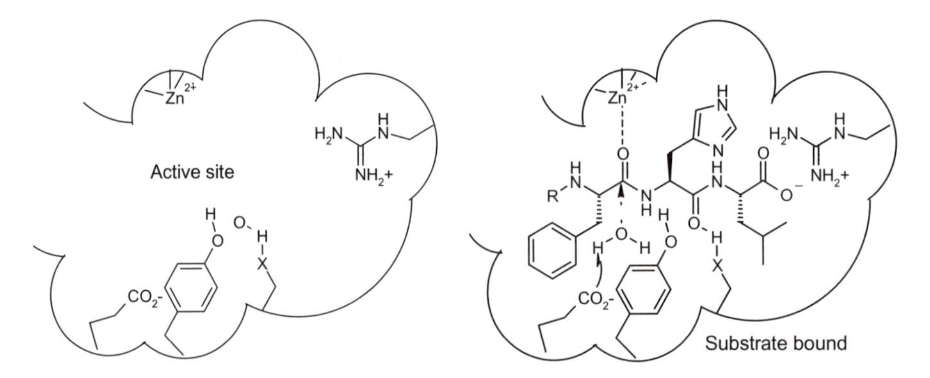


**Figure 12** *Representations of the ACE active site showing substrate binding and inhibition*

enzyme action to that illustrated is likely to be followed by many other zinc-dependent metalloproteases.

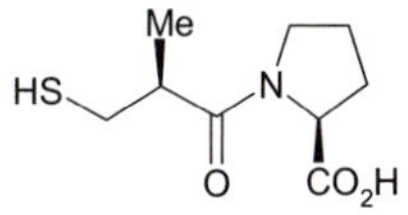

**10** 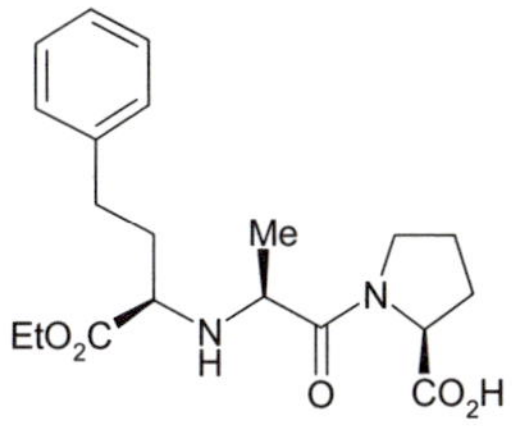 **11** 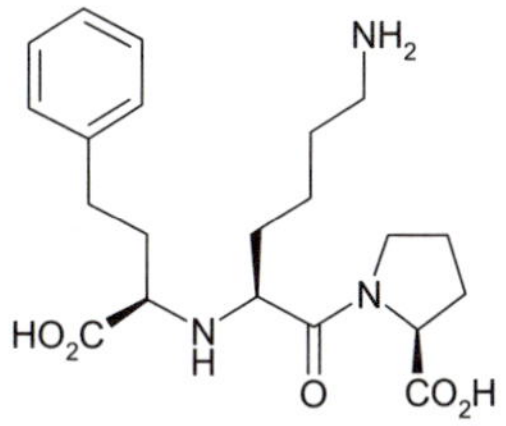 **12**

Despite their enormous clinical and commercial success, ACE inhibitors do suffer from a range of side-effects such as angio-oedema (mucosal lesions) and persistent cough. One possible explanation for this is that ACE is ubiquitous and unselective and will also cleave a number of other biologically important peptides such as bradykinin and substance P. It follows, therefore, that ACE inhibition may potentiate the actions of these peptides and lead to undesirable side-effects.

The conversion of angiotensinogen into AI by the enzyme renin is the first and rate-limiting step in the cascade. By contrast with ACE, renin has only one known substrate, being uniquely specific to angiotensinogen. Interruption of the cascade by renin inhibition could, therefore, be an extraordinarily specific pharmacological intervention, offering a valuable alternative to ACE inhibition with the potential for fewer side-effects.

Renin belongs to the aspartyl protease mechanistic class of enzymes and is related to pepsin and cathepsin D (mammalian enzymes), and penicillopepsin and endothiapepsin (fungal enzymes). Early inhibitor design strategies focused on computer modelling based on the substantial degree of homology displayed between renin and these two latter enzymes, for which high resolution X-ray crystal structures were available. This approach was further facilitated by the successful X-ray analysis at high resolution of the recombinant human protein. These X-ray structures show that the enzymes have active sites containing two catalytic aspartates, which can accommodate large peptidic substrates of eight amino acids or more. The postulated mechanism of action of this enzyme class (Figure 13) assumes that the ionised form of one of the aspartates acts as a general base to activate a water molecule to attack the scissile amide bond of the substrate, whilst the second protonated aspartate acts as a proton source for the developing tetrahedral transition-state.

One major difference between the substrate recognition motifs of the

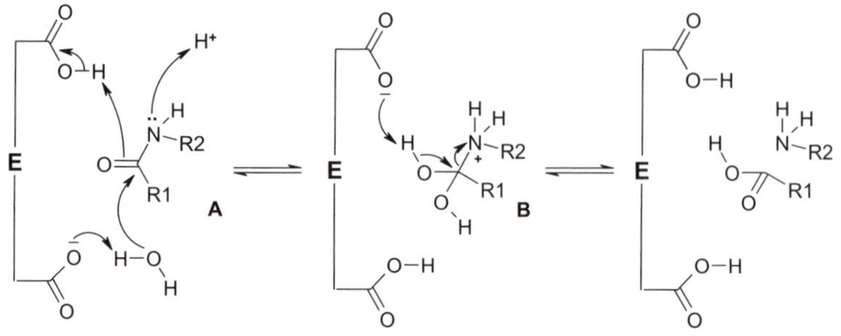

**Figure 13** *Schematic representation of the postulated mechanism of hydrolysis of substrate A by an aspartyl protease E* via *the transition-state B*

renin and ACE active sites is that renin lacks the dominant zinc ion. This means that the overall binding of inhibitors depends heavily on the multiple hydrophobic interactions which the enzyme makes with the side-chains of amino acid residues.

Although it is possible to prepare renin inhibitors based on the substrate structure, a much more attractive way to gain access to compounds with higher affinity is to make chemically stable, non-hydrolysable analogues which resemble the transition-state (Figure 14) in their geometry and electron distribution. This has been addressed by incorporation of the so called hydroxyethylene transition-state mimetic into structures such as the pseudo-peptidic inhibitor CGP 38 560 (**13**; $K_{is} = 0.7$ nM) and enalkiren (**14**; $K_{is} = 14$ nM), both of which also showed oral activity in primate models of hypertension. However, these compounds display poor bioavailability (<1% in humans) making them unsuitable for further development. More recent research efforts have led to the identification of compounds with much improved bioavailability, such as CP-108,671 (**15**; $IC_{50} =$ 4 nM; bioavailability 60% in dogs) and totally non-peptidic and orally active inhibitors such as compound **16** ($IC_{50} = 37$ nM) which apparently bind to the enzyme in a different fashion to the compounds above. Efforts to find agents with markedly improved oral absorption and pharmacokinetics continue.

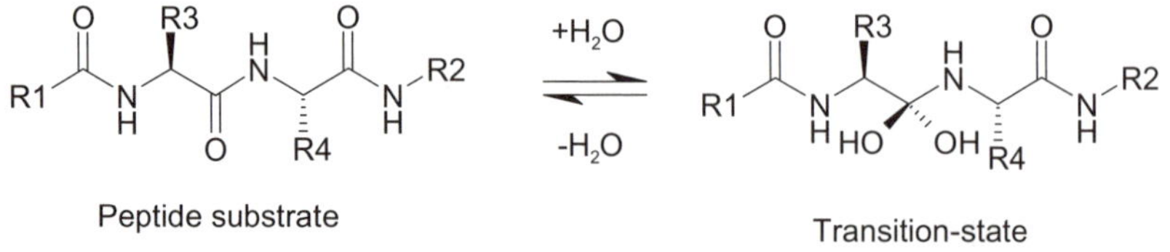

**Figure 14** *Schematic representation of the substrate and postulated transition-state in the hydrolysis step catalysed by renin*

A case history on the development of saquinavir, an inhibitor of another aspartyl protease, HIV protease, is the subject of Chapter 18.

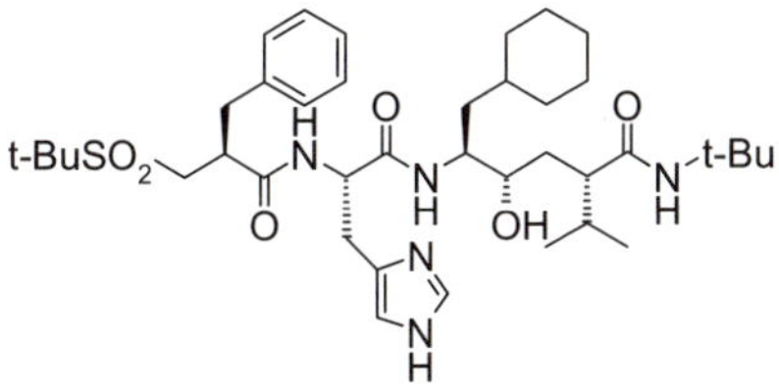

13

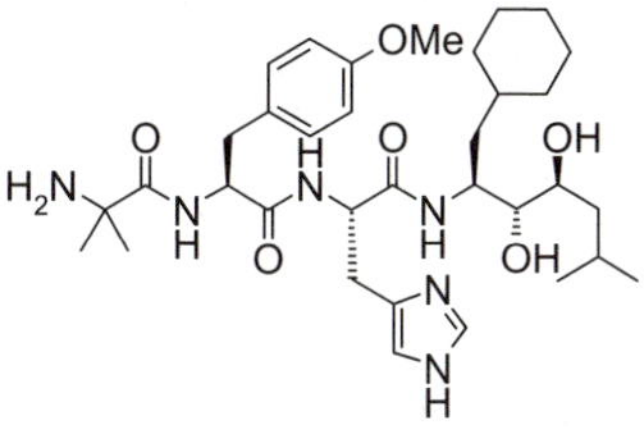

14

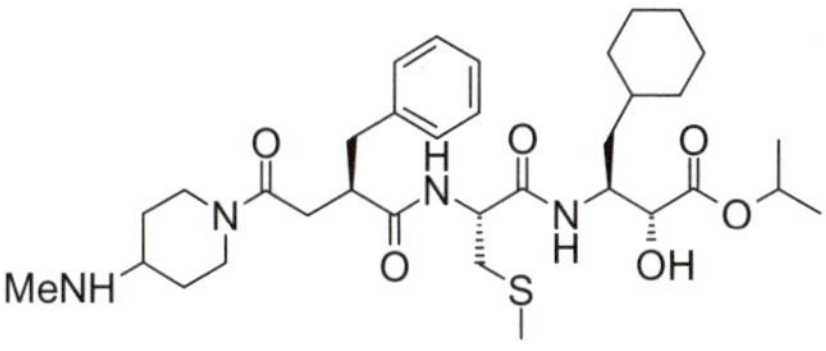

15

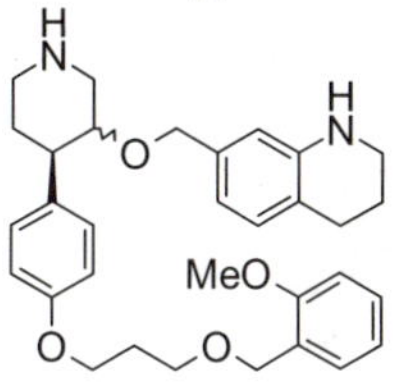

16

## 5 CONCLUDING REMARKS

Enzyme inhibition constitutes a powerful strategy for the design of novel drugs by enabling the selective blockade of specific biochemical cascades. We have entered a new era, where large numbers of pure enzymes have become more readily available through molecular biology. This growing target availability is being augmented by increasingly sophisticated protein NMR spectroscopic, X-ray crystallographic, thermodynamic and computational techniques, revealing more details of enzyme structure and mechanism and enabling the rational design of improved compounds. Informatics will allow increasing exploitation of accumulating databases. Within the pharmaceutical industry, there is a tendency towards increasing throughput, both in assays and in research projects. Enzyme kinetics show that $IC_{50}$s should not be taken at face value, but, rather, should be interpreted in the context of further information on the system. This approach will increase the rate of success in drug discovery projects.

### Acknowledgements

The authors are indebted to many colleagues at Alderley Park, especially Dave Timms, Geoff Holdgate and Pete Cook, for providing background material for this chapter.

## 6 REFERENCES

1. A.R. Fersht, *Structure and Mechanism in Protein Science*, W.H. Freeman and Company, New York, 1999.
2. W.W. Cleland, *Methods Enzymol.*, 1979, **63**, 103.
3. R.A. Copeland, *Enzymes. A Practical Introduction to Structure, Mechanism and Data Analysis*, Wiley-VCH, New York, 2nd edition, 2000.
4. B. Mannervik, *Methods Enzymol.*, 1982, **87**, 370.
5. W.H.J. Ward *et al.*, *Biochemistry*, 1999, **38**, 12514.
6. T.J. Franklin *et al.*, *Biochem. Pharmacol.*, 1999, **58**, 867.
7. P.N. Cook *et al.*, *Biochem. Pharmacol.*, 1995, **49**, 1043.
8. W.H.J. Ward *et al.*, *Biochem. Pharmacol.*, 1994, **48**, 659.
9. Y.-C. Cheng and W.H. Prusoff, *Biochem Pharmacol.,* 1973, **22**, 3099.
10. J.F. Morrison and C.T. Walsh, *Adv. Enzymol.,* 1988, **61**, 201.
11. L.M. Kauvar *et al.*, *Chem. Biol.*, 1995, **2**, 107.
12. W.H.J. Ward and G.A. Holdgate, *Prog. Med. Chem.*, 2001, **38**, 309.
13. S.B. Shuker *et al.*, *Science*, 1996, **274**, 1531.
14. A.A. Bogan and K.S. Thorn, *J. Mol. Biol.*, 1998, **280**, 1.
15. A.L. Jackman *et al.*, *Cancer Res.*, 1991, **51**, 5579.

### Additional Reading

1. *Comprehensive Medicinal Chemistry*, P.G. Sammes (ed.), Pergamon Press, Oxford, 1st edition, 1990, Vol. 2.
2. K.T. Douglas, *Chem Ind.*, 1983, 311.
3. J.K. Seydel and K.J. Schaper, in *Enzyme Inhibitors as Drugs*, M. Sandler (ed.), Macmillan, London, 1980, p. 53.
4. C. Walsh, *Tetrahedron*, 1982, **38**, 871.
5. J.R. Knowles, *Acc. Chem. Res.*, 1985, **18**, 97.
6. S.J. Cartwright and S.G. Waley, *Med. Res. Rev.*, 1983, **3**, 341.
7. M.A. Ondetti and D.W. Cusham, *CRC Crit. Rev. Biochem.*, 1984, **16**, 381.
8. M.J. Wyvratt and A.A. Pachett, *Med. Res. Rev.*, 1985, **5**, 483.
9. J. Cooper *et al.*, *Biochemistry*, 1992, **31**, 8142.
10. A.R. Sielecki *et al.*, *Science*, 1989, **243**, 1346.
11. J. Boger, *Annu. Rep. Med. Chem.*, 1985, **20**, 257.
12. R.H. Bradbury *et al.*, *J. Med. Chem.*, 1990, **33**, 2335.

CHAPTER 5

# Biological Evaluation of Novel Compounds

GARY W. PRICE, GRAHAM J. RILEY AND DEREK N. MIDDLEMISS

## 1 INTRODUCTION

The introduction of new drugs to the clinic is a protracted process, which involves a huge investment in time, resources and money. To make the selection of drug candidates as efficient and rapid as possible, a testing cascade is constructed. The definition of a testing cascade begins with the designation of the biological target. In the current era of genomics, this often starts with a knowledge of the gene encoding for a specific target protein. Five major classes of protein have produced practical targets for the discovery of drugs for the treatment of many illnesses. These classes of proteins can be broadly described as enzymes, ion channels, nuclear hormone receptors, transporters and 7-trans-membrane (7-TM) receptors. 7-TM receptors are characterised by their protein structure, which have 7 hydrophobic regions that allow them to insert in and across a plasma membrane. They are believed to make up in excess of 1000 genes in the human genome and have been the subject of considerable success in drug discovery. Examples of drugs which act at this class of proteins include $\beta$-blockers which are used in the treatment of hypertension, dopamine antagonists used in the control of psychotic symptoms in schizophrenics and 5-$HT_1$-like agonists, which effectively alleviate the symptoms of migraine.

Given the size of the human genome (there are estimated to be at least 40 000 genes), there are two possible approaches to the selection of which gene(s) to work on for a specific disease. If the gene causing the disease is known, for example in the case of Alzheimer's Disease (AD), where mutations of the $\beta$-amyloid gene are thought to be responsible for certain types of familial AD, the protein product of this gene may form the basis of a biological target in the hunt for a drug treatment. However, unfortunately, this is a rare example and the state of genomic research at the beginning of the 21st century is not sufficiently advanced to designate specific gene(s) abnormalities to specific diseases in many cases. Consequently, the pharmaceutical industry has used homology searches to define classes of genes that lead to protein products with a high probability for success in drug discovery. An example of these is the class of human 5-HT receptors which consist of at least 14 separate genes encoding for 14 distinct 5-HT receptors. Of these, the 5-$HT_{1B}$ receptor will be used throughout this article to illustrate the process of biological evaluation of new compounds. Such biological evaluation usually starts with a high throughput screen (HTS) of hundreds of thousands of compounds against the biological target to identify a lead molecule, which can give the medicinal chemist a rational starting point for the design of more potent and selective molecules.

Once the target protein and a lead molecule have been identified, a biological testing cascade is designed to identify potent, selective, biologically active candidate molecules for possible drug development. The general principles of such a cascade are given in Figure 1 below.

The primary screen is a medium throughput, *in vitro* assay which often incorporates a selectivity assay (*e.g.* 5-$HT_{1B}$ *versus* 5-$HT_{1D}$ receptor affinity – see below) and is followed by an *in vitro* functional assay which measures the biological efficacy of the compound; in the case of 7-TM

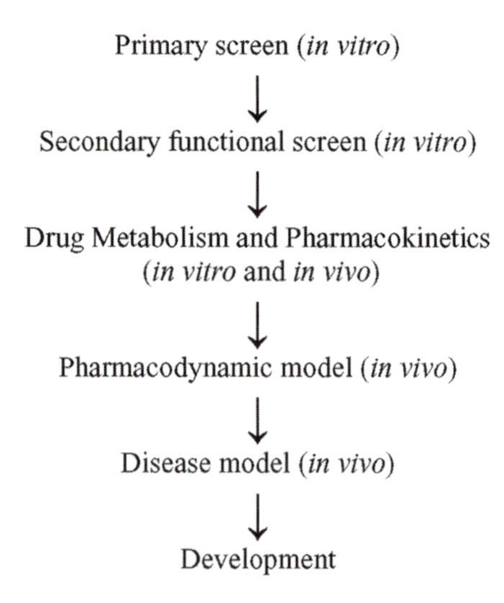


**Figure 1** *Example of a typical screening cascade*

receptor targets this would be whether the compound is an agonist, partial agonist, antagonist or inverse agonist at the receptor. After these *in vitro* tests, the candidate molecule is assessed for interactions with drug metabolising enzymes (cytochrome P450) and *in vitro* and *in vivo* intrinsic metabolic clearance. Once a compound has passed these basic criteria, it is next assessed in a pharmacodynamic assay for an action *in vivo* (see later for definition). This latter assay may be a simple physiological or behavioural read out of biological activity, such as effects on body temperature or locomotor activity. The final assessment before entry of the compound into development, is in an animal disease model. Each of these steps in the testing cascade will be discussed in more detail below and will be illustrated with the example of a selective 5-$HT_{1B}$ receptor antagonist, SB-236057.[1]

## 2 PRIMARY SCREENS

Primary assays must be robust, precise, reproducible, stable over long periods of time and have a capacity to test large numbers of compounds. They must also yield data that will predict activity in subsequent experimental procedures and also the intended therapeutic action. Many primary assays now use recombinant human proteins expressed in host cells, yet many of the secondary assays use animal, particularly rodent models. Therefore, it is important to confirm that the primary assay does predict an effect in the secondary assays. In most cases, it is possible to show that the protein target in an appropriate animal has the same functions as in the human derived protein, either by expressing the equivalent animal recombinant protein or by using native tissues and carrying out parallel experiments. If it is shown that the pharmacological profile of the intended animal tissue or recombinant-derived protein is significantly different from the human recombinant profile, then an alternative animal model must be sought.

The five classes of potential recombinant human protein targets are providing an increasing proportion of the targets for drug discovery. They have the advantage of relevance to the human disease condition and also availability. It is possible to derive quantities of biological material from cell growth in culture that are equivalent to many animals. This alone has extended the limits of the amount of experimentation well beyond the capacities of even 10 years ago. Human recombinant proteins from cell cultures also provide a relatively high level of purity and stability for *in vitro* use. Careful selection of the host cell can eliminate many problems that might have been present in native tissue preparations. The consequence is that assays tend to have fewer interfering factors that need to be controlled. Experimental designs can be simplified and coupled with new technologies. There is a move towards tests that require little more than mixing of the test compound with the biological reagent, incubate for a time to allow development of the signal then read it with an appropriate detector. Simple tests have fewer sources of error, providing the level of robustness, precision and reproducibility needed while allowing the volume of throughput required.

The general principle of a primary assay is to discover the concentration of a test compound that will achieve a defined end-point. For radioligand binding assays, the end-point is the concentration that inhibits 50% of the total binding, the $IC_{50}$. In order to achieve this, the test compound is diluted to provide a range of concentrations. In the ideal experimental design this should include concentrations high enough to give 100% inhibition and low enough to give little or no inhibition. From such curves good estimates of the $IC_{50}$ can be made. In practice, tested compounds have a wide range of activities and it is not possible to provide a range that will encompass them all. The concentration range is usually set to cover a window of expected concentrations. Occasionally, key compounds will have activities outside the expected range but they can be re-tested with more suitable concentrations.

Mixing the compound with the biological reagent and observing a relevant signal is the simplest experimental protocol. While methods are being developed to be able to observe receptor–ligand interactions directly, they are not yet widespread. For the majority of assays, a further reagent is necessary in order to be able to observe the interaction of the test compound with the target protein. In radioligand binding experiments, a radiolabelled compound that has a high affinity for the receptor protein is inhibited from binding by the test compound. The degree of inhibition can be used to determine the activity of the compound.

A key step in radioligand binding is the separation of the bound and unbound fractions of the radioligand. While centrifugation can be used it is generally slow and not amenable to automation. For many years the method of choice has been filtration. This is quicker because many more samples can be processed simultaneously by using semi-automated equipment. However, separation steps can be avoided by using proximity techniques. Scintillation Proximity Assay (Amersham) and Flashplates (NEN) work on the principle that the biological target is linked to a solid scintillation matrix. When the radioligand is bound to the receptor, it is brought into proximity with the scintillating matrix so that decay of the radio-isotope sends a ionising particle towards the solid scintillant, causing emission of a photon that can be detected. As the path-length of some ionising radiation, particularly from tritium and $^{125}I$, is short, only radiation from bound molecules is registered, while the unbound material is too far away. This is a simple and relatively quick procedure but the counting efficiency is low relative to filter assays and can preclude its use in some circumstances. Also, it can suffer interference from coloured or fluorescing test compounds.

When designing a radioligand binding assay, the key feature is that the components must be brought together in such a way that they react and reach equilibrium. Above all, the time of incubation must be long enough for equilibrium to be established but not so long that decomposition of any of the reactants becomes significant. Raising the temperature of incubation decreases the time but also increases the risk of decomposition. Incubation at 37 °C is desirable, because the data produced is more relevant to the therapeutic situation in mammals, but decomposition of the reagents may force the assay to be carried out at a lower temperature. Affinities of test compounds may be affected by this.

Advancing technologies are also allowing use of ligands that are tagged with a fluorescent moiety.[2,3] Fluorescence-based assays have considerable advantages over radiolabel-based assays in being more environmentally friendly, cheaper and having greater sensitivity and flexibility in detection (intensity, wavelength, polarisation). Thus fluorescence-based assays are adaptable to micro-assay systems and hence the major focus has been the development of ultra-high throughput assays using high density, 1536 well assay plates but are also equally valid for SAR screening. Assays have been developed that are similar to radioligand binding displacement assays. In this case, detection relies on a difference in fluorescence of a tagged ligand bound or unbound to the target protein. This difference can either be a change in polarisation (fluorescence anisotropy) due to a molecular-weight related rate of tumbling, or a quenching of the fluorescence. In the former case, there is a large difference in molecular weight between the fluorescent ligand and the ligand–protein complex, thus displacement of the ligand causes a change in the plane of polarisation and the ratio of bound:unbound can be measured. The medicinal chemist is often involved in the design of suitable ligands and one example within our laboratories was the identification of the labelled pleuromutilin, SB-452466 **1** which was used both for SAR and high throughput screening of the pleuromutilin binding site on the bacterial ribosome (see Chapter 3).[4,5] A second example was SB-477790 **2**, which was a ligand used in a ligand displacement primary assay for the ATP-binding site of B-raf kinase.[6] This ligand had exquisite sub-nM potency and, therefore, the assay could be configured to give us absolute values far below that achievable with the standard radiolabelled phosphorylation assay for our highly potent inhibitors.

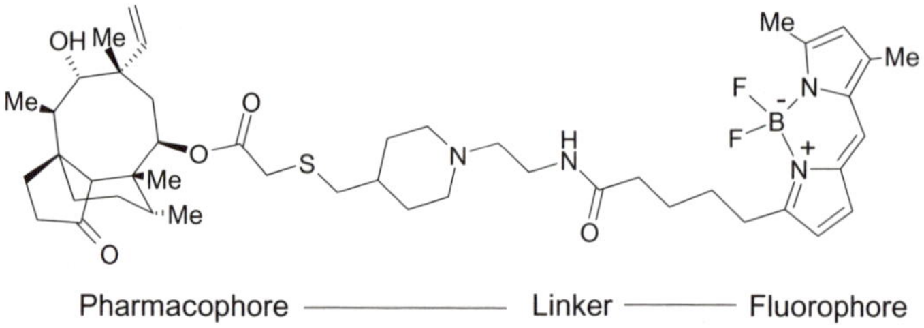

Pharmacophore —— Linker —— Fluorophore

**1** SB-452466

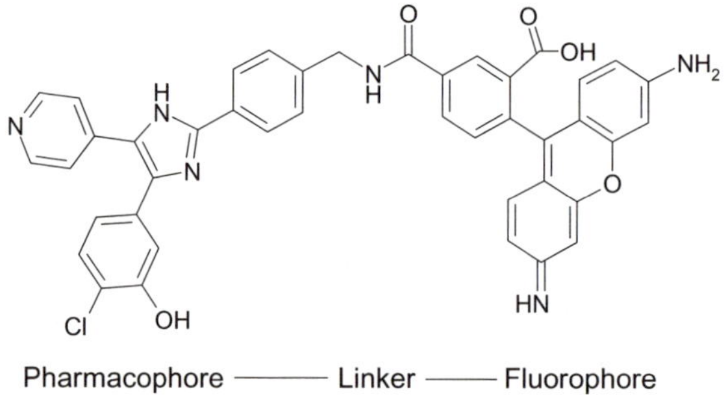

Pharmacophore —— Linker —— Fluorophore

**2** SB-477790

For detection by quenching, either the ligand or the protein may contain the fluorophore and its partner the quenching functionality. Thus, if the ligand contained the fluorophore and the protein the quencher, displacement of the ligand would result in an enhanced fluorescence signal. If the ligand were the quencher and the fluorescence of the protein (*e.g.* tryptophans) was being measured, displacement with a non-quenching ligand would again enhance the fluorescence signal. Both quenching and augmentation of the fluorescence has been used for enzyme cleavage assays. In this case, the substrate contains a fluorophore on one side of the cleavage site, and either a quencher or second fluorophore on the other side and relies on the close proximity of the quencher–fluorophore and fluorophore–fluorophore pairs. In the fluorophore–quencher pair, cleavage results in the two separating and hence the fluoresence is no longer quenched and an enhancement of signal is observed.[7] In the case of a fluoresence–fluoresence pair, the fluorophores are chosen so that the emission band of the first (donor) fluorophore overlaps with the excitation band of the second (acceptor) fluorophore. This enables "Fluorescence Resonance Energy Transfer" (FRET) between donor and acceptor. As long as the two fluorophores are in close proximity, excitation of the donor can be observed as fluorescence emission of the acceptor at a wavelength red-shifted from the donor spectrum. Cleavage then abolishes the acceptor fluorescence (Figure 2).

Fluorescence-based enzyme assays have also been set up where the enzyme reaction is liberating a fluorophore from a non-fluorescent substrate ("fluorogenic substrates") and this is the basis for most high throughput cytochrome P450 assays.[8]

A disadvantage of fluorescence-based assays is that finding a suitable ligand or substrate may be difficult because of the size of the fluorophores; thus it is not readily applicable to relatively small binding sites, for example within the trans-membrane region of a 7-TM receptor. In spite of this, currently approximately 70% of primary assays within GSK are fluoresence based.

The simplest measurement of affinity of the test compound is the concentration at which it inhibits 50% of the control binding or activity, the $IC_{50}$. However, this is not a constant as it is highly dependent on the experimental conditions; for example, the concentration of the radioligand or enzyme substrate. It does have the merit of being an empirical value that does not require the mechanism of the interaction to be known.

In pharmacologically clean experiments where the mechanism is known and the equilibrium constant for the radioligand ($K_d$) has been estimated previously, an estimate of the equilibrium constant for the test compound ($K_i$) can be made using:

$$K_i = \frac{IC_{50}}{\frac{[L]}{K_d} + 1}$$

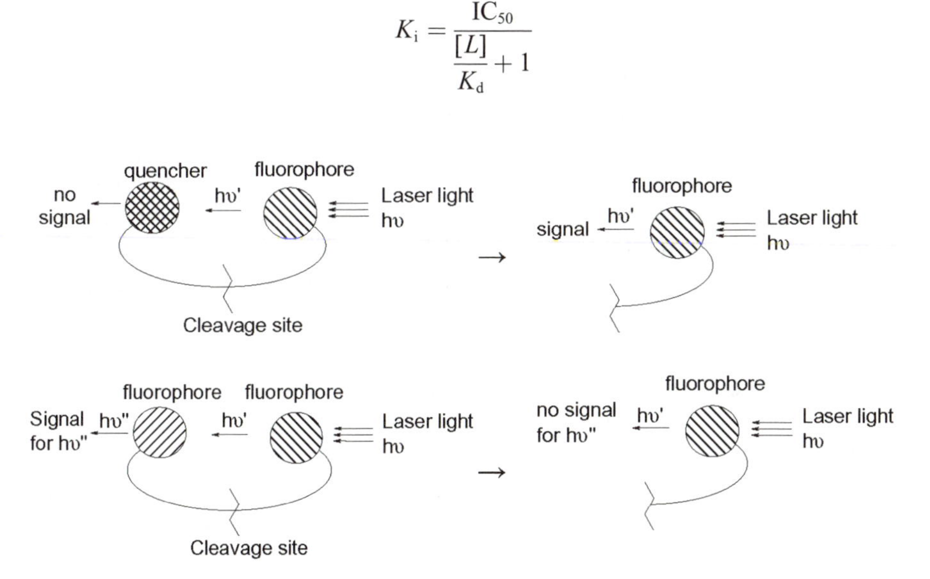


**Figure 2** *Diagrammatic representation of fluorescence quenched and transfer assays*

It must be appreciated that this correction contains approximations. The ligand concentration should be the unbound fraction of radioligand at 50% inhibition and the $IC_{50}$ should be the unbound concentration of inhibitor at 50% inhibition. The first of these is calculable and can be used, but often is not. The second is not measurable in radioligand binding experiments. The impact of these approximations increases as the $K_d$ and $K_i$ approach the concentration of binding sites ($B_{max}$) in the assay. This concentration varies between assays, depending on receptor expression, the specific activity of the radioligand and the affinity of the radioligand, but is usually in the range of 10 to 500 pM. In most test procedures, $K_i$ values greater than 1 nM are not influenced greatly by the approximations but sub-nanomolar $K_i$ may be.

In addition to the approximations in the conversion equation, there is an assumption that the $K_d$ value determined from, usually, a separate saturation analysis, often some time previously, applies on all subsequent assay occasions. Despite these assumptions and approximations, the correction is a useful tool to convert the $IC_{50}$ into a universal value. Since most experimenters try to use the same radioligand concentration every time, the correction factor approaches a constant value so it has little impact on the variation of the data. In our 5-$HT_{1B}$ receptor binding assay, utilising $^3$H-5-HT, the average correction applied in 1251 experiments was +0.26 p$K_i$ units, having a standard deviation of 0.08.

In presenting the affinity estimates, there is a choice between the $K_i$ as a molar concentration and the p$K_i$ which is the negative $\log_{10}$ of the molar $K_i$ (*cf.* pH). As the $IC_{50}$ (and therefore the $K_i$) is usually calculated from logarithmically related drug concentrations there is an impact on the statistical distribution of the data. Figure 3 shows how the p$K_i$ values for 5-HT in the 5-$HT_{1B}$ radioligand binding assay have a normal distribution compared with the skewed distribution for the $K_i$ values. We prefer to utilise p$K_i$ values but acknowledge that this is not universally adopted.

While there is no theoretical limit to the affinity of compounds for a receptor, the usual method of measurement of affinity constants in binding assays imposes an upper limit of affinity. This follows from the use of the $IC_{50}$ as the primary measurement. To achieve 50% inhibition of binding, the competing compound has to occupy about 50% of the available receptors. Therefore, the absolute minimum concentration of drug necessary to do this will be equal to half of the total receptor concentration. Depending on the assay, this usually falls between about 5 to 250 pM. In practice, there will be a fraction of unbound inhibitor that will increase the value of the total that is used to calculate the $IC_{50}$. If it were possible to plot percentage inhibition against the unbound fraction of the test concentration, this anomaly would be overcome. It is important to be

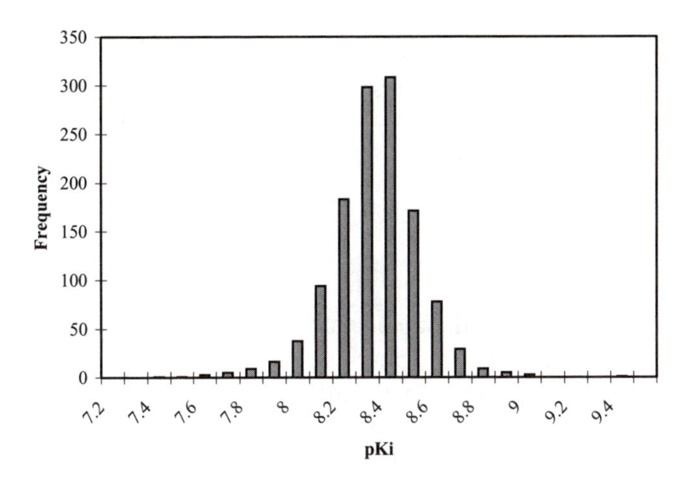


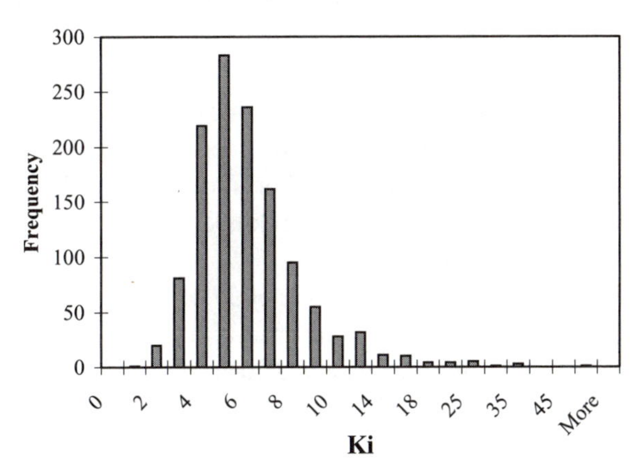


**Figure 3** *Frequency distribution of* $pK_i$ *and* $K_i$ *values for a 5-HT$_{1B}$ receptor binding assay*

aware of this limitation because it may affect the assessment and understanding of the target affinity.

## Broad Spectrum Evaluation of Compounds

All drug discovery departments within the pharmaceutical industry have a wide portfolio of assays through which compounds can be passed. Selected drug candidates can be circulated around these to identify any additional unexpected activities. Alternatively, there are specialist companies (*e.g.* CEREP, PanLabs or Novascreen) that provide a broad spectrum screening service that is often quicker and simpler to apply rather than sending to the multiple research sites of a large organisation. Either way, the purpose is to extend knowledge about the compound before committing to detailed further studies.

## SB-236057

When searching for new therapies that work by novel mechanisms, we cannot be sure what the ideal pharmacological profile is. Therefore, we have developed working criteria in order to help us through the process. For the initial stages of the screening process in the discovery of SB-236057 we used radioligand binding. We defined our ideal profile for this stage as an affinity (p$K_i$) of 8.0 or more and a selectivity over the other binding assays of 2.0 (100-fold).

The biological target was the 5-$HT_{1B}$ receptor subtype and the 5-$HT_{1D}$ receptor sub-type was used as the principal target at which activity was to be avoided. So, all compounds from the chemical programme were put through both of these assays and compounds that satisfied the primary criteria were passed to the next stage. The second stage was the evaluation of the compounds in a number of other aminergic radioligand binding assays. The objective was to identify unexpected activity in assays that might predict unwanted side-effects.

The profile that we found for SB-236057 is shown in Table 1. This profile met our working criteria as the affinity at 5-$HT_{1B}$ was a p$K_i$ of 8.14 and the selectivity was approximately 100-fold or greater over all other receptors tested.

**Table 1** *p$K_i$ values for SB-236057 in various aminergic radioligand binding assays*

| *Receptor*[a] | p$K_i$ *(s.e.m)* | n | *Receptor*[a] | p$K_i$ | n |
|---|---|---|---|---|---|
| 5-$HT_{1A}$ | <5.1 | 3 | 5-$HT_6$ | <5.1 | 6 |
| 5-$HT_{1B}$ | 8.14 (0.08) | 13 | 5-$HT_7$ | <5 | 6 |
| 5-$HT_{1D}$ | 6.29 (0.04) | 13 | Dopamine $D_2$ | <5.8 | 5 |
| 5-$HT_{2A}$ | <5.4 | 9 | Dopamine $D_3$ | <5.6 | 6 |
| 5-$HT_{2B}$ | <5.2 | 5 | Dopamine $D_4$ | <5.4 | 6 |
| 5-$HT_{2C}$ | <5.4 | 7 | Adrenergic $\alpha_{1B}$ | <5 | 3 |
| 5-$HT_4$ | 6.12 (0.14) | 3 | Adrenergic $\beta_2$ | <5.2 | 5 |

[a] All are human recombinant receptors except 5-$HT_4$ (guinea pig hippocampus).

## 3 SECONDARY ASSAYS

### *In Vitro* Evaluation of Compounds

Having selected compounds on the basis of high affinity for the target protein and selectivity over other related proteins, a process of elimination which typically screens out >90% of compounds, the next task is to confirm target protein affinity in native tissues and then evaluate functional activity of compounds.

### Binding Assays in Native Tissue *Versus* Recombinant Systems

As described in the previous section, there are many positive attributes of screening in recombinant systems. Compounds are selected on high affinity for the human protein, cloned cell lines are homogeneous and therefore greatly enhance reproducibility, sensitivity and stability of assays. They provide the ethical advantage of decreasing the use of animal tissues and offer very significant cost savings. However, it is prudent to remember that, in general, these proteins are expressed in an 'alien' cell, *e.g.* a neuronal receptor could be expressed in a Chinese hamster ovary (CHO) cell line, or in a human embryonic kidney (HEK) cell line. Secondly, these cells are grown in tissue culture, an artificial environment; therefore, post-translational modification of proteins such as phosphorylation and glycosylation which occur in the native tissue may not be present in the recombinant system. Trafficking of the protein to an appropriate part of the cell, such as the extracellular membrane, may require additional proteins which again may not be present and, finally, coupling of the target protein to its transduction mechanism may not occur. All of these factors can potentially affect the pharmacology of a protein. It is essential, therefore, that compound affinity for its target protein is confirmed in human native tissues, appropriate for the therapeutic indication. In addition, functional evaluation of compounds both *in vitro* and *in vivo* are likely to be determined in rodents. Therefore, confirmation of affinity in the relevant rodent species is also required.

Binding studies in native tissues are essentially carried out as for recombinant systems. However, with the possible presence in tissues of endogenous substrates for the target protein, a more extensive washing procedure is usually required during the preparation of membranes. Table 2 gives the affinity of the compound SB-236057 for rat and guinea-pig 5-$HT_{1B}$ receptors in brain tissue, which correlates well with the human receptor recombinantly expressed in CHO cells.

**Table 2** *Receptor binding affinity (p$K_i$ ± sem) for SB-236057 in human recombinant and rat and guinea-pig native tissues*

| *Human 5-$HT_{1B}$CHO* | *Rat 5-$HT_{1B}$ (cortex)* | *Guinea-pig 5-$HT_{1B}$ (striatum)* |
|---|---|---|
| 8.3 ± 0.1 | 8.3 ± 0.2 | 8.3 ± 0.2 |

## Evaluation of Functional Activity of Compounds

*Agonists/Antagonists/Inverse Agonists.* A limitation of binding assays is that usually the assay only provides data on a compound's affinity for a receptor but no information on a compound's efficacy, that is, the ability of a compound to stimulate the receptor and produce a functional response. An *agonist* binds to the receptor and so, induces a conformational change in the protein, which then activates a transduction pathway, for example, through activation of a G-protein or through the opening of an ion channel. An *antagonist* binds to the receptor but fails to induce receptor activation; however, the *antagonist* can prevent receptor activation by *agonists*. Between a full *agonist*, a compound with maximum efficacy and an *antagonist*, a compound with zero efficacy, lies a spectrum of possible efficacies or intrinsic activities for compounds known as *partial agonists*. A full *agonist* will always produce a maximal response in a receptor system, and an *antagonist* will never produce a functional response. However, the degree of intrinsic activity that a *partial agonist* displays depends entirely on the system being studied. In systems where there is a large number of receptors that are efficiently coupled to their transduction pathways, only a small fraction (<10%) of the receptors may need to be occupied in order to achieve a maximal response by a full agonist. This condition is described as exhibiting a high receptor reserve. Recombinant systems often present a high receptor reserve and, in such cases, a partial agonist can display maximal functional activity. In a system with no receptor reserve and poor coupling, a partial agonist can fail to produce any functional response. (See also Chapter 1.)

In recent years, this classification of the functional activity of compounds has acquired a third arm, *inverse agonists*, or compounds which display *negative intrinsic activity*. Though not fully accepted by all pharmacologists, the theory usefully explains phenomenon often observed in G-Protein Coupled Receptors (GPCR) in recombinant systems. In these systems (as observed for the human 5-$HT_{1B}$ receptors, see below), receptors can be constitutively coupled to their G-proteins, that is, they are in an actively coupled state in the absence of an agonist. A compound which uncouples these receptors from their activated state is known as an *inverse agonist*, because it, presumably, produces a conformational change in the receptor in the opposite direction to that produced by an agonist. An antagonist would inhibit the actions of either agonists or inverse agonists. It is still unclear, at this stage, whether such negative intrinsic activity can occur *in vivo*.

*Assays for Evaluating Functional Activity.* Prior to the advent of recombinant expressions systems, evaluations of functional activity in native tissues were usually time consuming, labour intensive and low in compound throughput. Intact native systems, such as brain tissues, are highly complex, and it can be difficult to tease out a functional signal that can be attributed directly to the target receptor of interest. Where possible, simpler surrogate experimental models in alternative biological preparations that had an easily quantifiable readout were employed. For example, it is difficult, though possible, to directly measure receptor activity of G-proteins or downstream effects such as adenylyl cyclase activity or phosphotidylinositol turnover in brain. If the same target receptor is also present on a blood vessel or present in the gut and activation of the receptor leads to a measurable contraction or relaxation of a strip of artery or gut, then this may be used instead of the more difficult and slower direct assay in the desired target tissue or organ.

Assessment of functional activity can occur at any stage in the transduction pathway. Cell-free preparations of membrane-bound receptors often appear as a mixture of high and low affinity states. Agonists display high affinity for one state of the receptor, low affinity for the other, whereas antagonists usually display similar affinity for both states of the receptor. This can be exploited by using two radioligands, one that is an agonist and another that is an antagonist. In these experiments, agonists will tend to have a high affinity for the inhibition of binding of the agonist radioligand but a lower affinity for the same effect with the antagonist radioligand. Expressing the difference between the $pK_i$ determined from agonist binding and that determined from antagonist, it is possible to get an indirect measure of intrinsic activity of compounds. It is usually preferable, however, to measure directly the function at the level of the G-protein (GTP$\gamma$S binding) at the *effector* (*e.g.* adenylyl cyclase activity/cAMP accumulation) or in terms of a change of a neuronal excitability or a consequence of such a change.

With recombinant systems, the receptor of interest is usually expressed at a very high level, usually 10 to 100 times greater than the density of sites expressed in the native tissue and at a much higher level than receptors endogenously expressed in the host cell lines. In these cells, once a transduction pathway is established, it is possible to study a receptor system virtually in isolation. In addition, it is possible to couple receptors artificially to alternative pathways or G-proteins and to enable a rapid functional activity readout and, therefore, greatly increase compound throughput. Such assays usually employ a fluorescent readout, such as fluorescing dyes or luminescent reporter gene assays. The former measures the levels of a second messenger generated by the effector, such as $Ca^{2+}$ mobilisation or increased $Ca^{2+}$ permeability. The introduction of the FLIPR (Flourometric Imaging Plate Reader, Molecular Devices, CA. USA) has enabled such assays to be performed in real time using live cells and, importantly, in a very high throughput format. The FLIPR can be used to measure functional activity in both G-protein coupled receptors and ligand-gated ion channels. Receptors not naturally coupled to $Ca^{2+}$, that is, receptors other than Gq-coupled, can be 'forced' into a $Ca^{2+}$ transduction pathway *via* phospholipase C by one of two methods. Cell lines expressing promiscuous G-proteins, such as $G_{\alpha 16}$, will couple to a range of receptors. An alternative is the co-transfection of cell lines with chimeric G-proteins, where carboxy terminal mutations of the $G_{\alpha}$ subunit can modify the receptor specificity of the G-protein; for example, replacement of a C-terminal portion of the $G_q$ subunit with that of $G_s$ ($G_{qs5}$ chimera) couples $G_s$-coupled receptors to phospholipase C instead of adenylyl cyclase. Thus this technology is potentially universally applicable to most GPCRs (though there are exceptions!). Another alternative is indirectly measuring cAMP levels through activation of a cAMP-dependent gene transcription pathway. Again, since these assays are amenable to high levels of automation this allows very high compound throughput. However, the further down a transduction cascade we proceed, the greater the increase in the complexity of the response. Measuring a functional response at the level of the G-protein with GTP$\gamma$S binding gives a simple direct-functional readout of receptor activity.

*GTP$\gamma$S binding.* On activation of a GPCR, the G-proteins undergo a cycle of conformational changes. GDP, which is bound to the $\alpha$-G-protein subunit of an $\alpha\beta\gamma$ trimer, is exchanged for GTP, triggering separation of the $\alpha$ and $\beta\gamma$ subunits. The latter stimulates the effector, such as adenylyl cyclase. The $\beta\gamma$ dimer is also thought to affect other effector systems, though their role, particularly *in vivo*, is still somewhat unclear. Meanwhile, the binding of GTP to the $\alpha$ subunit induces a conformational change in the receptor so that the receptor loses its high affinity for agonists, thereby accelerating dissociation of the bound agonist. Endogenous GTP-hydrolysing activity of the $\alpha$ subunit converts the G-protein into its GDP bound state, where the $\beta\gamma$ dimer re-associates with the $\alpha$ subunit, re-couples with the receptor and induces a high agonist affinity state of the receptor, thus completing the G-protein cycle.[9]

By introducing a non-hydrolysable form of radiolabelled GTP, typically [$^{35}$S]-GTP$\gamma$S into the cycle, it is possible to measure the accumulation of radioactivity into G-proteins. The amount of [$^{35}$S]-GTP$\gamma$S accumulated is directly proportional to the degree of receptor activation. Since the methodology employed is essentially that of a receptor binding assay, the GTP$\gamma$S binding assay is amenable to the same level of automation and throughput as the primary binding assay described earlier (Figure 4).

This assay was employed to measure the functional activity of SB-236057 in CHO cells expressing 5-HT$_{1B}$ receptors. As seen in Figure 5, 5-HT is a full agonist, GR127935 is a partial agonist and SB-236057 is an inverse agonist.

In Figure 6, the effect of SB-236057 on the 5-HT concentration–response curve is shown. SB-236057 shifts the 5-HT curve to the right in a concentration dependent manner with no suppression of the maximal response. The compound is acting as a competitive or surmountable antagonist, that is, 5-HT and SB-236057 are competing for the same site on the receptor. The magnitude of the rightward shift of the 5-HT curve is dependent upon the degree of receptor occupancy by SB-236057, which is governed by its affinity for the receptor and its free concentration. Knowing the free concentration of SB-236057 and by measuring the magnitude of the 5-HT curve shift, it is possible to calculate the affinity of the receptor for SB-236057. This can be determined in one of two ways.

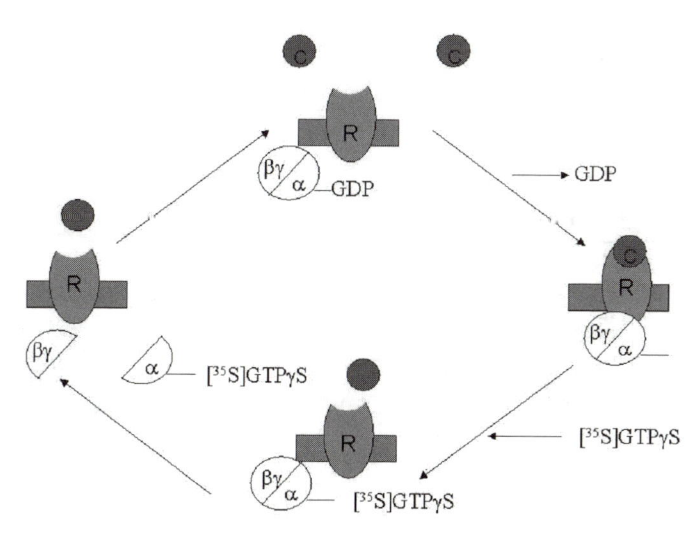


**Figure 4** *Schematic illustration of the [$^{35}$S]GTP$\gamma$S binding assay (C = compound, R = receptor, $\alpha$, $\beta$, $\gamma$ = subunits of the G-protein)*

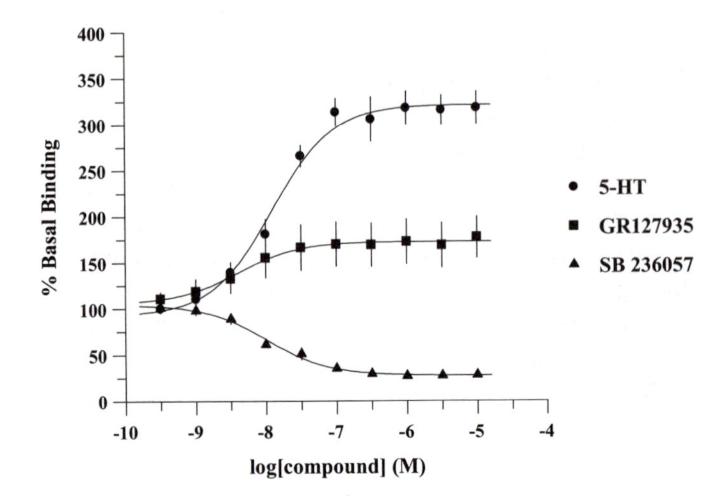


**Figure 5** *Functional profile of SB-236057 at h5-$HT_{1B}$ receptors as measured by [$^{35}$S]GTPγS binding in CHO cell membranes*

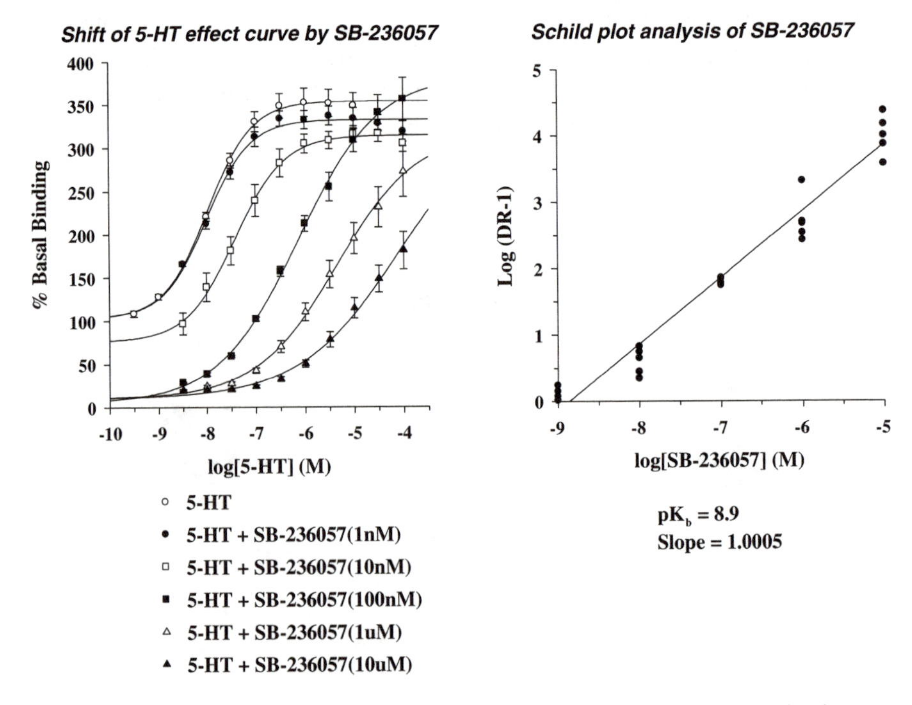


**Figure 6** *Measurement of the functional antagonist affinity of SB-236057 for h5-$HT_{1B}$ receptors as measured by [$^{35}$S]GTPγS binding in CHO cell membranes*

In Figure 6, the dose ratio is calculated for a concentration of antagonist; the dose ratio (DR) being, *e.g.* the $EC_{50}$ for the agonist in the presence of antagonists divided by the $EC_{50}$ for the agonist alone. From this value, using the Gaddum equation, the $pA_2$ can be determined. This is the negative logarithm of the molar concentration of antagonist that produces a dose ratio = 2, which equates to 50% occupancy of the receptor and the affinity of the antagonist for the receptor. If this is repeated with several different concentrations of antagonist and the log[antagonist] plotted against the log(DR – 1), if the resulting regression plot (Schild plot) is linear and has a slope of unity, then the antagonst is competitive and the intercept represents $-\log K_g(pK_b)$, which is again, an estimate of the equilibrium dissociation constant, $K_d$. Figure 6 shows such a Schild plot for SB-236057; the regression value is 0.90, the slope is 1.0, and the $pK_b = 8.9$, which correlates well with its affinity ($pK_i$) determined by radioligand receptor binding studies.

An alternative method for estimating antagonist potency is analogous to the receptor binding studies described earlier. A submaximal concentration of agonist is used (typically a concentration giving an 80% maximal response $EC_{80}$) and an inhibition curve is generated incubating in increasing concentrations of antagonist. From the inhibition curve the $IC_{50}$ is determined and, using the conversion equation described earlier for binding assays, the $pK_i$ is calculated.

These basic principles can be employed with any functional response, independent of the functional readout.

*Reporter Gene Assays.* Reporter gene assays have been configured in recombinant mammalian cell lines to identify novel ligands and for screening at many cell surface receptors, including G-protein coupled receptors and cytokine receptors, and nuclear receptors. Reporter gene assays have also been configured for compound screening in other cell types, including yeast and bacteria. In these assays, cell lines are generated containing the target protein of interest, together with a reporter gene specifically designed to detect ligand modulation of the target (Figure 7).

The interaction of a ligand with the target protein results in the upregulation of a signal transduction cascade to cause a change in the level of expression of a reporter gene in the cell nucleus. The reporter gene consists of two functional elements: a promoter element, the choice of which dictates the nature of the signal transduction event that will be detected, and a reporter enzyme. Stimulation of the promoter within the reporter gene causes an increase in the level of expression of the reporter enzyme. The reporter enzyme should offer a unique property to the cell such that it can be readily detected in either living cells or cell lysates.

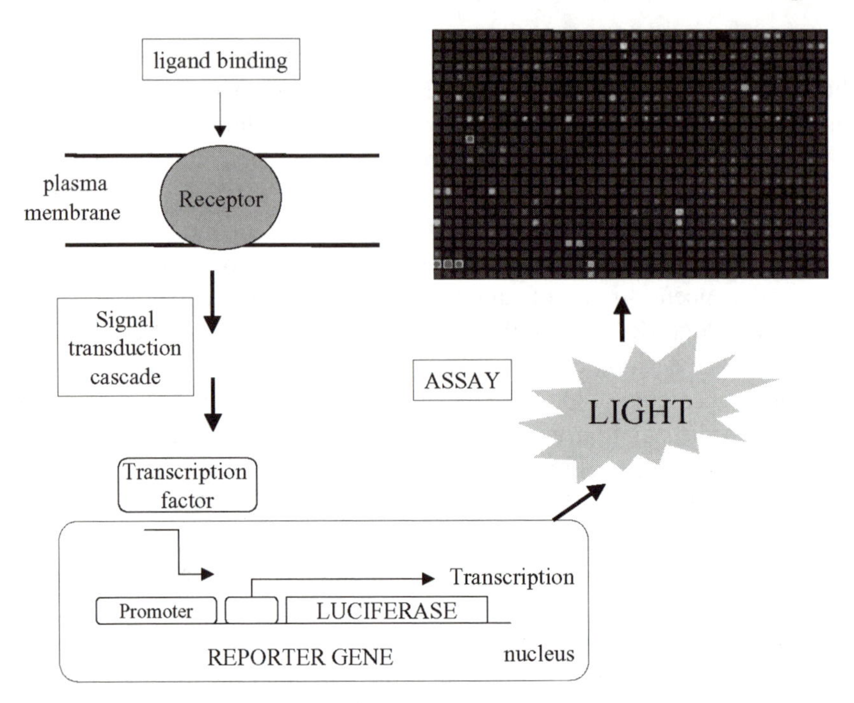


**Figure 7** *Schematic representation of a membrane receptor reporter gene assay*

Reporter enzymes such as firefly luciferase, secreted placental alkaline phosphatase, $\beta$-galactosidase and $\beta$-lactamase have been widely used in reporter gene assays. For each of these enzymes, assay reagents are available that generate either fluorescent or luminescent products. On colour enhancing imaging, these products can be readily detected in microtitre plate formats by giving a white/red image (lighter grey in Figure 7).

The availability of cheap, highly sensitive assay reagents has enabled the use of reporter gene assays for compound screening. Using image-based detection systems, reporter gene assays have been successfully run in 384 well and 1536 well microtitre plates in assay volumes as small as a few microlitres. The features of reporter genes and their use for both analytical pharmacology and high throughput compound screening have been extensively reviewed.[10,11]

*Measuring Function in Native Tissues.* As with receptor binding studies, functional studies in recombinant systems offer many advantages. However, care must be observed when interpreting recombinant data in relation to the functional response in a relevant native tissue. As shown in Figure 5, GR127935 displayed significant intrinsic activity in the GTP$\gamma$S binding assay at recombinant 5-$HT_{1B}$ receptors. However, in native tissues this compound behaves as an antagonist. Clearly the compound has some efficacy which is greatly amplified at recombinant h5-$HT_{1B}$ receptors and the G-proteins; that is, the system has a high level of receptor reserve. In brain tissue, where the density of 5$HT_{1B}$ receptors is approximately 100 times less than the transfected CHO cells, no intrinsic activity was observed. It is critical, therefore, that a compound's intrinsic activity is benchmarked in the relevant native tissue.

*[$^3$H]5HT Release from Brain Slices.* One of the established roles of 5-$HT_{1B}$ receptors is to function as terminal autoreceptors, receptors which, when activated by 5-HT, further inhibit 5-HT release. Therefore, blockade of this receptor should enhance synaptic levels of 5-HT when there is endogenous tone at this receptor. It is possible to measure release of 5-HT in the CNS both *in vitro*, in brain slices, and *in vivo*, in conscious freely moving animals using microdialysis (see *in vivo* models section).

Using HPLC and electrochemical detection, it is possible to measure endogenous 5-HT. Although this methodology is used for *in vivo* microdialysis samples, where sample numbers are limited, for large numbers of *in vitro* samples this method is time consuming and labour intensive. Therefore, as a surrogate model, brain slices (300 μm and 300 μm cross chopped) can be preloaded with [$^3$H]5-HT and released 5-HT levels are estimated from the levels of radioactivity in a superfusate. Both basal and stimulated (through electrical or chemical, *e.g.* high [$K^+$] stimulation) 5-HT release can be estimated. Inclusion of 5-HT in the superfusion stream inhibits stimulated release, inclusion of a 5-$HT_{1B}$ receptor antagonist such as methothepin potentiates release, neither treatment affects basal release. Figure 8 demonstrates the antagonism of the 5-HT inhibition of [$^3$H]5-HT release by SB-236057. In addition, under conditions of enhanced autoreceptor tone, that is, when the stimulation frequency is increased to elevate the biophase level of 5-HT, SB-236057 also blocks this increased inhibitory tone, thereby potentiating [$^3$H]5-HT release. This study therefore confirms that in native tissue SB-236057 acts as a terminal autoreceptor antagonist.

Ideally, such corroboration of functional activity should also extend to demonstration of activity in human native tissue; clearly this is rarely a possibility. During the evaluation of SB-236057, a collaboration with Professor Gothert from the Univeristy of Bonn allowed access to human cortex, transsected during removal of deep-seated cerebral tumours. So the above studies on [$^3$H]5-HT release were also performed in human tissue (Figure 9). Again, under conditions of high biophase 5-HT and, hence,

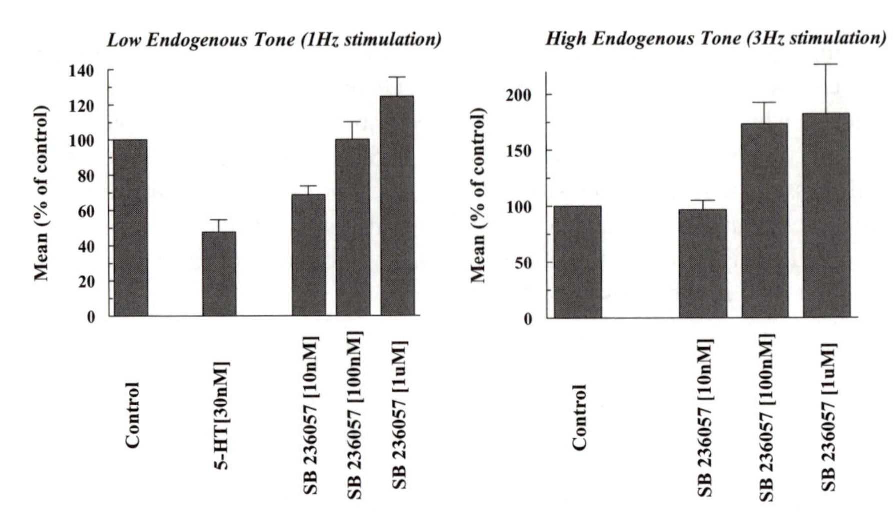


**Figure 8** *Demonstration of autoreceptor antagonism by SB-236057 as measured by [³H]5-HT release from guinea-pig cortical slices*

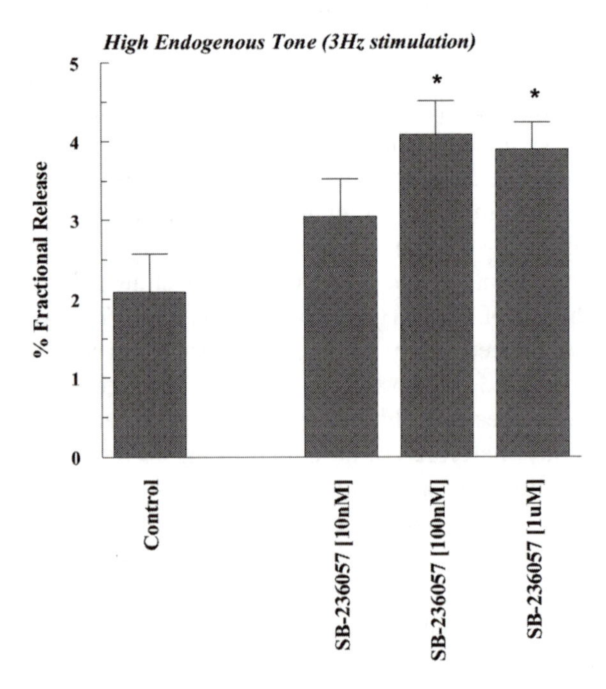


**Figure 9** *Demonstration of human autoreceptor antagonism by SB-236057 as measured by [³H]5-HT release from human cortical slices*

endogenous tone at the autoreceptor, SB-236057 blocked the terminal autoreceptor and potentiated [$^{3}$H]5-HT release.

## 4 DRUG METABOLISM AND PHARMACOKINETICS (DMPK)

Compounds that fulfil the criteria of affinity and efficacy set within the testing cascade progress to an assessment of their metabolic stability. This can be carried out both *in vitro* and *in vivo*, together with experiments to measure the penetration of the compound into the relevant biological compartment (*e.g.* blood, brain *etc.*).

*In vitro* methods for the measurement of the propensity of the compound to inhibit the various isozymes of cytochrome P450 have been developed and this allows an assessment of the liability for potential drug interactions. In addition, it is possible to measure the intrinsic clearance of the compound by liver microsomes from various species (*e.g.* rat, dog, human) in order to predict the metabolic liability of the compound.

*In vivo* the basic pharmacokinetic parameters of oral bioavailability, terminal half-life in the blood stream ($t_{1/2}$) and the degree of penetration into relevant biological compartments (*e.g.* brain/blood ratio after steady state infusion) can all be readily measured in animal species by the application of the sensitive mass spectrometry technology of the Sciex.

At this point in the testing cascade, it is possible to evaluate the DMPK data generated from the aforementioned tests and decide whether further biological evaluation of the compound *in vivo* is warranted. See Chapters 6 and 7 for further discussion.

## 5 PHARMACODYNAMIC ASSAYS

The purpose of a pharmacodynamic (P.D.) model in animals is to demonstrate that, after an acute dose, the compound is capable of reaching the target protein in concentrations sufficient to elicit a response. In some cases the P.D. model can be a model of the disease and, in these limited cases, drug pharmacodynamics and efficacy can be estimated simultaneously.

### Requirements of a P.D. Assay

The end point of test must be simple to measure, the assay must be robust and, ideally, the response should be through the target receptor, although use of a receptor mediating a side-effect at higher drug concentrations could also be utilised. In addition the P.D. assay must be capable of being executed in a small animal species such as mice, rats, guinea-pigs or gerbils.

## Examples of P.D. Assays

Several drug classes can cause reproducible falls in body temperature in experimental animals. Mice can survive substantial falls in core temperature of around 7–8 °C, larger laboratory animals like rats and guinea-pigs will experience falls of temperature of about 2 °C whereas humans at comparable doses do not experience noticeable changes. Although mice offer a large window in terms of possible drop in body temperature, it is preferable to use one species of animal for all tests and disease models tend to use rats or guinea-pigs. Body temperature is a readily applicable and highly reproducible measurement yielding reliable estimates of the pharmacodynamics of the test compound.

Measurement using a rectal probe is simple, quick and can be repeated a number of times to gain a time course of drug activity. It may be necessary to predose the animal with an agonist to provoke the fall in temperature before challenging with the test drug. In-dwelling telemetric devices can be used which allow remote monitoring of animals. There are several advantages to this refined methodology. Less handling of animals reduces the potentially confounding effects of stress and it allows for continuous monitoring. Additionally, it enables reuse of animals and hence a reduction in the number of animals being used in experimentation. Refinement of procedures, reduction in numbers being used and eventual replacement of animals are key aims of animal experimentation.

Locomotor activity models can also be useful pharmacodynamic assays, which have three potential advantages over hypothermic assays. Locomotor activity can be recorded automatically, allowing bigger experiments and no measuring probe is needed that might disturb or otherwise interfere with the animal and drugs may increase or decrease locomotor activity, thereby extending the type of possible observations. Movement of animals in activity chambers can be followed by counting the number of occasions in a time period that they break light beams set to cover the floor area of the chamber. By setting beams at different heights, rearing activity can also be measured and this can be useful in providing additional information. These two examples, body temperature and locomotor measurements, provide a major source of P.D. assays for CNS active drugs.

## SB-236057

The pharmacodynamic assay used in the discovery process for SB-236057 was typical of the general approach outlined above. We had shown that SKF-99101 was an agonist that caused whole body hypothermia mediated by the $5\text{-HT}_{1B}$ receptor subtype in the brain.[12] Experience of many

**Table 3** *The $ED_{50}$ of SB-236057 in blocking SKF-99101-induced hypothermia*

| *Time/h* | *$ED_{50}$/mg kg$^{-1}$* | *s.e.m.* |
|---|---|---|
| 1 | 0.27 | 0.09 |
| 2 | 0.24 | 0.04 |

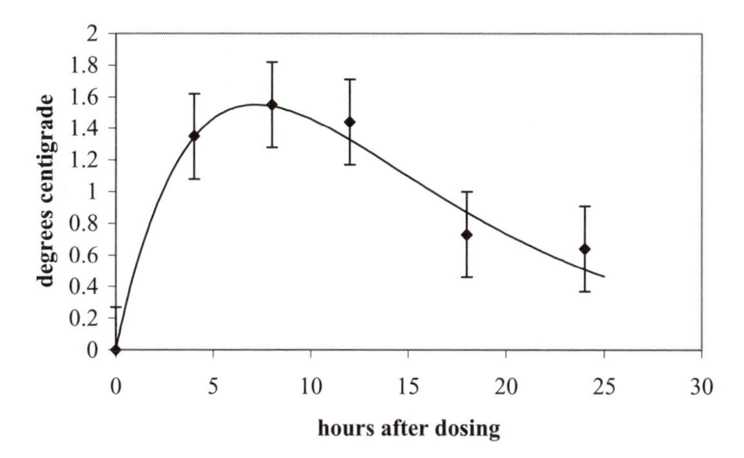


**Figure 10** *Time course of reversal of hypothermia induced by SKF-99101 by SB-236057*

compounds indicates that most compounds will have an effect on the CNS within 1–2 hours of an oral dose, although the effect may not be at a maximum at that time. Dosing SB-236057 orally to guinea-pigs 1 and 2 hours prior to testing its ability to block hypothermia induced by SKF-99101, allowed for the calculation of an $ED_{50}$ for SB-236057 at these times (Table 3).

Using these as a guide, guinea-pigs were dosed at 2.5 times the $ED_{50}$ and tested at intervals up to 24 hours. The reversal of the hypothermia over this time period showed that the compound was readily absorbed with satisfactory metabolism giving a maximum effect at around 8 hours after dosing (Figure 10). It could be concluded that SB-236057 was having a central effect at the 5-$HT_{1B}$ receptor.

## 6 ANIMAL MODELS – PRE-CLINICAL PROOF OF CONCEPT

Having selected the lead compound and achieved target protein affinity and function with an appropriate pharmacokinetic and metabolic profile, the compound moves on to evaluation in the animal models. The pharmacodynamic models will have confirmed the appropriate dose, route of administration and time to achieve the maximum pharmacodynamic effect (see previous section) thereby allowing the best possible chance of establishing pre-clinical proof-of-concept in animal models. Broadly these models are divided into two groups: proof of mechanism models and disease models.

Proof of mechanism models demonstrate that the compound achieves the profile set out in the biochemical rationale established at the outset of the drug discovery project, for example, *in vivo* blockade of a particular neurotransmitter system or inhibition of a specific metabolic pathway.

The disease model attempts to reproduce the disease itself, or at least some of the symptomatology associated with the disease, in an animal. Particularly with diseases and disorders of the CNS with a psychiatric component, modelling the disease itself is very difficult and few true disease models exist.

## Proof of Mechanism Models

In the current example of a 5-HT autoreceptor antagonist, the biochemical hypothesis behind the project is that current antidepressants, such as selective serotonin re-uptake inhibitors (SSRI), work through elevation of synaptic 5-HT, for example through blockade of the 5-HT re-uptake transporter. A selective 5-HT autoreceptor antagonist should increase 5-HT levels through potentiating the amount of 5-HT released on each nerve impulse. Therefore, proof of mechanism in this example is provided by the demonstration that systemic administration of the compound elevates 5-HT in the brain. Brain neurotransmitter levels can be assessed in conscious, freely moving animals with the technique of microdialysis. Here, a dialysis membrane is inserted into the brain and superfused with artifical cerebrospinal fluid. Through simple diffusion, free extracellular neurotransmitters pass through the membrane into the superfusion stream which then passes into a fraction collector. The transmitter levels are then concentrated and separated by chromotographic techniques and measured by highly sensitive methodologies such as electrochemical detection or mass spectrometry (see Figure 11).

Figure 12 demonstrates the increase in 5-HT levels in the dentate gyrus of the guinea-pig after systemic administration of SB-236057, an effect which can only be generated after 2–3 weeks treatment with the SSRI, paroxetine.[13] This, therefore, suggests that this compound could achieve therapeutic efficacy more rapidly than current antidepressants, which typically take 4–6 weeks to achieve therapeutic efficacy. Confidence in a compound and in the target is essential before embarking on a costly, time

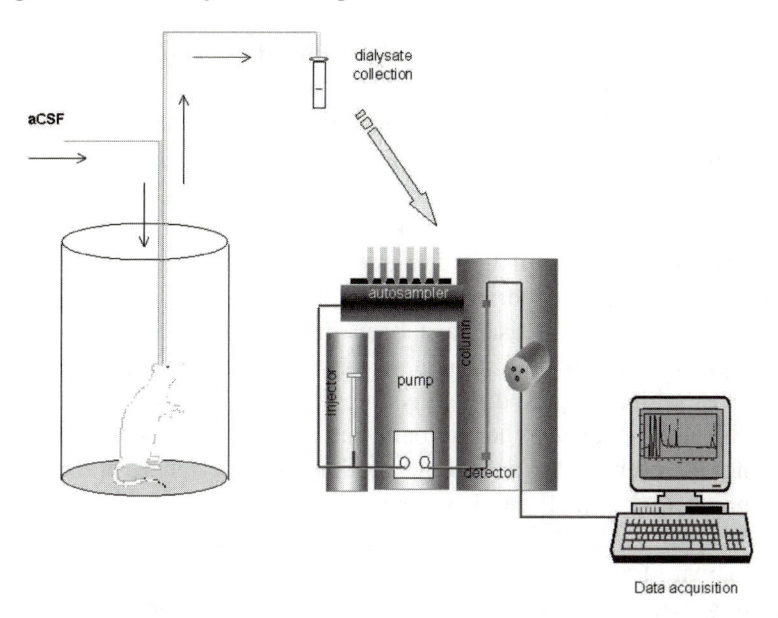


**Figure 11** *Schematic illustration of the microdialysis technique coupled to HPLC and electrochemical detection*

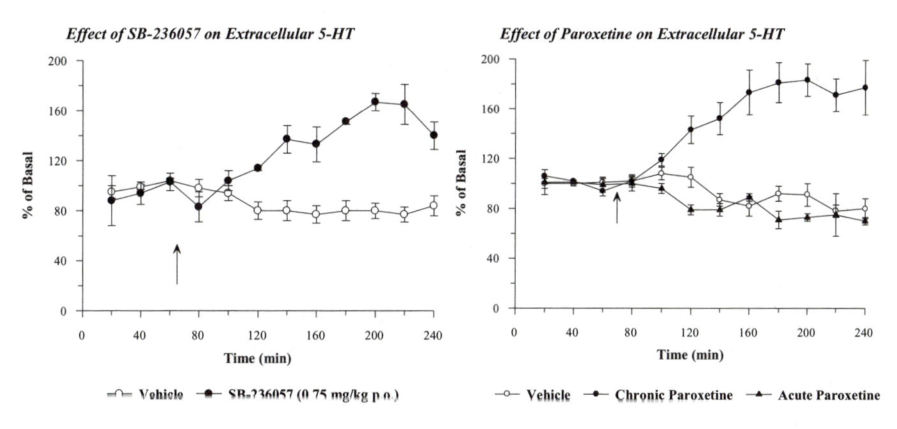


**Figure 12** *Effect of SB-236057 on extracellular 5-HT levels in the dentate gyrus of the guinea pig*

consuming programme of clinical development, so successful pre-clinical proof-of-concept is paramount.

## Disease Models

With neurological disorders, the pathophysiology of many of the disorders is reasonably well understood. Thus, in Parkinson's disease (PD) or ischeamic stroke, although models of the disorder *outcome* exist, such as lesions of the dopaminergic system for PD or inducing an ischaemic insult with a middle cerebral artery occlusion for stroke, there are no models for the disease itself. Therefore, models exist for the treatment of the symptoms but not for the treatment of the disorder or its prevention. In psychiatric disorders, where there are few instances of a demonstrated organic pathology, the situation is more difficult; how do we know if an animal is depressed or experiencing a psychotic episode?

Animal models of disease have been developed successfully in situations where it is reasonable to assume that an animal would share similar experiences with human and the physiological response to that situation is similar in both species. Whereas pain has a strong emotional component and hence is reserved for describing the human experience, nociception is the response to a potentially damaging stimulus, such as excessive heat or mechanical stress and is experienced by humans and animals alike. Physiological responses, such as elevation of blood pressure, heart rate and circulating catecholamines, as well as the rapid withdrawal from the stimulus, confirms that this is a shared experience. It is possible therefore to model at least certain aspects of human pain in animals. Anxiety can similarly be modelled in animals, since it is reasonable to assume that when faced with a potential predator or a conditioned stressful or painful stimulus, an animal's fight or flight reflex is activated as in humans. Models of pain and anxiety, therefore, have both face and construct validity and, in addition, respond to established therapeutics, hence are also pharmacologically validated.

Models of epilepsy and sleep disorders also have face, construct and pharmacological validity and greatly increase the degree of confidence in a target and a compound at a pre-clinical stage. For many disorders though this confidence can only be partly satisfied with proof of biochemical rationale and/or reduction in models of symptoms of disease. With such compounds, the final proof of concept will only be defined in the clinical situation.

## 7 ACKNOWLEDGEMENTS

We thank both Steve Rees and Frank King for their contributions to this chapter.

## 8 REFERENCES

1. D.N. Middlemiss *et al.*, *Eur. J. Pharmacol.*, 1999, **375**, 359.
2. R.P. Hertzberg and A.J. Pope, *Curr. Opin. Chem. Biol.*, 2000, **4**, 445.

3. S. Turconi *et al.*, *Drug Disc. Today*, 2001, **6**, S27.
4. E. Hunt, *Drugs Future*, 2000, **25**, 1165.
5. S. Turconi *et al.*, *J. Biomol. Screening*, 2001, **6**, 275.
6. F.D. King, personal communication.
7. N. Luciani *et al.*, *Biochem. J.*, 2001, **356**, 813.
8. J.C. Bloomer *et al.*, *Drug Metab. Rev.*, 2001, **33**, 80.
9. S. Lazareno *et al.*, *Life Sci.*, 1993, **52**, 449.
10. S. Rees *et al.*, in *Signal Transduction: A Practical Approach*, G. Milligan (ed.), Oxford University Press, 1999.
11. S.J. Hill *et al.*, *Curr. Opin. Pharmacol.*, 2001, **1**, 526.
12. J.J. Hagan, *et al.*, *Eur. J. Pharmacol.*, 1997, **331**, 169.
13. C. Roberts *et al.*, *N-S Arch. Pharmacol.*, 2000, **362**, 177.

CHAPTER 6

# Pharmacokinetics

PHILLIP JEFFREY

## 1 INTRODUCTION

A fundamental assumption of drug action is that the pharmacological response of the drug can be correlated with its concentration at the site of action. This relationship between drug dose giving rise to a drug concentration involves a number of processes, such as absorption, distribution, metabolism and excretion. The science of pharmacokinetics seeks to provide a mathematical basis for the description of these processes and predict the time-course of drugs in the body. Pharmacokinetics can, therefore, be defined simply as what the body does to the drug. However, with very few exceptions, there is no opportunity to measure drug concentration directly at the active site (*e.g.* the brain) and a fundamental assumption in pharmacokinetics is that there is a relationship between drug effect (either pharmacological or toxicological) and the drug concentration at an alternative but readily accessible site such as blood or plasma. The application of pharmacokinetics, therefore, is to obtain experimental data in the form of blood (or plasma) concentration–time curves and reduce these to an extent that will allow drug performance to be summarised and ranked. This process involves calculating a number of pharmacokinetic parameters (using equations, constants *etc.*), which, by the application of mathematical modelling techniques, may then be used to simulate and predict drug behaviour under different situations, such as increasing the dose or changing the dose interval.

Pharmacokinetics is very much a combined science with roots in chemical kinetics, physiology, pharmacology and biochemistry. The objective of this chapter is not to scare the reader with the complexities of mathematical modelling but to focus on obtaining an understanding of what these pharmacokinetic parameters mean, how they are derived (simply) and, perhaps most importantly, how the parameters are interrelated. Such an understanding will, hopefully, aid the reader in appreciating the importance of pharmacokinetics in the drug discovery and development process.

## 2 THE PROCESS AND TERMINOLOGY OF DRUG DELIVERY

The amount of drug administered and the inherent potency of the drug are not the sole factors that influence drug behaviour. The vast majority of drugs are administered orally and there are a number of variables and barriers that influence drug delivery to the active site in the tissues (Figure 1).

In order to be absorbed from the gastrointestinal tract, the drug must first be in solution; only then will it be able to cross the intestinal epithelia and enter the portal blood supply. The hepatic portal vein drains directly into the liver and, therefore, the next barrier the drug encounters is the liver, considered to be the major organ for drug metabolism. Here the drug may undergo metabolism and/or elimination in the bile prior to reaching the systemic (general) circulation (Figure 2). Once in the systemic blood

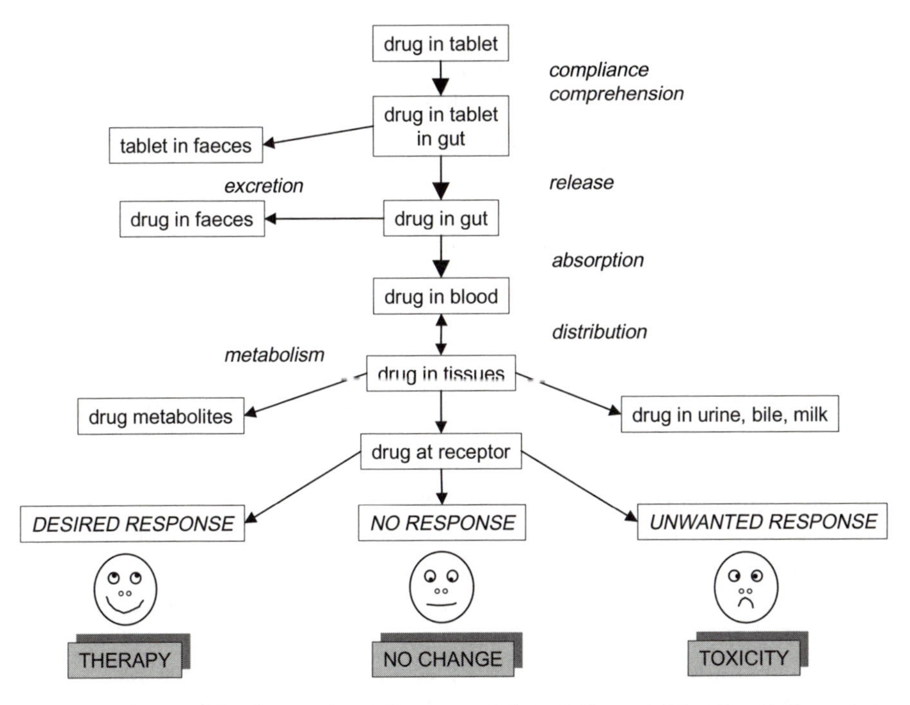


**Figure 1** *Some of the factors that influence oral drug delivery (with acknowledgement to Prof. G.T. Tucker)*

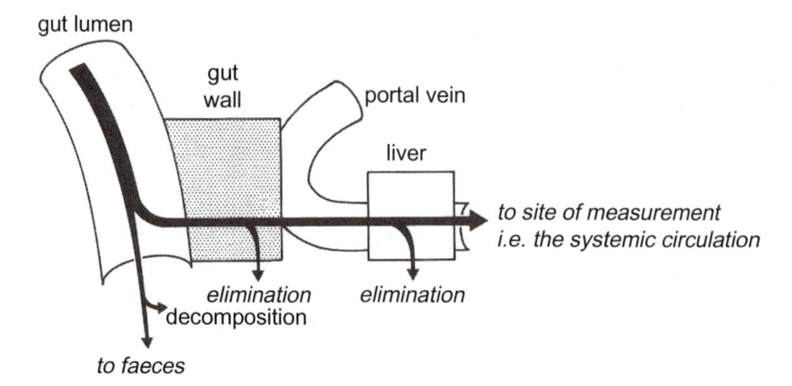


**Figure 2** *A schematic representation of the barriers associated with the passage of drugs from the intestinal lumen, across the gut wall into the portal vein, through the liver and into the systemic circulation (modified from Rowland and Tozer, 1995)*

supply, the drug is now available for distribution to the active site. This entire process is termed the fate of drugs and can be summarised by the following key descriptors:

- Absorption – all processes from the site of administration to the site of measurement
- Distribution – the reversible transfer of drug between the site of measurement and other sites within the body
- Elimination – the irreversible loss of drug from the site of measurement within the body
- Excretion – the irreversible loss of drug from the body
- Metabolism – the irreversible loss of drug from the body by biochemical conversion
- Disposition – elimination + distribution.

## 3 THE BLOOD (OR PLASMA) CONCENTRATION–TIME CURVE

The measurement of drug concentration in the blood, serum or plasma is the most direct approach used in assessing the pharmacokinetics of the drug. Advances in liquid chromatography (LC) and mass spectrometry (MS), coupled with automation of sample handling and preparation techniques, have resulted in the rapid development of sensitive, accurate, precise and robust assays suitable for the rapid determination of specific drug concentrations and subsequent pharmacokinetic data analysis.[1] Total radioactivity measurements are invariably non-specific, since they consist of all the radiolabelled drug related material and are, therefore, unsuitable for pharmacokinetic analysis. The blood (or plasma) concentration–time curve is generated by determining the drug concentration at various time points after drug administration and plotting these values against the corresponding time point.

Drugs may be administered by a variety of routes (Table 1) and the blood (or plasma) concentration–time curve is a useful method of evaluating the extent of absorption following administration from any site in the body including the subcutaneous, intra-muscular and intra-peritoneal routes. However, for the purposes of this introduction, only two routes will be considered; the intravenous (IV) route and the oral (PO) route. It should be noted that the systemic availability of a drug may be dependent on the route of administration, for example following IV administration of the potent and selective $5HT_{1A}$ agonist 8-OH DPAT, the blood levels in the rat

**Table 1** *Summary of the common routes of administration*

| | *Route* | *Comments* |
|---|---|---|
| ICV | Intra-cerebro vascular | Directly into the ventricle of the brain. Diffusion into the brain rapid, hence immediate effect, limited by low surface area. Elimination *via* the CSF and possibly local metabolism. |
| IV<br>IA | Intra-venous;<br>Intra-arterial | Guaranteed systemic exposure. Rapid blood levels and effect limited by protein binding, metabolism and distribution. Central activity limited by brain penetration. |
| SC | Sub-cutaneous | Systemic exposure produced more slowly than by IV. Only small dose volume appropriate with the possibility of precipitation at the dosing site. Other properties similar to IV but slower release may increase effects of metabolism *etc.* |
| IM | Intra-muscular | Similar to SC but more reproducible. |
| IP | Intra-peritoneal | Absorption into the systemic system or the hepatic portal vein, hence subject to first pass metabolism. Precipitation may occur in the peritoneum. Systemic exposure may be highly variable. |
| ID | Intra-duodenal | Gut transit time important. |
| IPV | Intra-hepatic portal vein | Subject to first pass metabolism. Avoids malabsorption through the gut wall and gut wall metabolism. |
| PO | Per Os (by mouth) | As for ID, systemic exposure limited by absorption from the gut, gut wall and first pass metabolism. Also limited by gastric emptying rates and gastric acid and enzyme stability. Differences may occur between fed and fasted states. |
| | Dermal/topical | Systemic exposure limited by surface area and absorption through the epidermal barrier. |
| | Buccal | Systemic exposure limited by surface area but avoids first pass metabolism. |
| | Rectal | Systemic exposure limited by absorption through the gut wall. Avoids first pass metabolism in the first 15 cm of the colon but absorption may be variable. |

are much greater than with the corresponding PO dose, reflecting the extensive hepatic metabolism of the drug.[2]

The relationship between the drug concentration–time curve, the various pharmacological, toxciological and pharmacokinetic descriptors and the concept of the therapeautic window is shown in Figure 3. When a drug is

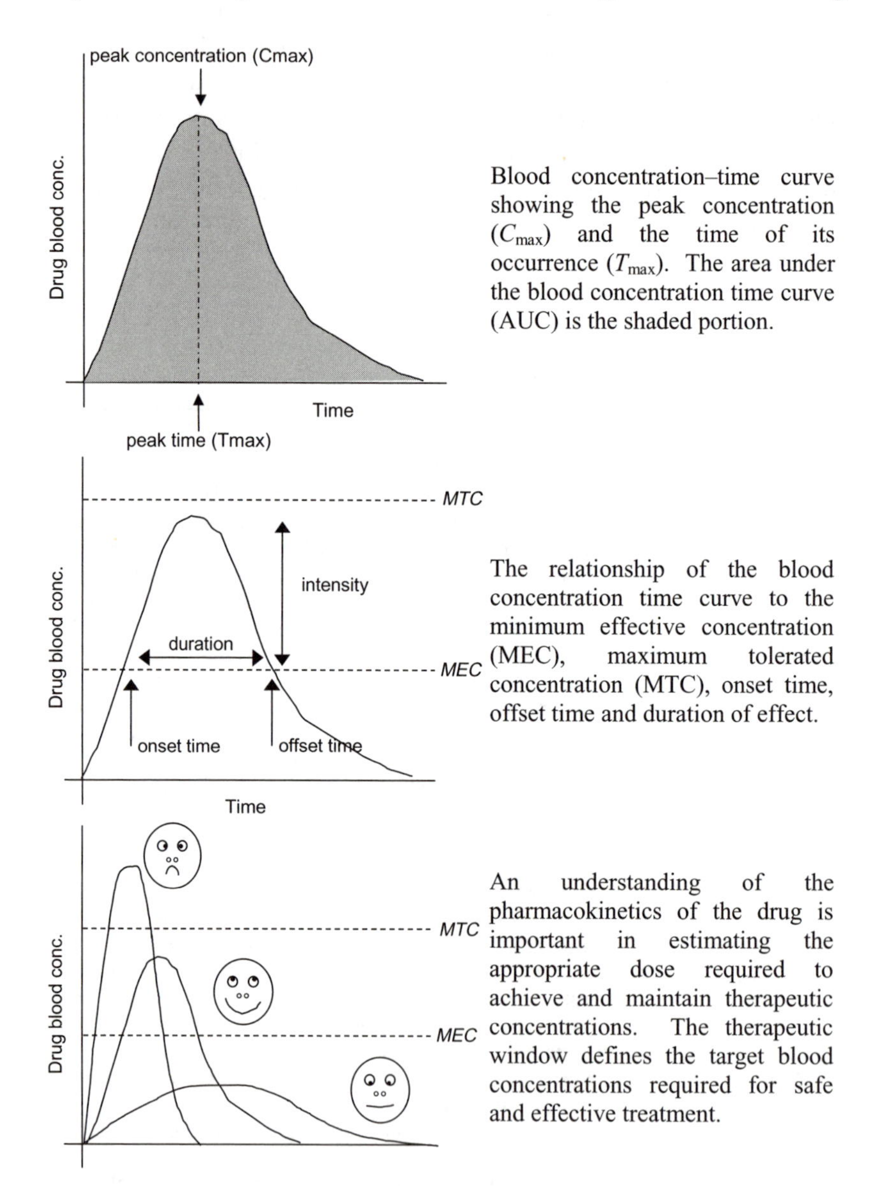


**Figure 3** *The relationship between the drug concentration–time curve, the various pharmacological, toxicological and pharmacokinetic descriptors and the concept of the therapeutic window*

administered into the body, it is absorbed and distributed into extravascular space and there is, therefore, an exchange of drug between the tissues and the blood. However, the drug may also be eliminated either by excretion of parent compound or by metabolism or by a combination of both. These processes do not function independently but simultaneously and the net effect is reflected in the blood (or plasma) concentration–time profile. Assuming the drug concentrations in the blood (or plasma) are in equilibrium with the tissues and, therefore, the site of action, the minimum effective concentration (MEC) represents the minimum drug concentration required at the receptor(s) to produce the desired effect.

The onset time is the time at which this concentration is achieved and the offset time is the time at which the systemic concentrations fall below this level. The duration time of drug action is the time difference between the onset and offset time and the magnitude of the pharmacological response is reflected by the increase in the systemic concentration, rising to a maximum. It should be noted that the pharmacodynamic response is not always reflected by the blood (or plasma) concentration–time profile, as sometimes this response is as a result of a cascade of effects rather than one single process. One such example would be in cancer treatment, where drugs act in an irreversible manner causing complex changes in protein synthesis resulting in eventual tumour cell death. Under these circumstances, the blood (or plasma) concentrations do not relate directly to the pharmacodynamic response and a more complex form of data analysis will be required.

As the drug reaches the systemic circulation, concentrations usually rise to a maximum ($C_{max}$) reflecting the fact that absorption is the more rapid process than either distribution or elimination, which will still occur during the absorption phase. At $C_{max}$, absorption may continue (as a result of residual drug still in the gastrointestinal tract) but not to the same extent as elimination and, therefore, the concentrations decline.

Pharmacokinetic data analysis is the numerical interpretation of the drug concentration–time curve and involves the calculation of certain pharmacokinetic parameters which provide an easier way to interpret the data rather than discussing individual blood (or plasma) concentrations. As already discussed, some of these parameters are relatively simple ones that are easily understood and interpreted as shown in Figure 3:

- $C_{max}$ – the maximum blood (or plasma) concentration
- $T_{max}$ – the time at which $C_{max}$ occurs
- Area under the curve (*AUC*) – the total area under the blood (or plasma) concentration–time curve is a measure of the quantity of

drug in the body, expressed in the unit of concentration multiplied by time.

However, these parameters only describe the pharmacokinetic behaviour of drugs in a limited manner and in order to provide a more detailed description, other pharmacokinetic parameters have been defined. The four most fundamental pharmacokinetic parameters are:

- Bioavailability – describes the rate and extent (amount) of drug reaching the systemic circulation and varies from 0 (no drug absorption) to 1 (complete drug absorption).
- Clearance – the measurement of the ability of the body to eliminate a drug, one of the most (if not the most) important pharmacokinetic parameters.
- Volume of distribution – the hypothetical volume of body fluid that would be required to dissolve the total amount of drug at the same concentration as that found in the blood, effectively a dilution factor representing the extent of drug distribution.
- Half-life – the time taken for a drug concentration to fall by one-half of its original value, used to describe elimination.

These parameters will now be described in more detail but the reader should note that it is not the calculation of these parameters that is the most important aspect of pharmacokinetic data analysis but the interpretation and the conclusions drawn from the data which allows for a more in-depth and thorough description of drug behaviour.

## 4 BIOAVAILABILITY

As discussed earlier, measurement of the blood (or plasma) *AUC* reflects the extent and, coupled with estimates of $C_{max}$ and $T_{max}$, the rate at which the drug reaches the systemic circulation. This is termed bioavailability or fraction of dose absorbed ($F$). After IV administration, all of the dose gets into the body, it is completely available and therefore $F$ equals unity. However, after PO administration several factors may combine to lower the bioavailability of the oral dose, such as incomplete release of the drug from the dosage form, poor permeability across the intestinal membrane and elimination by either the gut wall and/or the liver. Consequently, $F$ may vary from 0 (no drug absorption) to 1 (complete drug absorption). Note that the term drug absorption is used here (and throughout this introduction) to describe all processes from the site of administration (*e.g.* the mouth) to the site of measurement (the systemic blood), while permeability is used only to describe the passage of drug across the intestinal membrane. *AUC* is normally determined from time zero to infinity using a variety of mathematical methods, such as direct integration or the trapezoidal rule.

Absolute bioavailability of a drug is determined by comparing the respective *AUC*s after PO and IV administration (corrected for any difference in doses) thus:

$$\text{absolute bioavailability } (F) = \frac{AUC_{\text{po}}}{AUC_{\text{iv}}} \cdot \frac{\text{Dose}_{\text{iv}}}{\text{Dose}_{\text{po}}}$$

Sometimes it is necessary to compare other routes of administration with the PO rather than the IV route or the relative oral performance of a series of formulations (the test) compared with a solution (the reference). The relative biovailability is therefore:

$$\text{relative bioavailability } (F) = \frac{AUC_{\text{test}}}{AUC_{\text{reference}}} \cdot \frac{\text{Dose}_{\text{reference}}}{\text{Dose}_{\text{test}}}$$

Both instances assume that the pharmacokinetics are linear and do not change with dose, since for any extravascular route of administration, $F \leqslant 1$.

## 5 CLEARANCE

The concept of clearance is perhaps one of the most fundamental and, therefore, the most important in pharmacokinetic data analysis. In simple terms clearance, like volume of distribution, is a proportionality factor relating the concentration of drug in blood (or plasma) to the rate of drug elimination. Clearance is a flow parameter that is, under linear conditions, constant throughout the time interval. Drug elimination may take place at a number of different sites in the body (*e.g.* the liver, gut, kidney, intestine and brain) *via* a variety of mechanisms (*e.g.* elimination of parent compound in the bile or urine or by drug metabolism, or a combination of these processes) but whatever the site and mechanism of drug elimination, the net effect is a reduction in the drug concentration entering, compared with that leaving, the organ of interest. Since the overall rate of elimination will depend on the concentration of drug entering the organ then a parameter relating the rate of elimination to the measured drug concentration would be useful and this is the concept of clearance. However, it is not possible to measure all the individual organ clearances and since clearances are additive, it is usual to calculate the parameter total systemic clearance:

$$CL_{\text{total}} = CL_{\text{hepatic}} + CL_{\text{renal}} + CL_{\text{other}}$$

This apparent loss of drug as it passes through an organ of elimination uses the concepts of mass balance and is shown in Figure 4 using the liver as an example.

The rate of presentation of a drug to the liver is the product of organ blood flow ($Q$) and the drug concentration in the arterial blood flow entering the organ ($C_A$) hence:

$$\text{rate of presentation} = Q.C_A$$

and, similarly, the rate at which a drug leaves the organ is the product of blood flow and the drug concentration in the venous blood flow leaving the organ ($C_V$):

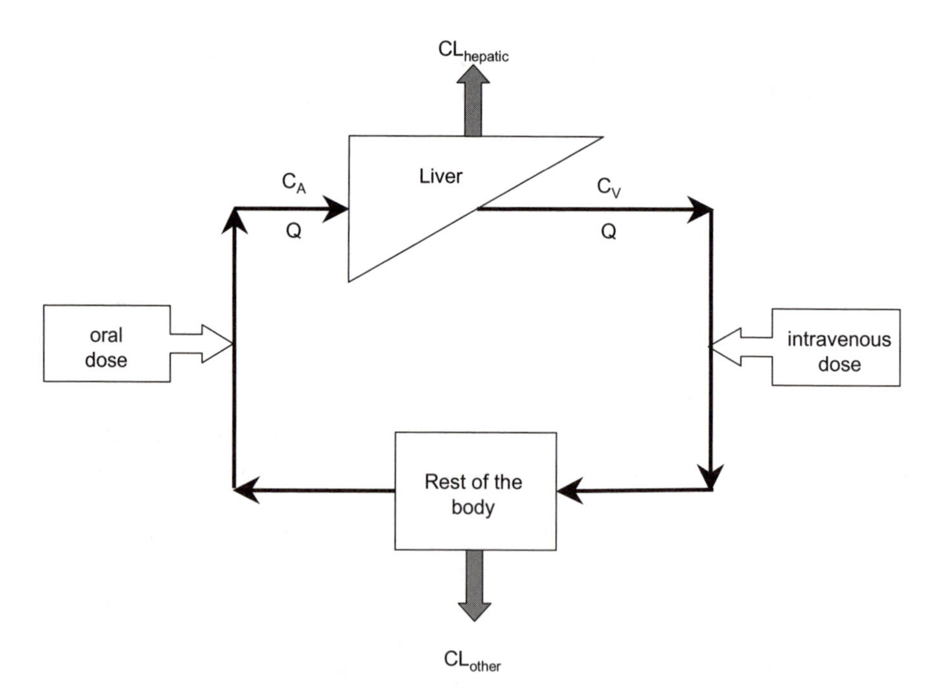


**Figure 4** *A schematic representation of hepatic clearance illustrating the relationship between oral administration, where all of the drug must first pass through the liver before it reaches the rest of the body, and intravenous administration, where all of the drug is distributed to the rest of the body before reaching the liver*

$$\text{rate of exit} = Q.C_V$$

The difference between the two rates is the rate of extraction by the organ and by mass balance accounts for the drug eliminated by the organ:

$$\text{rate of extraction} = Q.(C_A - C_V)$$

If the rate of drug extraction by the organ is related to the rate of drug presentation to the organ, then the parameter extraction ratio ($E$) may be derived:

$$E = \frac{\text{rate of extraction}}{\text{rate of presentation}}$$

$$\therefore\ E = \frac{Q.(C_A - C_V)}{Q.C_A}$$

$$= \frac{(C_A - C_V)}{C_A}$$

The numerical value of the extraction ratio can vary anywhere between zero, where there is no drug eliminated and hence there is no extraction since $C_V \approx C_A$ and, therefore:

$$E = \frac{(C_A - C_V)}{C_A} = \frac{(C_A - C_A)}{C_A} = \frac{0}{C_A} = 0$$

and 1, where all the drug is eliminated and hence there is complete extraction since $C_V \approx 0$ and therefore:

$$E = \frac{(C_A - C_V)}{C_A} = \frac{(C_A - 0)}{C_A} = \frac{C_A}{C_A} = 1$$

Drugs with an extraction ratio greater than 0.7 are classed as high extraction ratio drugs, those with an extraction ratio less than 0.3 are classed as low extraction ratio drugs and those with an extraction ratio between 0.7 and 0.3 are classed as moderate extraction ratio drugs.

The parameter clearance ($CL$) relates the product of organ blood flow to the extraction ratio and represents the volume of blood that is irreversibly cleared of drug per unit time, thus, for an organ such as the liver (with blood flow $Q_H$) the hepatic clearance ($CL_H$) can be defined as:

$$CL_H = Q_H E$$

or by substitution, since:

$$E = \frac{(C_A - C_V)}{C_A}$$

then:

$$CL_H = \frac{Q_H.(C_A - C_V)}{C_A}$$

As discussed earlier, following PO administration all of the drug will go through the liver before reaching the systemic circulation and, therefore, may be subject to hepatic elimination. This process is known as first-pass elimination (or first-pass metabolism) and drugs that undergo significant hepatic first-pass elimination will have high clearance rates since the extraction ratio ($E$) must approach 1 and therefore $CL$ will approach liver blood flow, *i.e.* since:

$$CL_H = Q_H.E$$

then as:

$$E \mapsto 1,\ CL_H \approx Q_H$$

total clearance ($CL$) is calculated by:

$$CL = \frac{F.\text{Dose}}{AUC}$$

where $F$ is the fraction of dose absorbed and $AUC$ is the total area under the blood (or plasma) concentration–time curve from time zero to infinity.

Clearance is normally determined following intravenous administration (since all of the dose gets into the body and therefore $F = 1$), whereas following PO administration, the value of $F$ is often unknown. During drug discovery and development, the pharmacokinetics of a drug is often determined in more than one preclinical species and it is useful to be able to compare compound performance, not only with other compounds, but also across other species. This can be achieved by normalising values of blood clearance (not plasma) to liver blood flow for the appropriate species as shown in Table 2, thus:

$$CL_{\text{normalised}} = \frac{CL}{Q_H}$$

**Table 2** *Theoretical clearance values in rat, dog and man normalised to liver blood flow*

| | *Rat* | *Dog* | *Man* |
|---|---|---|---|
| Liver blood flow ($Q_H$; mL min$^{-1}$ kg$^{-1}$) | 90 | 40 | 21.4 |
| High clearance (>70% $Q_H$; mL min$^{-1}$ kg$^{-1}$) | ⩾63 | ⩾28 | ⩾15 |
| Low clearance (<30% $Q_H$; mL min$^{-1}$ kg$^{-1}$) | ⩽27 | ⩽12 | ⩽6.4 |

After IV administration a simple "rule of thumb" can be used in estimating the maximum oral bioavailability ($F_{max}$). It is assumed that there is complete absorption of the drug from the intestinal tract and that elimination is solely hepatic ($E_H$), thus:

$$F_{max} = 1 - E_H$$

and that:

$$CL = Q_H . E_H$$

and therefore:

$$E_H = \frac{CL}{Q_H}$$

so, by substitution:

$$F_{max} = 1 - \frac{CL}{Q_H}$$

Thus by using the appropriate value of liver blood flow (see Table 2) and determining blood clearance after IV administration, the maximum oral bioavailability can be estimated. For high hepatic extraction drugs ($E_H \geqslant 0.7$), $F_{max}$ will be low, whereas for low hepatic extraction drugs ($E_H \leqslant 0.3$), $F_{max}$ will be high. This demonstrates the importance of clearance in limiting the oral bioavailability of drugs and clearance is therefore often used as a pharmacokinetic screen in drug discovery support.

## 6 VOLUME OF DISTRIBUTION

The concentration in blood (or plasma) will depend on the amount of drug in the body and the extent of distribution into areas of the body other than blood (or plasma). The value of volume of distribution ($V$) relates the amount of drug in the body to the concentration of blood (or plasma) and is therefore effectively a dilution factor:

$$V = \frac{\text{amount of drug in the body}}{\text{drug concentration in blood (or plasma)}}$$

In an average 70 kg man, the total volume of blood is 5 L, extracellular fluid is 12 L and total body water is 42 L, or since volumes of distribution are normally expressed as a percentage of total body weight (*i.e.* $\mathrm{L\,kg^{-1}}$), therefore blood, extracellular fluid and total body water represent approximately 7, 17 and 60% of body weight, respectively. Whilst the minimum volume of distribution must be at least equal to the blood volume ($0.07\ \mathrm{L\,kg^{-1}}$), few drugs have volumes of distribution that are physiologically relevant, highlighting the fact the volume of distribution is a hypothetical volume rather than a real volume. Many drugs have volumes of distribution in excess of physiological fluid volumes reflecting the hypothetical nature of the parameter (*e.g.* the antimalerial drug chloroquine has a volume of distribution of between 100 and $1000\ \mathrm{L\,kg^{-1}}$ in man) indicating that the drug is either distributed into a "deep" peripheral tissue compartment, pooled or stored in one particular tissue (*e.g.* fat) or bound to some specific biological matrices.

Volume of distribution can be considered as the volume of blood (or plasma) that would be required if the drug was distributed equally throughout all portions of the body and this concept is illustrated in Figure 5.

The volume of distribution depends on many factors, such as the blood flow to different tissues, the affinity of the drug for a given tissue and the type and capacity of the tissue, lipid solubility, pH and binding to other biological matrices such as plasma proteins and blood cells. Variations in blood cell association, plasma protein binding and tissue binding result in differences in volume of distribution, since only the unbound drug is available for distribution and capable of moving between blood (or plasma) and the tissue compartments as shown in Figure 6.

If we measure drug concentration in blood and $V$ is the apparent volume of distribution of the drug, then at equilibrium, the following mass balance relationship will exist:

$$V.C_{\mathrm{B}} = V_{\mathrm{B}}.C_{\mathrm{B}} + V_{\mathrm{T}}.C_{\mathrm{T}}$$

$$\text{amount of drug in the body} = \text{amount of drug in the blood} + \text{amount of drug in the tissue}$$

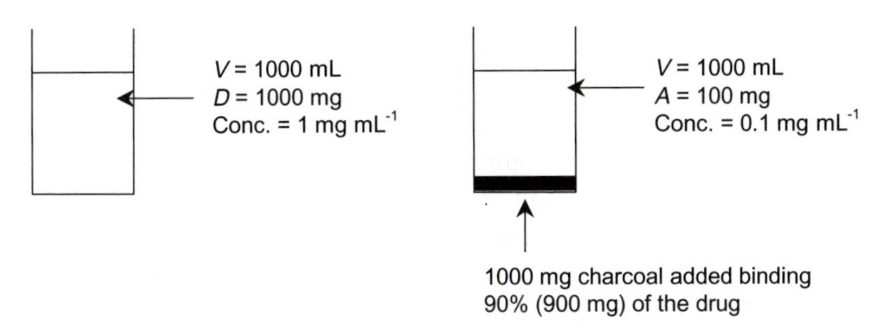


$$V = \frac{\text{amount of drug in the body}}{\text{drug concentration in blood (or plasma)}} = \frac{1000\ \text{mg}}{0.1\ \text{mg mL}^{-1}} = 10000\ \text{mL}$$

**Figure 5** *The concept of volume of distribution and the relevance of "deep" peripheral tissue binding to a hypothetical drug in a beaker. 1000 mg of drug is added to 1000 mL of liquid resulting in an initial drug concentration of 1 mg mL$^{-1}$. 1000 mg of charcoal, which adsorbs 90% of the drug (900 mg) is then added to the beaker and allowed to settle. The final drug concentration is now 0.1 mg mL$^{-1}$ and 10 000 mL of liquid would be required to account for all of the drug in the beaker, a volume much greater than the volume of the beaker*

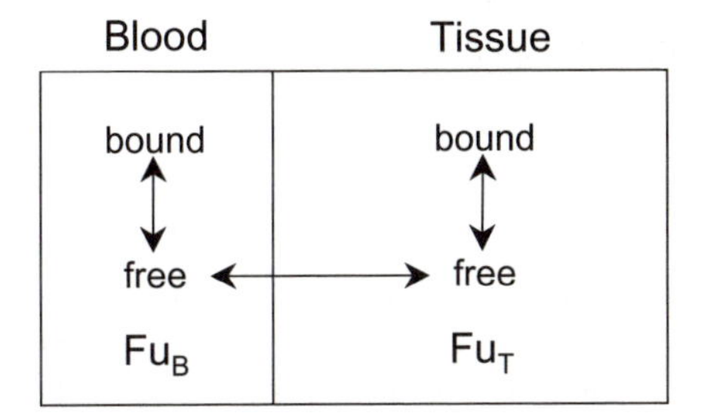


**Figure 6** *Only unbound drug is capable of moving between the blood and the tissues and at equilibrium the distribution of drug in the body depends on the binding to both blood components (blood cells and plasma proteins) and tissue components. $Fu_B$ and $Fu_T$ represent the unbound (or "free") fractions in blood and tissue, respectively*

where $C_B$ and $C_T$ are the total concentrations of drug in the blood and tissues, respectively. $V_B$ is the volume of blood and $V_T$ the apparent tissue volume outside of the blood. Incorporating the terms representing the unbound (or "free") fraction of drug in the blood, $Fu_B$, and the tissue, $Fu_T$, (see Figure 6), then the apparent volume of distribution, $V$, becomes:

$$V = V_B + V_T \frac{Fu_B}{Fu_T}$$

From this relationship, it becomes apparent that changes the volume of distribution will change depending on changes in $Fu_B$ and it is generally the case that drugs that are highly protein bound, resulting in a low $Fu_B$, will demonstrate low volumes of distribution, *e.g.* ibuprofen is >99% bound to plasma proteins and has a volume of distribution of $0.15 \, L \, kg^{-1}$ in man. Furthermore, protein binding can be important when highly protein bound drugs with narrow therapeautic windows are displaced as a result of either disease or co-administration of other highly protein bound drugs as shown in Table 3.

Blood and plasma concentration–time curves are normally reported as total concentrations and include both the bound and unbound ("free") drug. Whilst analysis of plasma samples is the more popular, and allows for therapeutic drug monitoring and the development of the appropriate dose in the clinic, measuring the concentrations in blood allows for a more in-depth examination and understanding of the interrelationships between extraction ratio, organ blood flow and blood clearance. It is important to appreciate that total concentrations and the appropriate unbound concentrations in blood and plasma are rarely the same, and that care should be taken when comparing pharmacokinetic parameters between different drugs and across different species.

**Table 3** *A comparison of the effect of a 5% reduction in protein binding of two hypothetical drugs; Drug A which is weakly protein bound and Drug B which is highly protein bound. Under normal circumstances only a small proportion of Drug B is available for distribution and therefore only a small proportion of the total amount of drug is active; however, a minor displacement of 5% causes a 100% increase in the free fraction of Drug B which may have toxicological in addition to therapeutic implications*

| | *Before displacement* | *After displacement* | *% Increase in free drug* |
|---|---|---|---|
| Drug A | | | |
| % bound | 50 | 45 | +10% |
| % free | 50 | 55 | |
| Drug B | | | |
| % bound | 95 | 90 | +100% |
| % free | 5 | 10 | |

## 7 HALF-LIFE

Half-life ($t_{1/2}$) describes the time it takes for the concentration in the blood (or plasma) to decline to half of its initial value and reflects the first-order nature of the processes of elimination (Figure 7). If the amount of drug decreases at a rate that is proportional to the amount of drug remaining, then the rate of disappearance is said to be exponential or first-order. Plotted on a linear scale, a typical blood (or plasma) concentration–time curve will have the following appearance; the blood concentrations of drug decline at an ever decreasing rate with increasing time and never appear to reach zero. By plotting the logarithm of drug concentration *versus* time, a straight line plot is obtained which indicates a first-order reaction. The half-life ($t_{1/2}$) may now be obtained from any two points on the plot describing a 50% decline in drug concentration. A useful 'rule of thumb' is that it takes approximately five $t_{1/2}$s for a drug to be completely eliminated from the body.

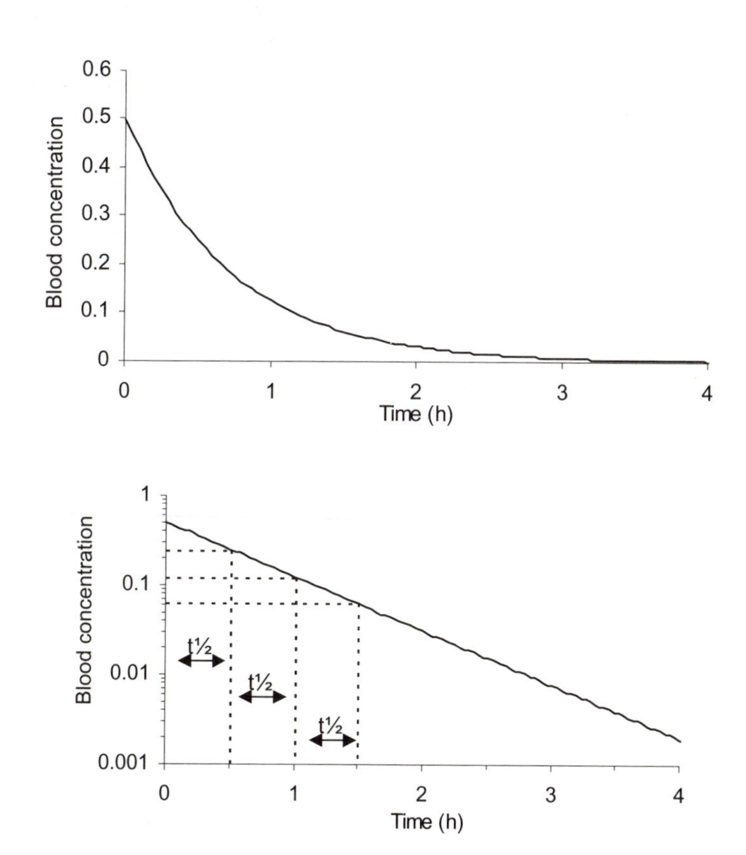


**Figure 7** *A description of half-life reflecting its first-order nature*

Of all the pharamacokinetic parameters discussed in this brief introduction half-life is probably the most familiar to the reader. To some people, it is the most important pharmacokinetic parameter since it reflects the length of time a drug exists in the body and, therefore, impacts on drug dose and dosage interval, peak and steady-state concentrations and drug accumulation (Figure 8).

However, half-life is a hybrid pharmacokinetic parameter and is a function of the volume of distribution and clearance of the drug in question. To understand the interrelationships between clearance, volume of distribution and elimination rate, we need to make some simple and fairly broad assumptions about the way drugs behave in complex biological systems, such as the body.

When a dose of drug is administered into the body it is distributed into the blood, extravascular space and cells. There is an exchange of drug between the blood (or plasma) and tissues. Drug is also eliminated as a result of hepatic or renal clearance and these elimination processes are assumed to occur at constant rate, that is a constant fraction of drug in the body is eliminated per unit time and this, therefore, is a first-order process. The blood (or plasma) concentration–time curve reflects these processes. Various mathematical models can be used to investigate and model drug absorption, distribution and elimination, the most common of which is

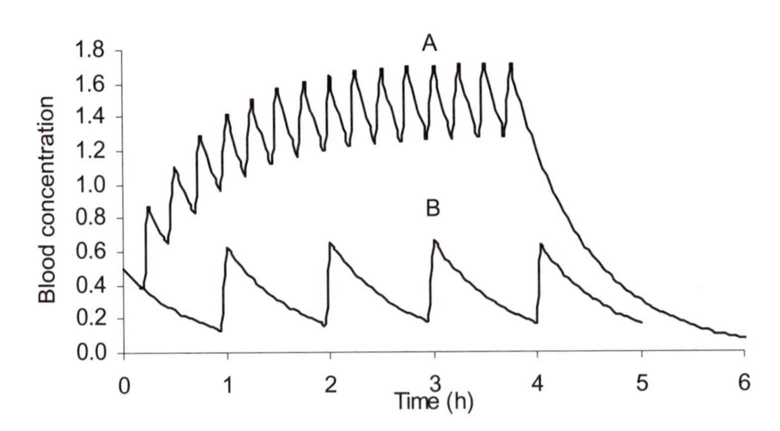


**Figure 8** *Theoretical blood concentration–time curves following an intravenous bolus dose of a drug with a terminal half-life of 0.5 h. Curve A represents drug administration twice every half-life (*i.e. *every 0.25 h) and Curve B represents the same drug dose administered once every two half-lives (*i.e. *every 1 h). Note that the frequency of drug dosing controls the maximum blood concentrations, the time it takes to reach a plateau (steady-state) and the amount of drug accumulation*

compartmental analysis. The body is represented as a series of compartments where drug moves freely from one compartment to another and mixing of drug within each compartment is instantaneous. If a drug is lost from one compartment it must appear in another compartment or be eliminated from the system, thereby obeying the principles of mass balance.

The simplest model is the one (or single) compartment model, which assumes that the drug is administered intravenously (as a bolus) into the blood stream and that the drug distributes and equilibrates rapidly with all the tissues in the body. The one-compartment model assumes that changes in blood (or plasma) concentrations will reflect the changes in overall tissue concentrations and is illustrated in Figure 9.

As discussed earlier, the rate of elimination for most drugs is a first-order process and is reflected in the terminal half-life ($t_{1/2}$) of the drug. However, the elimination rate constant ($k_e$) is a more useful parameter and can be obtained from $t_{1/2}$. Thus:

$$k_e = \frac{0.693}{t_{1/2}}$$

It should be noted that $k_e$ is a proportionality constant (having units of time$^{-1}$) and may be defined as the fraction of drug present at any time which would be eliminated in that unit of time. Thus, if $k_e = 0.5$ h$^{-1}$, then 50% of the drug present at any instant in time would be eliminated in one hour. However, it must be remembered that due to the first-order nature of the process, the instantaneous drug concentration is continuously declining and, therefore, although the fraction eliminated remains constant, the actual rate of elimination will decline with time.

In this simple system there are three variables; volume of distribution ($V$), clearance ($CL$) and drug dose ($D$). We have seen that drug elimination is dependent upon distribution and clearance and will proceed at a constant fractional rate. The relationship between $CL$ (a flow parameter with units of volume × time$^{-1}$), $V$ (units of volume) and $k_e$ (units of time$^{-1}$) is given by the equation:

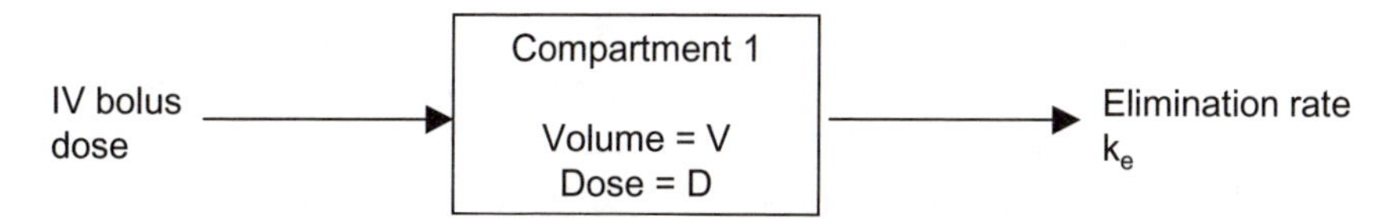


**Figure 9** *A simple illustration describing the one-compartment model with IV bolus input, the compartmental volume is equivalent to the volume of distribution and* $k_e$ *is the elimination rate constant*

$$k_e = \frac{CL}{V}$$

and since:

$$k_e = \frac{0.693}{t_{1/2}}$$

then by substitution:

$$t_{1/2} = \frac{0.693V}{CL}$$

Terminal half-life, then, is directly proportional to the volume of distribution and inversely proportional to clearance. It can be seen that since $t_{1/2}$ is dependent on both $CL$ and $V$, it can be doubled by either halving clearance or doubling the volume of distribution.

## 8 'ADVANCED' PHARMACOKINETICS

In reality, the blood (or plasma) concentration–time profiles of drugs rarely exhibit mono-exponential elimination profiles. Physiological reality is far more complex and most drugs display biphasic and triphasic disposition profiles. In these multi-compartment models, the drug distributes at different rates into the various different tissues, distribution is not instantaneous and takes a finite amount of time during which elimination may also be occurring. Drugs tend not to concentrate equally in tissues and different tissues will, therefore, accumulate the drug at different rates. The blood (or plasma) concentration–time curve reflects this complex and dynamic process and, in order to describe this, more complex mathematical models are required. For instance, whilst calculation of volume of distribution ($V$) is easy enough for a single compartment model, it becomes more difficult when there are two or more compartments and, therefore, the parameter volume of distribution at steady-state ($V_{ss}$) is used. Whilst the fundamental descriptions of $V$ and $V_{ss}$ are the same, both relate the amount of drug in the body to the concentration of drug in the blood (or plasma), in multi-compartmental models there will be several volumes of distribution associated with the different compartments. $V_{ss}$ reflects the differences in the distribution between the different compartments and is independent of the elimination rate constant, $k_e$. It is, therefore, the definition of choice used to correlate data on different drugs across different species and patients. For a more detailed description of pharmacokinetic data analysis and modelling techniques, the reader is referred to the texts listed in the bibliography.

## 9 CONCLUSIONS

A good pharmacokinetic profile is essential in drug selection and progression into development. Compounds with poor oral bioavailability, as a result of low permeability and/or high first-pass metabolism, and too short a half-life are not ideal drug candidates. Whilst it is not possible in this short review to detail all aspects of pharmacokinetics, the interrelationships between clearance, volume of distribution and terminal half-life have been discussed and the importance of clearance in limiting oral bioavailability highlighted. Knowledge of the pharmacokinetic properties of drugs is used to:

1. Predict blood (or plasma) concentrations at different doses and dose intervals
2. Optimise the drug dosage form for maximum oral bioavailibility
3. Estimate possible drug and/or metabolite accumulation after multiple dosing
4. Correlate drug concentrations with pharmacological and toxicological effects and establish a therapeutic window
5. Adjust drug dosage in special patient groups where the disease state or patient condition (*e.g.* aged, hepatically and renally impaired) alters the pharmacokinetics of the drug
6. Explain drug–drug interactions.

## 10 REFERENCES

1. A.P. Watt *et al.*, *Drug Discovery Today*, 2000, **5**, 17.
2. J.P. Mason *et al.*, *Xenobiotica*, 1995, **25**, 1371.

## 11 BIBLIOGRAPHY

D.W.A. Bourne, *Mathematical Modelling of Pharmacokinetic Data*, Technomic, Lancaster, PA, USA, 1995.

M. Rowland and T.N. Tozer, *Clinical Pharmacokinetics. Concepts and Applications*, Lea and Febiger, Media, USA, 1995.

L. Shargel and A. Yu, *Applied Biopharmaceutics and Pharmacokinetics*, Appleton and Lange, Stamford, USA, 1999.

J.G. Wagner, *Pharmacokinetics for the Pharmaceutical Scientist*, Technomic, Lancaster, USA, 1993.

CHAPTER 7

# Drug Metabolism

STEPHEN E. CLARKE

## 1 INTRODUCTION

The physicochemical properties of drugs that predispose them to good absorption, such as lipophilicity, are an impediment to their elimination. As a consequence, the elimination of drugs normally requires their conversion into water soluble compounds by a process of metabolism, which enables excretion *via* urine or faeces. Metabolism is often the major factor defining the pharmacokinetics of drugs, which in turn can influence the efficacy and side-effect profile of these compounds. The chemical nature and the means of identification of these biotransformations have been well known for many years, but in recent years major advances have been made in the understanding of the enzymes responsible for the metabolic pathways. Many of these enzymes are subject to genetic polymorphisms in man, demonstrate marked species differences and can be affected by diet and by the co-administration of other drugs. In addition to the widely recognised need to reduce metabolic liability, the avoidance of these issues represents the drug metabolism challenge to medicinal chemistry. An understanding of the enzymes and mechanisms responsible for the metabolism of drugs is essential to the success of this process.

## 2 DISTRIBUTION OF DRUG METABOLISM ENZYMES

Quantitatively, the most important source of drug metabolising enzymes is the liver. However, these enzymes are also found throughout the body, particularly at the interfaces where exposure to drugs or other xenobiotics might be anticipated, such as the gastrointestinal tract, skin, lung, kidney, nasal mucosa and the eye. Drug metabolising enzymes are also located in many other tissues, for example, the brain, adrenal gland, pancreas, testis, ovary, heart, spleen, placenta, plasma, erythrocytes and lymphocytes. Despite this wide distribution, only a few tissues have the metabolising capability, blood flow and strategic position between the site of absorption and site of action to influence therapy.

In general, the liver and intestine are the only significant organs of metabolism for most pharmaceutical endeavours. The liver has the highest metabolic capability, high blood flow and receives all the blood from the gastrointestinal tract. Overall, it is generally the major factor in the metabolic clearance of drugs and can limit the systemic bioavailability of orally ingested drugs. Extrahepatic sites are normally unimportant in the systemic metabolism of drugs due to their small size or inconsequential blood flow. However, depending on the route of administration, these tissues can exert a significant metabolic influence. For example, although the effective blood flow to the enterocytes on the villi tips of the small intestine is low, intestinal metabolism can still exert a significant first pass elimination for orally administered drugs, since the material must pass this barrier. Depending on the route of administration, there can be other exceptions and there can be important toxicological implications in terms of tissue specific chemical injury.

Within the cell, many drug metabolism enzymes are located in the lipid bilayer of the endoplasmic reticulum (microsomes) or the soluble fraction of the cytoplasm (cytosol). The S9 (supernatant following 9000 gav. centrifugation) is a combination of these two fractions. The drug metabolising capability of the mitochondria, nuclei and lysosomes is generally lower.

## 3 THE DRUG METABOLISING ENZYMES

The drug metabolising enzymes are usually classified by the reactions they catalyse, as either Phase I or Phase II. Phase I reactions introduce, or otherwise produce, a functional group (*e.g.* –OH, –SH, $–NH_2$, –COOH) into the molecule. These reactions include hydrolysis, reduction and oxidation and are performed by a wide range of enzymes. Oxidation is the commonest Phase I reaction, typically hydroxylation (aromatic or aliphatic) or *N*- or *O*-dealkylation. Hydrolysis of esters and amides produces acids, alcohols or amines and, given the common use of such groups to link two functional components, can lead to the cleavage of the molecule into those functional "halves". Reduction of nitro and carbonyl groups leads to amines and alcohols, respectively and disulfide, sulfoxide and quinone groups can also be reduced. Often these Phase I reactions precede Phase II biotransformations.

Phase II reactions involve the conjugation on a suitable chemical group of the molecule (parent compound or metabolite) and many drugs contain suitable functional groups without recourse to Phase I metabolism. Phase II reactions rarely precede Phase I reactions, as hydrophilicity of the products of Phase II reactions generally makes them poor substrates for the Phase I enzymes. Phase II reactions include conjugation with glucuronic acid, sulfate, glutathione or amino acids (*e.g.* glycine, taurine, glutamine), all of which increase the water solubility of the molecule. Conjugation reactions, such as *N*-acetylation of amines and *N*-, *O*- and *S*-methylation, generally result in more lipophilic products.

All of these biotransformation reactions are catalysed by a wide variety of enzymes (Table 1). They are frequently named for the reactions they catalyse, although this is not always the case (*e.g.* cytochrome P450; which were previously referred to as mixed function oxidases, and does describe their typical action well) and is occasionally misleading (*e.g.* aldehyde oxidase which also oxidises heterocycles such as substituted pyrimidines and purines). Many of the entries in Table 1 represent a large collection of related enzymes which frequently have broad and overlapping substrate specificity. These include isoenzymes, such as the family 1 members of

**Table 1** *The drug metabolising enzymes*

| *Enzyme* | *Phase* | *Reaction* | *Localisaton* | *No.*[a] |
|---|---|---|---|---|
| Alcohol dehydrogenase | I | Oxidation | Cytosol | 5 |
| Aldehyde dehydrogenase | I | Oxidation | Mitochondria, cytosol | 3 |
| Aldehyde oxidase | I | Oxidation | Cytosol | 1 |
| Carbonyl reductase | I | Reduction and oxidation | Cytosol | 3 |
| Carboxylesterase | I | Hydrolysis | Microsomes, cytosol | 9 |
| Cytochrome P450 | I | Oxidation or reduction | Microsomes | 36 |
| Diamine oxidase | I | Oxidation | Mitochondria | 1 |
| Epoxide hydrolase | I | Hydrolysis | Microsomes, cytosol | 3 |
| Flavin monooxygenase | I | Oxidation | Microsomes | 5 |
| Glucuronyl transferase | II | Conjugation | Microsomes | 15 |
| Glutathione S-transferase | II or I | Conjugation or reduction | Cytosol, microsomes | 26 |
| Monoamine oxidase | I | Oxidation | Mitochondria | 2 |
| *N*-acetyl transferase | II | Conjugation | Mitochondria, cytosol | 2 |
| Peptidase | I | Hydrolysis | Blood, lysosomes | 10 |
| Quinone oxidoreductase | I | Reduction | Cytosol | 2 |
| Sulfotransferase | II | Conjugation | Cytosol | 11 |
| Xanthine oxidase | I | Oxidation | Cytosol | 1 |

[a] Number of described human enzymes/forms.

the glucuronyl transferases and the superfamily of related enzymes of the cytochromes P450. It is not possible to provide complete coverage of this panoply of enzymes and their various complexities in an overview such as this; however, not all of these enzymes play a significant role in the metabolism of drugs. Many have predominantly endogenous substrates or are otherwise rarely involved in the metabolism of typical pharmaceuticals.

From a survey of the elimination pathways for approximately 450 drugs marketed in the US and Europe, the overall importance of the various clearance mechanisms has been determined. The elimination of unchanged drug *via* urine, bile, expired air or faeces represented, on average, approximately 25% of the total elimination of dose for these compounds. Of the 75% of clearance that is determined by metabolism, 75% of this is mediated by cytochrome P450 enzymes, with all other metabolic processes making up the remaining 25%. Thus, consideration of the major human drug metabolising cytochromes P450 and several other major enzyme systems enables coverage of the overwhelming majority of enzymes likely to be encountered in the metabolism of drugs.

## Phase I Metabolising Enzymes

*Cytochromes P450.*[1] Cytochromes P450 are ubiquitous throughout nature, are present in bacteria, plants and mammals and there are hundreds of known enzymes which show tissue and species specific expression. This diversity of enzymes has necessitated a systematic nomenclature system. The root name given all **CY**tochrome **P**450 enzymes is CYP (or *CYP* for the gene). Enzymes showing greater than 40% amino acid sequence homology are placed in the same family designated by an Arabic numeral. When two or more subfamilies are known to exist within the family then enzymes with greater than 60% homology are placed in the same subfamily, designated with a letter. Finally this is followed by an Arabic number representing the individual enzyme, which is assigned in the order of discovery. Only the 36 (almost certainly an underestimate) P450 enzymes described in man are likely to be of clinical relevance and only P450s in families 1, 2 and 3 appear to be responsible for the metabolism of drugs. Even of these 18 P450 enzymes, perhaps only 5 or 6 are quantitatively relevant and the relative significance of these is shown in Figure 1 and examples of substrate are given in Table 2.

The normal monooxygenase reaction mediated by cytochromes P450 can be represented by the following:

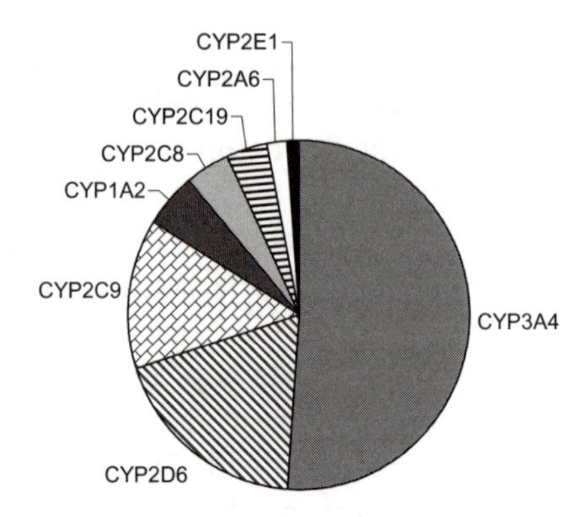


**Figure 1** *Relative significance of the major hepatic cytochromes P450 in the P450 mediated clearance of marketed drugs*

$$RH + O_2 \xrightarrow[2H^+]{2e^-} ROH + H_2O$$

where the oxidation is by oxygen at the formal level of peroxide. The electron transfer is mediated by two co-factors, one of which is flavin adenine dinucleotide (FAD) and the other being either flavin mononucleotide (FMN), iron–sulfur $Fe_2S_2$ redoxin or cytochrome *b5*. The reduction is mediated by the reduced forms of either nicotinamide adenine dinucleotide phosphate (NADPH) or nicotinamide adenine dinucleotide (NADH).

The expression cytochrome P450 activity is subject to modulation (inhibition and induction) from endogenous and exogenous factors as well as being subject to genetic variation, all of which can have a significant influence on the role these enzymes play in the metabolism of drugs.

*CYP3A4.* CYP3A4 appears to favour the metabolism of the more lipophilic drugs. However, its substrate specificity is very broad and it can metabolise drugs from the size of dapsone to cyclosporin. There are very many drugs that are predominantly eliminated by CYP3A4 and many others where CYP3A4 is a secondary mechanism of elimination. The most common reactions are carbon hydroxylation and *N*-dealkylation, with the positions of metabolism largely dictated by the ease of hydrogen abstraction for carbon hydroxylation and electron abstraction for *N*-dealkylation (Figure 2). The binding of substrates to CYP3A4 is relatively non-specific, seems to be essentially due to lipophilic forces and is, generally, relatively weak, which allows motion of the substrate in the active site. Thus, a single substrate may be able to adopt more than one orientation in the active site which results in the generation of several products. Moreover, there is considerable evidence that two or more substrate molecules may

**Table 2** *Examples of substrates of the major human drug metabolising cytochromes P450*

| *CYP1A2* | *CYP2C9* | *CYP2C8* | *CYP2C19* | *CYP2D6* | | *CYP3A4* | |
|---|---|---|---|---|---|---|---|
| Amitriptyline | Aceclofenac | Carbamazepine | Diazepam | Alprenolol | Mianserin | Alfentanil | Loratadine |
| Caffeine | Celecoxib | Cerivastatin | Diphenylhydantoin | Amiflamine | Nortriptyline | Alprazolam | Lovastatin |
| Clomipramine | Diclofenac | Paclitaxel | Hexobarbital | Amitriptyline | Paroxetine | Amiodarone | Midazolam |
| Clozapine | Dorzolamide | Pioglitazone | Mephobarbital | Aprindine | Perhexiline | Amlodipine | Nelifinavir |
| Fluvoxamine | Flurbiprofen | Retinoic acid | Omeprazole | Brofaromine | Perphenazine | Atorvastatin | Nifedipine |
| Haloperidol | Fluvastatin | Rosiglitazone | Proguanil | Cinnarizine | Propafenone | Benzphetamine | Omeprazole |
| Imipramine | Glimeripide | | *S*-Mephenytoin | Clomipramine | Sparteine | Budesonide | Paclitaxel |
| Paracetamol | Glyburide | | Teniposide | Codeine | Thioridazine | Carbamazepine | Quinidine |
| Phenacetin | Glypizide | | | Debrisoquine | Timolol | Clarithromycin | Rapamycin |
| Propranolol | Ibresartan | | | Deprenyl | Tomoxetine | Cyclosporin | Ritonavir |
| Riluzole | Ibuprofen | | | Desipramine | Tropisetron | Dapsone | Saquinavir |
| Ropinirole | Losartan | | | Dexfenfuramine | Venlafaxine | Diazepam | Sildenafil |
| Ropivacaine | Naproxen | | | Dextromethorphan | | Digitoxin | Simvastatin |
| Tacrine | Phenytoin | | | Encainide | | Diltiazem | Tacrolimus |
| Theophylline | Piroxicam | | | Flecainide | | Erythromycin | Tamoxifen |
| Zolmitriptan | Sildenafil | | | Flunarizine | | Ethinylestradiol | Teniposide |
| | Suprofen | | | Fluoxetine | | Etoposide | Terfenadine |
| | *S*-Warfarin | | | Fluphenazine | | Finasteride | Toremifene |
| | Tenoxicam | | | Guanoxan | | Flutamide | Triazolam |
| | Tienelic acid | | | Hydrocodone | | Ifosphamide | Verapamil |
| | Tolbutamide | | | Indoramin | | Indinavir | Zatosetron |
| | Torsemide | | | Metoprolol | | Lansoprazole | Zonisamide |
| | Citalopram | | | Mexiletene | | Lidocaine | |

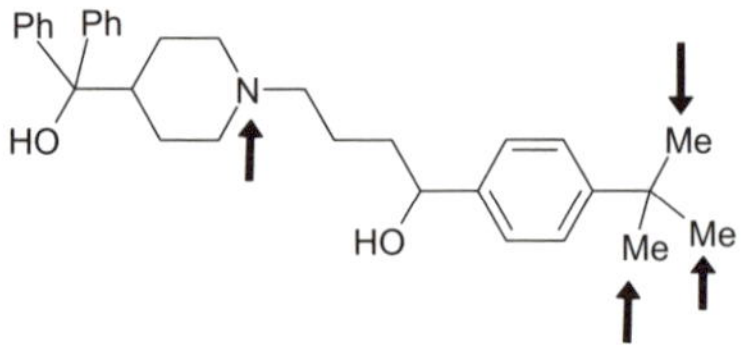

Figure 2 *Sites of CYP3A4-mediated metabolism of terfenadine*

bind simultaneously to the CYP3A4 active site. It is also thought that the CYP3A4 active site undergoes conformational changes upon the binding of substrates. CYP3A4 is the most complex human drug metabolising enzyme and there are no reliable models for its substrate specificity.[2]

CYP3A4 activity can vary considerably between individuals and can be modulated by dietary factors, hormones and other pharmaceutical agents. Significant genetic polymorphisms have also been identified in the regulatory region of the gene, all of which may contribute to this variability. CYP3A4 is inducible by several pharmaceutical agents, the majority of which either accumulate significantly upon multiple dosing, are given at the high doses of 100s of mgs or both, *e.g.* phenobarbital, felbamate, rifampin, phenytoin, carbamezepine and troglitazone. These inducers can profoundly effect the pharmacokinetics of other drugs with loss of efficacy being the most likely consequence, *e.g.* rifampin with the oral contraceptive pill. As the mechanism of induction is often *via* activation of a nuclear hormone receptor (pregnane X receptor), affinity against this receptor could be another component of lead optimisation.

Inhibition of CYP3A4 activity is of particular concern since so many drugs rely on this enzyme activity for their elimination. Thus co-administration of two drugs, one an inhibitor and the other a substrate, could result in a much higher exposure of the substrate. Several potent inhibitors of CYP3A4 are highly lipophilic and include an imidazole ring or some other means to bind to the haem of the P450. For example, oral ketoconazole and itraconazole are contraindicated with many CYP3A4 substrates as they can cause life threatening drug–drug interactions. Mechanism-based inhibitors or suicide substrates seem to be particularly prevalent with CYP3A4. Such compounds are substrates for the enzyme, but are believed to form reactive products that deactivate the enzyme. Several macrolide antibiotics, generally those with an unhindered tertiary amine function (*e.g.* erythromycin), are able to inhibit CYP3A4 in this manner. Due to the large number of drug molecules metabolised by CYP3A4, potent inhibition, by whatever mechanism, can have a detrimental effect on a compound's safety and marketability. Drugs have either not been approved or withdrawn from the market due to their inhibitory action on this enzyme, most recently mibefradil.

*CYP2D6.* The overwhelming majority of CYP2D6 substrates contain a basic nitrogen atom ($pK_a > 8$) which is ionised at physiological pH. All the models of CYP2D6 show essentially the same characteristics of a 5, 7 or 10 Å distance between this basic nitrogen atom and the site of metabolism. Despite the commonality of this ionic interaction, differences in binding affinity between substrates and inhibitors could be attributed to other types of interactions.[3]

CYP2D6 was perhaps the first and best characterised of the polymorphic cytochrome P450 enzymes. The poor metaboliser phenotype (PM) is characterised by an absence of any catalytically competent CYP2D6, resulting in a marked deficiency in the metabolism of such substrates, which can result in drug toxicity or reduced efficacy. The prevalence of the PM phenotype shows marked ethnic differences with a mean value of approximately 7% in Caucasian populations, but 1% or less in Orientals. Nearly 70 different CYP2D6 alleles have been identified and some individuals even have multiple copies of the gene giving them a super metaboliser phenotype.

CYP2D6 does not appear to be inducible. However, there are many known inhibitors, the most notable being quinidine. Although not metabolised by CYP2D6, quinidine conforms closely to the structural requirements for binding to the active site of the enzyme, as do several other inhibitors such as ajmalcine, fluoxetine and paroxetine. Not all CYP2D6 inhibitors have to have a strongly basic nitrogen atom. The HIV-I protease inhibitor, ritonavir (Figure 3), has a weakly basic centre but forms a relatively strong interaction with CYP2D6. However, the inhibitory potency is believed to be due to a number of hydrogen bonding interactions within the CYP2D6 active site.

*CYP2C9.* The majority of the CYP2C9 substrates are acidic or contain similar areas of hydrogen bonding potential (Figure 4). Models of the CYP2C9 typically produce schemes where the hydrogen bonding groups are positioned at a distance of approximately 8 Å from the sites of oxidation of these substrates.[4] It is also notable that almost no CYP2C9 substrates are oxidised at such a rate that first-pass metabolism is a barrier to their bioavailability.

There are allelic variants of CYP2C9, which, in general, show reduced rates of metabolism toward substrates, relative to the wild type. The extent of this can vary widely depending on substrate. (*S*)-Warfarin is metabolised predominantly by CYP2C9 and patients receive doses of between 4 and 8 mg of warfarin per day. However, patients with one of the mutant

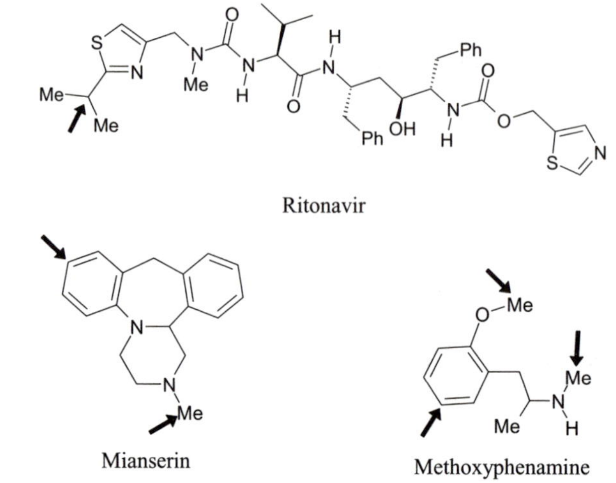

Ritonavir

Mianserin

Methoxyphenamine

**Figure 3** *Site of CYP2D6-mediated metabolism of ritonavir, mianserin and methoxyphenamine*

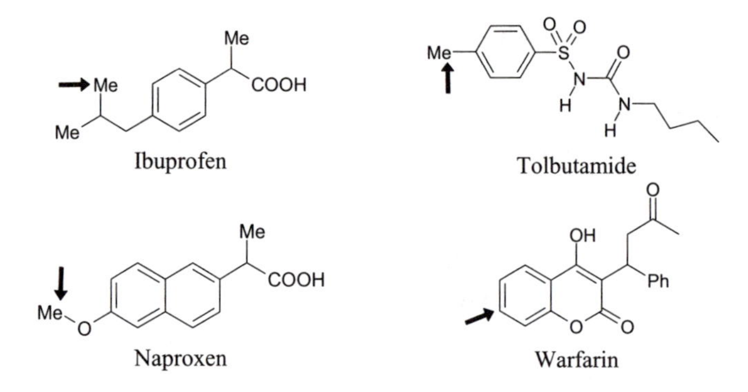


**Figure 4** *Sites of CYP2C9-mediated metabolism of ibuprofen, tolbutamide, naproxen and warfarin*

alleles are more sensitive to warfarin effects and one individual could not receive more than 0.5 mg of warfarin per day.

CYP2C9 is inducible by a number of the same drugs that induce CYP3A4, although usually to a lesser extent. A more significant problem is the inhibition of this activity, especially as its substrates include the narrow therapeutic index drug warfarin. Sulfaphenazole is perhaps the best known, most potent and selective inhibitor of CYP2C9. The mode of inhibition is *via* ligation to the haem, although which nitrogen atom from sulfaphenazole is involved is not known. Other significant inhibitors include fluvastatin and fluconazole which are far more widely used than sulfaphenazole.

*CYP1A2.* CYP1A2 substrates are mainly lipophilic planar polyaromatic or heteroaromatic molecules with a small depth and a large area/depth ratio, for example caffeine (Figure 5). Protein homology modelling suggests that the active sites of CYP1A2 are composed of several aromatic residues which form a cavity of restricted size and shape such that only planar structures are able to occupy the binding site.[5] Binding is made up of hydrophobic, defined hydrogen bonding and $\pi-\pi$ interactions. Unlike some of the other cytochromes P450, CYP1A2 does not have a clear preference for acidic or basic molecules.

Whilst CYP1A2 appears to be non-polymorphic in man, it is inducible by environmental factors such as cigarette smoking and dietary factors, such as consumption of cruciferous vegetables. Omeprazole is the best known inducer of CYP1A2, although this is only notable at higher doses or in poor metabolisers of omeprazole. The most clinically relevant inhibitors of CYP1A2 include the quinolone antibacterials enoxacin and pefloxacin, and the anti-depressant fluvoxamine. As with CYP2C9, few CYP1A2 substrates are highly cleared in man, but inhibition is a significant concern, since theophylline, a narrow therapeutic index drug, is metabolised by CYP1A2.

*CYP2C19.* Substrates for this enzyme include (*R*)-mephobarbital, moclobemide, proguanil, diazepam, omeprazole and imipramine, which do not show obvious structural or physicochemical similarities (Figure 6). CYP2C19 can bind compounds which are weakly basic like diazepam, strongly basic like imipramine or acidic compounds such as (*R*)-warfarin.

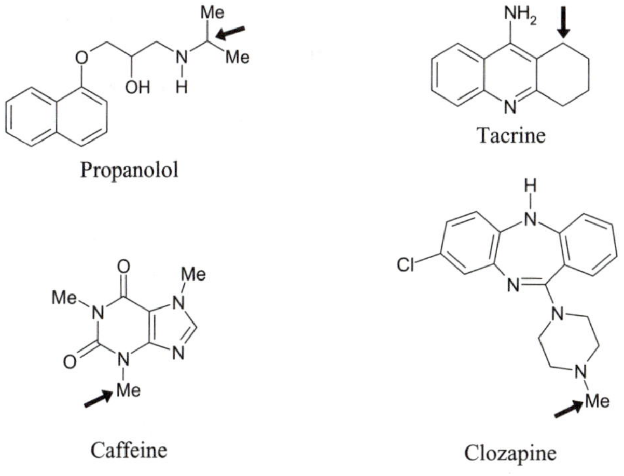

Propanolol

Tacrine

Caffeine

Clozapine

**Figure 5** *Sites of CYP1A2-mediated metabolism of propanolol, tacrine, caffeine and clozapine*

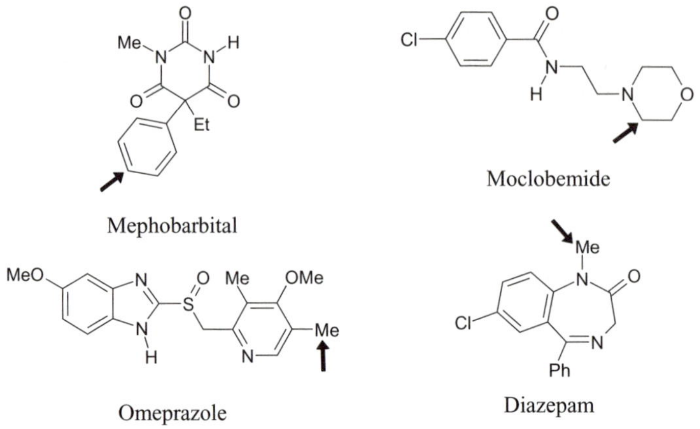

Mephobarbital

Moclobemide

Omeprazole

Diazepam

**Figure 6** *Sites of CYP2C19-mediated metabolism of mephobarbital, moclobemide, omeprazole and diazepam*

One possibility is that CYP2C19 binds substrates *via* a combination of hydrogen bond donor and acceptor mechanisms.[6]

CYP2C19 does exhibit a notable genetic polymorphism that shows marked inter-racial differences with an occurrence of approximately 3% in Caucasians and between 18 and 23% in Orientals. CYP2C19 poor metabolisers (PMs) lack any functional CYP2C19 activity and the mechanism of this polymorphism has largely been ascribed to two defects in the CYP2C19 gene. There are relatively few clinically relevant inhibitors of CYP2C19 and none that are entirely selective.

*CYP2C8.* Less is known about CYP2C8 than the other human drug metabolising P450 enzymes, as the appropriate tools for its study (specific substrates, sources of "pure" enzyme) have only become available or identified relatively recently. Its major substrates are often compounds that can be substrates for CYP3A4, like the glitazones and paclitaxel, or have the properties you might expect of CYP2C9 substrates like retinoic acid. No potent or selective inhibitors have been reported, although it has been shown to be subject to a genetic polymorphism, the incidence and consequences of which have yet to be fully evaluated.

## Other Oxidative Enzymes

*Flavin Monooxygenase.* The flavin monooxygenases (FMO) are microsomal enzymes and many of the reactions they catalyse can also be catalysed by cytochrome P450. Their reactions can be distinguished, *in vitro*, by the relative sensitivities of the two enzyme systems to heat (readily inactivates FMO) and non-ionic detergents (readily inactivates P450). The commonest FMO reaction is the oxidation of nucleophilic tertiary amines to *N*-oxides, although primary and secondary amines and several sulfur-containing drugs are also substrates. Nicotine, cimetidine and desipramine are all substrates of FMO, although this enzyme is not necessarily responsible for the major fraction of elimination of these drugs. Humans express five different FMOs (FMO1 to FMO5) and the major form in human liver (FMO3) is generally the one implicated in the significant metabolism of drugs, whereas FMO1 is the major FMO expressed in rats.

*Monoamine Oxidase.* Monoamine oxidase (MAO) is involved in the oxidative deamination of amines. Substrates include a number of endogenous amines, such as serotonin, as well as drugs like sumitriptan. There are two forms, MAO-A and MAO-B, and most tissues contain both forms of the enzyme. The oxidative deamination can be readily blocked by substitution on the *a*-carbon adjacent to the nitrogen, where hydrogen abstraction is thought to occur. Monoamine oxidase is most notable for its role in the activation of MPTP (1-methyl-4-phenyl-1,2,5,6-tetrahydropyridine) to a neurotoxic metabolite that causes symptoms of Parkinson's disease in primates, but not in rodents (Figure 7). The neurotoxic effects of MPTP can be blocked with *l*-deprenyl (a selective inhibitor of MAO-B), but not with clorgyline (a selective inhibitor of MAO-A). This may explain the species difference in sensitivity to MPTP, as rodents have very much lower levels of MAO-B compared to man.

*Aldehyde Oxidase.* Aldehyde oxidase can oxidise a number of substituted pyrroles, pyridines, pyrimidines and purines and its substrates include methotrexate, quinidine, azapetine, famciclovir and cyclophosphamide (Figure 8). It is particularly notable for the species difference in its level of expression. In rat and dog, levels of activity are low but the activity is highly expressed in human liver. Since rat and dog are frequently used to evaluate the pharmacokinetics of putative pharmaceuticals, for aldehyde oxidase substrates, these species can give a misleadingly promising outlook for likely success in man.

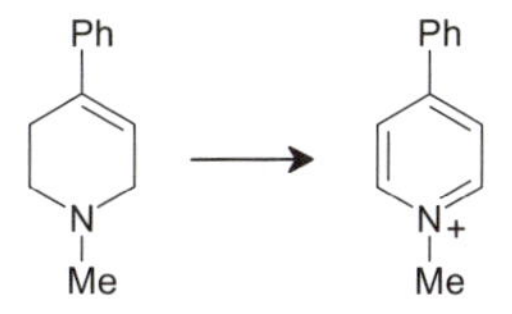


**Figure 7** *MAO-B-mediated oxidation of MPTP to a neurotoxic pyridinium*

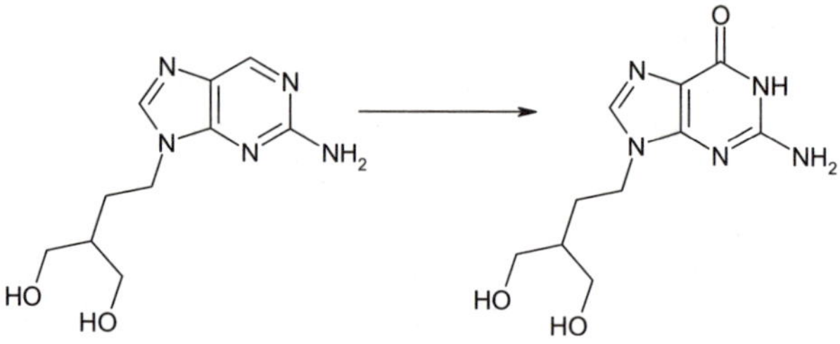

Figure 8 *Aldehyde oxidase mediated formation of penciclovir*

*Xanthine Oxidase.* Xanthine oxidase catalyses the oxidation of hypoxanthine to xanthine and to uric acid. It can also oxidise other purine derivatives, such as allopurinol and 6-mercaptopurine, as well as phthalazine. There is some overlap between the reactions catalysed by xanthine and aldehyde oxidase, although there are selective inhibitors that can be used, *in vitro*, to delineate the relative significance of each enzyme.

## Phase II Conjugation

*Glucuronidation.* UDP-glucuronosyl transferases (UGT) are a multi-gene family of enzymes that catalyse the conjugation of suitable functional groups (*e.g.* –OH, –SH, $-NH_2$, –COOH) with glucuronic acid (Figure 9). The resulting water soluble metabolite can then be excreted in bile or urine, where it is likely to be recognised by the biliary or renal organic acid transport systems.

Following excretion into the bile, the glucuronidated metabolite can be cleaved by $\beta$-glucuronidase activity in the intestinal microflora, releasing the parent compound that can then be reabsorbed. Such material is then available for metabolism by glucuronidation again and this process is known as enterohepatic recycling, although glucuronidation is not the only means by which this can occur. There are 15 human forms of UGT which are generally found in the highest concentration in liver, kidney or intestine. In addition to metabolising drugs, these enzymes play an

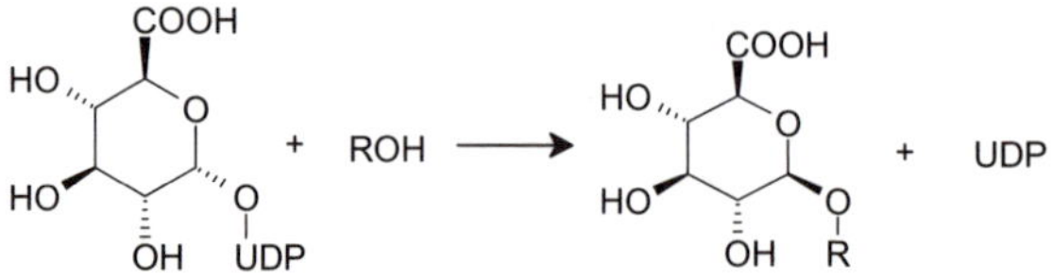

Figure 9 *Glucuronidation of a R–OH with UDP-glucuronic acid*

important role in the conjugation of endogenous compounds such as bilirubin. The key drug metabolising UGTs are UGT1A1, 1A4, 1A6, 1A9, and 2B7, but there are few drugs that are selectively metabolised by an individual enzyme and the substrate overlap is more significant than it is for the cytochromes P450 (Table 3).

*Sulfation.* Many of the drugs that undergo glucuronidation can also be sulfated by a reaction catalysed by sulfotransferases. Phenols and aliphatic alcohols are the preferred, but by no means only, substrates which include acetaminophen, chloramphenicol, ethinylestradiol, minoxidil, desipramine and ethanol. Sulfate conjugates of drugs can be excreted in bile or urine and, as described for glucuronidation, can be a mechanism of entero-hepatic recycling. The amount of co-factor for the reaction, 3′-phospho-adenosine-5′-phosphosulfate (PAPS), is limited by the concentration of free cysteine and as a consequence, the capacity of sulfotransferase is limited. Generally, sulfation is a high affinity, low capacity process in contrast to glucuronidation which is low affinity, high capacity. At low doses of acetaminophen, the sulfate is the main conjugate but at higher doses the relative amount of glucuronide increases. Although many forms of sulfotransferase have been identified in man, there has been little progress in identifying enzyme specific substrates or the extent of over-lapping substrate specificity.

*Acetylation.* *N*-Acetylation is a major route of metabolism for drugs containing an aromatic amine. However, aliphatic amines are occasionally substrates for *N*-acetyltransferase (NAT). Humans have two *N*-acetyltrans-ferases, NAT1 and NAT2. NAT1 is expressed throughout the body and NAT2 is expressed only in the liver. Sulfamethoxazole, *para*-aminobenzoic acid and sulfanilide are preferentially metabolised by NAT1 and isoniazid, hydralazine, procainamide, dapsone and sulfamethazine are preferential NAT2 substrates, although to some extent most can be metabolised by

**Table 3** *The major human drug metabolising UGTs*

| *UGT enzyme* | *Major localisation* | *Type of substrate* | *Example substrate* |
|---|---|---|---|
| 1A1 | Liver, intestine | Bilirubin, phenols, anthraquinones | Ethinylestradiol |
| 1A4 | Liver, intestine | Amines, phenols | Imipramine |
| 1A6 | Liver, intestine, kidney | Planar phenols | Acetaminophen |
| 1A9 | Liver, kidney | Bulky phenols, carboxylic acids | Propofol, ibuprofen |
| 2B7 | Liver, intestine, kidney | Morphine, steroids, carboxylic acids | Morphine, ibuprofen |

both forms. Genetic polymorphisms in NAT2 result in slow and fast acetylator phenotypes, which are evenly distributed in Western populations. However, slow acetylators are more common (~70%) in Middle Eastern populations and very much lower in Asians (<25%). As a consequence of this polymorphism, the pharmacological effect of the antihypertensive drug hydralazine is more pronounced in slow acetylators and these individuals are more predisposed to drug toxicities from compounds like isoniazid, dapsone and hydralazine. Genetic polymorphisms have been identified in NAT1, although the pharmacological and toxicological significance has yet to be fully determined.

*Glutathione.* Glutathione consists of glycine, cysteine, and glutamic acid and conjugates to electrophilic moieties of drugs or their metabolites (Figure 10). In addition to the electrophilic centre, the substrates for glutathione S-transferases are generally hydrophobic and can react non-enzymatically with glutathione. The concentration of gluthathione in the liver can be very high (10 mM) and this non-enzymatic conjugation can occur to a significant extent. The concentration of glutathione S-transferases is also very high and they are thought to account for ~10% of cellular proteins. They are also responsible for binding, storing, or transporting bilirubin, steroids and haem in addition to their role in conjugating glutathione to electrophilic species. Glutathione conjugates formed in the liver can be excreted intact in bile, or they can be converted into mercapturic acids by the sequential cleavage of glutamic acid and glycine from the glutathione, followed by *N*-acetylation in the kidney and excreted in urine. Glutathione conjugation is an important detoxification reaction, since electrophiles are potentially toxic species that can bind to and damage proteins, nucleic acids and other biological macromolecules.

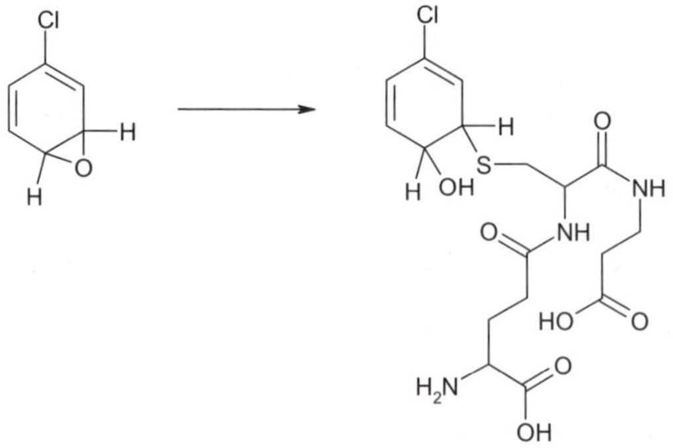

**Figure 10** *An example of glutathione conjugation with an electrophilic carbon*

## 4 CONCLUSIONS

The optimisation of the drug metabolism profile of a new pharmaceutical can potentially involve many different enzymes as the body brings to bear all of its tools to prevent any potential toxicological or pharmacological insult. Yet, unless the rate of metabolism is such that it results in poor pharmacokinetic properties (*e.g.* low bioavailability, short half-life), it may be considered not to be an issue. However, this may not be the case. For example, if a drug demonstrates a particularly high affinity for a drug metabolising enzyme, it may impair the clearance of other drugs that rely on that enzyme for clearance. The resulting drug–drug interactions could, at worst, prevent approval of the drug by the Regulatory Authorities and, at best, require labelling which may limit the therapeutic utility or marketability. Similarly, it is generally considered best if a compound is metabolised by multiple enzymes (multiple routes is not necessarily sufficient), as this will reduce the impact of any co-administered enzyme inhibitor or the influence of genetic polymorphisms. Other obvious targets include minimising the production of any reactive species produced as a consequence of metabolism, the presence of which is often indicated by the presence of glutathione related conjugates. Perhaps merely reducing the rate of metabolism is the most straightforward metabolism issue facing the medicinal chemist today.

## 5 REFERENCES AND BIBLIOGRAPHY

1. D.F.V. Lewis *et al.*, *Drug. Metab. Drug Interact.*, 1999, **15**, 1.
2. K.K. Khan and J.R. Halpert, *Arch. Biochem. Biophys.*, 2000, **373**, 335.
3. M.J. de Groot *et al.*, *J. Med. Chem.*, 1999, **42**, 4062.
4. V.A. Payne, Y.T. Chang and G.H. Loew, *Proteins*, 1999, **37**, 176.
5. J.J. Lozano *et al.*, *J. Comput.-Aided Mol. Des.*, 2000, **14**, 341.
6. V.A. Payne, Y.T. Chang and G.H. Loew, *Proteins*, 1999, **37**, 204.

### For Further Reading

G.G. Gibson and P. Skett, *Introduction to Drug Metabolism*, Chapman and Hall, London, 1986.
T.F. Woolf, *Handbook of Drug Metabolism*, Marcel Dekker, New York, 1999.
G.M. Pacifici and G.N. Fracchia, *Advances in Drug Metabolism in Man*, European Comission, Brussels, 1995.

G.J. Mulder, *Conjugation Reactions in Drug Metabolism: An Integrated Approach*, Taylor & Francis, London, 1990.
A. Parkinson in *Casarett & Doull's Toxicology. The Basic Science of Poisons*, C.D. Klaassen (ed.), McGraw-Hill, New York, 1996, pp. 113–186.

CHAPTER 8

# Toxicology in the Drug Discovery Process

SUSAN M. EVANS, ELISABETH GEORGE AND C. WESTMORELAND

## 1 INTRODUCTION

The prime concern in the development of new pharmaceutical agents is to ensure that the product is potent, effective and safe. European directive 65/65 states that the "primary purpose of any rules concerning the production and distribution of medicinal products must be to safeguard public health".[1]

Before administration of a pharmaceutical agent for the first time to humans (Phase I clinical trials), international regulations require that a detailed package of preclinical studies is undertaken to assess the potential toxicity of the new drug. Included in such a package are multi-species toxicology studies to detect any target organ toxicity and establish whether any toxic effects are reversible. The design and duration of such studies is specified by regulatory agencies such as the Food and Drug Administration (FDA) in the US. Recently there has been some harmonisation of the guidelines for toxicity studies by The International Conference on Harmonisation (ICH) which is a world-wide joint initiative involving both regulatory authorities and industry.[2]

A typical toxicology programme prior to Phase I studies with a new chemical entity (NCE) may consist of the following:

- Examination of the effects of three dose levels in repeat-dose studies in two animal species, including one non-rodent. The low dose level will be a small multiple of the predicted clinical dose and the high dose level should demonstrate some toxicity. An intermediate dose (often the geometric mean of the other two dose levels) will also be included to allow investigation of dose–response relationships. The duration of repeat-dose studies is dependent on the duration of the proposed clinical study.[3] Such repeat-dose toxicology tests assess whether a compound, which may be non-toxic in acute studies, accumulates and disrupts body function, resulting in increased toxicity after repeated administration.

- Testing the ability of the drug to damage genetic material (mutations in DNA and chromosome damage) using genetic toxicology tests. Prior to Phase I trials, genetic toxicology studies will usually include an *in vitro* test for mutation (*e.g.* microbial mutagenicity assay) and a second *in vitro* test for chromosome damage (*e.g.* cytogenetics assay in cultured mammalian cells). An *in vivo* test for genotoxic potential (*e.g.* rodent bone marrow micronucleus test) may also be performed.
- Examination of the overt pharmacodynamic effects of the drug on vital organ systems (cardiovascular, central nervous and respiratory systems) are carried out in safety pharmacology tests in two animal species (including one non-rodent). In addition, *in vitro* electrophysiological studies, such as Purkinje fibre preparations may also be carried out to predict a compound's potential to prolong the QT interval.[4]
- In the European Union and Japan, work must also be conducted to evaluate effects on embryo-foetal development if the proposed study is to include women of child-bearing potential. This is required to evaluate safety for the unintentional exposure of the developing embryo/foetus and would normally be carried out in rats and rabbits. In addition, it may be necessary to conduct *in vivo* studies to evaluate the effect of the drug on female fertility. At this stage in drug development, potential to affect male fertility is assessed during the repeat-dose studies by assessing testes weight and histology, including sperm staging.[5]
- During many of the *in vivo* toxicology studies described, assessment must be made of the systemic exposure to the drug (toxicokinetics). In addition, separate studies to assess the absorption, distribution, metabolism and excretion of the drug will be undertaken.

During later stages of drug development, other safety studies are carried out to support the clinical programme. Such studies can include acute toxicity studies, longer term repeat-dose studies, carcinogenicity studies, pre- and post-natal development studies and completion of the genotoxicity battery of studies. Regulatory toxicology studies are, and will continue to be, the definitive way to assess the pre-clinical safety of a new drug prior to and during clinical trials.

The quality of data generated in regulatory toxicology studies is of paramount importance and all studies are conducted in accordance with international Good Laboratory Practice (GLP) guidelines.[6] GLP is concerned with the organisational processes and conditions under which preclinical safety studies are planned, performed, monitored, recorded, archived and reported.

A considerable amount of time and resource is required to complete the GLP toxicology studies on a new drug prior to its first administration to humans. These resources will also include the chemical synthesis of the potential drug. For example, for a drug with low toxicity, a programme of work to support Phase I clinical trials as described above could use as much as 15 kg of drug substance during the preclinical development phase. When taking this into account, together with the numbers of animals used and the support work required to conduct fully GLP-compliant studies, it is clear that such work cannot be undertaken for a large number of compounds. Every effort should be made to screen out undesirable compounds prior to this stage of drug development.

The remainder of this chapter will concentrate primarily on the use of non-regulatory predictive tools in toxicity testing which have been developed more recently and which can be used for in-house decision making prior to GLP regulatory testing, *i.e.* during candidate selection or earlier during the drug discovery process. These tests are designed to eliminate candidate molecules with potentially unwanted side-effects at the earliest possible stage in drug discovery. The tests minimise both the quantity of drug substance and the number of animals used, such that those compounds that do pass through to the resource- and cost-intensive GLP studies have a greater chance of becoming new medicines.

## 2 OVERVIEW OF TOXICITY ASSESSMENT IN DRUG DISCOVERY

Within the pharmaceutical industry the selection and prioritisation of compounds with the highest potential to become drugs occur throughout the drug discovery and development process. Certain types of toxicity assessment can be implemented at most stages of this process as illustrated in Figure 1. The drug discovery and development process can be broadly divided into the following stages:

- Disease selection and target identification
- Lead series identification

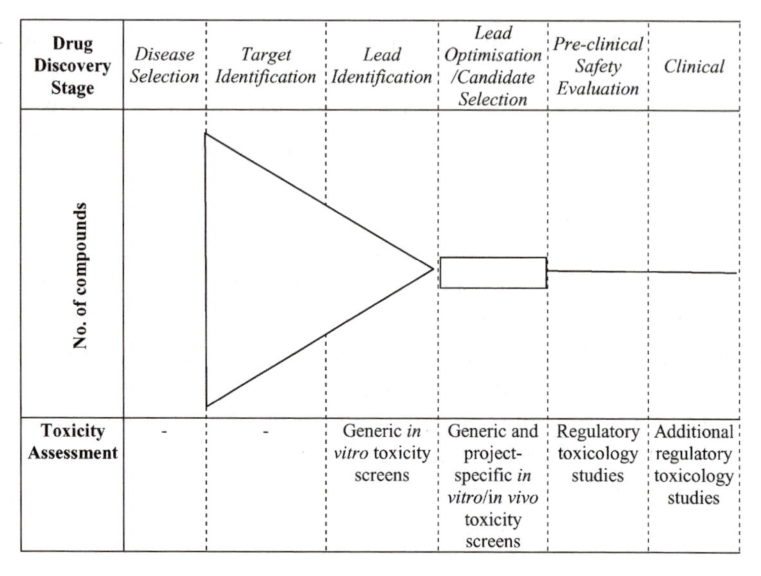


**Figure 1** *Overview of the use of toxicity studies during the drug discovery process*

- Lead compound optimisation/candidate compound selection
- Pre-clinical safety assessment
- Clinical.

## Disease Selection, Target Identification and Lead Series Identification

There is usually no practical involvement in terms of toxicity assessment during the disease selection and target identification stages of drug discovery. In certain cases, however, where specific toxic effects are known to be linked to particular types of therapies, *e.g.* nephrotoxicity of immunosuppressant drugs, contributions from toxicology can be made on a theoretical basis. Identifying potential toxicology issues that may be associated with a therapy or class of compound can speed up the later development of compounds from this project, *e.g.* developing *in vitro* screens representative of the specific toxic effects.

During the lead series identification stage, only small quantities of each compound are usually available (typically less than 20 mg). If no prior knowledge of any specific toxic effects of the new compound series exists it is appropriate at this stage to employ *in silico* and/or generic high throughput *in vitro* systems in the toxicity screening process. The aim of the assessment of toxicity at this stage of the drug discovery process would be to detect compounds that are most likely to fail during the later stages of the process, thereby selecting compounds with the highest potential to proceed through pre-clinical development.

### Lead Compound Optimisation/Candidate Compound Selection

The quantity of compound available at this stage of the process remains low (approximately 20 mg), therefore any assessment of toxicity needs to be *in vitro*. More knowledge may become available about toxicity issues, related to either the class of compound or the possible toxicological consequences of the pharmacological action in question, as well as the therapeutic target organ or cell population. If this is the case project-specific *in vitro* tests can be designed that help in the compound selection process. Examples of such project-specific *in vitro* tests will be described below (page 169). If the toxicity of the compounds is unknown generic *in vitro* toxicity screens can be employed to help prioritise with which compounds to proceed.

At the end of the lead compound optimisation stage a small number of promising candidate molecules are identified. The quantity of compound increases substantially, allowing the assessment of these compounds for prioritisation for further development using *in vivo* toxicity studies of short duration. The end result would be that only those compounds which are most likely to succeed are taken forward for GLP regulatory studies during the pre-clinical safety assessment and clinical development stages of the process.

### Pre-clinical Safety Assessment and Clinical Development

During the pre-clinical safety assessment stage, the toxicity of candidate molecules is assessed in regulatory toxicology studies, as outlined in Section 1. This stage also accompanies the remaining process through clinical development, where proof of concept investigations are undertaken to confirm the efficacy of the drug in humans, and where the safety of the drug in humans is assessed in clinical trials in healthy volunteer and patient populations. Other parameters, such as the potential side-effects, dose, pharmacokinetics, pharmacodynamics, interactions with other drugs or any contra-indications are also assessed during the clinical trials.

## 3 *IN SILICO* SYSTEMS FOR ASSESSING TOXICITY

*In silico* systems for assessing toxicity are computer systems that claim to predict the toxicological hazard associated with a compound. Computer-aided toxicity prediction has the potential to increase the speed and efficiency of toxicity testing, as well as to reduce the cost of the drug discovery process by prioritising compounds for further development. A number of systems are commercially available, such as DEREK (Deductive Estimation of Risk from Existing Knowledge), TOPKAT (Toxicity Prediction by Komputer Assisted Technology) and MULTICASE. These ‘expert’ systems are based on ‘rules’ derived from experimental data consisting of one or more toxicities associated with a chemical. Sets of compounds are used to derive these rules and these rules can be modified/improved on acquisition of more data. Some rules are based on mathematical approaches or QSAR[7] (induced rules), whereas other rules are derived from existing knowledge and expert human judgement (knowledge-based rules). Induced rules have the advantage of being unbiased towards a particular mechanism of toxic action and extending existing knowledge; however, they may be nothing more than empirical relationships devoid of biological meaning. On the other hand, knowledge-based rules are derived from existing knowledge, they do not extend existing knowledge but are based on specific mechanisms of toxicity. Both of these approaches have the potential to predict a range of toxicities from the physicochemical structure alone. The development and validation of the commercially available expert systems for predicting toxicity has been reviewed by Dearden *et al.*[8]

Some systems incorporate the mathematical induced rules *e.g.* MULTICASE, which divides a molecule’s structure into fragments (normally ranging from two to six heavy (non-hydrogen) atoms in length (*e.g.* $CH_2$–OH for an aliphatic hydroxy group). Other descriptors used in these systems include two-dimensional distances between atoms within a chemical structure, calculated electronic indices (molecular orbital energies, charge densities), calculated transport parameters (octanol–water partition coefficient, water solubility), various atom groupings such as hydrogen bond acceptors and donors and lipophilic centres. These descriptors are then statistically correlated with the toxicological activity of the compounds under study. The descriptor that has the highest probability of being responsible for the toxicological activity is identified (*e.g.* functional groups that are activated by metabolism to reactive sites, *e.g.* the bay region of polycyclic aromatic hydrocarbons which is transformed to a diolepoxide).[9] QSARs are then derived to predict the potency or extent of activity.

Due to the complexity of the toxicological response, the prediction of all aspects of toxicology in one system is not currently possible. *In silico* systems are currently most successful where a specific toxic endpoint is modelled. For example, the area of genetic toxicology (where a yes/no answer can be obtained for mutagenicity) is one of the main areas of focus for *in silico* toxicity prediction. An expert system such as DEREK can be used to predict the mutagenic potential of a compound at the very early stages of drug discovery. Mutagenicity of chemicals is influenced by many factors associated with the chemicals' structure, *e.g.* electrophilicity, size and shape of the molecules (which control access to target sites on DNA), ability of the chemicals or their metabolites to traverse the biological membrane. DEREK is a knowledge-based rule system that does not rely on algebraic or statistical relationships and makes qualitative and semi-quantitative predictions. The rules within DEREK are derived by human experts and reflect the current state of knowledge of structure–toxicity relationships with emphasis on the understanding of mechanisms and metabolism.[10] It also takes into consideration physicochemical properties such as $\log P$ and $pK_a$. Apart from mutagenicity, DEREK also covers a range of other specific toxicological endpoints which include carcinogenicity, neurotoxicity, teratogenicity, irritancy, skin sensitisation, respiratory sensitisation, corrosivity, lachrymation and methaemoglobinaemia.

The major hurdle with predicting general toxicity using *in silico* techniques is that compounds can be toxic for many different reasons such as the formation of toxic metabolites, the accumulation of compound in specific organs and species-specific effects. The accurate prediction of biokinetic factors is therefore a pre-requisite for the development of acceptable *in silico* systems of toxicity prediction. The generation of sufficient high quality, reliable toxicity data to validate completely the results produced by the computer-aided systems is also a necessity. The FDA are currently using *in silico* models as supplemental information for hazard identification.[11] The FDA does not currently consider computer-generated predictions as substitutes for any *in vivo*/*in vitro* studies.

## 4 INTRODUCTION TO *IN VITRO* SYSTEMS

*In vitro* systems can be defined as experimental systems using biological material other than intact animals. A large number of *in vitro* models exist, such as tissue slices, primary cell cultures and established cell lines, which, in combination with appropriate toxicity endpoints, can be used to investigate the potential toxicity of new compounds.

*In vitro* systems can be broadly classified into systems that work with primary material, and those that work with established cell lines (Table 1). Primary material is derived directly from a donor animal or human at the time of use. Examples of primary material used in toxicological investigations include isolated cells and tissue fragments or slices. In some circumstances, discarded material from surgery (such as excess tissue from

**Table 1** *Biological material used for* in vitro *systems*

| | *Biological model* | |
|---|---|---|
| | *Primary cells* | *Cell lines* |
| Definition | Directly derived from donor animal or human for each experiment | Obtained originally from human or animal but can proliferate indefinitely in culture. Often derived from tumours |
| Disadvantages | Often short-lived, still need tissue donors, inter-individual differences between human donors | De-differentiated (*i.e.* loss of tissue-specific functions) |
| Advantages | Functions close to *in vivo* situation | Long-term observations possible<br>No requirement for tissue donor for each experiment |
| Examples | Hepatocytes<br>Proximal tubule cells<br>Haemopoietic progenitor cells<br>Neutrophils<br>Keratinocytes<br>Skeletal myocytes<br>Tissue slices (liver, kidney, brain, *etc.*) | HepG2 , FAO (liver)<br>MDCK, LLPKC1, Vero (kidney)<br>HL60, TK6, L5178Y (haemopoietic system)<br>V79, CHO (fibroblasts) |

biopsies, breast surgery or liver transplants) can be used to provide primary human material. In most other cases, non-transplantable tissue must be used. The use of non-transplantable and surgical human tissue in *in vitro* toxicology raises important legal, ethical and safety issues.[12] The advantage of using primary *in vitro* systems in toxicological investigations is that the functions of the tissue are often closely related to the functions observed *in vivo*.

Primary material usually has a limited life-span (a few hours to several days). However, established cell lines have the ability to proliferate indefinitely in culture, and therefore have an unlimited life-span. Established cell lines are usually derived from tumour cells or are immortalised using genetic manipulation. These cell lines are often de-differentiated to varying degrees, which means that they have lost some, most, or all of their tissue-specific functions. In many cases, therefore, cell lines can provide only limited models for the *in vivo* situation. Such de-differentiation is thought to be the result of removing the cells from their hormonal and neuronal environment, and often cells will revert back to a functionally more embryonic state. The advantages of using established cell lines in toxicological investigations are that long-term observations can be made and that no new donor is needed to supply tissue for each experiment.

Cells can be cultured in a variety of culture vessels, ranging from large fermentors (for bulk production of cells), plastic Petri dishes and tissue culture flasks to multiwell plates (for high throughput screening). All cell types are grown in culture medium, which provides nutrients and must be renewed on a regular basis. Cells can be cultured as either monolayer or suspension cultures. Table 2 summarises the main differences between monolayer and suspension culture of cells.

Most cell types that are derived from solid tissues, such as the skin or kidney, adhere to the vessel surface to form a monolayer of flattened cells. If the cells proliferate, the monolayer grows to fill up the available culture surface, referred to as confluency. Once confluency is reached, many cell types stop proliferating due to a phenomenon called contact inhibition. In order to ensure continued growth, the cells must be subcultured (passaged). Solid tissues can also be cut into fragments or slices, which are then cultured on appropriate support surfaces. Such slices or fragments, however, degenerate inevitably in culture and can therefore only be used for a few hours. Cell types derived from blood or bone marrow are grown as suspension cultures, *i.e.* the cells maintain a spherical shape, float in the culture medium and do not adhere to the vessel surface.

When using *in vitro* systems for toxicological investigations, they are treated with the test compound and then they are assessed for compound-related toxicity. The endpoints used to measure toxicity *in vitro* can be broadly classified into cytotoxicity endpoints (non-specific markers) and functional endpoints.[13] Table 3 gives some examples of the type of toxicity markers that can be used *in vitro*.

**Table 2** *Difference between adherent and suspension cells*

| | *Cell type* | |
|---|---|---|
| | *Adherent cells* | *Suspension cells* |
| Definition | Derived from solid tissue | Derived from 'liquid' tissue |
| Disadvantages | Need to be trypsinised when subcultured | Can not be easily analysed microscopically, cells have to be centrifuged for medium changes |
| Advantages | Can be easily analysed microscopically, changes of medium and washes are easy to carry out | Can easily be 'subcultured' |

**Table 3** *Examples of toxicity markers available for studying toxicity* in vitro

| *Cytotoxicity markers* | | *Functional markers* | |
|---|---|---|---|
| Morphological measurements<br>Covalent binding to macromolecules<br>Lipid peroxidation<br>Levels of glutathione | • Cell necrosis, apoptosis | Hepatocytes | • Plasma protein production<br>• Triglyceride synthesis<br>• Urea synthesis |
| Plasma membrane integrity | • Dye uptake (*e.g.* Trypan blue)<br>• Ion leakage (*e.g.* $K^+$)<br>• Enzyme leakage (*e.g.* LDH, ALT, $\alpha$-GST) | Kidney proximal tubule cells | • Basolateral (PAH) transport<br>• cAMP response |
| | | Bone marrow cells | • Granulocyte/macrophage development |
| | | Adrenocortical cell function | • Cortisol production |
| Lysosomal integrity | • Neutral red uptake<br>• Levels of acid phosphatase | Cardiac myocyte function | • Cell contractility |
| | | Astrocytes | • Levels of GFAP |
| Mitochondrial activity/general redox balance of the cell | • Levels of ATP/nucleotide ratios<br>• Oxygen consumption<br>• MTT reduction<br>• Alamar blue reduction | | |
| Cell proliferation | • Cell growth<br>• $^3$H Thymidine incorporation | | |

LDH – lactate dehydrogenase; ALT – alanine aminotransferase; $\alpha$-GST – alpha glutathione-*S*-transferase; ATP – adenosine triphosphate; MTT – 3-(4,5-dimethylthiazol-2-yl)-2,5-diphenyltetrazolium bromide; PAH – *para*-amino hippurate; cAMP – adenosine 3′,5′-cyclic monophosphate; GFAP – glial fibrillary acidic protein.

Cytotoxicity endpoints quantitate dead cells or monitor biochemical changes common to all cells that are assumed to lead to or result from cell death. A wide variety of morphological and biochemical markers is available for measuring cytotoxicity.[14] Cytotoxicity in itself is not a useful or predictive parameter, as every compound will be cytotoxic if a sufficiently high test concentration is used. Results from cytotoxicity tests should be related to other parameters measured, such as the pharmacologically active concentration. Alternatively, specific cytotoxicity assays can be validated to predict another toxicologically relevant effect. It is also not possible to rank compounds across all mechanisms of toxicity with one cytotoxicity assay as different mechanisms of toxicity induce cell death *via* different pathways.

Functional endpoints look at impairment of a specific function in a particular tissue, for example albumin secretion in hepatocytes, transport functions in kidney cells or the ability of haemopoietic stem cells to differentiate into erythrocytes and granulocytes.

*In vitro* assays can never show all toxicities seen in the intact organism, and many toxicities are only seen when several organ systems are able to interact. However, *in vitro* models can be applied to augment compound selection along the drug discovery and development process. It is important to remember that the purpose of using such assays cannot be to predict all possible *in vivo* effects. Instead, *in vitro* assays are used to identify compounds with specific unwanted characteristics that are known to lead to failure during development.

Wherever *in vitro* assays are used, they have to be proven to be predictive for the particular purpose chosen in order to add value. It is therefore necessary that such assays are validated, both technically and scientifically. Technical validation addresses the question of whether the assay is reproducible, robust and reliable. Scientific validation needs to prove that the assay produces results that are relevant and meaningful for the chosen purpose. This is usually done with a set of reference compounds for which the *in vivo* effects are known. For example, the *in vitro* assay needs to eliminate compounds with a definite toxicological liability, but a limited number of false negatives (those compounds that appear clean *in vitro* but subsequently prove to be toxic *in vivo*) could be acceptable, provided they can be identified in early studies. However, false positives (those that are positive in the *in vitro* assay but are clean *in vivo*) would be less acceptable.

Validation can be a complex process when traditional *in vivo* assays used for regulatory purposes are substituted by *in vitro* assays.[15] However, for the purpose of compound selection during drug discovery, the situation is less complex.

## High Throughput Toxicity Screening

During lead series identification, often no prior knowledge of the toxicity of a compound exists. Therefore any toxicity assessment must be on a generic basis and needs to have a reasonably high throughput. The majority of high throughput toxicity screens are cytotoxicity screens which are used for the following three purposes:

- Prediction of *in vivo* toxicity
- Assessment of *in vitro* therapeutic indices
- Validation of potential drug candidates in cellular pharmacology screens.

*Cytotoxicity Screens Predictive of* In Vivo *Toxicity.* If a compound is acutely toxic to rodents at low doses, it is likely to fail early in the pre-clinical safety assessment process. Therefore, it is appropriate to try and predict, at an early stage, which compounds have the potential to be highly toxic following acute dosing in rodents *in vivo*. The purpose of such a screen would not be to predict the extent and nature of all possible effects *in vivo*, but rather to estimate the risk of failure when the compound is first administered in rodent studies. The rationale behind using cytotoxicity assays to predict *in vivo* toxicity stems from the concept of 'basal cell cytotoxicity' proposed by Ekwall.[16] It was suggested that, for most chemicals, toxicity is a consequence of non-specific alterations to cellular functions and that these may then lead to organ-specific effects. Evaluating the cytotoxic potential of compounds may, therefore, give an indication as to their toxic potential *in vivo*. Attempts to correlate acute toxicity *in vivo* with *in vitro* cytotoxicity have been made several times before with varying success.[17] Described below is an example from within Glaxo-Wellcome Research and Development where a high throughput toxicity assay capable of screening 500 compounds/week has been implemented to answer the question '*Does this compound have a high potential to be acutely toxic* in vivo?'

The validation of this high throughput cytotoxicity screen is detailed in Evans *et al.*[18] Briefly the validation involves assessing the cytotoxicity of a number of structurally unrelated compounds spanning a range of mechanisms of toxicity (*e.g.* mitochondrial poisons, alkylating agents, membrane disruptors) and varying degrees of *in vivo* toxicity. This was thought to mimic the range of compounds that would be under investigation in the finalised screen. A stable, fast growing cell line, Chinese hamster ovary (CHO) cells were treated with the test compounds and the degree of cytotoxicity measured using a fluorescent indicator of the cells' energy metabolism after a 24 h exposure period.

The experimentally derived *in vitro* $IC_{35}$ values were compared with the *in vivo* $LD_{50}$ values obtained from the RTECS (Registry of Toxic Effects of Chemical Substances) database.

By defining cut-off concentrations for both the *in vivo* and *in vitro* values it was possible to define a screen that was able to identify compounds with a high potential to be acutely toxic *in vivo* ($LD_{50}$ less than a certain value) based on the *in vitro* result ($IC_{35}$ less than a certain value) (Figure 2). An important factor to consider when defining the cut-off concentrations is the number of 'false positive' classifications. Results from these high throughput toxicity screens may influence decisions made on the progression of compounds therefore it is important to minimise the number of compounds that would be falsely classified as positive with respect to their potential to be acutely toxic *in vivo.* The number of false negative and false positive results translates into a sensitivity of 81.25% and a specificity of 100%.

It must be stressed that, as with most *in vitro* assays, the high throughput screen described above has limitations. For example, such a screen is unable to detect compounds that require metabolic activation to become

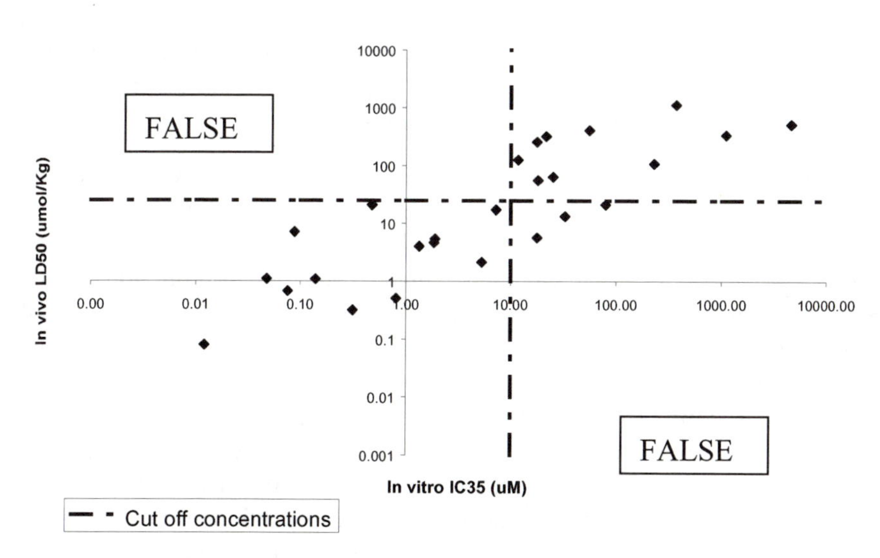


Sensitivity = Percentage of compounds correctly classified as positive = 81.25%

Specificity = Percentage of compounds correctly classified as negative = 100%

**Figure 2** In vitro–in vivo *plot of the high throughput cytotoxicity screen predictive of* in vivo *toxic potential, including cut-off concentrations and the resulting sensitivity and specificity*

toxic or are detoxified by xenobiotic metabolising enzymes. This is due to the limited xenobiotic metabolising capabilities of the cell line used, a fact which is true for most fast growing cell lines. This high throughput cytotoxicity screen has been used within GlaxoWellcome for in-house decision making at an early stage of drug discovery, to help prioritise which compound series or lead compounds to progress. To date, a total of 10 403 compounds have been tested from 38 different projects (1–1465 compounds per project). The rate of screen positives varied across projects from none (10 projects), <10% (20 projects), 10–20% (4 projects) to 20–37% (4 projects). There was no consistent trend between the number of compounds tested per project and the screen positive rate.

It is important to remember that this screen is not designed for regulatory acceptance and the results should be used in conjunction with an assessment of the other properties of a molecule, *e.g.* lipophilicity, solubility, absorption, metabolism and genetic toxicicity.

*Cytotoxicity Screens Used to Assess In Vitro Therapeutic Indices.* One of the beneficial uses of a cytotoxicity screen in the early stages of the drug discovery process is to assess the *in vitro* therapeutic index of a compound, *i.e.* the difference between the response of a compound in a pharmacological screen and its cytotoxicity. The aim would be to select compounds with the largest difference between the pharmacologically active and toxic concentration. For this, it is essential to perform both the pharmacological and cytotoxicity assays in the same cellular system because different cell types have different sensitivities to compounds.

*Cytotoxicity Screens Used to Validate Potential Drug Candidates in Cellular Pharmacology Screens.* During the drug discovery process compounds are tested in cellular pharmacology screens containing the target. A potential drug candidate may result in a reduction in the measured response due to its reaction with the target, *e.g.* if a pharmacology screen consists of cells that only emit a signal when the compound has not bound to the target. This decreased response may be due to a true effect (*i.e.* a compound that has bound to the target) or a decrease in cells caused by the cytotoxicity of the compound. A cytotoxicity screen can, therefore, be employed to establish the nature of the decreased response thereby confirming validity of the potential drug candidate. Once again it is essential to perform both the pharmacological and cytotoxicity screens in the same cellular system because different cells have different sensitivities to compounds.

It must be stressed that the uses of cytotoxicity screens described in this section (*i.e.* to predict the *in vivo* toxic potential of a compound, to assess the *in vitro* therapeutic index and to validate potential drug candidates in cellular pharmacology screens) are complementary. The results derived should always be interpreted in relation to the original purpose for which the screens were used and the limitations of the screens taken into consideration.

## *In Vitro* Screens for Specific Toxicities

High throughput (testing more than 100 compounds per week) *in vitro* toxicity screens can be used for screening compounds for non-specific toxicity. *In vitro* models can also be used to screen for particular types of toxicity. Screens can be used to detect toxicity towards a particular cellular sub-compartment (*e.g.* toxicity towards cellular DNA or mitochondria). Likewise *in vitro* models can also be used to study toxicity towards particular organs or organ systems. A number of papers have been published by ECVAM (European Centre for the Validation of Alternative Methods) which review the use of such models, *e.g.* reproductive toxicity,[19] neurotoxicity,[20] haematotoxcity,[21] respiratory toxicity,[22] nephrotoxicity,[23] skin irritation and corrosivity,[24] and hepatotoxicity.[25] Some of these *in vitro* screens (*e.g.* genotoxicity) can be applied routinely at an early stage in drug discovery as they can be run easily and require little compound. Others (*e.g.* Sertoli/germ cell co-cultures to study male fertility) are less applicable to routine screening and are used more commonly when investigating mechanisms of toxicity.

The remainder of this section will concentrate on the use of specific *in vitro* models for screening compounds in the drug discovery process.

*Genetic Toxicology.* Genetic toxicology is concerned with the effect of compounds on DNA and alterations in genetic material. When *in vitro* systems are used to study genotoxicity, tests are routinely carried out in the presence and absence of a mammalian metabolising system (S9 – a rat hepatic post-mitochondrial fraction). Studies to assess the genotoxicity of a compound can be divided into three categories:

1. Detection of mutagens (compounds which cause damage to DNA), *e.g.* microbial mutagenicity assays (including the 'Ames' test) or Mouse Lymphoma L5178Y mammalian cell gene mutation assay
2. Detection of clastogens (compounds that cause structural chromosomal damage), *e.g.* *in vitro* cytogenetic evaluation in cultured human lymphocytes[26]
3. Detection of aneugens (compounds that cause changes in chromosome numbers), *e.g.* the *in vitro* micronucleus test.[27]

As mentioned earlier, *in vivo* and *in vitro* GLP genotoxicity tests are required prior to administration of a new drug to humans. High throughput genotoxicity screens can reduce the number of compounds entering these costly, time-consuming tests by identifying genotoxic compounds early in the drug discovery process using significantly less test material. A number of advances have been made in the area of high throughput microbial mutagenicity screens. Assays such as the Ames mini well,[28] SOS/*umu* assay[29] and high throughput versions of the fluctuation test[30] have been suggested as high throughput screens to detect bacterial mutagens. As described earlier for cytotoxicity assays, these high throughput microbial mutagenicity screens must also be validated prior to implementation in drug discovery. In this case, the screens must be capable of accurately predicting the results of GLP microbial mutagenicity tests. These screens have limitations in that they are microbial systems and only detect mutagenic compounds, not clastogenic compounds. These limitations, however, can be addressed by additional screens employed later in the drug discovery process, *e.g.* Comet assay[31] or miniaturised versions of the Mouse Lymphoma L5178Y mammalian cell gene mutation assay.

*Target Organ Toxicity.* During the lead compound optimisation phase, more knowledge of the potential toxicity issues limiting the development of a class of compound is likely to become available. The toxicity information may relate to a type of lesion or specific target organ. If such prior knowledge exists, it is possible that assays for specific toxic effects can be implemented. Such assays may be used for ranking of analogues within that class of compounds, but have also been used for the investigation of species differences and have even contributed to risk assessment in later development stages.

The first step in developing tissue-specific assays is to model the specific effect *in vitro* with the compound that has previously been shown to produce a particular toxic effect *in vivo*. The next step is to prove that the effect seen in the *in vitro* model is specific to this compound/class of compound by using 'negative compounds'. Established cell lines or primary cultures can be used for these investigations. If established cell lines express the specific cellular function of interest, they can be used as the biological model. However, primary cultures are more frequently used for this purpose, more frequently as, in general, they are more likely to mimic the *in vivo* situation. The development and use of such primary culture systems is more time and resource consuming, but can give more predictive answers. The use of human-derived cells provides a tremendous opportunity. Toxicology data is routinely generated in animals *in vivo* but with the aim of predicting potential toxic effect in humans. If it is possible to model a lesion that is seen in animals *in vivo* using *in vitro* methods, and to investigate whether the same lesion is seen with human cells *in vitro*, this can add confidence to the prediction of the potential response in humans *in vivo* (Figure 3).

This approach may also provide evidence of species differences in toxicology where the toxicity seen in animals can be shown not to be relevant to humans. However, primary cells from humans are difficult to obtain for drug development purposes in many countries. As far as human established cell lines are concerned caution needs to be emphasised, as most human cell lines, similar to other animal cell lines, do not express their tissue-specific functions.

The following three examples demonstrate how target organ-specific *in vitro* models can help in compound selection and in establishing the relevance of toxicity observed in different species.

*Hepatotoxicity.* A widely quoted example for how *in vitro* methods have contributed to risk assessment is in the case of the peroxisome proliferators which are a structurally diverse group of compounds which are not uniformally genotoxic.[32] These compounds can cause both tumours and peroxisome proliferation in the livers of rats and mice. Ashby *et al.* reported that peroxisome proliferation was not observed in the livers of non-rodents, the question arose if there was a risk to humans. As an *in vitro* model, hepatocyte monolayer cultures were used from various species and the number of peroxisomes and levels of peroxisomal fatty acid $\beta$-oxidation were measured following treatment with peroxisome proliferators *in vitro*. *In vitro* peroxisome proliferation was induced in rat and mouse hepatocytes (as it was *in vivo* in these species) but not in non-rodent hepatocytes, thus providing an excellent correlation between the *in vivo* and *in vitro* findings. Using human hepatocyte monolayer cultures, no peroxisome proliferation was seen. This data formed part of a package of

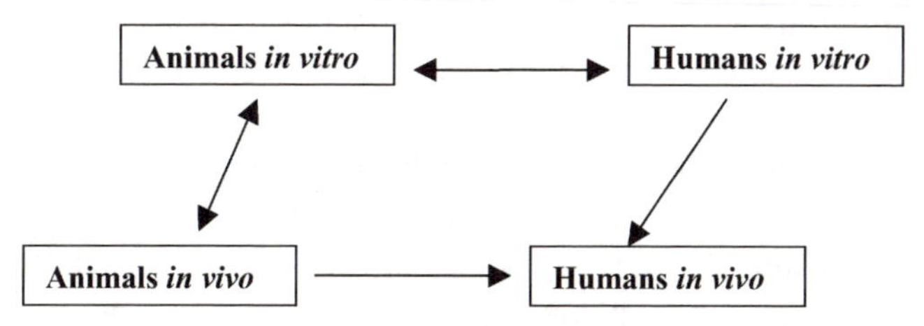


**Figure 3** *Schematic representation showing how the use of human cells can improve the predictive value of toxicology studies. If a lesion that is seen in animals* in vivo *can be modelled using* in vitro *methods, it can then be investigated whether the same lesion is seen with human cells* in vitro. *The results can then add confidence to the prediction of the potential response in humans* in vivo

work that led to the conclusion that the relevance to these compounds' hepato-carcinogenic effects for humans is considered to be negligible.[32]

Another example of the successful use of *in vitro* models is the assessment of hepatotoxicity associated with fibrinogen receptor antagonists.[32] During pre-clinical safety studies, several GlaxoWellcome fibrinogen receptor antagonists were found to cause centrilobular hepatic damage including the induction of mega-mitochondria. When primary hepatocyte cultures were treated *in vitro* with the same compounds, irreversible, concentration-related mitochondrial changes were also observed. A good correlation between *in vitro* and *in vivo* results was demonstrated by extending the treatment time with low concentrations *in vitro* which resulted in $IC_{50}$ values close to expected plasma levels which would cause hepatotoxicity *in vivo*. Changes were seen *in vitro* only after 2 days of treatment or longer, which emphasises the need to extend the treatment period in some *in vitro* investigations beyond the more traditionally recommended 24 h. The work shows that treatment of primary hepatocyte cultures may offer a suitable *in vitro* pre-screen for hepatotoxic fibrinogen receptor antagonists.[33]

*Nephrotoxicity.* It is well known that certain classes of antibiotics, some immuno-suppressants, analgesics and cancer therapeutics are nephrotoxic.[34] A large number of *in vitro* models exist to study kidney function, such as isolated proximal tubule fragments and cells, either in suspension or monolayer culture.[23] These models have successfully been used to rank series of compounds for their nephrotoxic potential.[35] In addition, *in vitro* models have been used with angiotensin-converting enzyme inhibitors to carry out mechanistic studies to identify which structural features were responsible for the observed toxicity, which gave the basis for the development of less nephrotoxic drug candidates.[35]

*Haematotoxicity. In vitro* tests have also been used extensively in the past to investigate bone marrow toxicity.[36] Haematotoxicity is a side effect of certain antivirals and most cancer therapeutics, and it is helpful early in the process to be able to rank compounds according to their haematotoxic potential. As an *in vitro* model for this, primary bone marrow from experimental species and humans can be used. For *in vitro* studies, the bone marrow is usually enriched in stem cells, and then the maturation of such stem cells into the various bone marrow lineages, such as granulocytes, erythrocytes, macrophages and megakaryocytes, is measured *in vitro*. Such approaches have been used in the past to rank compounds according to their therapeutic index, *i.e.* the ratio between the concentration that causes toxicity and the one at which efficacy is seen.[37] In other cases *in vitro* bone marrow cells have been used to determine which test species resembles most closely the response of human haematopoietic cells to treatment with a new compound. Furthermore, *in vitro* bone marrow tests have also been used as a tool in the planning of clinical trials.[37]

Whilst the above examples highlight occasions when *in vitro* models of target organ toxicity have aided in compound discovery and development, there are also examples where *in vitro* models for target organ toxicity have not been found to be predicitve. In an attempt to model the widely reported toxic effects of inhibitors of acyl-CoA cholesterol acyltransferase (ACAT) in adreanal glands experimental animals,[37] an *in vitro* screen was evaluated using two different cell types, a normal bovine adrenal cell line and primary cultures of adrenocortical cells. Cells were treated for 24 h with reference ACAT inhibitors (some of which caused adrenotoxicity *in vivo* and others which did not). Toxicity was assessed *in vitro* by cortisol production, morphology and reduction of MTT. Compound-related toxicity was apparent in all cell types and this did not correlate with *in vivo* data. The *in vitro* approaches used did not discriminate adequately between adrenotoxic and non-adrenotoxic ACAT inhibitors. Therefore, whilst adrenal cultures may be useful for studying the mechanism of ACAT inhibitor-induced adrenocortical toxicity an early *in vitro* screen to distinguish between adrenotoxic and non-adrenotoxic ACAT inhibitors was not adopted in this case.[39]

## 5 *IN VIVO* TOXICOLOGY IN CANDIDATE SELECTION

Whilst the *in vitro* approaches to toxicity screening described above can be used to screen out potentially toxic drug candidates, ultimately the compounds must be tested in the whole animal. Such regulatory toxicology tests in animals that are required prior to administration of a new drug to man are very resource-, animal- and cost-intensive. It is important that the drug candidates that enter this phase of development have the highest possible chance of progressing through the development programme.

*In vivo* studies to assess pharmacodynamics and pharmacokinetics are generally carried out on most drug candidates prior to the commencement of toxicology studies. Whilst the doses used in such studies are generally considerably lower than those which must be used in regulatory toxicology studies, it can often be very valuable to sample tissues from these animals for histological examination. Such additional work at an early stage of a project can sometimes highlight potential target organs which may be affected by a class of compound, thereby making subsequent design of toxicology studies, choice of animal species and dosages easier and less likely to need to be repeated.

During regulatory toxicology studies the toxicity of the drug is determined at high doses (see Section 1). It is also important to determine a 'no-effect' dose for toxicity in animal species. By demonstrating that toxic effects in animals occur only at high doses, but not at lower (more clinically relevant) doses, a margin of safety for the toxicity can be established. This is important when conducting subsequent risk assessments for potential clinical use of the drug. The regulatory tests that use the greatest numbers of animals and the largest amount of test compound are generally the repeat dose toxicity tests. To gain more information on the toxicology of new drug candidates in the whole animal prior to these regulatory repeat-dose studies, it is possible to design small experiments that can highlight unwanted toxicity. *In vivo* screens for repeat dose studies are therefore very beneficial. However, it must be accepted that any *in vivo* toxicity screening carried out during candidate selection is unlikely to predict all toxicity that will eventually be seen in the definitive regulatory studies.

Any *in vivo* screens to detect repeat-dose toxicity prior to regulatory studies should use the minimum number of animals and the minimum amount of drug substance, yet add information which can be used during candidate selection. The use of a single sex (*e.g.* male), single species (*e.g.* rat) and single route of administration (*e.g.* the proposed clinical route) clearly reduces the resource needed to identify toxicity with a particular compound. Likewise, the compound requirement can be minimised for these studies, by testing doses which equate to multiples of the proposed clinical dose (*e.g.* 30 and 100 times the proposed clinical dose) rather than using the maximum tolerated dose in the animal. In this way, some information on possible margins of safety and dose relationships for any toxicity observed can be obtained. The duration of screening studies to detect repeat-dose toxicity is also a compromise between the duration of a regulatory test (*e.g.* 1 month) which would involve the synthesis of a large amount of test compound and a sufficient duration of dosing to give meaningful information for candidate selection (*e.g.* animals might be examined following treatment for 2 to 7 days).

If carefully designed, these *in vivo* screens can provide an immense amount of information on a test compound from only a few animals. Early screens such as these will probably be the first occasion on which supra-therapeutic doses of a compound will be administered to animals and provide a good opportunity to collect some data on plasma levels of the drug. This will not provide definitive toxicokinetic data, which will be gathered later in the development process, *e.g.* maximum plasma concentration ($C_{max}$), area under the curve ($AUC$) and half-life ($t_{1/2}$). It will, however, give an early indication if it is going to be difficult to achieve high systemic exposure to the test compound in subsequent studies.

Many of the parameters measured in standard regulatory repeat-dose toxicology tests can be incorporated into *in vivo* toxicity screens such as monitoring for clinical signs of toxicity, ophthalmoscopy, electrocardiography, body weight measurement and food consumption. During routine repeat-dose studies clinical chemistry is performed on plasma or serum samples from the animals to identify any biochemical changes which may indicate drug-induced toxicity. Clinical chemistry may also be performed in screening studies. Such measurements can include activities of enzymes that may indicate damage to specific organs such as the liver (*e.g.* alanine aminotransferase) or muscle (*e.g.* creatine kinase). Damage to the kidney can often be detected as changes in creatinine and urea clearance as well as the salt balance of the plasma (*e.g.* sodium, potassium and chloride levels). Likewise, blood and bone marrow samples can be taken for haematological analysis which will provide information on whether exposure to the drug has altered the cellular components of the blood (*e.g.* erythrocyte counts, reticulocyte counts, haemoglobin content *etc.*). Some of the most important data from routine repeat-dose toxicology studies comes from macroscopic and microscopic examination of tissues at the end of the study. A full tissue list can be sampled from each animal (*e.g.* Table 4) and examined microscopically for any compound-induced changes.

**Table 4** *Organs that can be examined for drug-induced histological changes following* in vivo *toxicity screening studies*

| | | |
|---|---|---|
| Adrenals | Jejunum | Skin with mammary glands |
| Aorta | Kidneys | Spinal column with cord |
| Brain | Larynx and oropharynx | Spleen |
| Caecum | Liver | Sternum with marrow |
| Cervical lymph node | Lungs | Stifle joint/femur/bone marrow |
| Colon | Mesenteric lymph node | Stomach |
| Duodenum | Oesophagus | Testes |
| Epididymides | Ovaries | Thymus or thymic area |
| Eyes and optic nerves | Pancreas | Thyroids and parathyroids |
| Harderian glands | Peripheral nerve | Tongue |
| Middle/inner ear | Pituitary | Trachea |
| Nasal chambers | Prostate | Tracheal bifurcation |
| Nasopharynx | Rectum | Tracheobronchial lymph node |
| Zymbals gland | Salivary glands | Urinary bladder |
| Heart | Seminal vesicles | Uterus |
| Ileum | Skeletal muscle | Vagina |

An important feature of these *in vivo* toxicity screens is that the study design must be flexible to allow the maximum information to be obtained depending on the prior knowledge about the compound class. In addition, new technologies that may not be fully validated for use in regulatory studies (*e.g.* NMR analysis of urine – see section on page 178) can also be used in such investigative work.

## 6 THE USE OF NEW TECHNOLOGIES IN SAFETY ASSESSMENT

Over the last 10 years tremendous advances have been made in genomic, high throughput, robotic, biochemical, computing and bioinformatics techniques. The integration of these individual techniques has led to a number of new technologies that could be of great value in safety assessment in the future. It is now possible to monitor the effects of drugs on gene, protein and metabolite expression in biological systems. If specific gene, protein and metabolite expression changes can be mechanistically or correlatively linked with toxic effects, then new screens could be developed to identify toxic effects more quickly, easily and possibly more specifically.

### Toxicogenomics

The discipline of analysing gene expression changes is usually called genomics, or, in relation to safety assessment, toxicogenomics.[40] This term is slightly misleading, as in fact it is not the genome that is analysed, but the abundance of messenger RNA (mRNA) in biological samples. A number of different genomic techniques are available, which differ in the number of genes and in number of samples that can be assessed simultaneously. The most commonly used techniques use cDNA (DNA segments complementary to mRNA) or oligonucleotides (short nucleotide strands that are synthesised from DNA sequence information). These cDNA or oligonucleotides are then bound to either nylon or glass surfaces, to form so-called arrays (oligonucleotide glass arrays are frequently referred to as 'gene chips'). Then, mRNA (labelled either with radioactivity or fluorescent markers) from the sample of interest (*e.g.* liver from an animal treated with the test compound) is hybridised against the array (Figure. 4). The degree of hybridisation per gene on the array can be monitored and is proportional to the expression of that gene, *i.e.* the number of mRNA copies for that gene in the tissue under investigation. When comparing treated samples with control samples, it can be estab-

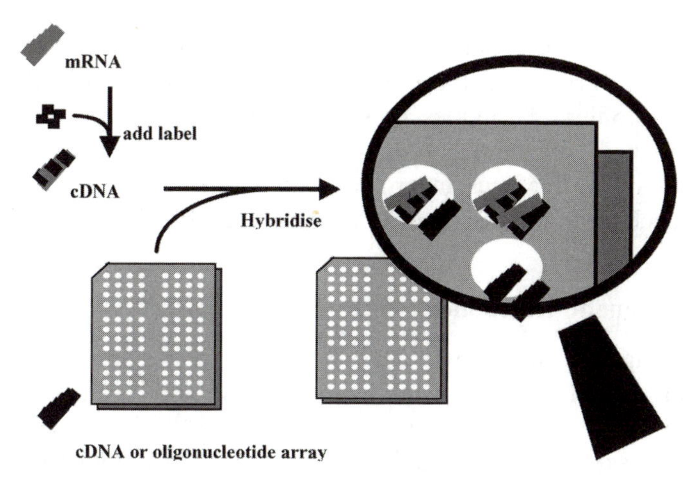


**Figure 4** *Principle behind toxicogenomics methods*

lished if the expression of a gene is increased or decreased by treatment with a test compound.

## Proteomics

Proteomics is a discipline which aims to assess systematically the total protein profile in a biological sample such as urine, serum or tissue. The major techniques used in the field of proteomics is two-dimensional polyacrylamide gel electrophoresis to separate the proteins and mass spectrometry in conjunction with databases of known proteins and genetic information[41] to identify any proteins of interest. These techniques allow the separation of hundreds of proteins simultaneously and give information on the abundance, but not the activity, of the proteins. Post-translational modification of proteins (*e.g.* phosphorylation, glycosylation) plays an important role in the functioning of biological systems and has been associated with certain disease states.[42] It is possible to identify these modifications by the position of the protein spots on the two-dimensional electrophoresis gel and the subsequent mass spectrometry analysis. Less comprehensive and more selective protein profiles can be obtained with 'protein chips', which selectively bind certain proteins from a mixture, based on chemical or biological surface characteristics.[43] Biological samples are applied onto these 'protein chips', the unbound material washed off and the bound material assessed by mass spectrometry. Proteomics has great potential in the field of toxicology where the protein expression profiles of samples of serum, urine, tissue from animals/cell cultures treated with a test compounds can be compared to control samples.[44] Any differences in protein expression may lead to an insight into the mechanism of toxicity of the compound administered.

## Nuclear Magnetic Resonance

Nuclear Magnetic Resonance ($^1$H NMR) is a technique which can identify any small molecule which contains a proton, even in a complex fluid such as plasma, urine or culture medium.[45] The use of biofluid NMR in the assessment of toxicity is based on the assumption that many toxic effects will lead to changes in the proportions of endogenous metabolites in biofluids and tissues, as a result of perturbed metabolic pathways in the affected individual or cell culture. Urinary biofluid NMR is a powerful tool for the assessment of toxicity, as all small molecules containing protons will be 'visible' to NMR, and no prior knowledge of the identity of metabolites is required.[46] As metabolites are accumulated in urine a continuous 'picture' of metabolic changes can be obtained, non-invasively, from the same individual or culture. Recovery from a toxic insult can also be monitored. The analysis of tissue samples by NMR has recently become possible through the use of Magic Angle Spin (MAS) $^1$H NMR.[47]

In all of the above technologies, the analysis of each sample can lead to thousands of data points, and the interpretation of such data is often difficult. The integration of complex data uses bioinformatics tools in order to make the information digestible to the scientist. Cluster analysis, principle component analysis, pattern recognition and multivariate data analysis can be employed to identify the important differences in data which result in some data being grouped together while other data may not.[48]

These technologies have developed rapidly in the last few years. However, due to the cost and resources involved in the use of these technologies, they are not used for regular high or even medium throughput screening. Moreover, due to the uncertainties regarding the interpretation of the complex results, these technologies are rarely used in regulatory toxicology. They can, however, be used to identify potential new markers of toxicity. Promising results have been published on the identification of disease or injury-specific patterns of gene pr protein markers[49–51] and of urinary metabolite markers for specific toxicities.[52] Once markers have been identified and validated, alternative methods of measurement, *i.e.* real-time PCR for gene, ELISAs for protein and HPLC/LC-MS for metabolite markers, could be used in routine screening. The other important strength of these new technologies is that they can give insight into mechanisms of toxicity and the pathways involved in the manifestation of toxicities, and can therefore be used for problem solving.

## 7 SUMMARY

In order to fully characterise a new drug's potential toxicity prior to its first administration to humans, resource-intensive GLP regulatory studies are required that necessitate large quantities of test compound and large numbers of animals. To improve the efficiency of drug discovery and development it is essential that only drug candidates with a high chance of success are progressed to these regulatory toxicology studies. If adequately validated *in silico*, *in vitro* and *in vivo* toxicity screens, such as those described above, can be incorporated throughout the drug discovery process, it is possible to identify and eliminate potentially toxic molecules using small quantities of test compound at an early stage in drug discovery. The use of new technologies in the field of toxicology offers the opportunity to make the screening process even more efficient in the future.

## 8 REFERENCES

1. Council Directive (EEC) 65/65, *Official J. Eur. Communities Engl. Special Edn.*, 1965–1966, p. 20.
2. D. Tweats in *Progress in the Reduction, Refinement and Replacement of Animal Experimentation*, M. Balls *et al.* (eds.), Elsevier, Amsterdam, 2000, p. 783.
3. http://www.mcclurenet.com/FedRegisterPDFs/S4a.
4. http://www.eudra.org/humandocs/PDFs/SWP/098696.pdf.
5. Anon, *Federal Register*, 1996, **61**, 15360.
6. Anon, in *The Good Laboratory Practice Regulations 1999*, The Stationery Office Ltd., London, 1999.
7. C. Hansch *et al.*, *Toxic Lett.*, 1995, **79**, 45.
8. J.C. Dearden *et al.*, *Alternatives Lab. Animals*, 1997, **25**, 223.
9. H.S. Rosenkranz *et al.*, *SAR QSAR Environ. Res.*, 1999, **10**, 277.
10. N. Greene *et al.*, *SAR QSAR Environ. Res.*, 1998, **10**, 299.
11. E.J. Mathews *et al.*, *J. Mol. Graph. Mod.*, 2000, **18**, 605.
12. R. Anderson, *Toxicol. In Vitro*, 1999, **13**, 729.

13. A. Guillouzo, in In Vitro *Toxicity Testing*, J.M. Frazier (ed.), Marcel Dekker Inc, New York, 1992, p. 45.
14. C.A. Tyson and C.E. Green in *The Isolated Hepatocyte: Use in Toxicology and Xenobiotic Biotransformations*, E.J. Rauckman, and G.M. Padilla (eds.), Academic Press, Orlando, Florida, 1987, 119.
15. M. Balls and W. Karcher, *Alternatives Lab. Animals*, 1995, **23**, 884.
16. B. Ekwall, *Ann. New York Acad. Sci.*, 1983, **407**, 64.
17. M.J. Garle, J.H. Fentem and J.R. Fry, *Toxicol. In Vitro*, 1994, **8**, 1303.
18. S.M. Evans *et al.*, *Toxicol. in Vitro*, 2001, **15**, 579.
19. N.A. Brown *et al.*, *Alternatives Lab. Animals*, 1995, **23**, 868.
20. C.K. Atterwill *et al.*, *Alternatives Lab. Animals*, 1994, **22**, 350.
21. L. Gribaldo *et al.*, *Alternatives Lab. Animals*, 1996, **24**, 211.
22. C.R. Lambré *et al.*, *Alternatives Lab. Animals*, 1996, **24**, 671.
23. G.M. Hawksworth *et al.*, *Alternatives Lab. Animals*, 1995, **23**, 713.
24. P.A. Botham *et al.*, *Alternatives Lab. Animals*, 1998, **26**, 195.
25. S. Coecke *et al.*, *Alternatives Lab. Animals*, 1999, **27**, 579.
26. D.J. Kirkland, *Basic Mutagenicity Tests: UKEMS Recommended Procedures*, Cambridge University Press, Cambridge, UK, 1990.
27. M. Kirsch-Volders *et al.*, *Mutat. Res.* 1997, **392**, 19.
28. D.A. Burke, D.J. Wedd and B. Burlinson, *Mutagenesis*, 1996, **11**, 201.
29. Y. Oda *et al.*, *Mutation Res.*, 1985, **147**, 219.
30. D.G. Gatehouse, *Mutation Res.*, 1978, **53**, 289.
31. R.R. Tice *et al.*, *Environ. Mol. Mutagen.*, 2000, **35**, 206.
32. J. Ashby *et al.*, *Human Experiment. Toxicol.*, 1994, **13 Suppl 2**.
33. C. Westmoreland *et al.*, *Human Experiment. Toxicol.*, 1997, **16**, 419.
34. S. Rosen, M. Brezis and I. Stillman, *Miner Electrolyte Metab.*, 1994, **20**, 174.
35. J.F. Sina and M.O. Bradley, in *In Vitro Methods of Toxicology*, R.R. Watson (ed.), CRC Press, Boca Raton, 1992, p. 81.
36. L. Garibaldo *et al.*, *Alternatives Lab. Animals*, 1996, **24**, 211.
37. R.E. Parchment *et al.*, *Toxicol. Pathol.*, 1993, **21**, 241.
38. B.D. Roth, *Drug Discovery Today*, 1998, **3**, 19.
39. G.H.I. Wolfgang *et al.*, *Toxicol. Methods*, 1994, **3**, 149.
40. W.D. Pennie, *Toxicol. Lett.*, 2000, **112–113**, 473.
41. D.F. Hochstrasser, *Clin. Chem. Lab. Med.*, 1998, **36**, 825.
42. K. Yamashita, *et al.*, *J. Biol. Chem.*, 1993, **268**, 5783.
43. G.L. Wright, Jr., *et al.*, *Prostate Cancer Prostatic Dis.*, 1999, **2**, 264.
44. S. Steiner and N.L. Anderson, *Toxicol. Lett.*, 2000, **112–113**, 467.
45. J.C. Lindon, J.K. Nicholson and J.R. Everret, *Annu. Rev. NMR Spectrosc.*, 1999, **38**, 1.
46. J.C. Lindon *et al.*, *Concepts Magn. Reson.*, 2000, **12**, 289.
47. J.K. Nicholson *et al.*, *Magn. Reson. Med.*, 1999, **41**, 1108.

48. K.P.R. Gartland *et al.*, *Mol. Pharmacol.*, 1991, **39**, 629.
49. E.F. Petricoin *et al.*, *Lancet*, 2002, **359**, 572.
50. L.J. Van't Veer *et al.*, *Nature*, 2002, **415**, 530.
51. R.S. Thomas *et al.*, *Mol. Pharmacol.*, 2001, **60**, 1189.
52. A.W. Nicholls *et al.*, *Biomarkers*, 2000, **5**, 410.

CHAPTER 9

# Chemical Development

PAUL SMITH

## 1 INTRODUCTION

The progression of new drug candidates from the point at which they are selected to the time they are marketed represents a very challenging period for the many chemists involved in 'Chemical Development'. These include specialists in Synthetic Chemistry, Analytical Chemistry, Reaction Hazard Assessment and Environmental Chemistry, who work with chemical engineers and pilot plant operators to ensure the provision of:

1. Timely supplies of drug substance (often called Active Pharmaceutical Ingredient – API) to allow safety, tolerability and efficacy studies to be carried out.
2. A robust, safe, cost effective process with minimal waste, leading to consistently high quality drug substance, which can be used to manufacture commercial supplies.

The drug development process spans many years (Figure 1) as evidence is gathered to demonstrate the safety and efficacy of a new medicine. Details of the activities at each phase of the development process are also outlined in Table 1. Of course, not all drug conditions survive this intense period of investigation, in fact of 10–15 compounds selected for development, only one is likely to make it through to become a marketed product.

The Discovery & Development exercise is therefore a very expensive activity and this is illustrated in Figure 2. Costs rapidly increase as development gathers pace and factors such as the pursuit of more difficult disease targets, the need for the use of more expensive modern technology and the pressure exerted by some Governments on drug prices all affect the return on investment. As a large part of the period of exclusivity

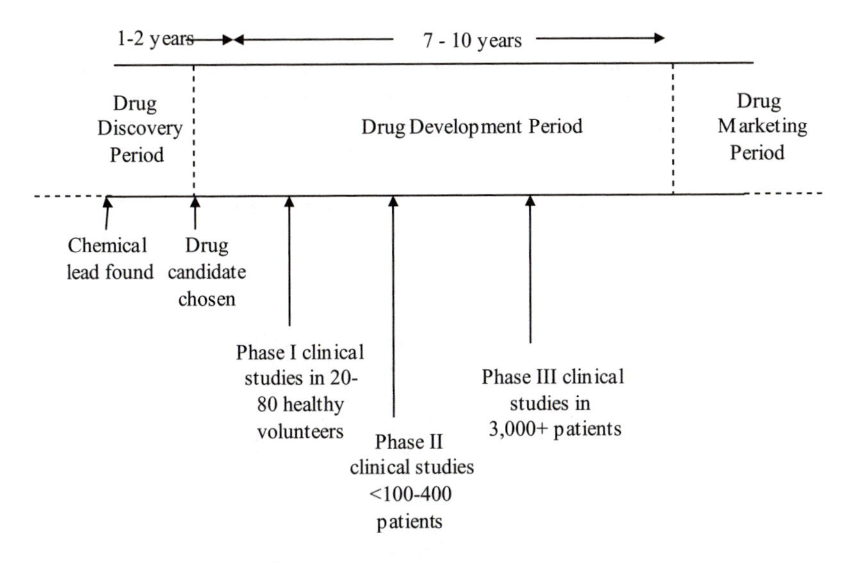


**Figure 1** *The stages of the drug development process*

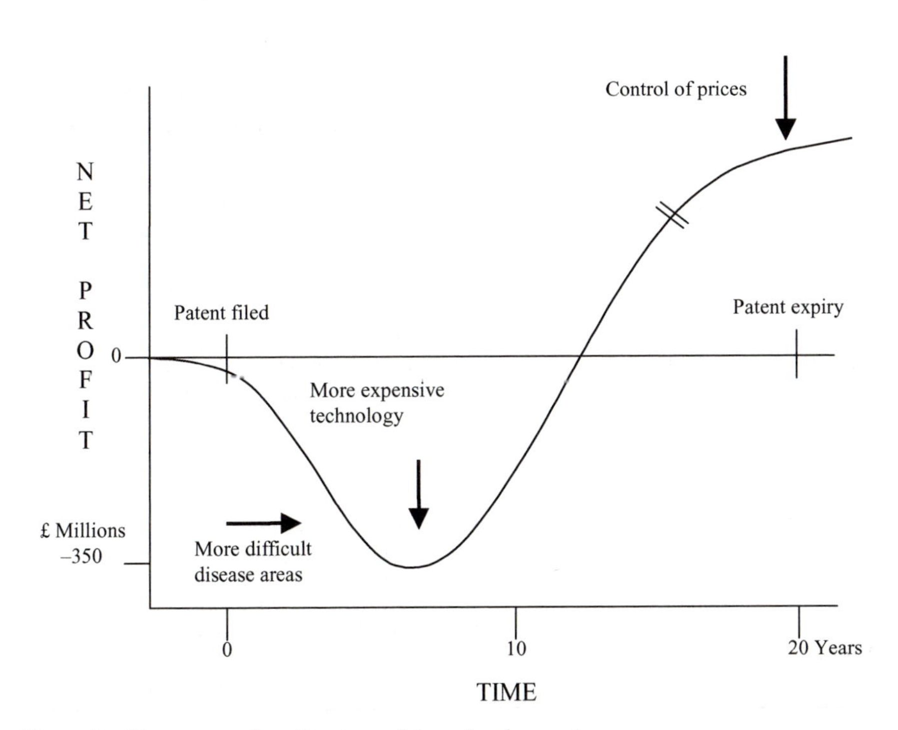


**Figure 2** *The costs and profit return of drug development*

**Table 1** *The stages of drug development for non-Chemical Development functions*

| *Stage* | *Definition* | *Major activities* |
|---|---|---|
| Pre-clinical | Up to first time in humans (FTIM) | 28 day toxicology in 2 species<br>Identify suitable formulations for FTIM<br>Prepare sufficient material to acceptable standard<br>Prepare documents for internal safety board review<br>Prepare protocols, identify centres, prepare regulatory documents (IND, CTX) to support Phase I studies |
| Phase I | ADME, safety and tolerability in humans | Single dose, dose escalation, repeat dosing in human volunteers (20–80 subjects)<br>Define drug metabolism and pharmacological activity and identify a dose range for Phase II studies<br>Identify suitable formulation for Phase II studies<br>Longer term toxicology to support Phase II<br>Prepare protocols, identify centres, prepare regulatory documents (IND, CTX) to support Phase IIA studies |
| Phase IIa | ADME, safety, tolerability and efficacy in patients | Relatively small studies (typically <100 patients) to identify side effect profile and proof of concept/mechanism<br>Initiate toxicology studies to support Phase III and registration (*e.g.* carcinogenicity studies)<br>Prepare protocols, identify centres, prepare regulatory documents (IND, CTX) to support Phase IIB studies |
| Phase IIb | Selection of dosage and regimen for Phase III studies | Dose ranging studies in sufficient number of patients (typically <100–400) to give adequate efficacy and safety data to justify the dose for Phase III<br>Identification of commercial formulation<br>Prepare regulatory documentation (IND & CTX) for Phase III studies |
| Phase IIIA | Definitive efficacy studies | Large scale studies to determine definitive efficacy and side effect profile in patients compared with placebo or current gold standard therapy<br>Prepare regulatory documents to support registration (NDA/MAA) |
| Phase IIIB | Post registration studies | Studies to support additional label claims and to support launch (*e.g.* additional Phase IIIA comparator studies) |
| Launch | Approval from regulatory authorities | Further efficacy/safety data may be required by regulatory authorities as a condition of launch |
| Phase IV | Post marketing surveillance | Further monitoring of efficacy/safety<br>Additional studies to support marketing |

provided by patent coverage has typically already been used up before a drug even reaches the market, there is a very limited amount of time available to recover costs and move into profit before the generic competitors, who have not had any major R&D costs, take away a large proportion of market share.

The key to future success is therefore to:

1. Minimise the time for drug development.
2. Reduce the rate of attrition of drug candidates, or at least identify causes of failure as early as possible.

As the timescale to market is measured in years, it is easy to forget that even a single day's delay in development can equate to the loss of £1 million of sales for a successful drug. Hence, achieving speed without compromising quality has become a key goal in all drug development activities.

In the Drug Development period, Chemical Development's activities are driven by the need to utilise the most appropriate route to a drug candidate at any given time and a typical progression is shown in Figure 3.

The challenge typically starts with the need to prepare a few hundred grams of drug substance for early safety assessment studies. A close working relationship with the Medicinal Chemists is very important to ensure that any issues associated with the 'Discovery Route' are highlighted early enough to allow them to be overcome without becoming rate limiting. In the majority of cases, the 'Discovery Route', or a modified version of it, will be used for the initial supplies and will be carried out in large laboratory scale (10–20 L) equipment.

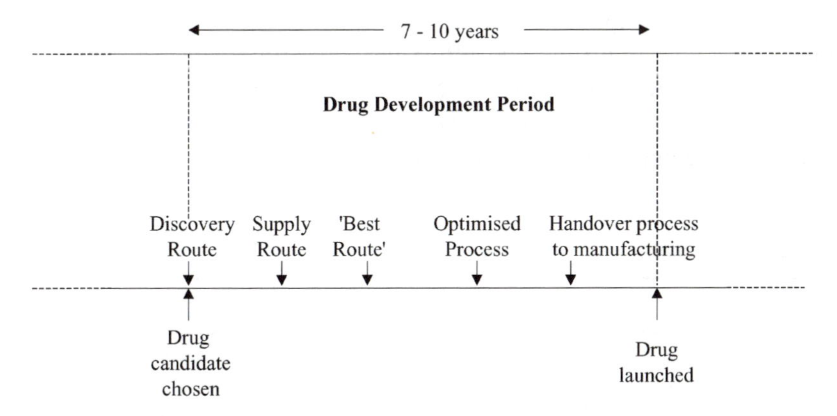


**Figure 3** *The role of development chemistry in the drug development process*

In addition to the viability of the synthetic route, it is also very important to consider the physical and chemical properties of the drug substance. Ideally a stable, water soluble, crystalline form will have been identified which can be prepared reproducibly. In practise, further work is often needed to determine whether to progress, for example, the parent compound or a pharmaceutically acceptable salt. It is also worth mentioning at this point the subject of polymorphism,[1] which is the ability of a solid compound to exist in more than one crystalline form. These forms can have different physical and/or chemical properties leading in some cases to changes in, for example, bioavailability, stability and processability. It has been known for a new polymorph to be produced a late as the manufacture of a marketed drug and the impact of that can be enormous as it is often difficult to go back to the original polymorph once a new one has been produced. Great efforts are now made to try to discover new polymorphs and control the formation of the selected one (usually the most thermodynamically stable) as early as possible.

The early laboratory work is hopefully laying the foundations for the next requirement, typically 4–5 kg for more extensive safety studies (*e.g.* 28 day repeat dose studies) and early Phase I studies in healthy human volunteers. In practice, this means that very early on, an analysis of each step of the 'Discovery Route' is carried out with respect to yield, scale-up potential (*i.e.* any factors likely to limit scale-up), isolation method and purification method. If these factors are all acceptable for scale-up (or can be quickly made acceptable) then the 'Discovery Route' can become the 'Supply Route' and be used to prepare supplies in the pilot plant. However, if there are major issues which cannot be overcome, *e.g.* unavailability of starting materials or uncontrollable reaction exotherms, then a new 'Supply Route' will be urgently needed.

In addition to the possibility of needing a new route for early supplies there is always a need to identify a commercially viable route. The rationale for investing time and effort into new route work is exemplified in Table 2. Thus for a 10 tonne year$^{-1}$ product, the optimised, best synthesis will provide a saving over a product lifetime of 10 years of £230M when compared with an original synthesis.

**Table 2** *Cost Saving: a typical example: £ kg$^{-1}$*

| | *Start* | *Optimised* |
|---|---|---|
| Original Synthesis | 5000 | 3000 |
| New Synthesis | 1800 | 1100 |
| Best Synthesis | 1300 | 700 |

A great challenge for Development Chemists is to discover and utilise the 'best route' as early as possible, so that all the knowledge and understanding of the chemistry that is built up as supplies are prepared will be useful at a later date. Where this does not happen, it is usually better to remain with a single 'Supply Route' rather than introduce several new routes into the supply chain. The key target is to select and demonstrate the 'best route' in time to be able to use the drug substance in definitive safety studies, such as 12 month repeat dose studies, and the carcinogenicity studies (this provides safety cover for any new impurities generated in the new process without the need for any additional toxicity testing).

In the early stages of development, the cost of preparing relatively small amounts of material, which are on the project's critical path, is typically not as important as the time it takes to prepare the material. Obviously, as soon as a future commercial process is being considered then the cost of goods becomes a major factor, but not the only one. Where there is more than one viable route to a drug candidate, an important choice has to be made as to which one to progress.[2] Some of the key factors are listed below:

- Raw material cost
- Labour cost
- Energy cost
- Raw material availability
- Robustness of the process
- Reproducibility of the process
- Capital investment required
- Safety of the process
- Environmental impact of the process
- Quality of the product
- Occupational risks
- Potential for further improvements to the process
- Patent situation

As no two processes will be at exactly the same level of development (*e.g.* a Supply Route may have been run in the plant on several occasions whereas a newly discovered route may have only been run in 20 L apparatus), sound judgement is required to determine whether sufficient information is available to make a sound choice, or whether more data needs to be collected.

Once a route has been selected as the basis for commercial production, the next phase of development can take place; namely, optimisation and determining boundary conditions. All of this work culminates in the preparation of 'Qualification Batches', which are run to demonstrate the reproducibility of the process and these are followed by the preparation of 'Validation Batches' which are run at the site of manufacture to demonstrate that the process is transferable to production equipment.

In order to function effectively, Chemical Development interacts with many other expert groups and Figure 4 outlines some of the principal ones. These interactions are explained in more detail below:

**Medicinal Chemistry:** Chemical Development obviously need to receive detailed information on the methods that were used to prepare initial quantities of drug substance. However, by working closely with the Medicinal Chemists, several other benefits are achieved, *e.g.* Development Chemists can provide early information on whether there are any potential development issues (such as cost of goods) with lead compounds. They can also provide feedback on improved synthetic methodologies to the Medicinal Chemists.

**Intellectual Property:** Without appropriate patent protection,[3] it is not commercially viable to develop new medicines. The most important patent is the one which claims the compound and its use, and this is very much the domain of Medicinal Chemistry. However, there are opportunities for Chemical Development to improve the patent protection of drug substances through their own discoveries, *e.g.* new processes, new solid state forms. It is also important to ensure that any new process under development is not going to infringe any existing patents, *e.g.* using an intermediate which is already covered in another company's patent.

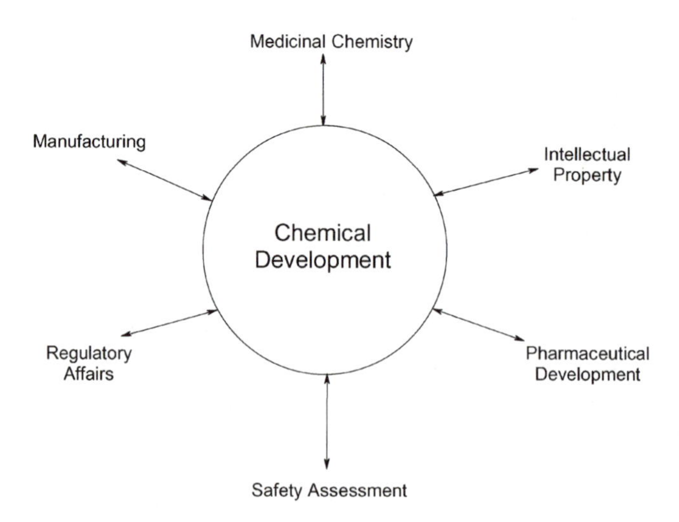


**Figure 4** *Chemical Development interactions*

**Pharmaceutical Development:** This group are responsible for converting the drug substance into a formulated product for clinical studies and are, therefore, a major customer for the supplies of drug substance prepared by Chemical Development. The provision of a drug substance that has good stability, good aqueous solubility and is of consistent quality (*e.g.* purity, solid state form, particle size) is highly desirable and it is important to work closely with Pharmaceutical Development as there is a large area of overlapping interest, *e.g.* Chemical Development can provide guidance on likely degradation products and likely interactions with excipients *etc.* Also, when as is inevitably the case, changes do occur in the physical parameters of the drug substance, *e.g.* particle size, level of hydration *etc.* Pharmaceutical Development need to be kept informed, as in some cases these changes could have serious implications on the way the drug substance performs in the formulation process and could also impact on bioavailability.

**Safety Assessment:**[4] Before any drug can be tested in humans it is a legal requirement that a variety of safety tests are first carried out in animals. The Safety Assessment group, who carry out these vital studies are Chemical Development's other major customer for drug substance supplies. It is important to ensure that the quality of the drug substance intended for use in human clinical trials is consistent with that used in these safety tests.

**Regulatory Affairs:** In order to obtain clearance to test drug candidates in humans and eventually to obtain a licence for marketing a new drug, it is necessary to satisfy the various national regulatory agencies (*e.g.* Food & Drug Agency in the USA) that the expected standards have been met. Chemical Development provide analytical batch data and synthetic details for inclusion in these submissions. Guidelines on these standards of working are available from the regulatory authorities and for Chemical Development, the recommendations on cGMP (current Good Manufacturing Practise) are very important to apply correctly. These guidelines[5] are aimed at helping to ensure that consistently high quality drug substance/product will be produced every time the process is run.

**Manufacturing:** In the very early phase of development there is little contact required with the manufacturing group (often referred to as Primary Manufacturing, as opposed to Secondary Manufacturing, who prepare the formulated product). However, a good knowledge of the manufacturing capabilities and limitations is useful at the stage of final route selection. Close contact is also needed to allow timely planning of future sourcing decisions and capital investments. Once a new synthetic route is selected for progression to become the basis of the commercial process, then discussions with manufacturing become important to ensure that any process improvements really are appropriate for the manufacturing environment.

## 2 ILLUSTRATIVE EXAMPLES

### The Importance of Quality

The pharmaceutical industry produces a product for which the patients have no way of being able to tell what the quality is like. It is, therefore, the responsibility of the industry to make sure that a quality product is delivered every time. A striking example of what can go wrong if proper controls are lacking occurred in the early 1980s when the illicit preparation of a new 'synthetic heroin', MPPP, resulted in several people being admitted to hospital with symptoms consistent with Parkinson's disease.[6] The problem was eventually traced to the presence of an impurity, MPTP, which is produced in large quantities if the reaction conditions are too vigorous (Scheme 1).

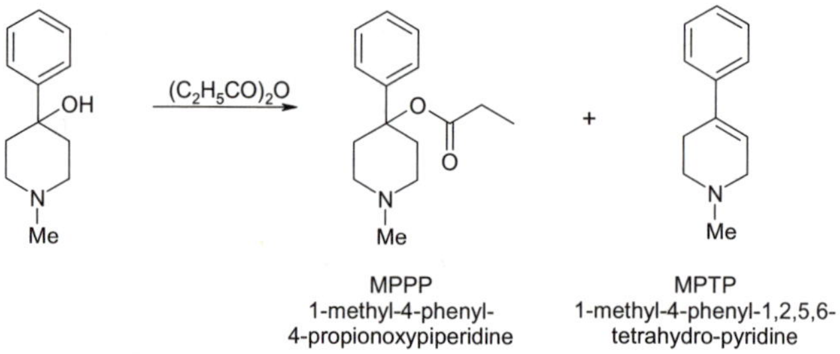
OH
N
Me
(C2H5CO)2O
O
O
N
Me
+
N
Me
MPPP
1-methyl-4-phenyl-
4-propionoxypiperidine
MPTP
1-methyl-4-phenyl-1,2,5,6-
tetrahydro-pyridine

**Scheme 1** *Preparation of MPPP*

### 3,3-Dimethylindoline. An Example of Route Discovery and Development

It is not appropriate to go into great detail in this chapter, but the example outlined below should provide a flavour of Route Discovery and Development.[7] 3,3-Dimethylindoline **2** was required for use in the preparation of a drug candidate. The Discovery Route[8] is shown in Scheme 2.

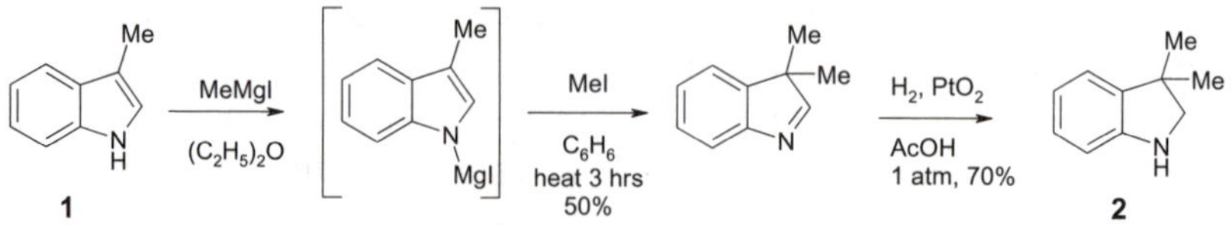
Me
N
H
MeMgI
(C2H5)2O
Me
N
MgI
MeI
C6H6
heat 3 hrs
50%
Me
Me
N
H2, PtO2
AcOH
1 atm, 70%
Me
Me
N
H

**1** **2**

**Scheme 2** *Discovery route to indoline 2*

This process was used to prepare a few grams of the indoline and was also useful in that, by substituting methyl iodide with other alkylating agents, a variety of analogues could be made prior to selection of the dimethylindoline. However, this process was not suitable for further scale up because of a number of drawbacks:

1. Diethyl ether and benzene are typically not acceptable in the pilot plant.
2. The low yield in the step 2 alkylation is due to poor selectivity between *C*- and *N*-alkylation.
3. The starting material, 3-methylindoline **1**, also known as skatole, has a very, very unpleasant smell!

A search of the literature[9] provided the basis for the 'Supply Route' shown in Scheme 3.

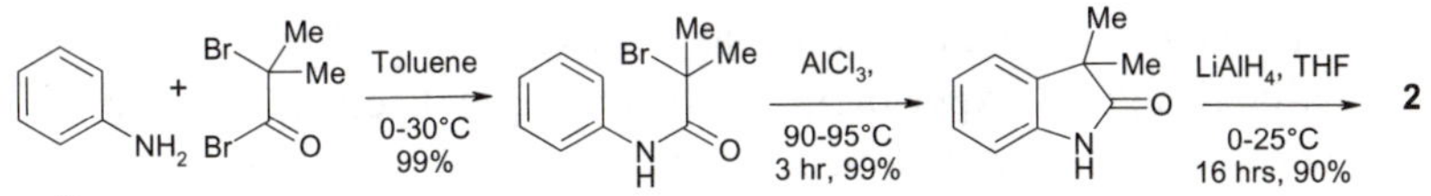

**Scheme 3** *Supply route to indoline 2*

This route was used successfully to produce tens of kgs of 3,3-dimethylindoline **2** from the cheap, readily available aniline **3**. Although the yields are very high, the process was not without problems. In particular, it would be highly desirable to avoid the reduction using lithium aluminium hydride (water sensitive reagent, releasing hydrogen) in the final step. A new route was investigated which replaced the reduction step with a simple hydrolysis and this is shown in Scheme 4.[10]

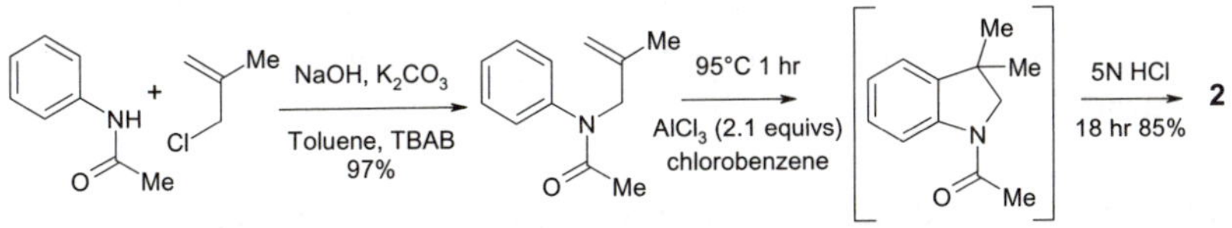

**Scheme 4** *Route of manufacture of indoline 2*

This route was chosen as the basis for the 'Route of Manufacture' and had been successfully run on tens of kgs. The project was terminated at this point and no further improvements, such as avoiding the use of the Friedel–Crafts 'catalyst' $AlCl_3$, were achieved.

## 3 FUTURE TRENDS

The commercial need to speed up the discovery and development of safe, efficient processes to produce high quality drug substances is providing an important driver for changing the way Chemical Development operates. Several initiatives are being pursued and some of the key ones are highlighted below:

**Automation:** has already had a major impact in the Discovery part of the Pharmaceutical Industry with the introduction of high throughput screening and automated parallel synthesis. It is now increasingly being introduced into the Development phase[11] and there are now several commercially available automated reactors (capable of carrying out up to 20 reactions at once) which allow rapid multiple screening, of for example, solvents, reaction conditions, catalysts *etc.*

The introduction of automation has also encouraged the greater use of 'statistically designed experiments' (for example Factorial designs),[12] for which there are some excellent, easy to use computer software packages available. Factorial designs alter several parameters simultaneously for each experiment, in contrast to the traditional approach of altering only one parameter at a time. The advantages of this approach are:

1. Better estimates of the effects of parameters (one level of a variable is tested several times).
2. Rapid identification of those parameters which have a big impact on the desired outcome of the experiments.
3. Identification of any important interactions between parameters (*i.e.* where the effect of one parameter is dependent on one or more other parameters). These are often not readily identified when varying one parameter at a time.
4. Greater confidence in results (confirmation over a wide range of parameter settings and the chemist is not misled by inherent variability).

Factorial and related designs are now increasingly being used to optimise reaction conditions (perhaps balancing yield *versus* certain critical impurities) and to assess boundary conditions and robustness.

There is also an increasing need to ensure that the amount of waste generated during the production of drug substances is kept to a minimum and to ensure that the waste that is inevitably produced can be readily recycled, or, if not, can be disposed of in an environmentally acceptable way. In order to achieve 'cleaner processes' the desired products have to be produced in as few steps as possible using high yielding, highly selective, atom economical chemistry.[13] There is now much work being carried out in academia under the heading 'Green Chemistry',[14] with particular emphasis on catalytic processes which can be applied in industry.

Traditionally, the manufacture of drug substances is based on batch processes and scale-up requires bigger and more costly vessels. Increasingly efforts are focussing on 'process intensification',[15] which includes miniturisation and continuous processing. This approach is very attractive as less capital investment is required since much smaller plant is needed and there are safety advantages as control of exotherms is easier

It is certainly true that the challenges for Chemical Development now and in the future are providing a very stimulating environment for those chemists responsible for delivering the goods.

## 4 ACKNOWLEDGEMENT

I am indebted to my colleagues at GSK for their helfpul suggestions and comments during the preparation of this chapter.

## 5 REFERENCES

1. *Org. Process Res. Develop*, 2000, **4**, 370; R. Byrn *et al., Pharm. Res.*, 1995, **2**, 945.
2. C.E. Berkoff *et al., Chemtech.*, 1986, **16**, 552.
3. P.W. Grubb, *Patents for Chemicals, Pharmaceuticals and Biotechnology*, Clarendon Press, Oxford, 1999.
4. J.H. Duffer and H.G.J. Worth (eds.), *Fundamental Toxicology for Chemists*, The Royal Society of Chemistry, Cambridge, 1996.
5. The following web sites provide useful information and guidance www.ifpma.org/ich1.html; www.fda.gov/cder/index.html.
6. J.W. Langston *et al., Science*, 1983, **219**, 979.
7. Many excellent examples of route development can be found in volumes of the journal *Org. Process Res. Develop.*
8. T. Hoshino, *Liebigs Ann. Chem.*, 1933, **33**, 500; A.H. Jackson, *Tetrahedron*, 1965, **21**, 989.
9. R. Stollé, *Chem. Ber.*, 1914, **47**, 2120.
10. T.W. Ramsay *et al., Synth. Commun.* 1995, **25**, 4029.
11. M.A. Armitage *et al., Org. Process Res. Dev.*, 1999, **3**, 189. D.F. Embiabata-Smith, *et al. ibid.*, 1999, **3**, 281. K.G. Gadamasetti (ed.),

*Process Chemistry in the Pharmaceutical Industry*, Marcel Dekker, New York, 1999, Ch. 23, p. 429.
12. L. Davies, *Efficiency in Research, Development and Production: The Statistical Design and Analysis of Chemical Experiments*, The Royal Society of Chemistry, 1993; K.G. Gadamasetti (ed.), *Process Chemistry in the Pharmaceutical Industry*, Marcel Dekker, New York, 1999, Ch. 22, p 411.
13. B.M. Trost, *Angew. Chem. Int. Ed. Engl.*, 1995, **34**, 259.
14. P.T. Anastas and J.L. Warner, *Green Chemistry: Theory and Practice*, Oxford University Press, New York, 1998.
15. C Ramshaw, *Process Intensification in the Chemical Industry*, Mechanical Engineering Publications Ltd., London, 1995. S.J. Haswell *et al.*, *Chem. Commun.*, 2001, 391.

## Further General Reading

S. Lee and G. Robinson, *Process Development: Fine Chemicals from Grams to Kilograms*, Oxford University Press, New York, 1995.
N.G. Anderson, *Practical Process Research and Development*, Academic Press, San Diego, 2000.

CHAPTER 10

# Physicochemical Properties

HAN VAN DE WATERBEEMD

## 1 INTRODUCTION

A successful drug candidate has the right attributes to reach and bind to its molecular target and has the desired duration of action. Binding to the target can be optimised by designing the proper three-dimensional arrangement of functional groups. Each chemical entity also has, through its structure, physicochemical and bio-pharmaceutical properties. These are generally related to processes such as dissolution, gastrointestinal absorption, uptake into the brain, plasma protein binding, distribution and metabolism.[1] Therefore, fine-tuning of the physicochemical properties by the medicinal chemist has an important place in lead optimisation. It is also important to note that many physicochemical characteristics of a compound are strongly inter-correlated (see Figure 1) and that modifying one most likely will affect others. An analysis of the World Drug Index (WDI) revealed that compounds are likely to have poor oral absorption when their molecular weight is over 500, the number of hydrogen bond

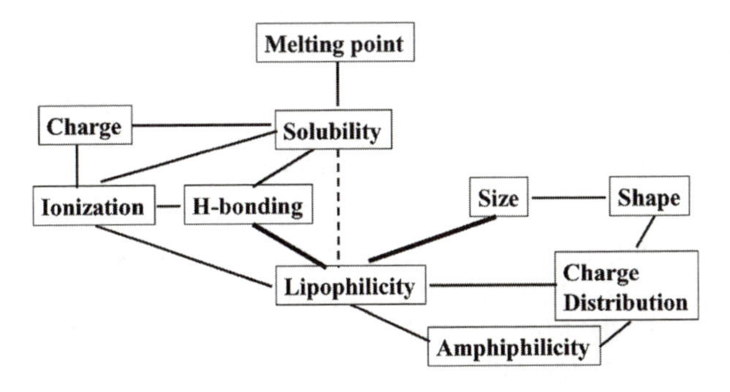


**Figure 1** *Relationships among various physicochemical properties*

donors is over 5, the number of hydrogen bond acceptors more than 10 and the calculated partition coefficient $C\log P$ more than 5.[2] This simple observation is known as the rule-of-5.

In this chapter we will discuss the key physicochemical properties of drugs, in particular with orally given, small molecule drugs in mind. For biologicals and drugs given by various other routes, other "rules" may apply. Further information can be found in reviews[3,4] and books.[5–7]

## 2 SOLUBILITY

### Dissolution and Solubility

For most drugs, the first step in the process of entering the body from a formulation is dissolution. Sufficient solubility and membrane permeability have been identified as important factors for oral absorption. For poorly soluble compounds, rate of dissolution is also an important factor. Modern formulation techniques can improve the rate of dissolution and, to a certain extent poor aqueous solubility, but the development of such compounds is likely to be long and difficult. But, what is sufficient solubility? Much is dependent on the dose. Estimates for a 1 mg kg$^{-1}$ dose give a minimum acceptable solubility of 5, 50 or 500 μg ml$^{-1}$ for low, medium and high permeability compounds, respectively.

The concept of maximal absorbable dose (MAD) is a useful concept for ranking potential drug candidates.

$$\mathrm{MAD} = (S) \times (K_a) \times (\mathrm{SIWV}) \times (\mathrm{SITT})$$

where $S$ = solubility (mg ml$^{-1}$) at pH 6.5; $K_a$ = transintestinal absorption rate constant (min$^{-1}$); SIWV = small intestinal water volume (~250 ml for man) and SITT = small intestinal transit time (~270 min for man). Using this equation, one can assess the minimum acceptable solubility (MAS) from:

$$\mathrm{MAS} = \frac{\mathrm{MAD}}{(K_a) \times (\mathrm{SIWV}) \times (\mathrm{SITT})}$$

Thus, for example if the MAD is 500 mg (typical for an antibiotic) for a 70 kg human, for a compound with high permeability ($K_a = 0.03$ min$^{-1}$) the solubility needs to be ~ 0.25 mg ml$^{-1}$, and for lower permeability ($K_a = 0.003$ min$^{-1}$), the solubility needs to be ~2.5 mg ml$^{-1}$. However, this assessment assumes no pH dependency on solubility.

The Food and Drug Administration (FDA) department in the US have recommended a classification for novel chemical entities (NCEs) based on dose strength, solubility and permeability, shown in Table 1, with comments on the consequences of drugs falling into each category. High permeability is defined as being at least 100-fold more permeable than mannitol and low permeability less than 10-fold that of mannitol. High solubility is defined as being >10 $\mu g\,ml^{-1}$, low solubility as <10 $\mu g\,ml^{-1}$ based on the lowest solubility in the physiological pH range (pH 1–8). Fast dissolution is defined as 85% of the drug dissolved in 0.1 M HCl within 15 min.

**Table 1** *FDA classification of novel chemical entities based on dose strength, solubility and permeability*

| | *NCE Classification for low dose strength* | |
|---|---|---|
| | *Low solubility* | *High solubility* |
| Low permeability | NCE1<br>Development extremely problematic<br>Probably not a drug | NCE 2<br>Absorption is dose and dissolution independent<br>Gastric residence time important<br>Formulation independent<br>Probable food effect |
| High permeability | NCE 3<br>Dissolution rate limited<br>Formulation significant – salt form and particle size important | NCE 4<br>Absorption is dose, permeability and dissolution independent<br>Formulation independent |
| | *NCE Classification for high dose strength* | |
| | *Low solubility* | *High solubility* |
| Low permeability | NCE1<br>Development extremely problematic<br>Not a drug | NCE 2<br>Absorption is dose and dissolution independent<br>Gastric residence time important<br>Formulation independent<br>Probable food effect |
| High permeability | NCE 5<br>Solubility and dissolution rate limited<br>Formulation significant – salt form and particle size and polymorphs important | NCE 6<br>Absorption is dose, permeability and dissolution independent<br>Formulation independent<br>Gastric residence time and dosing frequency important |

## Measurement of Solubility

The measurement of aqueous solubility seems, in principle, straightforward, but a number of pitfalls should not be overlooked, such as the effect of:

- Buffer- and ionic-strength (salting-out effect)
- Polymorphism and purity of the sample
- pH
- Supersaturation
- Thermodynamic (equilibrium) *versus* kinetic solubility.

Most compounds are less soluble in physiological salt solution (0.9% NaCl) than in pure water, due to the salting-out (or common ion) effect. Various salts have varying effects on the aqueous solubility. This can be seen when the solubility of different salts is compared, or solubility measurements are made in solutions of different ionic strengths.

The solubility–pH profile of a compound is a function of its ionisation constant(s). Zwitterions have the lowest aqueous solubility around their isoelectric point, but their log*D* (virtual 'log*P*') is then maximal.

Aqueous solubility (*S*, mostly expressed in $mg\,ml^{-1}$ or $g\,l^{-1}$, sometimes as mg (100 $ml^{-1}$ or %) is often inversely correlated to lipophilicity:

$$\log 1/S = a\log P + b \tag{1}$$

This equation holds only for series of congeneric compounds and has no general validity. Yalkowsky and colleagues derived an equation in which they relate aqueous solubility to octan-1-ol/water partition coefficients and the melting point (mp in °C) of the compounds:

$$\log S = -\log P - 0.01\,\mathrm{mp} + 1.2 \tag{2}$$

Thus, a lower melting point and low hydrophilic log*P* values led to more soluble compounds. This approach takes into account the effect of crystallinity upon solubility. The limitation of this approach is that it uses log*P* and mp values. In cases of a proposed, not yet synthesised compound, in particular the melting point is not known and cannot be estimated.

Traditional dissolution and solubility measurements are quite tedious. In today's discovery environment, higher throughput demands are extended to physicochemical measurements, such as assessment of solubility. Turbidimetric or nephelometric (kinetic) solubility evaluation has been reported as a reasonable first approximation to detect low solubility compounds,[2,8,9] and an automated potentiometric titration method for solubility measurement and solubility–pH profiles has been introduced.[10] Using a flow cell, turbidimetric solubility measurements can be set up for the range of 5–65 μg ml$^{-1}$.[2] For reasons of higher throughput, kinetic solubility is often measured on initially synthesised batches, whereas thermodynamic solubility is reserved for potential development candidates. Thermodynamic solubilities tend to be lower than kinetic, which is often due to higher purity of the batch of compound intended for development studies or to a different polymorphic crystalline form because another synthetic route or crystallisation procedure has been used.

Knowledge of aqueous solubility and dissolution rate may still be insufficient to predict *in vivo* performance, since food effects may have considerable and variable impact.[11] The selection of a proper salt is another important formulation-related issue.[12] Salt formation may be one approach to improve solubility problems. Small monovalent counter-ions mostly have better solubilities (*e.g.* $Na^+ = K^+ > Ca^{2+} = Mg^{2+} > Al^{3+}$). Organic counter ions (gluconic acid, lactic acid, *etc.*) are often better than inorganic ones (chloride, *etc.*).

## Calculation of Solubility

In addition to new experimental approaches, various efforts have been reported to estimate solubility from molecular structure, albeit with limited success. The reason is that most of these methods do not take into account the medium in which the solubility may be relevant. If one is interested in gastrointestinal absorption, this typically would be the fluid in the upper intestine.

Using the general solvation equation methodology or solvatochromic approach, Abraham and co-workers developed the following equation to predict solubility in an aqueous environment from molecular structure:[13]

$$\begin{aligned} \log Sw = {} & 0.52 - 1.00R_2 + 0.77\pi_2{}^{H} + 2.17\Sigma\alpha_2{}^{H} \\ & + 4.24\Sigma\beta_2{}^{H} - 3.36\Sigma\alpha_2{}^{H}\Sigma\beta_2{}^{H} - 3.99Vx \end{aligned} \tag{3}$$

$$n = 659;\ r^2 = 0.920;\ s = 0.56;\ F = 1256$$

where $R_2$ is the excess molar refraction, $\pi_2{}^{H}$ is the solute dipolarity/polarisability, $\Sigma\alpha_2{}^{H}$ is the overall or summation hydrogen-bond acidity, $\Sigma\beta_2{}^{H}$ is the overall or summation hydrogen-bond basicity, $\Sigma\alpha_2{}^{H}\Sigma\beta_2{}^{H}$ is a mixed term dealing with hydrogen-bond interactions between acid and basic sites in the solute, $Vx$ is McGowan's characteristic volume (for statistics, see pg. 227). These properties can be calculated using the program ABSOLVE.[13] No information is given on the potential intercorrelation of the terms used in this equation, which may limit its statistical significance. Future research should be directed to predict solubility in buffered media simulating the intestinal fluid.

Another approach in solubility prediction is based on molecular topology and neural network modelling.[14] The overall correlation between predicted and observed solubilities was $r^2 = 0.86$, which looks reasonable. However, the individual estimates may be 1 log unit or more wrong, which in the case of poorly soluble compounds, is unacceptable. Accurately measured values are still required to overcome shortcomings of current solubility predictors.

## 3 LIPOPHILICITY

### Definitions and Lipophilicity Scales

One of the key properties relevant to the biopharmaceutical and pharmacokinetic profile of a compound is lipophilicity. Very often lipophilicity is taken as equivalent to octan-1-ol/water partitioning or distribution, or seen as equivalent to hydrophobicity. Therefore, it is important to first repeat the recent IUPAC definitions here:[15]

*Hydrophobicity* is the association of non-polar groups or molecules in an aqueous environment, which arises from the tendency of water to exclude non-polar molecules.

*Lipophilicity* represents the affinity of a molecule or a moiety for a lipophilic environment. It is commonly measured by its distribution behaviour in a biphasic system, either liquid–liquid (*e.g.* partition coefficient in octan-1-ol/water) or solid–liquid (retention on reversed-phase high-performance liquid chromatography (RP-HPLC) or thin-layer chromatography (TLC) systems).

Partition coefficients ($\log P$) refer to the compound in its neutral state, while distribution coefficients ($\log D$) are measured at a selected pH, often 7.4. Octan-1-ol is the most widely used model of a biomembrane. Large compilations of octanol/water distribution data are available, *e.g.* in the MedChem database. Octan-1-ol is an H-bond donor and acceptor solvent and may not mimic the characteristics of very lipophilic membranes such as the blood–brain barrier (BBB). Therefore various other solvent systems have been used to estimate membrane transport (see Figure 2).[6] Although the octan-1-ol/water system is usually taken as the reference system to express $\log P$ values, many other systems may contain valuable information. It has been suggested that four different systems be used to characterise fully the lipophilic behaviour of a compound:

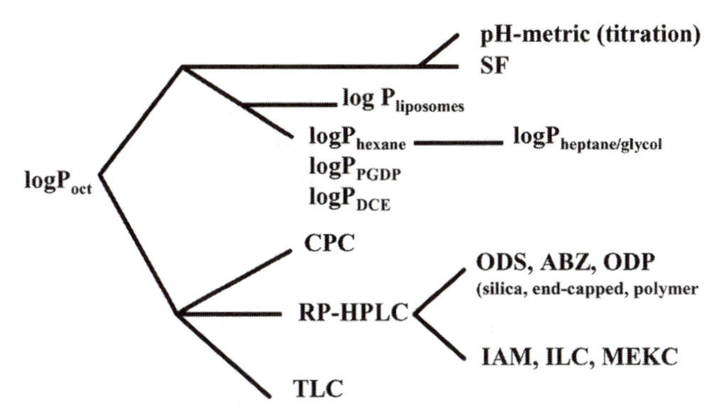


**Figure 2** *Experimental methods to measure lipophilicity.*[6] *$\log P_{oct}$: octan-1-ol/water partition coefficient, $\log P_{liposomes}$: partition coefficient between liposomes and buffer, $\log P_{hexane}$: 1-hexane/water partition coefficient, $\log P_{PGDP}$: propyleneglycol dipelargonate/water partition coefficient, $\log P_{DCE}$: 1,2-dichloroethane/water partition coefficient, $\log P_{heptane/glycol}$: a non-aqueous partitioning system, SF: shake-flask, pH-metric: logP determination based on potentiometric titration in water and octan-1-ol/water, CPC: centrifugal partition chromatography, RP-HPLC: reversed-phase high-performance liquid chromatography, TLC: thin-layer chromatography, ODS: octadecylsilane, ABZ: end-capped silica RP-18 column, ODP: octadecylpolyvinyl packing, IAM: immobilised artificial membrane, ILC: immobilised liposome chromatography, MEKC: micellar electrokinetic capillary chromatography*

- alkane/water (inert)
- octan-1-ol/water (amphiprotic = H-donor and acceptor)
- chloroform/water (proton donor)
- propylene glycol diperlargonate (PDGF) or di-1-butyl ether/water (proton acceptor).

More recently, a number of chromatographic systems have been suggested to measure lipophilicity. Among these, immobilised artificial membranes (IAM) have been well-studied.[16] Others include immobilised liposome chromatography (ILC)[17,18] and micellar electrokinetic chromatography.[19] Along the same lines several groups have suggested that studying partitioning into liposomes may produce relevant information related to membrane uptake and absorption.[20–22]

Often an optimum lipophilicity value is observed when lipophilicity is plotted against some form of 'activity'. This relationship between an 'activity' and $\log P$ can be described by either a parabolic (Hansch model) or bilinear (Kubinyi model) expression. The relationship between membrane permeability and lipophilicity is considered to be sigmoidal.

## The Information Content in log*P*

*The Major Contributions.* Partition coefficients contain information of at least two major contributions, one related to molecular size and the second on hydrogen bonding,[23] which can be expressed as:

$$\log P \text{ or } \log D = aV - \Lambda \quad (4)$$

in which $V$ is the calculated molar volume and $\Lambda$ a term for polarity, which is mainly the hydrogen bonding capability of the solute. The coefficient $a$ depends on the solvent system and on the program used to calculate the molar volume. $\Lambda$ depends on the solvent system and is related to the H-bonding capacity of the solute.

*Diff(log*$P^{N-I}$*).* More series-specific information can be obtained by considering the lipophilicity difference between neutral ($\log P^{N}$) and fully ionised ($\log P^{I}$) species,[24] termed diff($\log P^{N-I}$). This parameter expresses the influence of ionisation on the intermolecular forces and intramolecular interactions of a solute. Typically, in octan-1-ol/water the log*D* difference is *ca.* 3 log units, but may vary considerably due to different degrees of delocalisation of the formal charge.

*ΔlogP.* It has been observed that the difference between two different log*P* scales may be useful in drug absorption studies. Particularly brain uptake (see below in Estimation of Brain Penetration),[25] but also gastrointestinal absorption, can be estimated by this property.

$$\Delta\log P = \log P_{\text{octanol/water}} - \log P_{\text{alkane/water}} \quad (5)$$

Further studies on this property have demonstrated that it mainly encodes for the hydrogen-bonding capability of a compound, which is more conveniently assessed by calculation (see Section 4, pg. 204).

## Measurement of log*D*/log*P*

*From Shake-flask to High Throughput.* The traditional approach to measure a distribution coefficient is by shaking a two-phase system, such as a mutually pre-saturated octan-1-ol/aqueous buffer, until equilibrium is reached. This is known as the shake-flask approach. Variations have been developed. More recently there has been a trend to measure physicochemical properties including log*D* values using higher throughput methods. Such systems are now becoming commercially available.

*Difficulties with Alternative Solvent Systems.* Alkane water systems have been suggested as a model for the blood–brain barrier and a number of experimental values can be found in the literature.[26] An important practical limitation of an alkane is the poor solubility of most compounds in this solvent. As an alternative 1,2-dichloroethane (DCE) has been suggested,[27] but this solvent is carcinogenic and will most likely have little application in industry.

### Estimation of log*P* and log*D*

Octan-1-ol/water partition coefficients can also be calculated using an additivity scheme as suggested by Hansch and co-workers in the 1960s. The Hansch–Fujita substituent constant $\pi$ is defined as:

$$\pi_X = \log P_{RX} - \log P_{RH} \tag{6}$$

By this definition, the lipophilicity of a hydrogen atom is zero. Since this is not correct, later Hansch and Leo, as well as Rekker, have proposed fragmental values for lipophilicity contributions. Both have proposed an additivity scheme, using correction terms to account for certain intramolecular effects, such as the interaction between adjacent polar groups, to calculate log*P* values. Furthermore, atomic-based additivity schemes have been proposed, *e.g.* by Ghose and Crippen. The results of such calculations should be compared to experimental ones whenever possible. Calculated log*P* values reflect the molecule in its neutral uncharged state. More relevant for biological processes are distribution coefficients (log*D* values) at pH 7.4 (intracellular uptake) or pH 6.5 (gastrointestinal absorption).

Probably the best known algorithm to calculate the lipophilicity of a compound is CLOGP, which computes the calculated log*P* value of a compound in its neutral state.[28] Although in many cases useful, more relevant are log*D* values accounting for ionisation at a selected pH, often 7.4. Log*D* values can be estimated by combining estimates of log*P* and p$K_a$ (see Section 6) and using the appropriate Henderson–Hasselbach equations. Recent reviews of log*P* and log*D* computation can be found in the references.[29,30]

## 4 HYDROGEN BONDING

$\Delta$log*P* values, defined as the difference between two solvent system scales, particularly octan-1-ol/water and alkane/water, appear to encode for the H-bonding capacity of a solute and uptake in the brain.[25] Further research showed that H-bonding can be more conveniently assessed by computational methods.[31–34] H-bonding capability of a drug was demonstrated to be

**Table 2** *Hydrogen bond descriptors*[43]

| *Calculated* | |
|---|---|
| Number of donors and acceptors | HD, HA |
| Total number of heteroatoms | HT |
| Total number of possible H-bonds | HBOND |
| Free energy factors | $C_a$, $C_d$ |
| Total of free energy factors | $C_{ad}$ |
| Polar surface area | PSA |
| Dynamic polar surface area | $PSA_d$ |
| *Experimental* | |
| From partitioning in different solvent systems | $\Delta \log P$ |
| Polarity contribution in log*P* | $\Lambda$ |

an important property for crossing membranes and to correlate well with human intestinal absorption. This property can be simply calculated by summing up the polar surface area of all nitrogen and oxygen atoms in a molecule. Thus, conformational flexibility is taken into account. There is debate whether the minimum energy conformation is sufficient or whether a wide range of conformations should be considered and factored into a dynamic polar surface area. Even more simple, but also more crude, is to count total numbers of H-bond acceptors and donors in a molecule. Various computational approaches are commercially available such as the program HYBOT, which computes H-bond factors for donors and acceptors based on a large database of experimental data on H-bonding.[35] In Table 2 an overview is given of currently available descriptors for hydrogen bonding.

## 5 MOLECULAR SIZE

The most convenient way to express the size of a compound is by calculating its molecular weight (*MW*). Very closely related to *MW* are properties such as surface area, molar volume, parachor, McGowan's characteristic volume, molar refractivity, and others. In various studies, it was demonstrated that large size is detrimental to membrane permeation. However, cyclosporin A (*MW* 1203) is a reasonably well-absorbed drug and is often cited as an example of a large molecule. Difficult in such debates is the lack of knowledge of the role of molecular shape on membrane permeability, or the putative role of yet unidentified transporters facilitating transport. In Table 3 a number of size descriptors are collected.

**Table 3** *Molecular size descriptors*[42]

| *Calculated* | | *Experimental* | |
|---|---|---|---|
| Molecular weight | *MW* | Molecular radius | |
| Molar refractivity | *MR* | (Stokes–Einstein radius) | $r$ |
| Van der Waals volume | *V* | Hydrodynamic radius/size | $r$ |
| Total surface area | *S* | Hydrodynamic volume | $K_d$ |
| Water accessible volume | *VW* | Cross-sectional area | $A_D$ |
| Water accessible surface | *SW* | | |
| Ovality | *O* | | |
| Non-polar part of the volume | *VNP* | | |
| Non-polar part of the surface area | *SNP* | | |
| Principal axes | $X, Y, Z$ | | |
| Ratio length/width 1 | $X/Y$ | | |
| Ratio length/width 2 | $X/Z$ | | |

In QSAR studies, sometimes only the contributions of substituents at various sites are considered. In that case, substituent-specific descriptors such as Taft and Verloop parameters can be used.[36]

## 6 IONISATION CONSTANTS

### Ionisation/Protonation State

Since neutral compounds display better passive transport properties than ionised ones, the ionisation state of a drug is an important property. Indeed, most drugs on the market are acids or bases with p$K_a$ values 2–12.

The p$K_a$ of a compound is defined as the state at which 50% of the species are protonated. Without further notation, the p$K_a$ is measured in water. However, poorly soluble compounds may be measured in mixtures, *e.g.* water/methanol. Electron-accepting neighbouring groups (*e.g.* nitro) have a p$K_a$-lowering effect, while electron-donating groups (*e.g.* alkyl) have a p$K_a$-increasing effect.

Compounds with a basic and an acidic functional group may exist as four different species: zwitterion, anion, cation and neutral. The four microscopic protonation constants can be calculated from the two measured macroscopic p$K_a$s and one of the micros.

The protonation state has an influence on the distribution properties of a compound. The partition coefficient measured at a certain pH is called the distribution or apparent partition coefficient (log$D$ or log$P'$). Log$P$ values refer to neutral molecules. Log$P$ and log$D$ values can be related to each other using the compound's p$K_a$ value(s), for example; for monoprotic compounds:

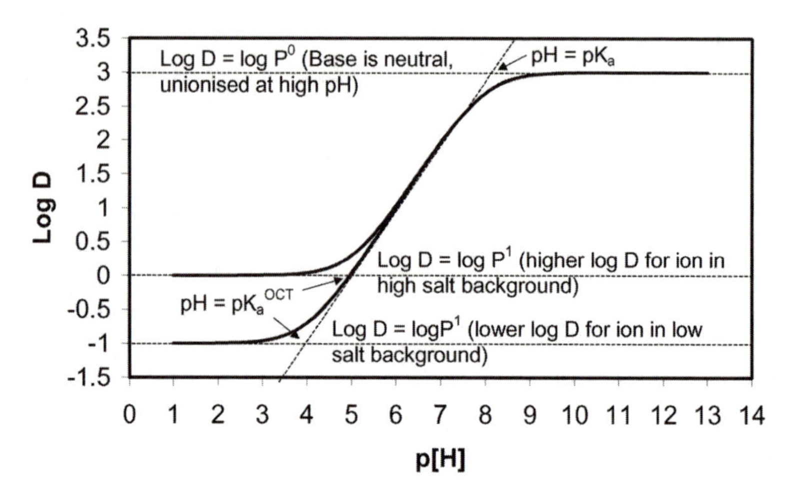


**Figure 3** *Typical log*D *vs. pH profile for a monoprotic base* (copyright John Comer, Sirius Analytical Instruments)

$$\text{acids:} \quad \log P = \log D + \log(1 + 10^{\mathrm{pH}-\mathrm{p}K_\mathrm{a}}) \tag{7}$$

$$\text{bases:} \quad \log P = \log D + \log(1 + 10^{\mathrm{p}K_\mathrm{a}-\mathrm{pH}}) \tag{8}$$

These equations are only valid close to the p$K_a$ (± two units). A graphical representation of a log$D$–pH curve for a simple monoprotic base is given in Figure 3.

An empirical formula which gives a better fit to experimental data for a wider pH range is the tanh function, *i.e.*

$$\text{Mono-protic acids:} \quad \log P = \log D + (1 - \tanh(\mathrm{p}K_\mathrm{a} - \mathrm{pH} + 1)) \tag{9}$$

$$\text{Mono-protic bases:} \quad \log P = \log D + (1 - \tanh(\mathrm{pH} - \mathrm{p}K_\mathrm{a} + 1)) \tag{10}$$

## Estimation of p$K_a$

A number of programs, such as ZPARC, ACD/p$K_a$ and PALLAS, are available to calculate p$K_a$ values. In Table 4, a comparison is made for the antihistaminic cetirizine. Clearly the computed results are not identical for each of the programs and deviate from the experimental values.

The quality of such calculated data depend on the complexity of the structures. More innovative structures, or non-precedented chemotypes, are more likely to be difficult cases. One approach to calculate p$K_a$ values is based on the use of Hammett constants. In 1938 Hammett reported $\sigma_{para}$ and $\sigma_{meta}$ substituent constants for benzoic acids. It was found that $\sigma$

**Table 4** *Comparison of the calculation of* $pK_a$ *values for cetirizine*[50]

| | *Cetirizine* | | |
|---|---|---|---|
| ACD | 2.1 | 3.3 | 6.4 |
| ZPARC | 6.2 | 3.6 | 6.7 |
| PALLAS | 0.5 | 3.3 | 7.6 |
| Experimental | 2.2 | 2.9 | 8.0 |

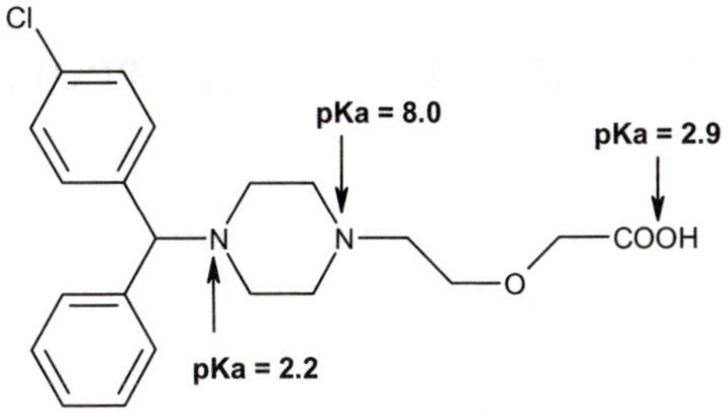

values for a given substituent in the *para*- and *meta*-position need to be different, since the observed electronic influence of a substituent in these positions is different. Electron-releasing substituents have negative $\sigma$ values and electron-withdrawing substituents have positive $\sigma$ values. Thus, for *meta*- and *para*-substituted benzoic acids their $pK_a$ can be calculated from:

$$pK_a = 4.20 + \rho\Sigma\sigma \quad (11)$$

Where 4.20 is the $pK_a$ of benzoic acid at 25 °C and $\rho$ is a susceptibility constant in a particular solvent ($\rho = 1$ in water at 25 °C). Many similar equations have been derived and published for other functional groups.

## 7 ELECTRONIC PROPERTIES

The electronic characteristics of a molecule can be described in various ways. The formal charge, due to full or partial ionisation of an ionisable functional group, as discussed above, is the best known. Using molecular orbital calculations a wide range of different atomic and molecular properties related to the electron distribution within the molecule can be computed.[36]

Many surface properties are related to the electronic structure of a compound. In molecular modelling studies, 3D surface descriptors, such as molecular electrostatic potentials (MEP) or molecular lipophilic potentials

(MLP) have often been successfully used to model (dis-)similarity among related compounds as well as the interaction between ligand and target.

## 8 ESTIMATION OF OTHER MOLECULAR PROPERTIES

### Computational Properties

Molecular structures and substituents (substructures or fragments) can been classified using the following properties:

- lipophilic/hydrophobic/amphiphilic
- steric/topologic/size/shape
- electronic/electrostatic
- hydrogen bonding

(Physico-)chemical descriptors used in QSAR studies and in the design of virtual or real combinatorial libraries can be subdivided as:

- local (substructural) or global (structural)
- experimental or calculated
- unique or composed

In the previous sections, the computational calculation of various properties has been briefly discussed. Programs such as Tsar, Cerius2, Sybyl, SciQSAR, MOLCONN-Z, HYBOT, VolSurf, MolSurf, MS-WHIM, EVA and others, are commercially available to compute these and a wide range of other properties. In particular, topological descriptors appear to be a rich vein for the creation of more and more descriptors, many of which are strongly intercorrelated and of questionable use to drug design.

Many different substituent parameters have been tabulated in the literature and can be used for series design and structure–activity relationship studies. One can either use the original variables or derived ones, which are called principal properties or principal components. The problem of redundancy of many of these descriptors in QSAR studies can be overcome by using suitable variable selection techniques, or statistical methods such as partial least-squares (PLS), but this is beyond the scope of this chapter.

### Ligand–Receptor Interactions

The potency of a drug is often expressed by it's $K_i$ or $IC_{50}$ value. The Cheng–Prusoff equation relates $K_i$ to $IC_{50}$ values:

$$\mathrm{IC}_{50} = K_\mathrm{i}(1 + F/K_\mathrm{d}) \tag{12}$$

where $K_\mathrm{i}$ is the inhibition constant or equilibrium dissociation constant of the drug–ligand complex, $F$ is the concentration of the free radioligand used in the binding study and $K_\mathrm{d}$ its dissociation constant. Care should be taken by comparing $\mathrm{IC}_{50}$ values between laboratories. In contrast to $\mathrm{IC}_{50}$ values, $K_\mathrm{i}$ values are independent of the receptor and ligand concentration used in the assay. The free energy of binding of a ligand to its target (receptor, ion-channel, enzyme or nucleic acid) is given by:

$$\Delta G = 2.303RT\log K_\mathrm{i} = \Delta H - T\Delta S \tag{13}$$

Since $2.303RT$ is *ca.* 1.4 kcal mol$^{-1}$, this allows us to estimate differences in binding energies. A factor of 10 in $K_\mathrm{i}$ is thus 1.4 kcal mol$^{-1}$; a factor of 100 is 2.8 kcal mol$^{-1}$.

Energy terms involved in ligand–receptor binding can be classified as

| | | |
|---|---|---|
| hydrophilic | ion–ion | translational |
| hydropobic | dipole–dipole | rotational |
| Van der Waals | | vibrational |

The prediction of binding affinities, knowing the ligand and even the target structure, is still quite difficult. One approach is based on summarising contributions of functional groups[37] using the equation:

$$\Delta G = T\Delta S + n_\mathrm{DOF}E_\mathrm{DOF} + n_\mathrm{x}E_\mathrm{x} \tag{14}$$

where $T\Delta S$ represents the unfavorable entropy term for a ligand binding to its receptors, assumed to be constant and estimated to be 14 kcal mol$^{-1}$; $n_\mathrm{DOF}$ are the internal degrees of conformational freedom, rotatable bonds in the ligand, while $E_\mathrm{DOF}$ is the average entropy loss on binding per rotatable bond; $n_\mathrm{x}$ is the number of occurrences of functional group x in the ligand, and $E_\mathrm{x}$ represents the average intrinsic binding energy for group x (see Table 5).

Another approach to calculate binding energies is to use programs such as LUDI, which is based on the summation of various types of contributions, such as Van der Waals interaction, hydrogen bonding, *etc.*

**Table 5** *Intrinsic binding energies* (kcal $mol^{-1}$)[37]

| | | | | | |
|---|---|---|---|---|---|
| $N^+$ | 11.5 | CO | 3.4 | DOF[a] | −0.7 |
| $PO_4^{2-}$ | 10.0 | OH | 2.5 | | |
| $COO^-$ | 8.2 | Halogen | 1.3 | | |
| | | N | 1.2 | | |
| | | O, S | 1.1 | | |
| | | C ($sp^3$) | 0.8 | | |
| | | C ($sp^2$) | 0.7 | | |

[a] Degrees of freedom or number of rotatable bonds.

## 9 RELATIONSHIPS TO DRUG DISPOSITION

### Estimation of Gastrointestinal Absorption

Prediction of human intestinal absorption direct from molecular structure has been studied by several groups.[38] The above-mentioned correlation with polar surface area is an example of a correlation with a single descriptor (Figure 4). Others have used combinations of calculated properties,[39] sometimes mixed with experimental physicochemical data such as log*D* values.[40]

All of these methods currently do not take into account the role of the biochemical barrier of a membrane. In the future, such models need to be complemented by terms for P-gp efflux and possibly contributions of other transporters, and for gut wall metabolism, mainly by CYP3A4.[3]

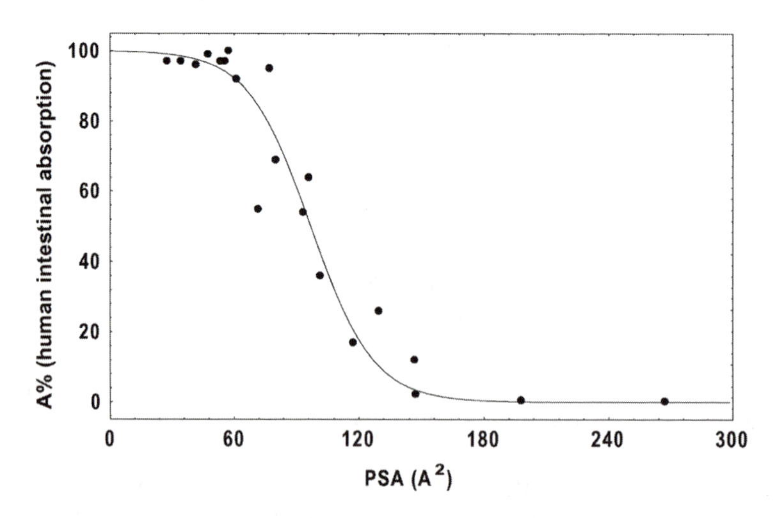


**Figure 4** *Polar surface area (PSA) for the prediction of human gastrointestinal absorption*[32,33]

Using almost identical approaches as for the prediction of human gastrointestinal absorption, the permeation through Caco-2 cells,[41,42] uptake into the brain[34,43–47] (see below) or percutaneous permeability[48,49] can be estimated.

## Estimation of Brain Penetration

The influence of log*P*/log*D* on brain uptake for CNS agents has been widely studied and is well-documented.[43,44] Most CNS drugs are relatively small in size, *MW* around 350. Octan-1-ol/water distribution coefficients are typically in the range 0–4, also reflected by a well-known rule-of-thumb that optimal CNS drugs have a log*D* of ~2 (Figure 5). Based on a comparison of CNS *versus* non-CNS drugs, it was concluded that the physicochemical constraints on compounds targeted to the brain are slightly more restrictive than for oral absorption. *MW* should be below 450, while for GI absorption a limit of 500 has been suggested. The total polar surface area, a measure for hydrogen bonding capacity, should be below 90 $Å^2$, while for GI absorption this limit can be somewhat higher.[6,44] Using surface activity measurements, it was established that compounds with a cross-sectional area >80 $Å^2$ are unlikely to cross the blood–brain barrier.[45] Further discussion on this topic is included in Chapter 12.

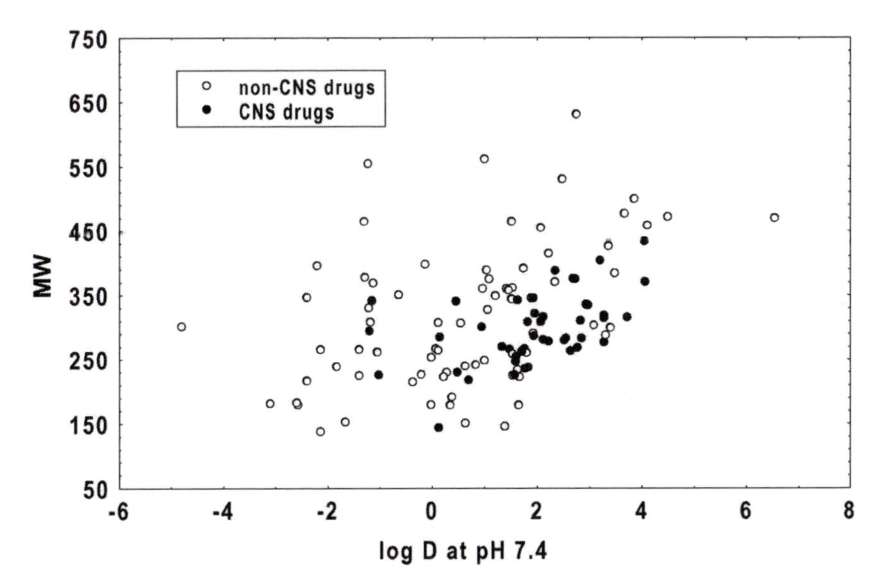


**Figure 5** *LogD against molecular weight for a series of CNS and non-CNS drugs*[44]

**Estimation of Pharmacokinetic Properties**

Increasing lipophilicity results in higher plasma protein binding. The resulting effect on the unbound volume of distribution ($V_{du}$) is demonstrated in Figure 6, in which the unbound volume of distribution (volume of distribution at steady state corrected for plasma protein binding) is plotted against log$D$ values. The trend for basic and acidic compounds to have higher values than neutral compounds is clearly visible. The increase in free volume of distribution with increasing lipophilicity largely reflects association with plasma proteins (albumin) for acidic compounds and association with tissues and lipoproteins for neutral and lipid compounds. Basic compounds bind mainly to albumin and partially to $\alpha_1$-acid glycoprotein. With increasing lipophilicity, unspecific binding to red blood cells, leukocytes and platelets probably also increases. The high association for acidic compounds with albumin is due to both ion-pair and hydrophobic interactions. Similar reasons apply for the affinity of basic compounds for tissues. Here the ion-pair and hydrophobic interactions are due to the head and tail groups of phospholipids present in cell membranes. Similar to the prediction of intrinsic metabolic clearance, such associations can be studied with *in vitro* systems, such as plasma protein and cell membrane binding.

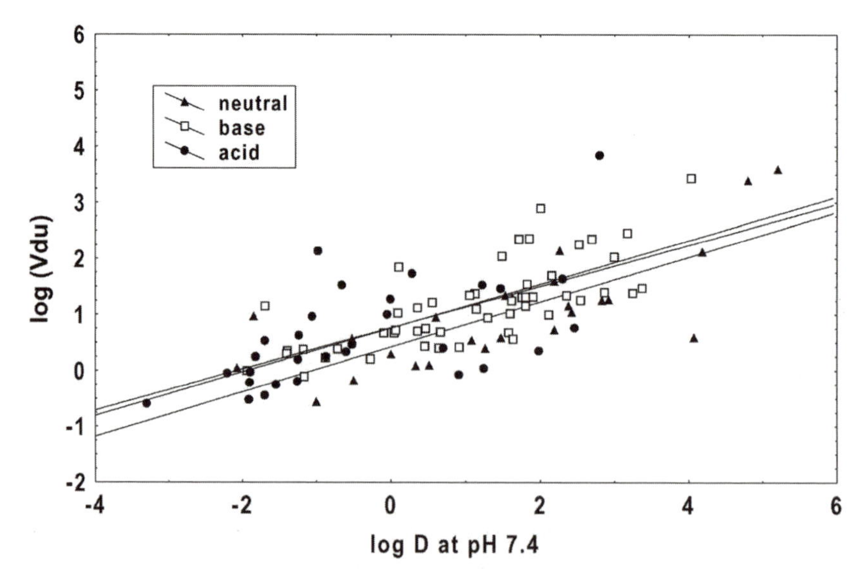


**Figure 6** *Effect of log*D *on unbound volume of distribution for different types of compounds*

## 10 REFERENCES


1. O.H. Chan and B.H. Stewart, *Drug Disc. Today*, 1996, **1**, 461.
2. C.A. Lipinski *et al.*, *Adv. Drug Deliv. Rev.*, 1997, **23**, 3.
3. H. van de Waterbeemd *et al.*, *J. Med. Chem.*, 2001, **44**, 1313.
4. D.A. Smith, B.C. Jones and D.K. Walker, *Med. Res. Revs.*, 1996, **16**, 243.
5. H. Kubinyi, *QSAR: Hansch Analysis and Related Approaches*, VCH, Weinheim, 1993.
6. H. van de Waterbeemd, in *Methods for Assessing Oral Drug Absorption*, J. Dressman (ed.), Dekker, New York, 2000, p. 31.
7. H. van de Waterbeemd, in *Modern Methods of Drug Discovery*, R. Hilgenfeld and A. Hillisch (eds.), Birkhäuser, Basel, in press.
8. M. Kansy *et al.*, in *Molecular Modeling and Prediction of Bioactivity*, K. Gundertofte and F.S. Jørgensen (eds.), Kluwer Plenum, New York, 2000, p. 237.
9. C.D. Bevan and R.S. Lloyd, *Anal. Chem.*, 2000, **72**, 1781.
10. A. Avdeef, *Pharm. Pharmacol. Commun.*, 1998, **4**, 165.
11. V.L. Stella *et al.*, *J. Pharm. Sci.*, 1999, **88**, 775.
12. P.L. Gould, *Int. J. Pharmaceut.*, 1986, **33**, 201.
13. M.H. Abraham and J. Le, *J. Pharm. Sci.*, 1999, **88**, 868.
14. J. Huuskonen *et al.*, *J. Chem. Inf. Comput. Sci.*, 1998, **38**, 450.
15. H. van de Waterbeemd *et al.*, *Pure and Appl. Chem.*, 1997, **68**, 1137; *Annu. Rep. Med. Chem.*, 1998, **33**, 397 (http://www.iupac.org/reports/1997/6905vandewaterbeemd/index.html).
16. B.H. Stewart and O.H. Chan, *J. Pharm. Sci.*, 1998, **87**, 1471.
17. P. Lundahl and F. Beigi, *Adv. Drug Deliv. Rev.* 1997, **23**, 221.
18. U. Norinder and T. Österberg, *Perspect. Drug Disc. Des.*, 2000, **19**, 1.
19. M.D. Trone *et al.*, *Anal. Chem.*, 2000, **72**, 1228.
20. C. Ottiger and H. Wunderli-Allenspach, *Pharm. Res.*, 1999, **16**, 643.
21. R.P. Austin *et al.*, *J. Pharm. Sci.*, 1998, **87**, 599.
22. K. Balon *et al.*, *J. Pharm. Sci.*, 1999, **88**, 802.
23. H. van de Waterbeemd and B. Testa, *Adv. Drug Res.*, 1987, **16**, 85.
24. G. Caron *et al.*, *Pharm. Sci. Technol. Today*, 1999, **2**, 327.
25. R.C. Young *et al. J. Med. Chem.*, 1988, **31**, 656.
26. N. El Tayar *et al.*, *J. Phys. Chem.*, 1992, **96**, 1455.
27. F. Reymond *et al.*, *Chem. Eur. J.*, 1999, **5**, 39.
28. A.J. Leo, *Chem. Rev.*, 1993, **93**, 1281.
29. P. Buchwald and N. Bodor, *Curr. Med. Chem.*, 1998, **5**, 353.
30. R. Mannhold and H. van de Waterbeemd, *J. Comput.-Aid. Mol. Des.*, 2001, **15**, 337.
31. H. van de Waterbeemd and M. Kansy, *Chimia*, 1992, **46**, 299.

32. K. Palm *et al.*, *J. Med. Chem.*, 1998, **41**, 5382.
33. D.E. Clark, *J. Pharm. Sci.*, 1999, **88**, 807.
34. D.E. Clark, *J. Pharm. Sci.*, 1999, **88**, 815.
35. O.A. Raevsky and K.J. Schaper., *Eur. J. Med. Chem.*, 1998, **33**, 799.
36. P.C. Jurs *et al.*, in *Chemometric Methods in Molecular Design*, H. van de Waterbeemd (ed.), VCH, Weinheim, 1995, p. 15.
37. P.R. Andrews *et al.*, *J. Med. Chem.*, 1984, **27**, 1648.
38. H. van de Waterbeemd, in *Pharmacokinetic Optimization in Drug Research: Biological, Physicochemical and Computational Strategies*, B. Testa *et al.* (eds.), Verlag HCA, Zurich, 2001, p. 499.
39. U. Norinder *et al.*, *Eur. J. Pharm. Sci.*, 1999, **8**, 49.
40. S. Winiwarter *et al.*, *J. Med. Chem.*, 1998, **41**, 4939.
41. G. Camenisch *et al.*, *Eur. J. Pharm. Sci.*, 1998, **6**, 313.
42. H. van de Waterbeemd *et al.*, *Quant. Struct. Act. Relat.*, 1996, **15**, 480.
43. S.P. Gupta, *Chem. Rev.*, 1989, **89**, 1765.
44. H. van de Waterbeemd *et al.*, *J. Drug Target.*, 1999, **6**, 151.
45. H. Fischer *et al.*, *J. Membr. Biol.*, 1998, **165**, 201.
46. U. Norinder *et al.*, *J. Pharm. Sci.*, 1998, **87**, 952.
47. A.M. Ter Laak *et al.*, *Eur. J. Pharm. Sci.*, 1994, **2**, 373.
48. W.J. Pugh *et al.*, *Int. J. Pharm.*, 1996, **138**, 149.
49. W.J. Pugh *et al.*, *Int. J. Pharm.*, 2000, **197**, 203.
50. B. Testa *et al.*, *Clin. Exp. Allerg.*, 1997, **27**, S13.

CHAPTER 11

# Quantitative Structure–Activity Relationships

DAVID J. LIVINGSTONE

## 1 INTRODUCTION

Many useful medicinal compounds have been discovered by the simple, albeit time consuming, process of administering a mixture of naturally occurring compounds to a sick person. The mixtures were often derived from parts of plants (*e.g*. bark, root or seeds), fungi, insects or animals, or from extracts of these organisms and the 'volunteer' patients from the general human population. Some drugs still in current use today were discovered millennia ago by this process, for example, morphine ($\sim$4000 BC), reserpine ($<$1000 BC), aspirin ($<$200 BC) and ephedrine ($\sim$1 AD). Changes in attitude to the value of human life, the development of sciences and medicine, commercial pressures and the effect of regulators have all led to the abandonment of this strategy for drug discovery.

So, what has taken its place? In the middle of the century before last Crum-Brown and Frazer realized that the curare-like paralysing properties of a set of quaternised strychnines depended on the nature of the quaternising group.[1] In their paper they proposed the following equation:

$$\Phi = f(C) \tag{1}$$

in which $\phi$ is a measure of biological activity ('physiological action') and $C$ is a measure of chemical structure ('chemical constitution'). The major problem in obtaining an accurate definition of $f$ in the equation was attributed to the difficulty of expressing changes in $\phi$ and $C$ with sufficient 'definiteness'. This situation has changed but has by no means been resolved as is discussed later.

The development of synthetic organic chemistry and methods of structure determination, coupled with the recognition that changes in chemical structure led to changes in biological activity, had a profound effect on the search for new medicinal compounds. The source of compounds altered from complex mixtures derived from natural products to pure, well-characterised molecules produced by synthetic methods. Testing procedures took longer to change, for example in the early 1900s an antimalarial research programme at the Bayer research institute used as their 'guinea-pigs' patients who had been rendered insane and paralysed by the final stages of syphilis and who were then deliberately infected with malaria. The antimalarial pamaquine emerged from these studies and was marketed in 1926. The trend, however, in biological testing was towards simpler systems, often using isolated organs, tissues or cells and as a result it became possible to draw up structure–activity relationship (SAR) tables such as the one shown in Table 1, where R is the substituent, $C$ is the concentration (mM) required to produce a desired level of effect in a biological test.

In Table 1 the potency of the molecules is expressed as a concentration ($C$) required to produce some standard effect, *e.g.* 50% inhibition of an enzyme, thus the phenyl analogue is the most active with the hydroxyl substitution the least active. It is important to ensure when constructing a table of this type that we are comparing 'like' with 'like' hence the need for standard effects such as $ED_{50}$, $IC_{50}$, $LD_{50}$, and so on. An SAR table provides information about the effect of change in chemical structure on biological properties and, in principle, allows the comparison of these effects for multiple positions or parts of the structure. In practice, however, most comparisons are made in a pair-wise fashion and one of the major

**Table 1** *A typical structure–activity relationship table*

| *R* | *C* |
|---|---|
| H | 300 |
| $CH_3$ | 80 |
| Cl | 15 |
| OH | 790 |
| $NO_2$ | 450 |
| $OCH_3$ | 350 |
| $CH_2OC_2H_5$ | 400 |
| $C_6H_5$ | 4.0 |

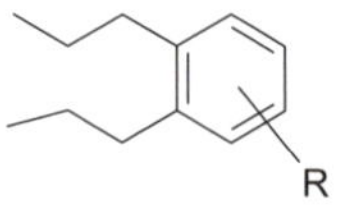

R

disadvantages of an SAR table is that it is only possible to assess the contribution of chemical structures which are represented in the table.

The SAR approach is an obvious development from Crum-Brown and Fraser's observations although the results are not usually expressed in a mathematical form such as Equation (1). A more mathematical approach was developed from SAR, as discussed in the next section, but what of the present? In the final decade of the last century, the technique known as combinatorial chemistry was born. This approach allows the efficient creation of large sets of compounds, either singly or as mixtures and as a result, dramatic changes in biological testing took place to create high throughput screening (HTS) systems which were capable of handling these large numbers of compounds. At first it was thought that these techniques might take the place of the so-called 'rational drug design' methods which had been in place for twenty or so years; indeed, the use of the term 'screening' suggested that this might become a 'high tech.' replacement for the ancient process of random screening described earlier. It soon became clear, however, that numbers which appear large on the human scale, thousands or hundreds of thousands of compounds, are still tiny compared with the vast numbers of compounds which might be made (estimates vary from $10^{50}$ to $10^{180}$). Rather than replacing QSAR and other drug design methods combinatorial chemistry and HTS have, if anything, made these techniques even more valuable.

## 2 BACKGROUND TO QSAR

At about the same time as Crum-Brown and Frazer's paper, other workers reported that the properties of molecules were important determinants of biological activity. Richardson showed that the toxicities of ethers and alcohols were inversely related to their water solubility, Richet demonstrated a relationship between the narcotic effect of alcohols and their molecular weight and Meyer and Overton independently showed that the narcotic action of many compounds was dependent on their oil/water partition coefficients. In the 1930s, chemists were beginning to explore the effect of changes in chemical structure on the rates and equilibrium constants of chemical reactions, resulting in the birth of physical organic chemistry. Perhaps the most famous of these early studies was the work of Hammett who devised a scale of electronic effects of substituents using Equation (2):

$$\rho\sigma_{X} = \log K_{X} - \log K_{H} \tag{2}$$

where $K_X$ is the equilibrium constant for a reaction involving an X substituted compound and $K_H$ is the equilibrium constant for the unsubstituted (or H substituted) parent. The left-hand side of the equation contains two constants $\rho$, the reaction constant which takes a characteristic value for a particular reaction, and $\sigma_X$, the substituent constant for the substituent X. In the original defining equation for $\sigma$ scales Hammett chose the ionisation of benzoic acids as the standard reaction and assigned a value of 1 to $\rho$. The Hammett equation applies to both equilibrium and rate constants and many different chemical reactions have been used to create $\sigma$ scales in order to characterise different types of electronic effects.[2]

In the 1960s Corwin Hansch made a seminal contribution with the suggestion of a chemical model system for hydrophobicity based on octanol/water partition coefficients[3] (log$P$), a system now almost universally adopted. Partition coefficient is simply defined as the ratio of concentrations of a compound (Y) in octanol and water:

$$P = \frac{[Y]_{OCT}}{[Y]_{aqu}} \tag{3}$$

Octanol was chosen as the reference organic phase because it was felt that it might simulate the lipid components of biological membranes, while water modelled the aqueous phases of a biological system. The original proposal involved a substituent constant, $\pi$, defined using Equation (4):

$$\pi_X = \log P_X - \log P_H \tag{4}$$

The similarity between this equation and Equation (2) is clear, the main difference being the lack of a reaction constant for the hydrophobicity parameter $\pi$. With these two substituent constants it is now possible to rewrite Table 1 as a quantitative structure–activity relationship (QSAR) table as shown in Table 2, where R is the substituent and $C$ is the concentration (M) required to produce a desired level of effect in a biological test.

In Table 2 the biological potency is now expressed as the logarithm (base 10) of the reciprocal of the concentration to produce a standard effect. This has a number of advantages; taking a reciprocal means that 'big' numbers are 'good', a natural pattern to recognise, and taking the logarithm improves the distribution of the data (making it more 'normal'), puts it on an easily handled scale and makes it suitable for comparison with free energies.[4] The changes in chemical structure are now represented by changes in the values of the two substituent constants, $\pi$ and $\sigma$, with the result that structure is now represented quantitatively. This is the

**Table 2** *A typical quantitative structure–activity relationship table*

| *R* | π | σ | *Log(1/C)* |
|---|---|---|---|
| H | 0.0 | 0.0 | 0.5 |
| $CH_3$ | 0.56 | −0.17 | 1.1 |
| Cl | 0.71 | 0.23 | 1.8 |
| OH | −0.67 | −0.37 | 0.1 |
| $NO_2$ | −0.28 | 0.78 | 0.35 |
| $OCH_3$ | −0.02 | −0.27 | 0.45 |
| $CH_2OC_2H_5$ | −0.24 | 0.03 | 0.4 |
| $C_6H_5$ | 2.0 | −0.01 | 2.4 |

R

meaning of the Q in QSAR, quantitative refers to the way that chemical structure is represented not, as commonly thought, to the activity or to the relationship between activity and structure.

One major advantage of a QSAR table is that it is now possible to consider the effect of chemical changes, which are not included in the original table, simply by looking up the substituent constant values for any new group. Consultation of a table of substituent constants shows that some other common hydrophobic substituents which may be used to replace phenyl are $SCF_3$ ($\pi = 1.44$), $C_3H_7$ (1.55), 3-thienyl (1.81), $C(CH_3)_3$ (1.98) and so on. The descriptors (substituent constants) shown in Table 2 may otherwise be called physicochemical properties, this is a more general term used to refer to quantitative parameters based on chemical structure.

Another major contribution made by Hansch and his co-workers was the recognition that the 'explanation' of biological potency might require the use of more than just one chemical property, such as log*P*, and thus the 'Hansch equation' was born,[5] shown in its generalised form below:

$$\log(1/C) = a\pi + b\pi^2 + c\sigma + dE_s + \text{constant} \quad (5)$$

The terms in Equation (5) are the hydrophobic and electronic substituent constants $\pi$ and $\sigma$ as before and a steric substituent constant $E_S$, due to Taft. This equation is known as a multiple linear regression equation, or model, because it consists of a linear combination of terms, even though one of those terms is a square ($\pi^2$). The process of fitting such a model to a set of data is called multiple linear regression analysis, often abbreviated to MLR, and this is discussed later in this chapter.

The 'state of the art' in QSAR at the beginning of the 1970s consisted of compound selection (see next section) and MLR modelling using substituent constants obtained from a look-up table such as the one compiled by Hansch *et al.*[6] QSAR was often called the 'Hansch approach' and it suffered from a number of limitations. Molecular descriptors, *i.e.* substituent constants, were often missing from the tables either due to gaps or because a particular substituent was not listed. For complex structures it was often not easy to decide which substituent constant value to use and there was a general dearth of data for non-aromatic systems. MLR requires quite precise measurements of the dependent variable (the biological response) and this was often not easy to obtain. As a result of these and other limitations, drug designers began to examine alternative means of characterising molecules and other ways to create models, which would link biological activity to chemical structure. In the following sections of this chapter some important features of the modern QSAR approach will be briefly discussed.

## 3 COMPOUND SELECTION

One of the earliest consequences of the use of physicochemical descriptors to characterise molecular structure was the recognition by Craig[7] of the fact that a plot of two such parameters represented a chemical property 'space'. Figure 1 shows a plot of $\sigma$ against $\pi$ for a set of common substituents.

Construction of a useful set of compounds involves the choice of substituents which span this space and, as activity results are obtained,

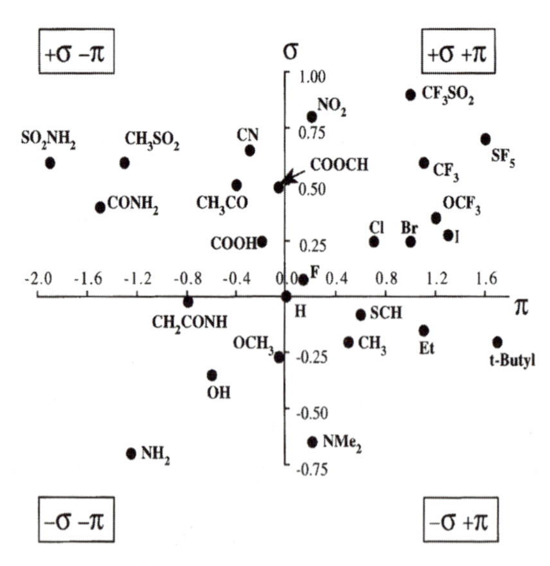


**Figure 1** *Plot of* $\sigma$ versus $\pi$ *for a set of common substituents*
(From reference 7, copyright (1971) American Chemical Society)

further compounds can be chosen which are similar or different to the existing set depending on the outcome of the testing. Reference to the choice of compound sets here introduces some useful jargon from the field of pattern recognition. A set of compounds that is chosen for biological testing or, where test results are already available, for the production of a QSAR model such as a MLR equation is known as a 'training' or 'learning' set. A set of compounds that will be made or chosen from an existing collection for further testing is known as a test set. Test sets are used to check the performance of QSAR models or examine the correctness of some mechanistic or chemical feature hypothesis. Appropriate choice of these sets is crucial to the success of a research programme since the application of any method of analysis can only extract as much information from a set as it contains. Judicious selection of the set(s) in the first place can help to ensure that this information content is maximised.

Set selection based on plots of pairs of parameters, known as Craig plots, has an obvious intuitive appeal but there are disadvantages to this approach, perhaps the most important being that only two physicochemical descriptors are involved in the choice. The importance of compound selection was recognised at a very early stage in the development of modern QSAR techniques and a number of approaches were reported which offered improvements over the use of Craig plots.[8] The concept of a chemical property space is useful in the explanation of how most of these methods operate. Figure 2 shows a plot of 9 compounds described by three physicochemical properties labelled A, B and C.

These three properties are known as 'factors' in classical experimental design terms and the compounds at the vertices of the cube represent design points at the extremes of these factors. Compound 8, for example, is a compound that has high values of all three factors whereas compound 6 has high values of factors A and C with a low value of factor B. A full factorial design[8] at two levels for these three factors would consist of the eight compounds at the vertices of the cube. A means of reducing the number of experiments, and hence compounds required, in a factorial design is to use a method known as fractional factorial design. Compounds 5, 2, 3 and 8 in Figure 2 would constitute a half-fraction factorial design for this set, compounds 6, 1, 7 and 4 would constitute another. The full design 'explores' the three-dimensional chemical property space by means of a cube, the half-fraction designs explore it by means of a tetrahedron, a well-known shape to most chemists. What about the compound shown at the centre of the cube, labelled as m? This corresponds to the mid-point of the factor space and is often considered in factorial and fractional factorial experimental designs. If the three

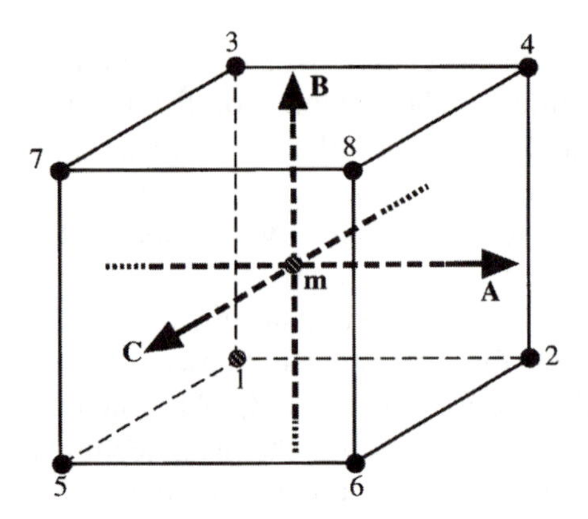


**Figure 2** *Representation of a factorial design in three factors (A, B and C)*
(From Austel,[9] with permission of the *Eur. J. Med. Chem.*)

physicochemical properties are substituent constants such as $\pi$, $\sigma$ and MR (a steric parameter, see next section) which are scaled to $H = 0$, then this midpoint is the unsubstituted parent.

Factorial and fractional factorial designs can be calculated to consider several different physicochemical properties simultaneously. Another technique that works in a high-dimensional physicochemical property space, using multiple properties in other words, is cluster analysis. Figure 3

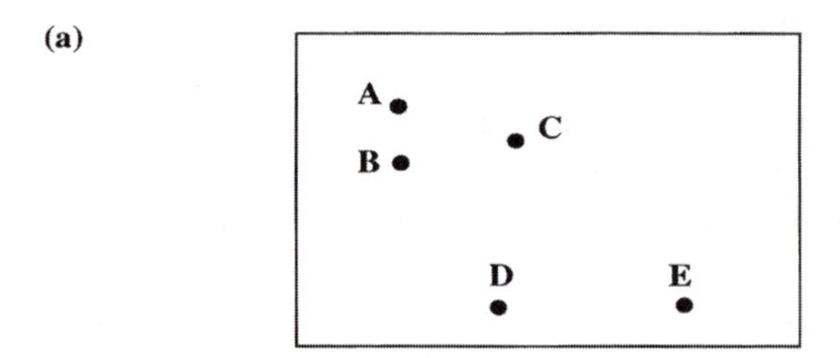


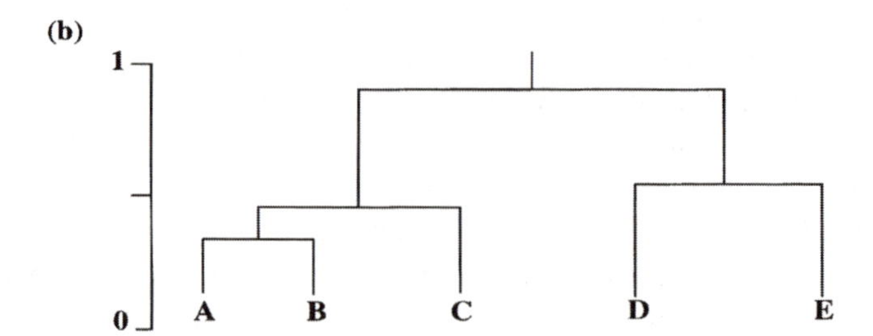


**Figure 3** *Illustration of the production of a dendrogram for a simple two-dimensional data set*
(From reference 8, with permission of Oxford University Press)

**Table 3** *Examples of substituents clustered from a six dimensional property space* (From Hansch and Leo[6], with their kind permission)

| *Cluster set number* | *Number of members* | *Examples of substituents* |
|---|---|---|
| 1 | 26 | $-Br$, $-Cl$, $-N_3$, $-CH_3$, $-CH_2Br$ |
| 2 | 17 | $-SO_2F$, $-NO_2$, $-CN$, $-$1-tetrazolyl, $-SOCH_3$ |
| 3 | 2 | $-IO_2$, $-N(CH_3)_3$ |
| 4 | 8 | $-OH$, $-NH_2$, $-NHCH_3$, $-NHC_4H_9$, $-NHC_6H_5$ |
| 5 | 18 | $-CH_2OH$, $-NHCN$, $-NHCOCH_3$, $-CO_2H$, $-CONH_2$ |
| 6 | 21 | $-OCF_3$, $-CH_2CN$, $-SCN$, $-CO_2CH_3$, $-CHO$ |
| 7 | 25 | $-NCS$, $-$pyrryl, $-OCOC_3H_7$, $-COC_6H_5$, $-OC_6H_5$ |
| 8 | 20 | $-CH_2I$, $-C_6H_5$, $-C_5H_{11}$, $-$cyclohexyl, $-C_4H_9$ |
| 9 | 21 | $-NHC{=}S(NH_2)$, $-CONHC_3H_7$, $-NHCOC_2H_5$, $-C(OH)(CF_3)_2$, $-NHSO_2C_6H_5$ |
| 10 | 8 | $-OC_4H_9$, $-N(CH_3)_2$, $-N(C_2H5)_2$ |

shows a two-dimensional illustration of the way that cluster analysis operates to produce a tree-like diagram, known as a dendrogram, of the degree of similarity between objects in the two-dimensional space. Similarity in this case is defined simply as the Euclidean distance between the points (objects) A–E although other distance measures may be used.

The number of distinct clusters produced by cluster analysis is dictated by the similarity structure of the data set and also by the level of similarity chosen to define clusters. Use of a high level of similarity (low value of the similarity scale in Figure 3) will lead to a large number of clusters each containing a small number of compounds. A low value of similarity, on the other hand, will yield a small number of clusters—one ultimately if the similarity value is set low enough. Hansch and Leo used this property of cluster analysis to create tables of sets of substituents, which could be used to make sets containing 5, 10 and 20 compounds.[6] Table 3 illustrates the clustering of substituents for a 10 cluster aromatic set.

## 4 DESCRIBING CHEMICAL STRUCTURE

Reliance on tabulated values of substituent constants imposed restrictions on the application of QSAR, particularly in the case of sets of diverse compounds. A solution to this problem is to use calculated rather than experimental descriptors and, indeed, some very early, published QSAR models used parameters such as superdelocalisability and the energy of the highest occupied molecular orbital. Molecular connectivity is a descriptor, which can be calculated simply from the two-dimensional representation of a molecule. Connectivity indices in their simplest form are computed from the hydrogen-suppressed skeleton of a compound by the assignment of a degree of connectivity, $\delta_i$, to each atom ($i$) representing the number of atoms connected to it. Figure 4 shows the degree of connectivity for each of the four heavy atoms in *iso*-butanol.

For each bond in the structure, a bond connectivity, $C_k$, can be calculated by taking the reciprocal of the square root of the product of the connectivities of the atoms at either end of the bond. For example, the bond connectivity for the first carbon–carbon bond (from the left) in the structure is:

$$C_1 = \frac{1}{\sqrt{(1 \times 3)}} \tag{6}$$

More generally the bond connectivity of the $k$th bond is given by:

$$C_k = \frac{1}{\sqrt{(\delta_i \delta_j)}} \tag{7}$$

where the subscripts $i$ and $j$ refer to the atoms at each end of the bond. The molecular connectivity index, $\chi$, for a molecule is found by summation of the bond connectivities over all of its $N$ bonds.

$$\chi = \sum_{k=1}^{N} C_k \tag{8}$$

For the butanol shown in the figure, the four bond connectivities are the reciprocal square roots of $(1 \times 3)$, $(1 \times 3)$, $(2 \times 3)$ and $(2 \times 1)$ which gives a molecular connectivity value of 2.269. This simple connectivity index is known as the first-order index because it considers only individual bonds, in other words paths of two atoms in the structure. Higher order indices may be generated by the consideration of longer paths in a molecule and other refinements have been considered, such as valence

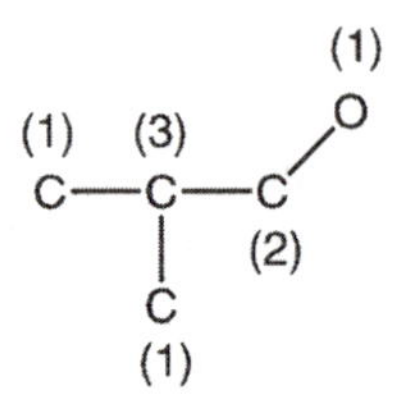


**Figure 4** *Atom connectivities for* iso-*butanol*
(From reference 8, with permission of Oxford University Press)

connectivity values, path, cluster, and chain connectivities. Connectivity indices have proved useful in the correlation of a number of simple chemical properties and also biological activities.[10] They have the disadvantage that there is no simple way to assign a chemical interpretation to them and for this reason many QSAR practitioners have avoided them. They have the distinct advantage, however, that they can be calculated rapidly and unambiguously for any structure; recent reports, for example, have shown that they may be used to compute solubility[11] and partition coefficient[12] values for libraries of compounds.

The increasing accessibility of computational chemistry and molecular modelling packages during the 1980s, due in part to the rapidly decreasing cost of computing power, led to the widespread use of these systems to calculate molecular descriptors. The advantage of this approach is clear since it no longer requires the use of look up tables, with their often missing values, and the parameters produced are, in the main, easily interpreted. There are problems in that a "reasonable" 3-D structure of the molecules is usually required and the very diversity of the properties available can be confusing.[13] The use of molecular modelling packages resulted in the creation in the late 1980s of a new form of QSAR known as 3D QSAR. This technique is discussed separately in Section 6.

Of the 'traditional' descriptors used in QSAR, hydrophobicity is probably the single most successful type as shown in a survey of the QSAR journal in 1998 where over 50% of the QSAR models involved a hydrophobicity term.[13] Many different parameters have been proposed for this property, based on experiment and theoretical calculations, but the original concept of octanol/water log$P$ remains the firm favourite. The substituent constant $\pi$ is infrequently used now, partly because candidate molecules tend to more complex than simple substituted aromatics but mainly because there are a number of reliable methods for estimating log$P$.[14] The steric substituent constant, $E_S$, has largely been replaced by a parameter called molar refractivity (MR). This descriptor may be calculated from tables of fragment values and a number of computer programs are available for this purpose. Electronic effects remain the most difficult to parameterise and the search for successors to $\sigma$ has given rise to what is arguably the largest group of computed physicochemical properties.[13]

## 5 BUILDING QSAR MODELS

Building and interpreting QSAR models is a process which deters many medicinal chemists from the QSAR approach, since it involves a knowledge of mathematics and statistics which is seldom taught in a chemistry course. There are many excellent statistics textbooks but they can be impenetrable to those with little or no training in the area. This is unfortunate since modern statistics packages are very easy to use and require only a modest investment of time in order to understand their output and use them effectively. The book by Kubinyi,[4] a chemist himself, gives a thorough introduction to the field of QSAR and presents many useful examples of its application. This author has written a practical guide[8] to the use of many of the commonly used methods in QSAR, with the emphasis being on the application of the techniques rather than their mathematical or statistical basis.

For much of the 1960s and 70s, MLR was the only method used in QSAR modelling, with one or two notable exceptions. Recognition of the restrictions imposed by the MLR approach, proliferation of molecular descriptors due to the use of computational chemistry and the need to handle different forms of response data (*e.g.* classified) led to the gradual adoption of a wide variety of other techniques. There is insufficient space here to do more than briefly mention some features of the more important techniques but the two books (refs. 4 and 8) and the references in them will provide plenty of further reading.

## Multiple Linear Regression

Examination of the data in Table 2 will probably reveal a correlation between the response and $\pi$ by eye but what happens if we fit a line to the data:

$$\log(1/C) = 0.96(9.2)\pi + 0.65(7.6) \tag{9}$$

$$n = 8 \qquad R^2 = 0.93 \qquad F = 84 \qquad s = 0.23$$

The slope of the line (0.96) and the intercept (0.65) are known as regression coefficients and these are fitted to the data using a technique known as least squares. This simple linear regression equation contains all the statistical quantities that are necessary to evaluate the quality of any regression model, however complex. They are also all the statistics that would normally be quoted when such a model is published, although extra quantities may also be quoted and sometimes they are quoted in a different form (see later). So, what do the numbers mean? The equation for the line is quite clear. The $\log(1/C)$ value for a new compound in the set may be found by looking up its $\pi$ value, multiplying this by 0.96 and adding on 0.65. The numbers in the brackets after the regression coefficients are the values of the $t$ statistics for the coefficients. Without going into detail, the

$t$ statistic tells us whether the coefficient is significantly different from zero and is used by comparing the value with a tabulated value. A good rule of thumb for judging the significance of $t$ statistics is that they should usually be greater than 2.

What about the line of numbers below the equation? The letter $n$ refers to the number of data points (compounds) used in the fit. It is important to know this, as the value of many statistical quantities depends on the number of degrees of freedom (no. of data points – fitted coefficients). The next quantity in this list is the square of the correlation coefficient which has values in the range from zero (no correlation) to 1 (a perfect correlation). The square of the correlation coefficient multiplied by 100 gives the percentage of variation in the dependent variable (the response), explained by the regression model. In this example, the equation accounts for 93% of the variance in $\log(1/C)$, which is clearly quite a reasonable fit. The next entry in this list is the $F$ statistic, which is used to assess the overall significance of the fit. Briefly, $F$ statistics are used by comparing the calculated value with a tabulated value at a particular level of significance or confidence, although some statistics programs do this "look up" themselves. The tabulated value at the 95% confidence level for this data set is 10.13, a value comfortably bettered by the computed $F$ statistic so, again, this fit would be judged significant. The final quantity, $s$, is the standard error of prediction and it is important to compare this with the experimental standard error, if known. Clearly, a regression model should not be able to predict the data with greater precision than it can be measured. If $s$ is lower than the experimental $s$ then this is a good sign that the model is "overfitting" the data and should be reduced in complexity by removing terms.

This has been necessarily, a brief introduction to regression modelling and for a better understanding the reader should consult the two books cited earlier or, indeed, any standard statistics text. There is, however, one further important point that should be made here and that is the matter of correlations between the descriptor variables, a situation known as co-linearity. Consider the following two regression equations which describe the $pI_{50}$ for the inhibition of thiopurine methyltransferase by substituted benzoic acids in terms of calculated atomic charges.[15]

$$pI_{50} = 12.5(\pm 5.2)q_{2\pi} - 8.3(\pm 5.1) \tag{10}$$

$$n = 15 \qquad r = 0.757$$

$$pI_{50} = 12.5(\pm 4.8)q_{6\pi} - 8.4(\pm 4.8) \tag{11}$$

$$n = 15 \qquad r = 0.785$$

The two equations are almost identical which suggests that the two physicochemical descriptors should be correlated with one another. This is, indeed, true as the correlation coefficient between $q_{2\pi}$ and $q_{6\pi}$ is 0.997 ($r^2 = 0.994$ or 99% explained variance). Chemically, this is not surprising since the numbers 2 and 6 refer to substitution positions on the aromatic ring of benzoic acid, in other words the two *ortho* positions. Combination of these two descriptors into a multiple linear regression model gives Equation (12):

$$\mathrm{p}I_{50} = -74(\pm 57)q_{2\pi} + 84(\pm 56)q_{6\pi} - 6.3(\pm 4.5) \tag{12}$$

$$n = 15 \qquad r = 0.855$$

Here the correlation coefficient has increased, although that is to be expected since the equation contains an extra term, but notice that the regression coefficients have changed considerably with one even changing sign. The statistics quoted are in a different form to that used earlier, the correlation coefficient is shown un-squared and instead of reporting $t$ statistics, the author quotes the standard errors of the regression coefficients. These are easily obtained, however, by division of the regression coefficient by the standard error and, remembering the rule of thumb of 2 for significance, it can be seen that none of the terms in Equation (12) are significant. It should also be noted that in this case the original paper did not report values of the $F$ statistic.

Co-linearity in the descriptor set will lead to effects such as that seen in Equation (12) and thus should be avoided. It should also be avoided as it can be misleading in chemical terms, *i.e.* interpretation of a QSAR; reduction of co-linearity is another one of the aims of set selection.

## Principal Component Methods

Principal component analysis (PCA) forms the basis of a number of methods which are useful in QSAR, including a technique called partial least squares (PLS) which is an integral part of a popular 3-D QSAR method called CoMFA (see later). So what is PCA? In brief, the method is a mathematical transformation of a data set into a new data set, which contains a smaller number of variables made up from linear combinations of the original variables. These new variables, which may be called PCs or latent variables, have two important properties:

1. The first principal component explains the maximum variance in the data set, with subsequent components describing the maximum part of the remaining variance subject to the condition that:
2. All principal components are orthogonal to one another.

The structure of the PCs may be expressed in the form of equations as shown below:

$$\begin{aligned} PC_1 &= a_{1,1}v_1 + a_{1,2}v_2 + \ldots a_{1,N}v_N \\ PC_2 &= a_{2,1}v_1 + a_{2,2}v_2 + \ldots a_{2,N}v_N \\ &\ldots \\ PC_Q &= a_{Q,1}v_1 + a_{Q,2}v_2 + \ldots a_{Q,N}v_N \end{aligned} \tag{13}$$

This representation may appear daunting at first sight but is, in fact, easily explained. The subscript 1 to $Q$ for the PCs simply represents the number of that principal component. The subscript to $v$ represents the original variable number from a set of $N$ variables and the coefficient, $a$, represents the contribution of that variable to each PC. Thus, $a_{1,1}$ is the contribution of the first variable to the first PC, $a_{3,2}$ is the contribution of the second variable to the third PC and so on. These coefficients are called the loadings of the variables on to the PCs, their sign gives the direction of the loading (also the correlation of the variable to that PC) and their magnitude describes their contribution to the variance explained by that PC. The new variables, the PCs, are called PC scores and so compound 1 which was originally described by the values of the $N$ original variables will now be described by the values of $Q$ PC scores. PCA will generate as many principal components as the smaller of $N$ (variables) or $P$ (data points or compounds). This process is shown diagrammatically in Figure 5.

So, how is PCA useful? The new data matrix is smaller than the original, the new variables, the PCs, are orthogonal to another and they explain the variance (information) in the original data set in a known way. One useful application of this technique is to use the PC scores as a means of representing the original high-dimensional data ($N$ dimensions where $N$ is the number of original variables) as a two- or three-dimensional plot as shown in the next section. Another application is to use the PC scores in building regression models. Equation (14) shows the $ED_{50}$ data for a set of benzodiazepines described by three principal components ($X_1$, $X_2$ and $X_3$) computed from a set of quantum chemical descriptors.

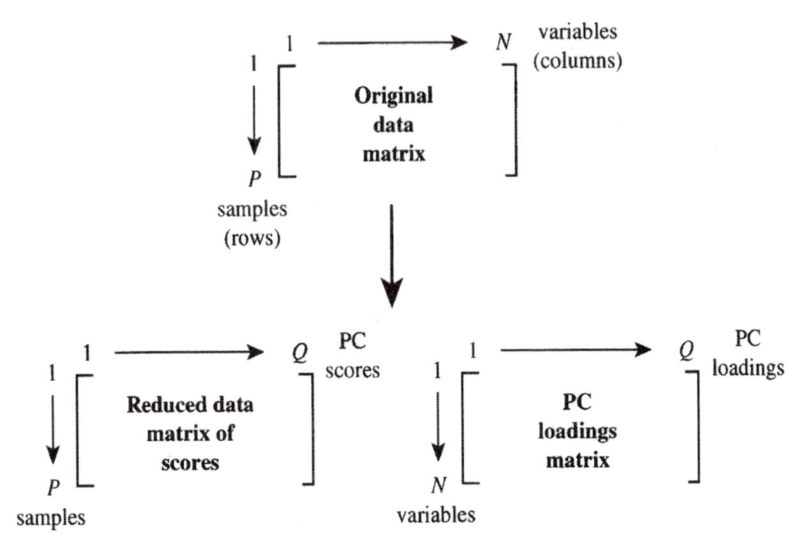


**Figure 5** *Illustration of the process of principal components analysis to produce a 'new' data matrix of* Q *scores for* P *samples where* Q *is equal to (or less than) the smaller of* N *(variables) or* P *(samples). The loadings matrix contains the contribution (loading) of each of the* N *variables to each of the* Q *principal components*
(From reference 8, with permission of Oxford University Press)

$$\log \mathrm{ED}_{50} = -0.25(\pm 0.07)X_1 + 0.2(\pm 0.08)X_2 + 0.34(\pm 0.16)X_3 + 1.34 \qquad (14)$$

$$n = 19 \qquad R^2 = 0.89 \qquad F = 39.26$$

## Data Display

The use of multiple physicochemical properties, measured, calculated or theoretical, to characterise a set of molecules generally provides a better, or at least more complete, description of the compounds than the use of just one or two. The disadvantage of using such high-dimensional or multivariate data sets is that we cannot take advantage of our natural ability to recognise patterns visually. PCA provides one means to overcome this by plotting the scores (for compounds) or loadings (for variables) on the axes of two or three principal components.

In Figure 6, the loadings of a set of biological tests and two physicochemical properties are shown plotted against the first two factors from a factor analysis (factor analysis is related to PCA, the two factors have similar meaning to the first two principal components).[16] The two points Ke* and Ke-pred* are experimental and predicted values of a measure of chemical reactivity and, since they are close together on the

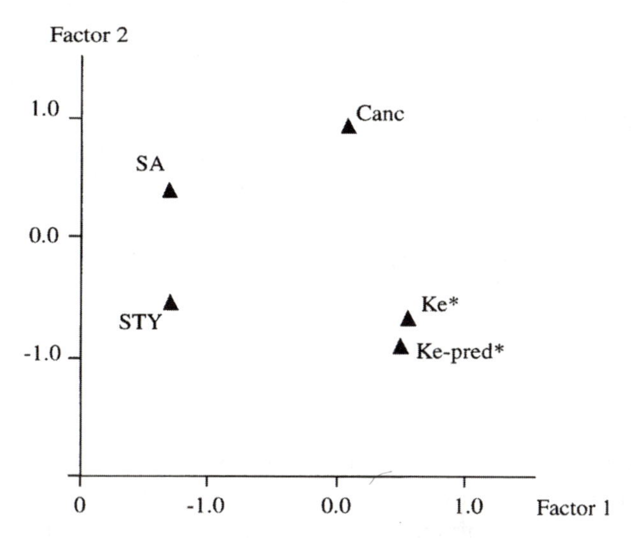


**Figure 6** *Loadings plot from a factor analysis of* in vivo *and* in vitro *tests, and measured and calculated physicochemical descriptors*
(From reference 16 by permission of Oxford University Press)

plot, are seen to be correlated for this set of compounds. The other three points are Canc, an *in vivo* measure of carcinogenicity, STY, an experimental measure of mutagenicity and SA, the result of a manual method of alerting for carcinogenicity based on chemical structure. This plot shows that none of the other four descriptors are particularly well related to the *in vivo* Canc results, with the structural alerts (SA) being the most similar.

Cluster analysis is another means of visualising high-dimensional data sets since the dendrogram is a two-dimensional illustration of the relationship between data points in the original *N*-space. A technique known as non-linear mapping (NLM) or multidimensional scaling can be used to produce an analogous plot to a PC scores plot using, as the name implies, a non-linear procedure. The process can be easily described as follows. For any given data set of points in *N* dimensions, it is possible to calculate the distances between pairs of points by means of an equation such as that shown in (15).

$$d_{ij} = \sqrt{\sum_{k=1,N} (d_{i,k} - d_{j,k})^2} \tag{15}$$

This is the expression for the Euclidean distance where $d_{ij}$ refers to the distance between points $i$ and $j$ in an $N$-dimensional space given by the summation of the differences of their co-ordinates in each dimension ($k = 1, N$). The distances for the starting data set are fixed and labelled $d^*_{ij}$ to indicate that they are $N$-space distances. The next step is to randomly (usually) assign positions to the data points in a two- or three-dimensional space and calculate distances between them, again using Equation (15). The two sets of distances, $d_{ij}$ in two dimensions and $d^*_{ij}$ in $N$-dimensions, may be compared using an error function such as that shown in Equation (16).

$$E = \sum_{i>j}(d^*_{ij} - d_{ij})^2/(d^*_{ij})^\rho \tag{16}$$

Minimisation of this error function can be achieved by altering the positions (the coordinates) of the data points in the lower dimensional space and, once a minimum has been reached, the corresponding co-ordinates may be used to produce a two- or three-dimensional plot of the data. This plot, known as a non-linear map, is a 'best' representation of the data in a low-dimensional space whilst retaining the relative inter-point distances from the $N$-dimensional space. Figure 7 shows a non-linear map of 166 substituents described by six physicochemical properties[17] (a 6-dimensional space), with the same data set used by Hansch and Leo to produce tables of clusters of substituents as described earlier (Table 3).

### Classified Data

The biological response data shown so far, such as $IC_{50}$, $ED_{50}$ and so on, has all been measured on a continuous scale but what if the results are only available as a classification, *e.g.* active/inactive, toxic/non-toxic, *etc.*? A variety of methods exist for the treatment of classified data and they can yield as much information about a QSAR as the techniques which operate on continuous data. Indeed, in some cases where, for example, the response is poorly distributed, it may be better to convert continuous data into a classification.

One technique, which operates with a classified response is known as $k$-nearest-neighbour (KNN). The starting point for KNN is the calculation of a distance matrix just as for non-linear mapping using an equation such as Equation (15). KNN operates with multiple physicochemical descriptors, just like PCA and NLM, and works by comparison of the classification of the $k$ nearest neighbours of an unknown sample. The value of $k$ is usually taken to be an odd number so there can be a majority vote with no ties, but choice of the value of $k$ is problematic and is usually chosen empirically as that number which gives the best results on the training set.

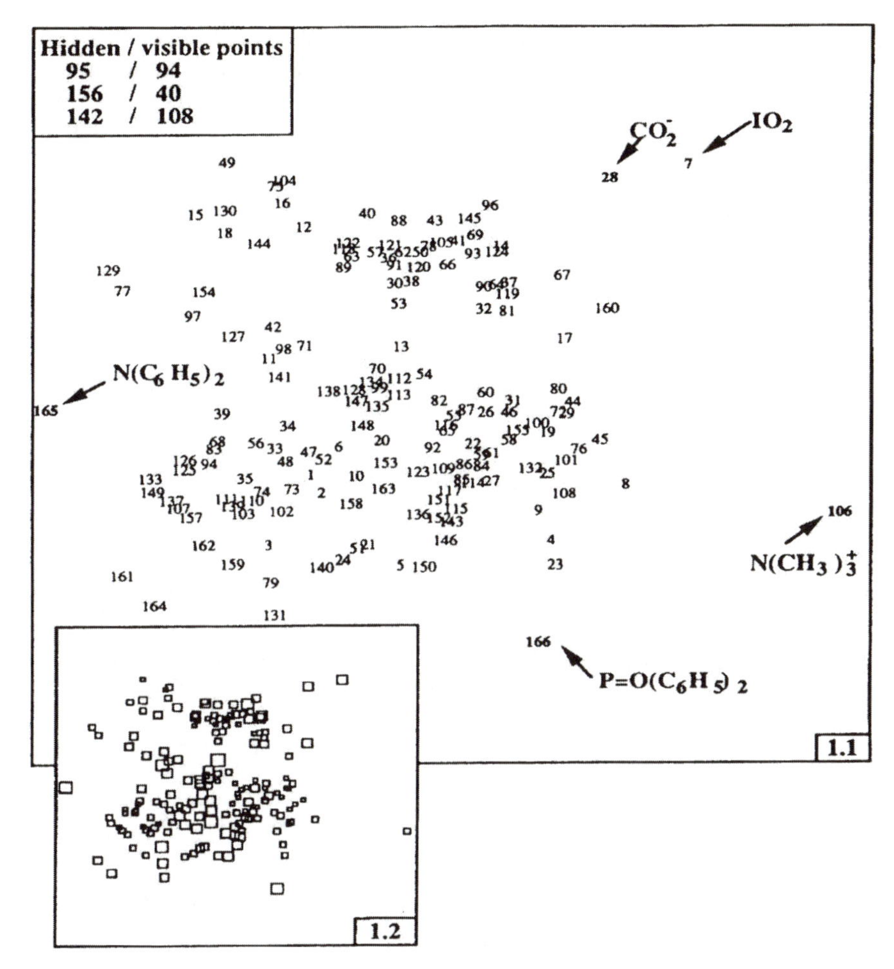


**Figure 7** *Non-linear map of substituents*
(From reference 17, copyright (1994) American Chemical Society)

An impressive example of the application of KNN has been reported for the prediction of antineoplastic activity.[18] The training set consisted of 138 structurally diverse compounds tested in a mouse brain tumour system. The compounds were described by a large number of sub-structural descriptors, 421 in all, and various procedures were adopted to reduce this to smaller sized sets. The results of the application of KNN to a test set of 24 compounds are shown in Table 4.

Discriminant analysis is the classified data equivalent of multiple linear regression. Where regression aims to fit a line or surface *through* a set of data points, discriminant analysis fits a line or surface *in between* two classes of points. This is shown in Figure 8 where the points represent

**Table 4** *Comparison of predicted and observed antineoplastic activities* (From reference 18, copyright (1975) American Chemical Society)

| | *Non-active* | *False negative* | *Active* | *False positive* |
|---|---|---|---|---|
| KNN | 14 | 1 | 6 | 3 |
| Experimental | 17 | 7 | | |

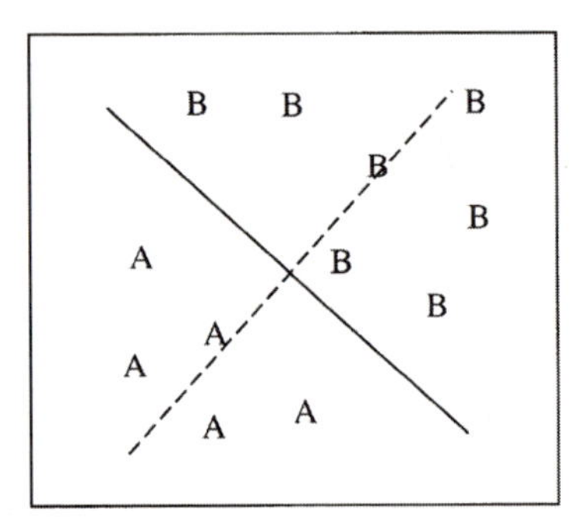


**Figure 8** *Two-dimensional representation of discriminant analysis. The dashed line represents the discriminant function and the solid line a discriminant surface which separates the two classes of samples*
(From reference 8, with permission of Oxford University Press)

compounds belonging to one of two classes, A or B, and the line represents a discriminating surface.

The discriminant function may be represented by an equation, which has a similar form to an MLR equation:

$$W = a_1x_1 + a_2x_2 + \dots a_nx_n \tag{17}$$

Here the $\chi$s represent the physicochemical properties used to describe the compounds and the *a*s are their coefficients computed by the fitting procedure. Table 5 shows the results of the application of stepwise discriminant analysis to mutagenicity data collected on a set of 90 structurally diverse compounds. The compounds were characterised by three different sets of physicochemical descriptors (see Table 5 for details) and three different variable selection procedures were used to "prune" the sets.[19,20]

As can be seen from Table 5, this discriminant analysis was quite successful, giving fitting results of 90 to nearly 98% correct and 'jack-knife' results of 86 to nearly 97% correct for various combinations of the descriptors. The term 'jack-knife' refers to a means of evaluation of a

**Table 5** *Summary of discriminant analysis results for a set of 90 mutagens* (Reproduced from reference 13, copyright (2000) American Chemical Society)

| *ID analysis* | *Pooled data set*[a] | *No. of variables*[b] | *Classification function* (% correct) | *Jack-knifed validation* (% correct) |
|---|---|---|---|---|
| Pool 1 | EVA 342 SMC + WHIM SMC | 6 | 92.2 | 90.0 |
| Pool 2 | EVA 342 PDRv2.0 + WHIM SMC | 9 | 96.7 | 96.7 |
| Pool 3 | EVA 288 CORCHOP + WHIM SMC | 9 | 97.8 | 95.6 |
| Pool 4 | EVA 342 SMC + TSAR 58 | 9 | 92.2 | 90.0 |
| Pool 5 | EVA 342 PDRv2.0 + TSAR 58 | 9 | 93.3 | 91.1 |
| Pool 6 | EVA 288 CORCHOP + TSAR 58 | 5 | 90.0 | 86.7 |
| Pool 7 | EVA 342 SMC + TSAR stand. | 9 | 92.2 | 90.0 |
| Pool 8 | EVA 342 PDRv2.0 + TSAR stand. | 8 | 93.3 | 91.1 |
| Pool 9 | EVA 288 CORCHOP + TSAR stand | 5 | 90.0 | 86.7 |
| Pool 10 | WHIM SMC + TSAR 58 | 8 | 97.8 | 94.4 |
| Pool 11 | WHIM SMC + TSAR stand. | 8 | 94.4 | 92.2 |
| Pool 12 | EVA 342 SMC + WHIM SMC + TSAR 58 | 9 | 97.8 | 94.4 |
| Pool 13 | EVA 342 PDRv2.0 + WHIM SMC + TSAR 58 | 9 | 97.8 | 94.4 |
| Pool 14 | EVA 288 CORCHOP + WHIM SMC + TSAR 58 | 9 | 96.7 | 93.3 |
| Pool 15 | EVA 342 SMC + WHIM SMC + TSAR stand | 10 | 96.7 | 92.2 |
| Pool 16 | EVA 342 PDRv2.0 + WHIM SMC + TSAR stand | 10 | 96.7 | 92.2 |
| Pool 17 | EVA 288 CORCHOP + WHIM SMC + TSAR stand | 8 | 95.6 | 92.2 |

[a] *EVA, WHIM and TSAR refer to variables computed using the EVA and WHIM methods and the structure activity program TSAR. CORCHOP, SMC and PDRv2.0 refer to variable selection methods.*
[b] *The number of variables included in the discriminant function, see reference 31 for details.*

model, any model, by the process of omitting a compound, building the model with the remainder and then predicting the left out compound. This procedure is repeated until every compound in the set has been left out once. This is also referred to as cross-validation or leave-one-out (LOO). Cross-validation can also mean the omission of groups of compounds but, unless otherwise stated, it usually refers to leave-one-out. The statistics computed by this process are often labelled with a subscript CV to distinguish them from their fitting equivalents. An evaluation of the potential predictive performance of a model by this means is not perfect, for a number of reasons, but it does give an indication and should result in lower performance statistics.

## 3D QSAR

There are a number of techniques which may be described as '3D QSAR' but this section will just address two closely related methods, comparative molecular field analysis[21] (CoMFA) and GRID. Both of these techniques arose from molecular modelling studies. CoMFA was designed at the outset to characterise sets of small molecules. GRID, on the other hand, was initially described as 'a computational procedure for determining energetically favourable binding sites on biologically important macromolecules'.[22] The two techniques both involve the calculation of interaction energies between a probe and a small molecule, or part of a macromolecule, at points in space defined by a grid placed around the target structure. Figure 9 shows a molecule surrounded by a 10 × 10 × 10 grid of points and it can be seen from this how a data matrix consisting of interaction energies can be built up.

In the case of CoMFA the steps involved in the modelling procedure can be summarised as follows:

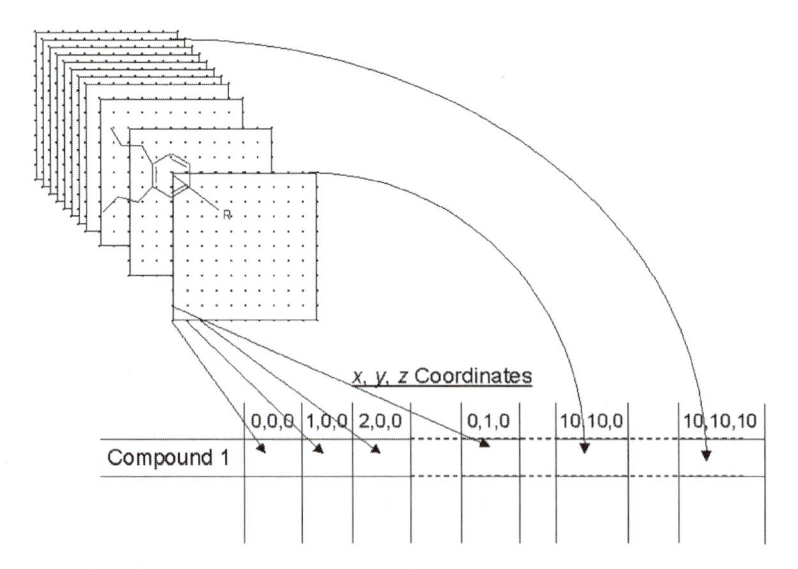


**Figure 9** *Illustration of the CoMFA and GRID procedures for a 10 × 10 × 10 matrix* (From reference 13, copyright (2000) American Chemical Society)

- Obtain a suitable 3D structure for each molecule in the training set.
- Derive partial atomic charges so that an electrostatic field can be generated.
- Align the molecules using some suitable alignment strategy (after conformational analysis if required).
- Create a cubic lattice of points around the molecules (usually larger than the largest member of the set).
- Compute interaction energies using a probe such as a pseudo methyl group with a unit positive charge. This generates a steric interaction energy based on a Lennard-Jones potential and an electrostatic interaction energy based on a coulombic potential.
- Fit a PLS model to the biological response and the interaction energies.
- Make predictions for a test set, visualise the results as contour plots on displays of the individual molecules in the set.

The last step but one mentions fitting a PLS model, but what is this? As mentioned in the section on principal component analysis above, PLS is related to PCA. It is also related to MLR, as the PLS process involves the calculation of new variables which:

- Explain maximum variance in the X set.
- Are orthogonal to one another.
- Are maximally correlated with the Y set (usually only 1 variable).

The first two properties of these PLS variables (usually called latent variables or LVs) are the same as for principal components but the last property is a characteristic of PLS. This process results in models, which describe the variance in the response as a regression equation made up of a small number of LVs and which can be used for prediction in much the same way as an MLR equation.

The CoMFA approach is only available in the SYBYL molecular modelling package but CoMFA type models may be constructed using the program GRID along with a PLS modelling algorithm. So, what are the principal differences between CoMFA and GRID? A major difference is that GRID uses more probes for the determination of interaction energies, presumably as a result of the original design aim of GRID to investigate macromolecular binding sites. In addition to probes such as $CH_3$, $NH_2$, $NH_3^+$ and $O^-$, considerable effort has been concentrated on the parameterisation of hydrogen bonding in the GRID force-field and the current version of GRID offers 56 probes from the regular menu with nine multi-atom probes available from an auxiliary menu.

The popularity of CoMFA and related approaches can be seen in a book dedicated to 3D QSAR[23] and progress since that book has been remarkable; there have been hundreds of reported applications of the technique and those published between 1993 and 1997 are nicely collated by Kim.[24]

## 6 ARTIFICIAL INTELLIGENCE

It may, at first sight, appear strange to have a section headed 'Artificial Intelligence' in a chapter on QSAR or, indeed, in a book about medicinal chemistry. There are, in fact, a surprising number of artificial intelligence (AI) based methods employed in QSAR and computer-aided drug design techniques. Expert systems, for example, are used to calculate log*P*, estimate 3D chemical structure, plan reaction routes, provide structural alerts for carcinogenicity and estimate toxicity for various endpoints. Molecular modelling packages may be thought of as expert systems since the "force field" used in molecular mechanics essentially consists of a set of rules governing atomic radii, bond lengths, angles and torsion angles. The calculation of a molecular mechanics energy and its subsequent minimisation to produce a low energy structure is achieved by the application of these rules, originally devised by humans. Although there are various ways in which the systems operate, the heart of any expert system consists of a set of rules, sometimes referred to as a rule base or knowledge base, which has been put together by "experts". The definition of an expert is problematical but usually means anyone who has an opinion on the problem!

One AI technique, which has found many applications in situations where a solution is sought in a very large answer 'space', is the genetic algorithm (GA). Although there are various types of GA they mostly share common features in that the solution to a particular problem is represented by a genetic vector made up of 'genes', which are the variables or parameters of the problem. The GA algorithms then make alterations to the genetic vector using the evolutionary techniques found in biology, *i.e.* mutation, crossover, mating. Typically, a GA will start with a population of solutions, which may have been created randomly, or by variation of some known starting solutions. These solutions are assessed using some fitness function and a number of the best solutions are kept with others being created from them. This new population of solutions is assessed, the worst discarded and a new population created. The GA cycles until some particular target value of the fitness function is achieved or a set number of iterations is reached or some other stopping criterion is applied.

An example may serve to illustrate this process. Figure 10 shows the structure of a simple molecule, a pyrethroid insecticide, with five rotatable

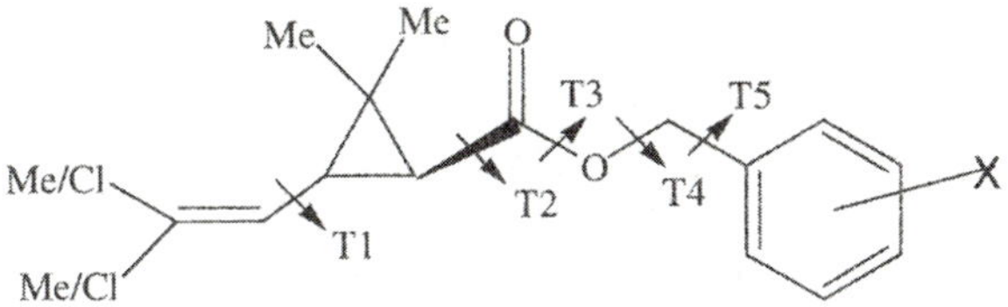

**Figure 10** *Illustration of the rotatable bonds in the pyrethroid parent structure* (From reference 25, copyright (1992) Wellcome Foundation Ltd)

bonds labelled $T_1$ to $T_5$. Actually one of these ($T_3$) is effectively fixed as it is part of the ester group. The structure of any particular conformation of this compound may be represented by a genetic vector made up from the values of the torsion angles of these four bonds, $T_1$, $T_2$, $T_4$, $T_5$. The fitness function, in this case, may be taken as the molecular mechanics energy, which is computed using an equation such as (18):

$$E = \tfrac{1}{2}\sum K_b(b - b_0)^2 + \tfrac{1}{2}\sum K_\theta(\theta - \theta_0)^2 + \tfrac{1}{2}\sum K_\phi(1 + s.\cos(n\phi)) \qquad (18)$$

In this simple example of a molecular mechanics force field the three terms represent the energy contribution due to bond lengths, bond angles and torsion angles, respectively. The summations are over all bonds and the constant terms, $K$, are the force constants for the distortion of these terms from their standard values (denoted $b_0$ and $\theta_0$ for the bond lengths and angles). At each full cycle of the GA, the energy of the solutions in the population are evaluated, the best solutions kept and a new population 'bred' from them. By this means a population of 'good' conformations may be very rapidly discovered in an enormous solution space.

GA algorithms and their variants have been applied to molecular modelling, experimental design, overlay and docking and the generation of QSAR models such as PLS and MLR equations. An example of the latter will perhaps illustrate the power of the GA approach. Kubinyi has examined[26] a well-known data set which consists of 31 molecules described by 53 computed molecular descriptors. It was calculated that there are $7.16 \times 10^{15}$ possible regression models that can be computed for this data set; at a rate of generation of one regression equation every second it would take 226 million years to evaluate them all. Using a GA approach called MUSEUM, Kubinyi was able to discover a set of good regression models in seconds from this solution space.

Another popular AI method is known as Artificial Neural Networks (ANN). These algorithms have arisen from AI research, which aimed to replicate the structure of the brain, rather than emulating the functions of

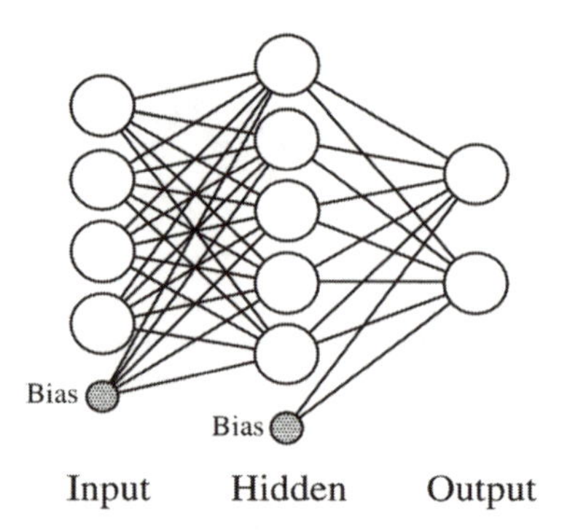


**Figure 11** *Diagram of an artificial neural network consisting of three layers of neurons – an input layer, hidden layer and output layer. Each neuron in a layer is connected to every neuron in the next layer*
(From reference 8, with permission of Oxford University Press)

the brain. ANN consist of layers of artificial neurons connected together as shown in Figure 11.

Training a neural network typically consists of presenting descriptor data to the input neurons and comparing the output with some known target, *e.g.* $\log(1/C)$. The connection weights between the neurons are altered to produce a better fit between the calculated and observed data and this process is conducted through a series of iterations until some desired accuracy is achieved. One of the advantages of the ANN approach for model building is that the underlying model may be non-linear. One disadvantage of this technique is that the actual model is 'hidden' amongst the connection weights and although methods have been proposed to discover the models, interpretation is not straightforward. Applications of ANN in medicinal chemistry have been recently reviewed.[27]

## 7 SUMMARY

The QSAR approach to drug design has much to offer the medicinal chemist. The techniques may appear daunting at first sight but with a little effort even the most complex methods may be readily understood and applied. The field has developed rapidly over the last 30 years or so, both in terms of the properties used to describe molecular structure[13] and the techniques used to relate activity to these properties.[8,28] Applications of QSAR in drug design are frequently reported in the medicinal chemistry literature, in reviews[4,23,27,29] and in the proceedings of conferences.[30] Recent advances in medicinal chemistry, such as combinatorial synthesis and high throughput screening, coupled with the established and developing methodologies of QSAR and molecular modelling will continue to provide useful and life saving medicines.

## 8 REFERENCES


1. A. Crum-Brown and T. Frazer, *Trans. Roy. Soc. Edinburgh*, 1868–9, **25**, 151.
2. M. Charton, in *Advances in Quantitative Structure–Property Relationships*, M. Charton (ed.), JAI Press, Inc., Greenwich, 1996, pp. 171–219.
3. C. Hansch, *et al.*, *Nature*, 1962, **194**, 178.
4. H. Kubinyi, *QSAR: Hansch Analysis and Related Approaches*, VCH, Weinheim, 1995, pp. 15–16.
5. C. Hansch, *et al.*, *J. Am. Chem. Soc.*, 1963, **85**, 2817.
6. C. Hansch, *et al.*, *Exploring QSAR. Hydrophobic, Electronic, and Steric Constants*, American Chemical Society, Washington DC, 1995.
7. P.N. Craig, *J. Med. Chem.*, 1971, **14**, 680.
8. D.J. Livingstone, *Data Analysis for Chemists. Applications to QSAR and Chemical Product Design*, Oxford University Press, Oxford, 1995, Chapter 2.
9. V. Austel, *Eur J. Med. Chem.*, 1982, **17**, 9.
10. *Topological Indices and Related Descriptors in QSAR and QSPR*, J. Devillers and A.T. Balaban (eds.), Gordon and Breach Science Publishers, Amsterdam, The Netherlands, 1999.
11. J.J. Huuskonen *et al.*, *Eur. J. Med. Chem.*, 2000, **35**, 1081.
12. J.J. Huuskonen *et al.*, *J. Chem. Inf. Comput. Sci.*, 2000, **40**, 947.
13. D.J. Livingstone, *J. Chem. Inf. Comput. Sci.*, 2000, **40**, 195.
14. R. Mannhold and K. Dross, *Quant. Struct.-Act. Relat.*, 1996, **15**, 403.
15. O. Kikuchi, *Quant. Struct.-Act. Relat.*, 1987, **6**, 179.
16. R. Benigni *et al.*, *Carcinogenesis*, 1993, **13**, 547.
17. J. Devillers, in *Chemometric Methods in Molecular Design*, H. van de Waterbeemd (ed.), VCH, Weinheim, 1995, pp. 255–263.
18. K.C. Chu *et al.*, *J. Med. Chem.*, 1975, **18**, 539.
19. D.J. Livingstone and E. Rahr, *Quant. Struct.-Act. Relat.*, 1989, **8**, 103.
20. D.C. Whitley *et al.*, *J. Chem. Inf. Comput. Sci.* 2000, **40**, 1160.
21. R.D. Cramer *et al.*, *J. Am. Chem. Soc.*, 1988, **110**, 5959.
22. P.J. Goodford, *J. Med. Chem.*, 1985, **28**, 849.
23. *3D QSAR in Drug Design: Theory Methods and Applications*, H. Kubinyi (ed.), ESCOM, Leiden, 1993.
24. K.H. Kim, *Perspect. Drug Discov. Des.*, 1998, **12/13/14**, 257.
25. B.D. Hudson *et al.*, *J. Comput.-Aid. Mol. Design*, 1992, **6**, 191
26. H. Kubinyi, *Quant. Struct.-Act. Relat.*, 1994, **13**, 393.

27. D.T. Manallack and D.J. Livingstone, *Eur. J. Med. Chem.*, 1999, **34**, 195.
28. D.J. Livingstone in *Methods in Enzymology*, J.J. Langone (ed.) Academic Press, San Diego, 1991, Vol. 203, pp. 613–638.
29. D.J. Livingstone in *Structure–Property Correlations in Drug Research*, H. van de Waterbeemd and R.G. Landes (eds.), Austin, USA, 1996, pp. 81–110.
30. See, for example, proceedings of the 12th European Symposium on QSAR – *Molecular Modelling and Prediction of Bioactivity*, K. Gundertofte and F.S Jørgensen (eds.), Kluwer Academic/Plenum Publishers, New York, 2000.
31. M.D. Smith, PhD Thesis, University of Portsmouth, Portsmouth, UK, 1999.

CHAPTER 12

# Computational Chemistry and Target Structure

COLIN EDGE

## 1 INTRODUCTION – THE BASIC TOOL KIT

Computational chemistry can be defined as the use of calculations to describe and predict the properties of molecules. This review will concentrate on computational chemistry applied to pharmaceutical research, particularly techniques to aid drug discovery. Computational chemistry has spawned modern sub-disciplines such as bioinformatics and cheminformatics and is a wide area in its own right, so we shall visit key areas in brief detail and highlight recent review articles to which the interested reader can refer for more information.

A basic tool kit of computational chemistry for drug discovery work is shown in Figure 1. This can be broken down simply into programs for drug-sized molecules, programs for target sized molecules and programs to combine the two. These programs tend to run on Unix graphics work stations, but are increasingly available for PCs, running either Windows NT or Linux.

### Graphics Programs

The first program one needs for looking at drug-sized (small) molecules is a graphics package and all the major software vendors supply such packages. These programs can be used to build computer models of molecules, to modify existing models and to calculate simple properties, such as geometries and energies. They usually contain a force field representation of a molecule, which is a simple classical mechanics description of a molecule in terms of bond lengths, angles, torsions and

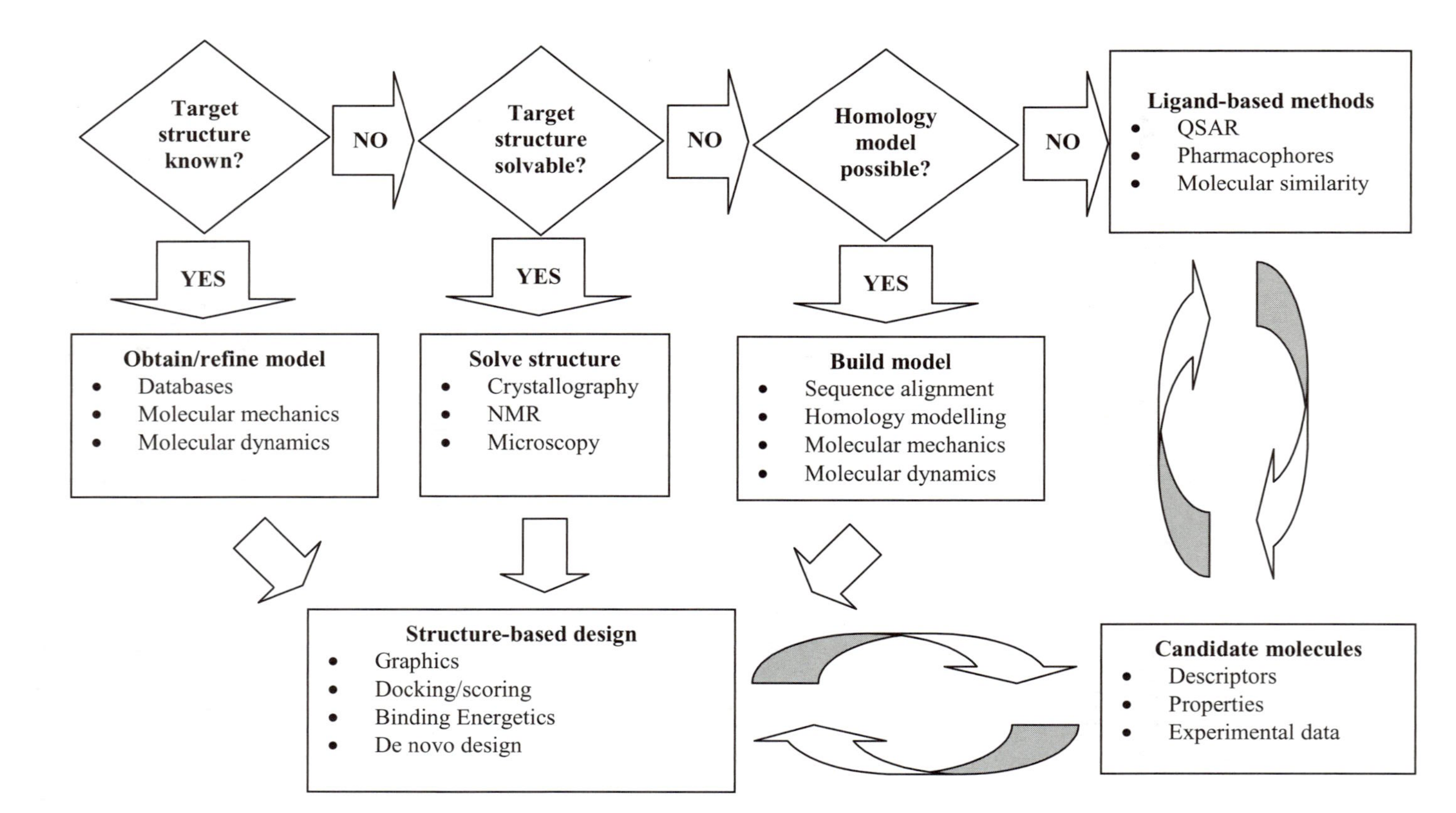


**Figure 1** *A basic tool kit of computational chemistry for drug discovery work*

simple, non-bonded forces such as electrostatics and van der Waals repulsion and attraction.

The graphics programs also act as central organisers of project work, as there are often associated programs that can be run from the main console. These ancillary programs perform further characterisations of the molecule of interest, such as the determination of additional properties or the exploration of molecular conformation. One important set are the quantum chemistry programs that calculate properties of isolated molecules as properties derived from the wave function. The properties fall into two broad camps: energy based, such as preferred conformation, interaction energy, vibrational frequencies and thermochemical stabilities, and those based on the electronic distribution, such as charge transfer, dipole moment and electrostatic potential. There are three broad classes of theoretical method used. The most commonly used are the semi-empirical methods found in programs like MOPAC, AMPAC and VAMP. Then there are the *ab initio* molecular orbital (MO) codes, such as Gaussian and Spartan. Although valence bond formalism programs exist, in practice they are not used in pharmaceutical studies. The final category are the density functional methods. These rely on an abstraction of the overall electron density of the molecule. As the programs have become faster and more widely available, particularly by their incorporation into commonly used packages such as Gaussian, they have become direct competitors to the low level MO calculations.

## Protein Modelling Programs

If one has structural information on the target of interest, perhaps an NMR or X-ray crystal structure, it is advisable to use a specialist protein modelling package rather than the general purpose ones mentioned above. These are often tightly coupled to a program that investigates protein structure and conformation *via* a specialised polypeptide force field, such as CHARMM or AMBER. This force field is the source of the power and also the Achilles heel of these modelling packages as they are heavily tuned towards the observed behaviour and properties of typical proteins. Whilst this is fine for most small to medium sized soluble proteins, one runs into problems if one encounters amino acids that are not in the standard basis set of the twenty commonly occurring amino acids. For instance, the hydroxyproline residues found in collagen often have to be built by modifying standard prolines either by applying a modification script (a patch in the jargon) or even by writing one's own description.

Another major feature of protein modelling software is the provision of the requisite tools for the building of model structures based on related proteins. This is termed homology modelling. This is an important activity for many computational chemists as there are 'only' 14055 protein structures available as of the end of 2000 and these represent about 5000 different templates for homology modelling.[1]

### Programs that Combine Receptor and Ligand

The third category of computational chemistry programs are those that combine ligands and receptors. These either analyse the binding interaction, for instance docking programs, or suggest novel molecules that might fit a cavity, *i.e. de novo* design programs.

### Abstract Site Models

As well as the programs that use atomic models of the target site, there are also programs that use an abstract model of the site. This may be because structural information is not available, but there are identifiable structure-activity relationships (SAR). In this case, one can build a pharmacophore model of the salient features needed for efficacious binding and search a database of candidate molecules.

## 2 STRUCTURAL INFORMATION FOR COMPUTATIONAL CHEMISTRY

The use of molecular structure in medicinal chemistry can be traced back a century to predictions of simple physicochemical properties dependent on molecular composition. However, computational chemistry, in the modern sense of a discipline where molecular properties and behaviour are modelled using computers and computer graphics, is much younger. This author, along with many others, would argue that the earliest triumph of molecular modelling was the determination of the double helical structure of DNA by Watson, Crick and Wilkins in 1953.[2] The structure was solved by using X-ray crystallographic data to compare proposed models of the DNA chain. The models were laboriously built by hand using wood and metal. Today, models that exist only within computers, on disk or even in memory, are the norm.

### X-ray Crystallography

Since the time of Watson and Crick, X-ray crystallography has played an important part in supporting computational chemistry.[3] Most medium to large pharmaceutical companies possess X-ray crystallography groups and there are a growing number of start-up concerns concentrating on structure-based drug design.

The major bottleneck in structure determination by crystallography is the production of viable crystals from purified protein, so high throughput protein crystallisation is an area of current research.[4] An interesting alternative is the possibility of using extremely powerful X-ray lasers on single molecules, collecting reflections in the brief instant before the molecule is destroyed by radiation, thus negating the need for a crystal altogether![5]

## Nuclear Magnetic Resonance

Nuclear magnetic resonance is the other major method used by the pharmaceutical industry to determine three-dimensional structure.[6] There are roughly 100 new structures in the literature each year.[7] Admittedly there are limitations on the use of NMR; size, solubility and provision of labelled material spring to mind, but the technique offers insights that are missing from crystallographic analyses. One recent application that is pertinent to drug discovery is the so-called SAR by NMR approach reported by Fesik's group.[8] In this, one maps the target site of a nitrogen15-labelled protein by analysing a collection of small molecules and determining the binding constants by heteronuclear single quantum correlation (HSQC) spectroscopy. Structurally related analogues of any hits are then tested, to try to optimise the binding affinity. Once this has been done for multiple sites in the protein, linkers can be designed to connect the optimal fragments, with the hope that the resulting molecule will have much improved affinity. Computational chemistry plays a part in this process both in the identification of structurally related test molecules and in the design of the linker.

## Structural Databases

At the end of 2000 there were 14055 crystal structures deposited in the Protein Data Bank (PDB, http://www.rcsb.org/pdb). These have been organised by various groups into hierarchical databases, based on structural characteristics of the proteins. The databases are publicly available *via* the Internet and are an excellent source of structural information. Notable examples are the SCOP database of Murzin *et al.*[9] and the CATH database of Orengo *et al.*[10] Both organise protein structures into families and superfamilies on the basis of structure and sequence. Proteins which are thought to share a common evolutionary ancestor (homologues) will be clustered together. In addition to these related proteins, one also finds unrelated proteins (analogues) that share a common fold. These databases are obviously a good source of basic material for homology modelling, that is building models of proteins of unknown structure.

## 3 THE USE OF STRUCTURE IN DRUG DESIGN

Until recently, structure-based drug design has been seen to be in competition with the 'less rational' combinatorial chemistry/high-throughput screening approach. Happily, this is no longer the case, as can be seen from recent reviews in the literature dealing with the synthesis of combinatorial chemistry and rational design.[11–13] The main use of structure in pharmaceutical research has been the characterisation of ligand–receptor binding modes and the subsequent prediction of new pairings. This has resulted in several success stories, including inhibitors of HIV protease, cathepsin K, neuraminidase and thymidylate synthase.

### Docking and Virtual Screening

Lately, reports have appeared in the literature of using protein structures, particularly enzymes, as templates for virtual screening exercises. In an analogous fashion to high throughput screening, thousands of candidate structures are docked into the protein and scored for the 'goodness' of fit. Although virtual screening seems to have been successful in the case of fairly rigid enzyme structures, such as proteases and synthetases, one must appreciate that the current scoring regimes are not at all accurate. The best one can hope for in virtual screening with today's technology is an enrichment of the number of molecules that might interact, compared to a random selection. This is fine for a true screening exercise, where one is not concerned about what one misses, as long as one finds useful leads for further chemistry. However, the faster docking algorithms, such as DOCK and ICM cannot be recommended for virtual SAR work.

How good are today's docking programs? There are a plethora available, for example Dock, Flo, Growmol, Flexx, Gold, Ludi, *etc.*[14,15] The fact that no one program has prevailed over the others, and that new docking programs are still being reported in the literature, suggests that the automatic insertion of ligands into a target site is not a mature discipline or a solved problem.

In an interesting and entertaining paper, Davis and Teague[16] show many examples in which, in their words, 'experimental determination of the structure of drug–receptor complexes often result in surprises'. They attribute this to the important part that hydrophobic interactions play in drug–receptor binding. Whilst hydrogen bonds are easy to identify, both by eye and by computer program, they do not contribute much to the overall binding of a complex unless at least one of the donor or acceptor are charged. Neutral hydrogen bonds are quite weak. The authors describe cases where hydrogen bonds have been successfully replaced with hydrophobic interactions, giving stronger binding, often at the expense of some lost conformational energy as the receptor moves to accommodate the ligand. This induced fit of receptor to ligand is also described by Williams and Bardsley in a recent paper, again in terms of the failure of the rigid receptor hypothesis, though it is unclear who still defends this theory![17] Nevertheless, both these papers show that hydrogen bonds have been accorded an undeserved prominence in the characterisation and prediction of ligand–receptor interactions.

This can be put in simple terms by examining the role of solvent water. Consider a hydrogen bond donor group, an $NH_2$ group, say, in a ligand. In aqueous solution, this will hydrogen bond to solvent water molecules surrounding it. As water is such a good hydrogen bonding medium, it will be easy to 'satisfy' the bonding requirements of the ligand. Now let us rip the molecule out of the solvent and place it in a protein pocket. If we do not create a hydrogen bond to the $NH_2$ group we shall be losing free energy as the water can rearrange to establish its own hydrogen bond requirements, but the ligand hydrogen bonds have been lost; if we create a good hydrogen bond, we shall just be breaking even. This is a simplistic argument, as it does not take account of the other energies gained or lost in the filling of the cavity, equilibration of the solvent or in the expulsion of solvent water from the protein cavity. Nevertheless, one can see from this simple model that one cannot expect too much free energy of binding by just considering hydrogen bonds.

The limitations of current docking methods have also been highlighted by Murray *et al.*[18] They found that their docking software, PRO_LEADS, could identify the correct binding conformation 76% of the time if the enzyme conformation was that of the crystal structure of the docked ligand. This is an impressive result. Unfortunately the program was only successful 49% of the time when the enzyme structure was in the conformation derived from the crystal structure of a different complex. This second case is the one that the computational chemist is generally interested in: one has a crystal structure, often with a ligand bound and one wants to predict the binding mode of another ligand.

## Prediction of Binding Energies

There have been several recent studies on the efficacy of binding energy predictions.[19–21] So and Karplus compare the performance of seven methods: five QSAR based methods and two structure-based methods. In this case, the structure-based methods performed quite poorly in correlating with the observed biological activities, reflecting the current state of the art. The QSAR methods fared better on the whole and a consensus method using results from all five worked better, on average, than any one method.

A slightly different approach to the prediction of binding affinities has been pioneered by Aqvist.[22] This uses a thermodynamic cycle approach which is fitted to experimental results. The method, known as the linear interaction energy approach, is much cheaper to calculate than the free energy perturbation method that was in favour ten years ago.[23] However, it is a semi-empirical approach, relying on experimentally determined free energies of binding and might be considered to be a glorified QSAR approach with all the attendant problems of extrapolation to uncharacterised systems. If one has planned a series of fairly similar molecules from a traditional medicinal chemistry exploration of a lead or a combinatorial array expansion of a template, then this method has some promise.[24]

### *De Novo* Design

The therapeutic target structure can also be used as a foundation for the invention of completely novel entities. Computer programs are available to devise drug-sized molecules within a protein pocket.[25] These are known as *de novo* design programs, as one is supposedly designing a molecule from scratch. In practice, one often modifies an existing lead to optimise interactions.

There are two types of *de novo* design program. Firstly there are the fragment based methods, such as Ludi. These have a database of small molecular fragments which are joined in some combinatorial fashion to produce the putative ligand. Obviously these are limited by the size and quality of the fragment library. The second class are the atom-based methods, such as Sprout. These build up molecules from a much smaller database of atoms. The limitation, in this case, is the huge combinatorial space that is available to the user. As a (slightly obscure!) example, there are at least 23 different $C_4$ hydrocarbons possible.[26] Thus the atom-based methods tend to run quite slowly and also produce many synthetically unfeasible candidates.

## 4 RELATED (HOMOLOGOUS) STRUCTURES

Often in drug discovery work, one has incomplete information on the target structure. One may have available a structure of a protein from the wrong species, an orthologue, for instance a bovine or murine crystal structure rather than the human isoform. This is particularly true in antimicrobial studies, where one is unlikely to have structural information on all target species. Even worse, there may be no structure of an isoform available. In this case, the best one can hope for is the structure of a related protein, which one hopes possesses the same three-dimensional fold. These orthologous and homologous proteins can be used as a basis for the construction of a model of the desired protein. Thus this is known as homology modelling.

## Protein Sequence Alignment

How does one identify homologous proteins? This is commonly achieved by searching databases of protein sequences, using the sequence of the protein that one wants to model. There are large publicly available databases of sequences on the Internet, for instance the Entrez system at NCBI (http://www.ncbi.nlm.nih.gov/entrez). These can searched by keyword to identify the protein of interest. Once one has the protein sequence, one can use a program to align this sequence to those of related proteins. Again, there is a site maintained by the NCBI which allows one to align sequences from a large database *versus* a target sequence.

Currently one of the best alignment tools is the PSI-BLAST program devised by Altschul *et al.*[27] This improves on the original BLAST methodology in various ways, including the modification of the scoring system depending on the position in the sequence.[28] This and its sister program, IMPALA,[29] are now the methods of choice for sequence database searching and multiple sequence alignment.

## Homology Modelling

If one can identify one or more related proteins of known structure in a multiple sequence alignment, one can attempt to build a homology model. There are three main steps: copying the coordinates of identical regions, modelling the coordinates of the rest of the structure and adjusting the whole model structure to relieve strain. All the commonly available protein modelling packages have methods for these activities. A typical protocol would be:

(a) Copy the backbone (*i.e.* the amide and alpha-carbon part) and side-chain of amino acids that are identical matches in the sequence alignment;

(b) Copy the backbone of amino acids that are similar in the sequence alignment (*e.g.* all are hydrophobic at that position);
(c) Construct the side-chains of the similar amino acids using standard geometries; construct the backbone and side-chain of any remaining amino acids using standard geometries;
(d) Relieve strain from 'hot spots' at insertions and deletions in the sequence alignment using some form of conformation generator, such as molecular dynamics or a Monte Carlo method.

An optional final step often taken is to relieve strain from the whole protein model by subjecting it to energy minimisation or molecular dynamics protocols.

Homology modelling works reasonably well when the similarity between the two protein sequences is above the so-called 'twilight zone' of 20–35% sequence identity.[30] What can one do below this threshold? One practice that can be used in some circumstances is to build a partial model. If the site of interest has greater similarity than the surrounding protein, then one can build a model of the site itself. This is only useful if one is careful to use the model appropriately. For instance, a molecular dynamics refinement of a partial model is likely to produce nonsense unless the model is of a whole sub-domain or structural unit of the parent protein. On the other hand, ligand docking studies can be attempted unless there is good reason to suspect that the missing part will affect the behaviour of the ligand.

## Membrane Proteins – Difficult Cases

Partial models have been used extensively in the modelling of the membrane-bound proteins. A good example is the trans-membrane regions of G-protein-coupled receptors (GPCRs).[31] G-protein-coupled receptors are good drug targets; about 30–40% of the drugs on the market interact with GPCRs. Homology models of GPCRs have been built ignoring the N-terminal region of the protein before the first trans-membrane helix, the C-terminal domain after the seventh helix, the large third intracellular loop and often all the other loops. This leaves just the seven trans-membrane helices themselves. The rationale for this partial model is that small ligands, such as the monoamine neurotransmitters 5-hydroxytryptamine and dopamine, bind to the central cavity contained by these helices. Thus, agonist and antagonist drug candidates might be expected to bind in the same place.

GPCR models have become more accurate as time goes by. Bacteriorhodopsin has been supplanted by bovine rhodopsin as the template for homology modelling and two-dimensional crystallographic data has been replaced by three-dimensional results.[32,33] The models are still relatively imprecise as the template structures are not of high resolution.

Other membrane proteins also lend themselves to partial homology modelling. Structures of two ion channels have been reported, MscL mechanosensitive, non-selective channel and KcsA potassium-selective channel.[34] The first is a pentameric structure which is opened by pressure in the lipid bilayer of the membrane; the second is a slightly simpler tetrameric design which does not seem to be gated, but is relatively specific for potassium flow. Although both these channels are from prokaryotes, they have been seen as templates for models of more therapeutically relevant channels.[35,36]

## 5 THE ABSENCE OF TARGET STRUCTURAL INFORMATION

A major theme of drug design in the pharmaceutical industry throughout the latter half of the 1990s has been the development of high throughput methods. Genomics projects have provided data on thousands of potential targets for therapeutic intervention, both in humans and in microbes; high throughput screens have provided (crude) biological data on thousands of sample compounds; combinatorial chemistry has provided large numbers of related molecules for testing. How does the computational chemist make sense of all these data? More importantly, how can computational chemistry support the subsequent drug discovery process?

### Quantitative Structure–Activity Relationships

In the absence of any structural information on the target structure, one is forced to consider just the ligand. This sort of study has a long history, going back as far as 19th century experiments on the relationship between oil/water partitioning of simple alcohols and their narcotic effects on tadpoles! The establishment of a quantitative structure–activity relationship (QSAR) or, more generically, a quantitative structure–property relationship (QSPR) relies on experiment. One must have a set of molecules for which a biological activity or physicochemical property has been measured. One can then try to establish some sort of relationship between this property and other properties of the molecule, which are themselves either measured or calculated. The potential of the method is then that one can predict the activity/property of a yet to be synthesised molecule.

QSAR relies heavily on having a wide range of molecular descriptors available. Traditionally used parameters are described in more detail in

Chapter 11. We shall concentrate here on molecular descriptors that have some relationship to structure.

### Molecular Descriptors

A commonly used structural descriptor is the molecular fingerprint. This encodes structural (and sometimes other property) information as a string, that is a list of characters. The advantage of this representation is the rapid comparison of molecular structures that can be achieved. If the string is a bit-string, that is contains just 1s and 0s, one can do very rapid matching of thousands of molecules using simple metrics such as the number of bits in common. Many pharmaceutical companies have adopted this approach using the Daylight system,[37] although some aspects of the Daylight fingerprints have been criticised.[38]

## 6 WHEN THE TARGET STRUCTURE IS ALMOST IRRELEVANT

Knowledge of the pharmacological target structure is important and useful in drug design; it is, however, only the start of the process. A drug must be reasonably potent at its target site, but it must also possess a host of other properties, which fall under the umbrella of ADME (absorption, distribution, metabolism, excretion). Twenty years ago, medicinal chemists strove to find potent molecules using data from whole animal or whole tissue experiments. Nowadays, the primary biological data is usually from isolated targets *in vitro*. The focus has shifted from potency and selectivity to an almost holistic approach, often termed 'developability', where ADME properties are also considered.

### Drugs *Versus* Leads

A lead molecule is usually far from being a drug candidate. Let us consider an (idealised) antibiotic. The molecule must be potent against its target, say a component of the protein synthesis pathway, such as a tRNA synthetase enzyme. It will have affinity for a range of isozymes from different bacterial species, but will show little or no affinity for the human isoform, as we do not want a toxic agent. So far, we have been in the realm of structure-based design. It is conceivable that we would have crystal structures of human and bacterial enzymes, the latter with bound inhibitors present. We could use the docking and *de novo* design programs mentioned above to devise new chemical entities which were potent and selective. These are not drugs, however, until we also consider ADME factors. An orally bioavailable compound will have to be stable to stomach acid, diffuse across the gut lumen, survive first-pass metabolism in the liver, partition between aqueous and fatty phases in such a way as to provide sufficient concentrations at the site of infection, penetrate the bacterial cell walls, inhibit the enzyme and finally be metabolised and excreted from the patient. Some of these steps may depend on specific protein interactions, for example hydroxylation by a cytochrome P450 metabolic enzyme, or expulsion by an ATP-driven pump. Many, however, are dependent on simple passive diffusion and are thus independent of any particular target structure. Our antibiotic must have gross physicochemical properties, such as log*P* and charge, such that it will survive long enough inside the human body to arrive at its target in sufficient concentration and also have the required properties to be able to penetrate the bacterium in order to reach its target. This overall mix of required properties is not just the purview of antibiotics: for instance, centrally acting compounds, such as antidepressants, must be bioavailable and also cross the blood–brain barrier.

## Drug-like Qualities

The determination of a molecule's physicochemical properties is discussed in depth in Chapter 10. These overall properties of a molecule have proven to be very useful in characterising the ADME performance of drug-like molecules. A well-known example is the Rule of Five, proposed by Lipinski and co-workers.[39] Contrary to popular belief, this is not a set of five rules, but a set of rules based on multiples of five. The rule states that a molecule is unlikely to be bioavailable if two or more of the following conditions hold:

(a) Molecular weight more than 500;
(b) *C*log*P* more than 5;
(c) more than 5 hydrogen bond donors;
(d) more than 10 hydrogen bond acceptors.

The hydrogen bond donors and acceptors are not true donor/acceptor counts, but simple summations of OH and NH groups and of O and N atoms respectively. It is important to realise that these rules are not predictive of oral bioavailability, but define the boundaries outside of which oral bioavailability is unlikely to be achieved.

The Rule of Five can be seen as an attempt to define drug-like molecular properties. Other researchers have made similar attempts using a variety of methods. Bemis and Murcko analysed sets of known drugs to define commonly found structural parameters, looking first at ring systems[40] and then at side-chains.[41] They showed that one only needs to specify 32 different scaffolds to characterise more than half of the drugs that they examined. Thus, one could expect that a drug-like scoring system could be devised. Sadowski and Kubinyi trained a neural network to distinguish drug-like compounds (members of the World Drug Index database) from non-drug-like compounds (members of the Available Chemicals Data-base).[42] An interesting feature of their work was that they discovered that an atomic descriptor system that had been designed to predict log*P* performed well as the input descriptor for the neural network.[43] In the same issue of *J. Med. Chem.*, Ajay, Walters and Murcko, that is the group who originally analysed the structures of known drugs, also reported a neural network that could recognise 80% of the molecules in the MACCS Drug Data Report to be drug-like.[44]

## Blood–Brain Barrier

Another barrier that concerns some medicinal chemists is the blood–brain barrier (BBB). This is at the junction between blood vessels and astrocytes, comprising the interface between the nerve cells and the epithelium of the capillaries. The BBB prevents hydrophilic molecules from entering most of the central nervous system, although this can be effected by carrier proteins in special cases. A simple model of the blood–brain barrier is thus a lipid membrane with no pores, through which lipophilic molecules can diffuse passively.

The rules of thumb that govern blood–brain barrier penetration have been summarised by Pardridge[45] as follows: the molecular weight between 400 and 600; the number of hydrogen bonds not to exceed 10. The rules rely on the following scoring system: 4 H-bonds counted for terminal amide; 3 for primary amine or amide; 2 for hydroxyl; 1 for ether or carbonyl and 0.5 for ester. Thus one strategy to gain BBB penetration is to 'trade down' the above list, for instance blocking hydroxyls by converting them into alkoxy groups.

Although the simple rules of thumb may work for many xenobiotics, care must be taken with entities that one might consider to be less drug-like and more like the molecules typically found in the blood. For example, glucose is the prime fuel of the brain and is efficiently extracted from the bloodstream. Peptides are also good examples. Although many peptides get into the brain by passive diffusion, there are some that are transported by quite specific transport proteins. For instance, interleukin-1 is actively taken into the brain, but similar cytokines, such as interleukins-2 and -6 and tumour necrosis factor $\alpha$, are not. Similarly leptin is actively taken up, but other peptides that affect feeding, such as orexin A and B, MCH, neuropeptide Y and agouti-related protein, do not seem to be.[46]

Some attempts have been made to arrive at more general descriptions of BBB permeation, using QSAR techniques. A recent example is that reported by Crivori *et al.*,[47] in which a series of molecular descriptors were used to provide a predictive model. The descriptors included simple quantities such as molecular weight, but the authors also attempted to use three-dimensional descriptors, including various volume and surface properties. The authors used a training set of 44 molecules and a test set of 108 drugs and claimed better than 90% discrimination between compounds that do, and do not, cross the blood–brain barrier.

## 7 CONCLUSION

The use of computational chemistry to aid in the discovery of drugs is routine in the 21st century. There are many techniques available to address problems whether one has a target structure or not. There are also methods which allow one to construct a target model in some cases. This review has highlighted some of the commonly available and commonly used methods. Obviously many interesting efforts have not been discussed. Any reader who needs more detail should look at the review articles mentioned here, or inspect relevant journals, such as *J. Comput. Chem.*, *J. Comput.-Aided Mol. Design* and *J. Med. Chem*.

All methods used in computational chemistry are approximations, as are all experimental results. I have tried throughout to cite estimates of quality, where these are known. An understanding of the limitations of a technique help greatly in its use. For instance there is little hope of predicting aqueous solubility to an accuracy better than 0.6 log units, since this is the variation in the experimental data. Also, there is little point in running long molecular dynamics simulations on large libraries of compounds in target binding sites, since the calculations will take longer to run than the compounds take to make. Nevertheless, I hope I have shown that computational techniques have recently played an important part in drug discovery efforts and will continue to do so for the near future.

## 8 REFERENCES


1. S.K. Burley, *Nat. Struct. Biol.*, 2000, **7** (suppl.) 932.
2. B. Alberts *et al.*, *Molecular Biology of the Cell*, 2nd edn., Garland, New York & London, 1989.
3. A.J. Oakley and M.C.J. Wilce, *Clin. Exp. Pharmacol. Physiol.*, 2000, **27**, 145.

4. R. Stevens, *Curr. Opin. Struct. Biol.*, 2000, **10**, 558.
5. J. Hajdu, *Curr. Opin. Struct. Biol.*, 2000, **10**, 569.
6. D.J. Craik and M.J. Scanlon, *Annu. Rep. NMR Spectrosc.*, 2000, **42**, 115.
7. G.C.K. Roberts, *Curr. Opin. Biotechnol.*, 2000, **10**, 42.
8. S.B. Shuker *et al.*, *Science*, 1996, **274**, 1531.
9. A.G. Murzin *et al.*, *J. Mol. Biol.*, 1995, **48**, 443.
10. C.A. Orengo *et al.*, *Structure*, 1997, **5**, 1093.
11. H.-J. Bohm and M. Stahl, *Curr. Opin. Chem. Biol.*, 2000, **4**, 283.
12. H.-O. Kim and M. Kahn, *Comb. Chem. High Throughput Screening*, 2000, **3**, 167.
13. T. Hermann and E. Westhof, *Comb. Chem. High Throughput Screening*, 2000, **3**, 219.
14. M.A. Murcko, in *Reviews in Computational Chemistry*, K.B. Lipkowitz and D.B. Boyd (eds.), Wiley-VCH, New York, 2000, Vol. 10, p. 1.
15. D.E. Clark *et al.*, in *Reviews in Computational Chemistry*, K.B. Lipkowitz and D.B. Boyd (eds.), Wiley-VCH, New York, 2000, p. 2.
16. A.M. Davis and S.J. Teague, *Angew. Chem. Int. Ed.*, 1999, **38**, 736.
17. D.H. Williams and B. Bardsley, *Perspect. Drug Discovery Des.*, 1999, **17**, 43.
18. C.W. Murray *et al.*, *J. Comput.-Aided Mol. Des.*, 1999, **13**, 547.
19. H. Gohlke *et al.*, *Perspect. Drug Discovery Des.*, 2000, **20**, 115.
20. S.-S. So and M. Karplus, *J. Comput.-Aided Mol. Des.*, 1999, **13**, 243.
21. D.E. Clark *et al.*, *Rev. Comput. Chem.*, 1997, **11**, 67.
22. J. Aqvist *et al.*, *Protein Eng.*, 1994, **7**, 385.
23. W.L. Jorgensen, *Acc. Chem. Res.*, 1989, **22**, 184.
24. T. Hansson *et al.*, *J. Comput.-Aided Mol. Des.*, 1998, **12**, 27.
25. M.A. Murcko, *Pract. Appl. Comput.-Aided Drug Des.*, 1997, 305.
26. F.R. Burden, *J. Chem. Inf. Comput. Sci.*, 1989, **29**, 225.
27. S.F. Altschul *et al.*, *Nucleic Acids Res.*, 1997, **25**, 3389.
28. S.F. Altschul and E.V. Koonin, *Trends Biochem. Sci.*, 2001, **23**, 444.
29. A.A. Schaffer *et al.*, *Bioinformatics*, 1999, **15**, 1000.
30. B. Rost, *Protein Eng.*, 1999, **12**, 85.
31. G. Muller, *Curr. Med. Chem.*, 2000, **7**, 861.
32. T. Okada, *et al.*, *J. Struct. Biol.*, 2000, **130**, 73.
33. K. Palczewski *et al.*, *Science*, 2000, **289**, 739.
34. H.L. Li *et al.*, *Jpn., J. Mol. Biol.*, 1998, **282**, 211.
35. S.R. Durell and H.R. Guy, *Biophys. J.*, 1999, **77**, 789.
36. G.M. Lipkind and H.A. Fozzard, *Biochemistry*, 2000, **39**, 8161.
37. Daylight, Computer Program, Daylight CIS, Mission Viejo, CA, 2000.
38. D.R. Flower, *J. Chem. Inf. Comput. Sci.*, 1998, **38**, 379.

39. C.A. Lipinski *et al.*, *Adv. Drug Delivery Rev.*, 1997, **23**, 3.
40. G.W. Bemis and M.A. Murcko, *J. Med. Chem.*, 1996, **39**, 2887.
41. G.W. Bemis and M.A. Murcko, *J. Med. Chem.*, 1999, **42**, 5095.
42. J. Sadowski and H. Kubinyi, *J. Med. Chem.*, 1998, **41**, 3325.
43. V.N. Viswanadhan *et al.*, *J. Chem. Inf. Comput. Sci.*, 1989, **29**, 163.
44. Ajay *et al.*, *J. Med. Chem.*, 1998, **41**, 3314.
45. W.M. Pardridge, *J. Neurochem.*, 1998, **70**, 1781.
46. A.J. Astin *et al.*, *Brain Res.*, 1999, **96**, 848.
47. P. Crivori *et al.*, *J. Med. Chem.*, 2000, **43**, 2204.

CHAPTER 13

# Patent Medicine

BILL TYRRELL

## 1 INTRODUCTION

Since writing my original chapter in the previous edition of this book[1] patent law has moved on so much that a mere cosmetic update could not do justice to the subject. What follows is another dose of Patent Medicine, but completely reformulated.

Recently Euroforum, a firm specialising in organising conferences, held a two-day seminar for patent lawyers aptly entitled 'Patents in the Age of Genomics'. This underlined just how much the 'gene to screen' methodology used in modern drug discovery has sent shock waves through the patent profession, just as it has revolutionised the way pharmaceutical companies organise their research effort. The vast increase in patent filings that has resulted from companies trying to capture the fruits of their DNA research, much of it *in silico*, is well known – and has led to heated debate to which I will return later. These days, patent practitioners not only have to be familiar with depositing microorganisms in culture collections, but also with bioinformatics, mathematical algorithms and providing electronic listings of DNA sequences.

This shake-up in practice, of course, affects only a part of the patent profession. Those who deal in mechanical inventions such as improved corkscrews, humane mousetraps or sock presses continue much as they have always done. Fortunately small inventors continue to provide entertaining examples of 'off-the-wall' patents. I will refer to one or two of these later to provide light relief and to demonstrate that patents do not necessarily have to relate to ground-breaking inventions.

Despite a heightened awareness of intellectual property rights over the last few years, there still persists a tendency in some quarters to think that patents are boring, esoteric documents with which the medicinal chemist need not be troubled. Nothing could be further from the truth! Admittedly, some aspects of a patent attorney's life are not a thrill a minute (but the high spots can be very high indeed) and it is fair to say that patent specifications can be rather dry. But it is indisputable that patents are vital to the research-based pharmaceutical industry. The fortunes of companies are sharply influenced by the patent protection they have on their key products. To take just one example, the *Financial Times* of 11 August, 2000 printed an article headed "Prozac ruling hits drug companies", noting the successful challenge to one of Eli Lilly's US patents. The article states:

> *"Shares in Lilly fell more than 30 per cent . . . after a US appeal court said its 'method of use' patents were so similar to the original composition patent that they were effectively an unfair extension."*

The article concluded that Prozac could face generic competition in the US three years earlier than had been thought, cutting as much as $5 billion from Lilly's sales over four years. The above case illustrates a growing trend: generics companies will often not wait for patent expiry, but will take their chances in litigation, arguing, as they are fully entitled to do, that the patent is invalid and should never have been granted in the first place.

Many patents are upheld in the courts but some are not and when it comes to litigation, everything is raked over in fine detail. Imagine that you were the medicinal chemist responsible for inventing a blockbuster drug. Would you not feel a bit sick if some blunder on your part, such as a conference presentation before patenting (thereby destroying novelty in most countries), prejudiced the validity of a patent covering a billion dollar plus product?! That is why this chapter is important. It is **vital** for scientists to gain a good working knowledge of the patent system, both to avoid making costly errors and to help flag inventions when they are made.

I have dwelt on the fact that patent protection is vital to the profitability of research-based pharmaceutical and biotech companies, but in doing so I do not want to overlook the more fundamental point that patent protection is vital to **patients**, all over the world. It is only through strong patent protection that R&D investment will be forthcoming to provide new medicines for presently incurable diseases. It can now cost some $500 million to develop (over a period of 10–12 years) a new medicine.[2] The largest pharmaceutical companies spend as much as $4 billion per annum on R&D, much of it abortive. Absent the prospect of getting a return on this investment by securing an all-too-short period of patent protection, funding will dry up and the research-based industry will cease to exist. As the Association of the British Pharmaceutical Industry (ABPI) neatly puts it: 'No patents, no cures'!

By and large, this message has got home and we have seen in the last 7 years a strengthening of patent laws around the world, as part of the so-called GATT-TRIPS initiative (see Section 7). Nevertheless there still exists a powerful anti-patent lobby which argues that patenting is unethical. The protestors have seized particularly on recent developments in biotechnology and the issue of access to medicines in developing countries to knock the system. Patents continue to be a hot political issue!

In the first edition of this book, I mentioned that patents were stranger than fiction. That is just as true today – perhaps even more so. Patents and patent cases continue to take on bizarre twists and turns, which provide a constant source of fascination – in the pharmaceutical arena just as in other fields of technology. A taste of this will, I hope, come through in the examples below.

Here, then, is a completely revised chapter on medicinal chemistry patents at the start of the new millennium. In fact, I had better put a precise date on it: May 2002. It is astonishing that a system that began in the 15th century is still changing so rapidly and is probably evolving today faster than at any time in its history. Let us hope it will not be too much out of date by the time you read it.

## 2 WHAT ARE PATENTS?

While patents are the primary means of legal protection in the pharmaceutical industry, it is worth re-iterating that they are an important part of a whole gamut of rights known as **Intellectual Property (IP) rights**. The medicinal chemist should at least be aware of other forms of IP. These include ***Trade Marks***, intended to protect a word or logo associated with a particular product (TAGAMET and ZANTAC are examples), ***Registered Designs*** (intended to protect the 'look' of a particular article which has eye appeal, such as a pill-dispenser) and ***Copyright*** (which can be relevant to prescribing information on pack inserts). ***Trade Secrets,*** sometimes referred to as proprietary ***know-how***, represent another important aspect of IP which should always be considered as an alternative to patenting. The fundamental difference between patenting and know-how is that in the former case, the invention is inevitably published, whereas with the latter the intention is to keep it deliberately a secret. Another important right in the pharmaceutical industry is the ability to prevent access to one's regulatory file data for a limited period; this is known as ***Regulatory Exclusivity*** (RE).[3]

Patents, or 'Letters Patent', are legal documents designed to protect inventions by affording to the inventor a limited period of exclusivity, the basic idea being to stimulate innovation.[1] The patent system has been in existence for many years and has been adopted in, one form or another, in most countries. The first British patent with a medical application was granted in 1698 for a cheap method of making Epsom salts. Figure 1 shows the front page of a patent granted recently in the Russian Federation.

Patent term can vary but in major countries is now fairly uniformly 20 years from the date of application. During the life of the patent the inventor (or more usually a company to which the inventor has assigned the patent rights as part of a contract of employment) can exclude others from exploiting the patented invention without permission. To enforce the patent it may be necessary to sue the alleged infringer by bringing a civil action in the courts. As an alternative, the patent proprietor may choose

РОССИЙСКАЯ ФЕДЕРАЦИЯ

ПАТЕНТ

НА ИЗОБРЕТЕНИЕ

№ 2160120

Российским агентством по патентам и товарным знакам на основании Патентного закона Российской Федерации, введенного в действие 14 октября 1992 года, выдан настоящий патент на изобретение

**КОМБИНИРОВАННАЯ ВАКЦИНА НА ОСНОВЕ ПОВЕРХНОСТНОГО АНТИГЕНА ВИРУСА ГЕПАТИТА В, СПОСОБ ЕЕ ПОЛУЧЕНИЯ И СПОСОБ ПРЕДУПРЕЖДЕНИЯ ИНФЕКЦИИ ГЕПАТИТА В У ЧЕЛОВЕКА**

Патентообладатель(ли):

*СМИТКЛАЙН БИЧАМ БАЙОЛОДЖИКАЛС С.А.* (BE)

по заявке № 94046145, дата поступления: 15.05.1993

Приоритет от 23.05.1992

Автор(ы) изобретения:

*Жан ПЕТР* (BE), *Пьер ОЗЕР* (BE)

Патент действует на всей территории Российской Федерации в течение 20 лет с **15 мая 1993 г.** при условии своевременной уплаты пошлины за поддержание патента в силе

Зарегистрирован в Государственном реестре изобретений Российской Федерации

г. Москва, ***10 декабря 2000 г.***

*Генеральный директор*

*А.Д. Корчагин*

**Figure 1** *Front page of a patent granted in the Russian Federation*

(on agreed terms) to grant a licence to allow others to practice the invention. Such licenses can be exclusive or non-exclusive.

It is important to note that you can only exclude someone from using your invention *commercially*; pure research on the subject matter of the patented invention is not an infringement. Nevertheless, this is a grey area and there is a developing area of the law dealing with what does, or does not, constitute an 'experimental use' exemption to infringement, a frequently asked question with regard to conducting clinical trials on drugs which are covered by patents.

Pharmaceutical inventions are somewhat different to those in other fields, *e.g.* those relating to tin openers or telephones, in that a long period of time is required to satisfy regulatory authorities, such as the FDA, of a drug's safety and efficacy. Since patents are generally applied for while a new chemical entity is at the stage of early research, there may only be 10 years or less of 'effective patent life' remaining once a drug is put on the market. This has led to patent term extensions (or the equivalent, such as 'Supplementary Protection Certificates') being granted in some countries, which is at least some compensation. In return, non-research-based companies have lobbied for the right to obtain regulatory approval for generic copies of drugs during the lifetime of the relevant patent so they are ready to launch immediately the patent (or extension) expires. This is allowable in the US in certain circumstances and is a growing practice elsewhere.[4]

The fact that a patent gives only a right to *exclude* others from making use of a patented invention means that a patent holder (patentee) cannot assume that he or she has the right to commercialise the invention to which the patent relates. To do so might infringe a patent of broader or overlapping scope belonging to another. A licence may have to be negotiated in order to put a pharmaceutical product on the market. Establishing one's own patent position is desirable but checking for the existence of dominating patents is essential: failure to carry out an infringement search is a sure recipe for litigation!

When patent lawyers talk about infringing a patent, they are referring to trespassing on the 'claims', the numbered paragraphs at the end of the specification, which are meant to set out the metes and bounds of the invention. Other things being equal, the most valuable patents are 'broad' ones because they provide the greatest area of exclusion, so preventing 'me-too' competition. In the chemical arena, typical patent claims rarely relate to one specific compound but to thousands of compounds with different groups R, $R^1$, $R^2$ *etc.* carefully defined. An even broader and more powerful patent claim is the so-called 'mechanism of action' claim covering the use of *any* compound which interacts with (*e.g.* inhibits) a receptor shown to be associated with a particular disease state. A claim of this type might be written (in Europe) as follows:

> *"The use of an inhibitor of [receptor X] in the manufacture of a medicament for the prophylaxis or treatment of [disease Y]."*

In assessing the validity of patents, the prior work of others is obviously critically important. The term 'prior art' is generally used in patent circles for this body of existing knowledge. One can obtain patents in an already crowded field if a genuine contribution has been made, but in general, the closer the prior art, the narrower the allowable scope of the patent. The question of what is, and is not, patentable is dealt with in the next section.

## 3 WHAT IS PATENTABLE?

Granting, or 'issuing', patents is not a rubber-stamping exercise. It is a mistake to imply that one simply 'takes out' a patent as though this were an afternoon's work. In fact, patents have to be fought for, by convincing patent examiners in different countries around the world that the patent claims are valid. To do that it is necessary to satisfy some basic criteria for patentability, which are addressed below. One can apply these criteria to any potentially patentable subject matter. In the pharmaceutical field that could, for example, be a new chemical entity, pharmaceutical composition, synthetic method or intermediate, or a new use for a known compound. Those engaged in more upstream research might, typically, wish to patent DNA, expression vectors, hosts, drug targets or screening methods.

### Novelty

In most countries (the US is a notable exception) there is an 'absolute novelty' requirement, meaning that if your invention has been publicly disclosed either in writing or orally before a patent application is filed, then your claims will lack novelty and accordingly be unpatentable. The rationale for this is simple: a patent should not prevent the public from having the opportunity to practice something they already (in principle) know about (the fact that most members of the public will not, in fact, have read or heard the disclosure is irrelevant). Except in the US, this rule also applies to disclosures made by the inventor as well as to disclosures by others. Even in the US, a patent application must be filed within a one year of the inventor's own publication (the 'grace period', a provision which, while hotly debated, has not yet caught on in Europe). The golden rule is: **do not disclose your invention, except in confidence, to anyone before a patent application has been filed**.

It is worth noting that there are certain points that this simplified discussion has not addressed (patent law is not noted for simplicity). To pick just two, the legally-minded reader might well ask:

- What happens if I make a disclosure in confidence and that confidence is broken?
- What happens if someone else invents the same thing as me and files an earlier patent application, which is only published *after* I have filed my application?

For the keen reader I have put the answers to these and some other more complex questions in an optional Advanced Module (Section 9).

### Inventive Step

To be patentable, an invention (as set forth in a patent claim) must involve an 'inventive step', that is to say be non-obvious to a 'person skilled in the art' over any disclosure (written or oral) which has been 'made available' to the public. The reason for this is similar to that discussed above in relation to novelty. Members of the public should not be barred from practising an invention which, whilst not actually disclosed in the prior art, is an obvious extrapolation of what is already available to them. This was the reason given by a UK judge in November 2000 for revoking a patent held by Pfizer on the oral use of sildenafil (Viagra) for the treatment of male erectile dysfunction (MED).[5]

The concept of the person skilled in the art (the patent equivalent of the man on the Clapham omnibus) is a 'legal fiction', created in order to assess an inventive step. He or she is assumed to have read (or heard) everything in the field in question (but little outside it), yet to have no inventive capability. As the judge in the above-mentioned Viagra case stated:

> *"The question of obviousness has to be assessed through the eyes of the skilled but non-inventive man in the art. This is not a real person. He is a legal creation. He is deemed to have looked at and read publicly available documents and to know of public uses in the prior art. He understands all languages and dialects. He never misses the obvious nor stumbles on the inventive. He has no private idiosyncratic preferences or dislikes. He never thinks laterally."*

Mr Justice Laddie in *Lilly Icos v Pfizer (2000)*

Probably more has been written on the subject of inventive step than anything else in patent law because, unlike novelty, the matter tends to be subjective and it is difficult to judge what is, or is not, obvious to the skilled person. It is well accepted that patent offices responsible for granting patents such as the European Patent Office (EPO) and national patent courts which adjudicate patent disputes take a pragmatic view. Rather than consider the impact of a prior disclosure on one person skilled in the art, they normally consider the impact on an appropriately staffed multidisciplinary team.

On the other hand the EPO has been criticised for its rather formalistic approach to inventive step. This involves identifying the 'closest state of the art', usually a single document or other disclosure, and then formulating an artificial 'problem' to which the inventor can be considered as having provided a 'solution'. Both the EPO and the US Patent and Trademark Office (USPTO) seem to be in step on whether an invention should be deemed obvious from a legal standpoint when it is accepted that the steps leading to it are 'obvious to try'. In such cases patentability turns on whether or not there was a 'reasonable expectation of success'. However, the EPO and USPTO are rather at odds on the question of the obviousness of DNA sequences, an important consideration for biotech inventions.[6]

**Sufficiency**

The criterion for sufficiency is that the invention must be described sufficiently clearly and completely for it to be carried out by a person skilled in the art. This sounds straightforward but, as with inventive step, is rarely black and white. Again one can go back to first principles to understand this requirement. It would be unfair if the inventor cheated by enjoying the exclusive rights provided by a patent, whilst not providing in the specification a proper disclosure which interested parties could put into practice once the patent had expired. It is for this reason that lack of sufficient disclosure (Americans call this 'non-enablement', a term which has caught on even in the UK courts) is both a ground for objection pre-grant and a ground for revocation post-grant.

The medicinal chemist can readily understand that providing full details of reagents, reaction conditions *etc.* is essential and if you provide the same sort of write-up in a patent application as you would submit to a learned journal you will not go far wrong. One thorny problem, however, has concerned biotech inventions relying on a key microorganism, which is not publicly available. How then can the sufficiency requirement be met? This difficulty has been overcome by allowing the applicant to deposit, in a specially approved International Depositary Authority, a sample of the microorganism, which will be available in due course to anyone requesting a sample (certain terms and conditions apply).

For patent lawyers, the concept of sufficiency is actually far more complex than is indicated by the above summary. For example, insufficiency arguments can be invoked to attack over-broad claims or to deny a claim to priority. A little more will be said about this in the advanced module (Section 9).

### Utility or 'Industrial Application'

Put simply, an invention has to be *useful*, but in the past Patent Offices and Courts have been rather relaxed about this criterion and have hardly considered it. The utility requirement (in Europe 'industrial application') is nevertheless worthy of mention. That is because it has raised its head in relation to patenting of partial DNA sequences (ESTs or expressed sequence tags) of genes having no known function. The attempted patenting of ESTs has in fact caused such a furore that patent office practice has been tightened up in the USA, making it now much less likely that DNA sequences with no credible utility can be patented. In the EPO, adoption of the EU Biotech Directive (see section 8) has meant that the industrial applicability of a DNA sequence (or partial sequence) has to be clearly indicated in the patent specification.

Provided some utility can be shown, an invention does not have to be a blockbuster to be patentable provided the other criteria discussed above are satisfied. That is quite clear from the selection of weird and wonderful inventions which can be found in the patent literature, now easily accessible through a growing number of web sites devoted to 'wacky patents' (see Section 10). To illustrate the point, here is a claim from US Patent Number 5,971,829:

*A hand-held apparatus for rotating a common edible pastry cone comprising:*

*(a) a housing adapted to be grasped and supported by a person's hand;*
*(b) a cup rotatably supported within and substantially surrounded by said housing, said cup being adapted to receive and vertically support in a stable manner a common edible pastry cone; and*
*(c) a drive mechanism supported by said housing, said drive mechanism including rotating means for imparting a rotary motion upon said cup, said rotary motion providing feeding means for rotationally*

> *feeding the contents of said common edible pastry cone against a person's outstretched tongue.*

The rather verbose language for describing what most people would simply call 'a motorised ice cream cone' leads us naturally into the next section on patent terminology, or 'patentese'.

## 4 'PATENTESE'

Patents tend to be viewed as being rather impenetrable documents, full of obscure and repetitive language, the only bright spot for scientists being the examples at the end, probably written by fellow scientists. As for the claims attached to the specification: what about putting in a little punctuation? A few full stops wouldn't go amiss! Why can't patent attorneys write sensible English like anyone else?

By way of explanation, the primary objective of the patent attorney is to try to formulate claims capturing the actual **invention**. In relation to the *prior art*, the claims should be as broad as possible (to keep copiers out), yet valid (satisfying all the criteria mentioned in Section 3 above). The patent attorney is not employed to write a piece of elegant literature, but a legal document which, if needs be, will withstand scrutiny in the courts. Because of the need to capture the invention as broadly as possible, **generic** language is frequently used in patent documents. This is often awkward, just as it sounds a little odd to ask for one's drink in a "receptacle" (thus covering any container, whether it be glass, tankard or even egg-cup). In the example above, the invention relates to rotating the edible pastry cone; there is no need to limit the contents of said cone to ice-cream!

For an invention in the mechanical field, a coiled spring might be just one way of providing 'springiness' and to cover this the patent attorney will typically refer to 'resilient means'. The fundamental idea is to put beyond doubt infringement by a similar device, even if it uses a different type of spring. Similarly, in the chemical or biotech field, where possible, the draftsperson uses generic terms such as an acyloxy group or a non-inducible promoter. If a functional or generic term is not available, other ways can be found to extrapolate from a given example. For example, a substituent is frequently defined as C1–6 alkyl when perhaps only one or two compounds with alkyl groups in that range have actually been made and tested. However, great care should be taken to check that such extrapolation is well-founded (see Section 9).

Patent specifications do not make great reading because once such generic language has been used, it has to be used over and over again.

There should be no elegant variation, as that creates legal uncertainty. For the same reason, by convention a patent claim is written as just one sentence, no matter how long it is! Naturally this tends to produce a high 'Fog Factor'![7]

Another feature of patentese that is difficult for the non-specialist to grasp is the idea of sub-claims. Normally claim 1 is the 'main claim' which will contain the broadest definition of the invention. For example, a main claim could relate to a process for preparing a compound of defined formula (I), which comprises treating a compound of formula (II) with a reducing agent in a non-ionic solvent. By comparison, Claim 2 may typically claim: 'A process according to claim 1 in which the reducing agent is a metal hydride'. Claim 3 may claim: 'A process according to either claim 1 or claim 2 in which the solvent is hexane'. Claim 4 might state: 'A process according to any preceding claim which additionally comprises irradiating the reaction mixture with ultraviolet light.'

This is a strange way of writing English but it is a clever way of providing fall-back positions should the main claim be held too broad. For example, if a prior art reference were cited which disclosed the reduction of the compound of formula (II) by hydrogenation, claim 1 would lack novelty. However, claim 2, in which it is specified that the reducing agent is a metal hydride, would not and neither would claim 3 if the hydrogenation reaction had been done in a solvent other than hexane. Notwithstanding that claims 2 and 3 might be novel, they would most likely be held obvious over the cited reference as metal hydrides are well known reducing agents. However, claim 4, specifying irradiation with ultraviolet light, could well be valid. The test of obviousness is not whether a skilled chemist *could* have done the experiment under conditions of UV irradiation, but whether he or she, in the light of the prior art, *would* be motivated to do it.

This example also helps to illustrate the use of the word 'comprising' in patentese. This by convention has an open-ended meaning, unlike the phrase 'consisting of' which is closed. If my cup of tea comprises milk it may also contain sugar even if that is not stated. Patent claims frequently have the word 'comprising' in them as a way of specifying only the essential elements of the invention. If patent claims were more restricted, and in particular if they were limited to claiming only the precise conditions used in a working example, they would be too readily circumvented to be of any commercial value.

For all this strained language and supposed legal certainty, patent disputes frequently turn on the correct construction of the claims and what exactly they are supposed to cover. In the UK there has traditionally been a fairly literal claim interpretation, but there is some scope for elasticity based on the notion of 'purposive construction'.[8] In the US there is a well-established 'doctrine of equivalents'. Under that doctrine something falling close to, but not literally within, the claims can still be held to infringe, particularly if, with respect to what is claimed, it performs the same function in the same way. That is, however, unless the patent applicant has put on the record a statement which would lead the public to believe the claim excluded the alleged infringement. In that case 'file wrapper estoppel' will prevent the patentee arguing for a broader interpretation, another piece of patent jargon!

One could go on and on about the language patent people use amongst themselves: continuation-in-part, Markush group, the Protocol on Article 69 EPC, double patenting, interlocutory injunction, Gillette defence, Swiss claims, Budapest Treaty, Paris Convention *etc.* There is no doubt patentese can be an impenetrable language, but if the medicinal chemist knows what is set out in this section he or she should know more than enough to get by.

## 5 APPLYING FOR PATENTS

The process of obtaining (and subsequently nurturing) patents is the job of patent attorneys who are usually both scientists (so they more or less understand what you are doing) and specialised lawyers. However, this process can involve the research scientist in several important ways.

One of the most fundamental contributions the scientist can make is to provide a sufficient description of the invention, usually by writing up several working examples prior to the filing of a patent application. There is no requirement in most countries for a patent application to describe the best way of carrying out the invention. However, it is important to be aware that in the US there is a statutory requirement to disclose fully the best method or best compound known to the inventor at the time of filing (the **'best mode'** requirement). Failure to disclose best mode can seriously jeopardise the validity of a US patent. The inventor(s) should be aware of this when providing patent examples.

The actual mechanics of applying for patents need not greatly concern the medicinal chemist. Suffice it to say that the normal procedure is to apply, in the first instance, for a patent in the country in which the invention originated. For a UK company, one would normally file a patent application at the UK Patent Office in Newport in Wales, although there is a receiving office in London and other major cities. Nowadays it costs nothing to file a UK patent application: nothing that is in terms of initial filing fees but search, examination and maintenance fees will all have to be paid subsequently.

An important requirement, which the patent attorney has to think about at this early stage, is identifying one or more inventors – the names of which ultimately have to be supplied to the patent office and which appear on the front of the specification when the patent application is published. Determining the identity of the true inventor(s) is a legal question and, unlike submitting a paper for publication, one cannot include everyone involved in the work. The critical question is: who actually devised the invention? The person who carried out the work on instruction is unlikely to be an inventor unless particular difficulties had to be surmounted. No hard and fast guidelines can be given here on this delicate question except to note that every case will depend on the precise circumstances.

For most inventions, especially those in the pharmaceutical field, obtaining patent protection in just one's home country will not be enough. For a blockbuster drug, patent protection in a large number of countries will be required. However, the beauty of the system is that, thanks to an agreement known as the Paris Convention, a period of 12 months is available from the date of first application (the 'priority date') before a decision needs to be taken about filing abroad. Within that 12 month period, the inventors can refine the invention and the company can decide the next steps. At the end of the 'Convention year' there are three main choices:

- The priority application may be abandoned if there is no further interest
- The priority application can be abandoned and refiled (with or without modification), so creating a new priority date
- The priority application can act as a 'priority document' for a 'final' application, which is then 'foreign-filed'.

The second option is appropriate if the invention has merit but more time is required to work on it. However, the penalty for refiling is loss of the original priority date, making it more probable that you could be pipped to the post in a competitive field. It is also important to note that the abandon and refile option will not be available if in the priority year you have published your invention. That is because the reset priority date will be later than the date of your own publication, so the latter will count as prior art against you, a classic case of 'shooting oneself in the foot'! One of the more unpopular tasks your patent department has to perform is to monitor proposed publications to stop precisely that sort of thing happening, as well as to check generally that no statements are being made which could be detrimental to the progress of other applications.

If it is decided to proceed with foreign filing, medicinal chemists may well be called upon to assist. Typically the foreign filing text will 'cognate' several applications which have been filed in the preceding 12 months and you may be asked to add to, and perhaps tabulate, some of the examples. The true scope of the invention will have to be determined; in other words the breadth of the inventive concept. For example, do only substituted benzimidazoles with a thiophene ring in the substituent have the desired biological activity? Could the thiophene ring be replaced by a furan ring? If the ethyl analogue works can it be assumed that any lower alkyl substituent will afford similar activity? What is in the prior art and why is what you have done inventive over it? These are the sort of questions on which you may be required to comment. Finally, although this is a decision normally taken with input from senior management, you may be asked to make a recommendation as to the breadth of geographical filing.

The last decision is critically important because once a filing has been carried out it is not normally possible to add countries later.[9] It would be unfortunate if, owing to an error of judgment at this point, a drug which proved to be a blockbuster could be copied without infringement in most parts of the world. A filing of broad geographical scope is the safest option but can be extremely costly, so a balance has to be struck. Fortunately, a way of making applications overseas is available, which not only solves the problem of having numerous translations made at the last minute but also helps to keep options open in relation to geographical coverage. This is the PCT (Patent Cooperation Treaty) route, administered by the World Intellectual Property Office (WIPO) in Geneva.

The PCT system has enjoyed spectacular success since it was first introduced in 1978 and is now the route generally used in industry for making foreign filings. What is particularly useful is that it is possible to file, in English, a single application designating over 100 countries.[10] Provided the PCT application is made within 12 months of the first priority application, priority can be claimed, potentially providing a way of 'getting behind' any relevant publications which have been made during that 12 month interval. (Warning: this glosses over the question of priority entitlement for subject matter not present in the priority document but only in the final application filed, see Section 9.)

At 18 months from the claimed priority date the PCT application will be published with a WO prefix. The publication shows on its face applicant information, the title and an abstract of the invention and the designated countries. By way of illustration, I have selected an application which is unlikely to give you or your company a headache, although the same could not, perhaps, be said for the individual depicted (see Figure 2)!

The initial cost of a PCT application is not as much as one might think. The filing fee and designation fees amount to no more than about £2,500, although there is a page charge for specifications greater than 30 pages.[11]

PCT WORLD INTELLECTUAL PROPERTY ORGANIZATION
International Bureau

INTERNATIONAL APPLICATION PUBLISHED UNDER THE PATENT COOPERATION TREATY (PCT)

| (51) International Patent Classification 6 : A42B 1/02 | A1 | (11) International Publication Number: WO 99/39598<br>(43) International Publication Date: 12 August 1999 (12.08.99) |
|---|---|---|

(21) International Application Number: PCT/US99/02785

(22) International Filing Date: 9 February 1999 (09.02.99)

(30) Priority Data:
09/020,605 9 February 1998 (09.02.98) US

(71)(72) Applicant and Inventor: FLANN, Randall, D. [US/US]; Room 7, 413 West Mineral Street, Milwaukee, WI 53204-1741 (US).

(74) Agents: RYAN, Daniel, D. et al.; 633 West Wisconsin Avenue, Milwaukee, WI 53203 (US).

(81) Designated States: CA, European patent (AT, BE, CH, CY, DE, DK, ES, FI, FR, GB, GR, IE, IT, LU, MC, NL, PT, SE).

Published
*With international search report.*

(54) Title: SUBSTANCE DISPENSING HEADGEAR

(57) Abstract

A headgear (12) for dispensing a substance has a container (28) to carry the substance. A spigot (14) is secured to the container (28). The spigot (14) can be opened to dispense the substance by gravity, suction, pressure or levity flow from the container (28). The spigot (14) can be closed to retain the substance in the chamber (20). A hat-like recess (26) is formed within the bottom wall (24) of the container (28) sized for wearing on an individual's head, and for maintaining the container (28) in a freestanding condition during hands-free ambulation of the individual.

**Figure 2** *A representative PCT publication*

The rub comes later, typically at 30 months from the first priority date (usually about 18 months from filing the PCT application) when it is time to 'enter the national phase'. At this point the real costs cut in, as what one is doing is effectively filing individual patent applications in each of the designated countries (or regions). Costly translations may have to be provided and filing and examination fees will have to be paid to each of the designated patent offices. It is clearly an appropriate time to review whether to proceed with national phase entry in all the countries originally designated and, perhaps, reduce costs by cutting back on geographical coverage.

A further advantage of the PCT system is that it is possible to obtain a non-binding international preliminary examination report (an IPER) before this critical 30 month stage. That report will show whether the invention is likely to be patentable and worthwhile pursuing. It is also possible to amend the claims prior to entering the national phase so as to improve the chances of a favourable review by a national patent examiner later.

The inexperienced scientist will often make the mistake of treating a competitor's PCT patent publication as 'a patent'. It is not a patent, merely an ungranted application. There is no guarantee that the claims will be granted by national patent offices later, or that the competitor will even pursue the application. However, that does not mean that relevant third party patent applications can be ignored. Claims could ultimately be issued by national or regional patent offices that might block your company's commercial activities, so the progress of troublesome third party applications should be monitored by your patent department. A second point is that published PCT and other patent applications can (depending on the date) form part of the prior art and thus, are potentially relevant to the patentability of one's own work.

I have spoken of the advantages of the PCT route. Where, you may be thinking, does the European Patent Office fit in? The EPO, based in Munich, is the body responsible for granting European Patent Applications. It is possible to obtain a European Patent, designating 20 European Countries, by making a direct application to the EPO.[12] This could be done in a manner similar to that described above, by filing a European Patent Application claiming priority from one or more national applications filed up to 12 months previously. A European Patent can also be obtained, *via* the PCT route, by designating the EPO as a regional patent office.[13]

A European Patent application is published at 18 months from the earliest priority date. But in the age of genomics, we may be entering an era when, for some inventions, conventional publication does not occur. In order to search DNA sequences, it is a requirement that these are supplied to the Patent Office in electronic version on a floppy disk, but it is also a requirement that a paper copy is filed. In July 2000, the President of the European Patent Office published a notice saying that he was refusing to publish, in the normal way, a copy of a patent application, which amounted to 50 000 pages in length. However, a copy can be purchased on CD-ROM!

This section has dealt with applying for patents, but that is only the start of the task. If patents are to have any value, they have to be granted, with claims which are enforceable, if needs be, in the courts. These activities, prosecution and litigation, are dealt with in the next section. It will be seen once again that the scientist can play a crucial part.

## 6 PROSECUTION AND LITIGATION

The word 'prosecution' suggests some criminal activity, but in the patent world the word is used to describe the process of getting a patent granted after it has been filed. Except in a very few countries (for example South Africa), patents rarely go through 'on the nod' and frequently have to be fought for by written and sometimes oral argument. This is where the patent attorney's advocacy skills come to the fore, but it is also an opportunity for the scientist to contribute.

The scientist will know far better than the patent attorney what publications already exist in the field of interest. If the inventor is aware of relevant prior art, in most countries there is no obligation to bring it to the attention of the patent offices, but **that is not the case in the USA**. In the US the applicant has a 'duty of candor' and **must** bring all relevant prior art to the attention of the US Patent Office, usually in the form of an Information Disclosure Statement (IDS). If you do not do so you run the risk that any patent ultimately obtained will be declared unenforceable. Accordingly, if you are an inventor you will almost certainly find that your patent attorney periodically sends you a memo reminding you to disclose all relevant prior art of which you are aware.

Frequently the scientist can help during prosecution of the patent application before various patent offices by providing additional data, sometimes in the form of a declaration or sworn affidavit. These additional data cannot be added to the patent specification (that would be 'adding matter' after filing which is forbidden) but they can be used to overcome objections, for example that the invention cannot be performed over the whole range claimed or that the invention lacks inventive step.

To take a simplified example, imagine there is a known compound that is structurally similar to the compound you have invented. If it has the same pharmacological effect, then it would be standard patent office practice to reject your application on the grounds of obviousness. One way round that problem is to provide a showing (if possible), which demonstrates that your compound has 'surprising activity', *e.g.* is unexpectedly ten times more active than the prior art compound. Following the submission of such data, I would expect the obviousness rejection to be withdrawn and a patent to be granted.

It is not only patent examiners who get involved in assessing patentability. In nearly all countries, patents can be challenged during their lifetime by third parties seeking revocation on the grounds that they should never have been granted in the first place. In the European Patent Office, this can be done centrally if an opposition is filed within 9 months of grant. Challenging validity is the opposite side of the coin to alleging infringement in patent litigation. If your company sues someone for infringing a patent, you can be sure that the alleged infringer will claim that the patent is invalid and therefore cannot be infringed. Rather than getting the pirate off the market, the plaintiff could find the precious patent revoked and the door opened to a flood of generic competition!

Full-blown patent litigation is extremely expensive, especially in the USA where a patent action could cost $5–10 million. Therefore it should not be entered into lightly. Furthermore, in the US a patent trial, at first instance, normally takes place in front of a jury. The jury members will definitely not be familiar with medicinal chemistry or biotechnology, making the outcome something of a lottery! If one is found guilty of 'willful infringement' of a US patent (normally as a result of blatantly ignoring it) an award of triple damages can be made. Negotiating a licence before litigation commences, or coming to an out of court settlement while it is still ongoing, may often be the preferred option for both parties.

Patent litigation in Europe is comparatively less costly but judicial systems can vary dramatically from one country to another. For example, in Germany infringement and validity are treated in quite separate proceedings. Since infringement is usually tried first, it is possible that a defendant could suffer an injunction, yet find later that the patent was not worth the paper it was written on. Until recently, Holland was a popular forum for patent litigation as the Dutch court was famed for granting injunctions in rapid proceedings. Furthermore, in appropriate cases, such injunctions were pan-European in scope. This threat led resourceful defendants to devise 'the Belgian torpedo', the idea being to strike first by bringing a counter-suit in Belgium, a country with a notoriously slow judicial system. Having launched the 'torpedo' it was argued that, pending the outcome of the case in Belgium, litigation initiated elsewhere could not proceed.

All this is a far cry from the rather more predictable system in the UK, where to enter the High Court in London one cannot help feeling one is stepping back in time. Echoes of old cases such as Rex *v* Arkwright (1785), still cited today, can almost be heard in the corridors and stories of the bar in former times still abound (see below).

**Judge (to barrister):** *"After hearing your submission Mr Smith I am none the wiser."*

**Barrister:** *"None the wiser perhaps, M'Lud – but far better informed!"*

Nevertheless the UK system is not slow and is overseen by some quite excellent specialist patent judges who were once formidable patent barristers. Even so, the judgment of the lower court is often reversed by the Court of Appeal. Once or twice in a decade, a patent appeal finds its way to the House of Lords and is decided by five Law Lords. In the 1990s, the Biogen case (see Section 9) was one such case. The Merrell Dow *v* Norton case relating to the anti-histamine terfenadine and its *in vivo* conversion into an active metabolite was another.

All these cases make marvellous reading. Where else but in a patent case would you be asked to consider the implications of following a cooking recipe by Elizabeth David? And in the same judgment find a discussion of what the Amazonian Indians knew about the 'magic spirit of the bark' (per Lord Hoffmann in the terfenadine case). In keeping with this literary tradition, one leading patent solicitor decided to abandon the law in favour of becoming a novelist.[14]

Let me conclude this section with a word of warning. In the US, UK and some other countries, litigation goes hand in hand with 'discovery', that is the compulsory delivery up of any documents having a bearing on the case. The discovery process can include scientific memoranda, minutes of team meetings and handwritten notes. Virtually any document not enjoying legal 'privilege' (such as an opinion provided by a qualified patent specialist) is fair game. Deleted e-mail messages can even be retrieved! Time and time again, one comes across damaging documents that threaten to compromise the company's patent position. I can only repeat the advice I gave in the previous edition of this book: ensure that every single document you write (or even annotate) would not be an embarrassment if one day it was produced in a court.

In any litigation, and I include in this opposition proceedings, scientists are frequently asked to attend patent hearings as expert witnesses and can play an invaluable role in convincing a court or appeal board of a patent's validity or otherwise. Anyone who has ever been involved in patent litigation will know that it is a thrilling, if nerve wracking experience.

## 7 TRIPS AND US PRACTICE

Patent law can differ quite considerably from country to country. National patent law is established by the highest levels of government (in the UK by an act of parliament) and what goes into the law is highly political. One would not think dusty legal documents could raise such a stir, but they do! Another major chapter in patent law, which has unfolded since the first edition of this book has been the "GATT-TRIPS" agreement. Readers will be familiar with the General Agreement on Tariffs and Trade (GATT) which established the World Trade Organisation (WTO) in 1995.

What they may not know is that an important part of the GATT talks related to Trade-Related Aspects of Intellectual Property Rights (TRIPS).

The TRIPS agreement calls for all member countries of the WTO to amend their IP laws (if needs be) to provide certain minimum standards. Historically, a cause for concern has been the poor standard of patent protection, in particular for pharmaceutical inventions, in developing countries. The TRIPS provisions promise to change all that. TRIPS sets out:

- Minimum term of patents will be 20 years from filing date.
- 'Product' protection will be available for pharmaceuticals (formerly many countries would grant only 'process' claims which are harder to police and more easily circumvented).
- Patents will be available without discrimination as to field of technology.

Post-TRIPS, all signatory countries have to provide these and other minimum standards of IP protection. So-called 'developed' countries had only 1 year to bring about amendments and many 'developing' countries should by now have changed their laws, although they were given until 2005 if they did not provide product protection for pharmaceuticals.[15] However, the TRIPS agreement is still contested and was central to the much-publicised dispute between several pharmaceutical companies and the South African Government in relation to generic AIDS drugs.[16]

From the pharmaceutical company perspective, the TRIPS agreement is a huge step forward but is by no means perfect. For example, there are no provisions dealing with the growing problem of international 'exhaustion of rights'. Exhaustion of rights, already a fact of life within the EU, limits the rights of a patentee once he has put his product on the market. If the exhaustion doctrine were adopted internationally, the patentee could not sue a third party who buys the product legitimately in one country and then imports it for sale into another country where a patent right exists. Clearly this represents a considerable threat to pharmaceutical companies striving to make medicines available at low prices in developing countries. Despite its shortcomings, TRIPS has rightly been described as 'the single most influential factor in determining the future of patent law world-wide'.[17]

Paradoxically, it is the TRIPS agreement, aimed primarily at parts of the world with weak patent laws, that has prompted a number of notable changes in US practice. The US patent system has always been strong and renowned for being different to that of other countries! At the time the GATT agreement was reached, the term of a US patent was 17 years from grant. That was clearly contrary to TRIPS because if a US patent were granted quickly the patent term would be less than the agreed minimum of 20 years from application. Accordingly, in June 1995 a new law took effect providing a patent term of 20 years from application for all US applications filed after that date, but maintaining the 17-year term for all 'pre-GATT' cases. This led to a huge increase in the number of applications filed just before the cut-off date: patents normally take longer than 3 years to grant and one is usually better off with a 17 year term!

Nevertheless, for all cases filed after 8 June 1995, the term of a US patent is a predictable 20 years from the filing date. In time this tactic will put a stop to US 'submarine' patents, that is patent applications surfacing as granted patents many years after filing and still enjoying a 17 year term. Until recently, US patent applications were not published before grant, unlike the system in the EPO, for example, which is completely transparent. That is partly why submarine patents were so dangerous: one could never tell what was lurking in the depths of the US Patent Office. However, in a complete *volte face*, US patent applications are now published before grant, provided there is an equivalent foreign application.

Significantly, the TRIPS agreement was silent on the famous US 'first-to-invent' system, according to which being first to the Patent Office, while desirable, is not essential to steal a march on a competitor. Not surprisingly, therefore, the US have seen no reason to harmonise with the first-to-file system found in Europe, Japan and most other countries in the world. US attorneys are still happily engaged in 'interference practice' – that is going through a formalised procedure before a special department in the Patent Office to determine who conceived the invention first and whether they were diligent in 'reducing it to practice'. However, nowadays it is no longer necessary to have carried out the invention in the USA (or imported it) to establish 'conception': a laboratory notebook record made anywhere in the world suffices. Keeping good notebook records, with each page countersigned as read and understood by a colleague, is a discipline to which the scientist should pay particular attention.

## 8 THE BIOTECH REVOLUTION

The story of modern biotech patenting can be traced back to 1980 when the famous case of Diamond *v* Chakrabarty was decided by the US Supreme Court. This was not an infringement case, but the final chapter of an appeal by Chakrabarty against the refusal of the US Patent Office to grant him a patent (Diamond was the then Commissioner of Patents and Trademarks). Chakrabarty discovered a process by which plasmids capable of degrading different oil components could be transferred to a *Pseudo-*

*monas* bacterium. The resulting transformed microorganism was willing and able to gobble up oil slicks! Despite the clear utility and ingenuity of the invention, the objection was raised that the microorganisms were alive and were not statutory subject matter, despite the fact that Louis Pasteur had been granted a US patent in 1873 on 'yeast, free from organic germs of disease ...'.[18] The Chakrabarty case incidentally saw the first appearance of 'No Patents on Life' protesters, although that term has since become a much more common soundbite in Europe. Chakrabarty won the case, but only by the narrowest of margins: 5-4. It was famously held that 'anything under the sun that is made by man' was potentially patentable subject matter in the US.

If, instead of '1066 and all that', Sellar and Yeatman had written 'Biotech and all that' the decision by the Supreme court would no doubt have been classified as a GOOD THING. Certainly it paved the way for more patenting in the same vein and the start-up genetic engineering companies of the early 1980s such as Biogen, Amgen and Genentech benefited greatly from the positive patent climate that the Chakrabarty decision engendered. Today the golden age of genetic engineering has given way to the even more golden age of genomics.

Genomics, the study of genes and their contribution to disease, is likely to provide the key to the development of treatments for presently incurable diseases in the future. Alzheimer's disease, breast cancer, schizophrenia and even obesity are all now clearly associated with a genetic component. It seems that only if you die of a gunshot wound can it really be said that you are suffering 100% from environmental causes – and not even then if you have a genetic predisposition to become a gangster! Predictably, the consequence in the patent world has been a huge effort directed to patenting human genes. Encouragingly, the US administration has continued to recognise the importance of stimulating R&D investment by the continued provision of a strong patent system. The US Patent Office regularly grants patents on isolated genes. However, Europe has been different and a whole book could be written on the events of the last 10–15 years. The story began with the introduction of a proposal for an EU Directive intended to harmonise patent protection amongst the member states in relation to biotech inventions. This was known generally as the draft 'Biotech Directive'.

It is unfortunate for the pharmaceutical industry that this directive was being debated at the same time as the European Patent Office granted a patent on the famous 'oncomouse' (based on the Chakrabarty precedent, the US Patent Office had granted a corresponding US patent in 1988 with barely a murmur). In addition, other advances in biotechnology such as genetically engineered plants were becoming of increasing public concern, as was the whole issue of biodiversity and alleged ‘biopiracy’ (making inventions from genetic resources belonging to others). The term ‘bioethics’ quickly entered the vocabulary of the patent profession.[19]

The first version of the EU Biotech Directive was unsurprisingly voted down by the European Parliament in 1995. However, the EU Commission rapidly put a new draft back on the table which, after much modification and debate, was accepted by the European Parliament in May 1998. It was critical to the pharmaceutical and biotech industries that patent protection for biotech inventions was not weakened by the Directive. Fortunately, the final form of the Directive was positive. Article 5 is of particular note as this confirms that the DNA which constitutes human genes, provided it is claimed in isolated form and not *in situ*, is in principle patentable, and provided that the ‘industrial application’ is set forth in the specification.

The furore over the patenting of ESTs, already alluded to in Section 3, has now partly subsided, as it is clear that patent offices will find ways of refusing patents, either on utility or obviousness grounds. Concern has instead focussed on the patenting of ‘research tools’ (*e.g.* targets and screening methods) for fear that these could prove to be a disincentive to research. It is also argued that if companies are allowed to establish proprietary positions on key genes (such as the BRCA breast cancer genes) this will lead to an increase in the price of medicines and diagnostic tests.

There is no easy answer to any of these questions and the debate continues. However, it is the position of the Association of the British Pharmaceutical Industry, the BioIndustry Association and many other industry groups that it does no good to place a bar on patenting what is clearly patentable subject matter. This would be the slippery slope to weakening the patent system, so undermining the incentive for investment on which the industry depends. There is no reason in principle why research tool patents should be treated any differently from other patents and licences negotiated where necessary.

A claim to a method of screening is one thing, but where most practitioners would draw the line is extrapolation to any compound identified by such a method, the so-called ‘reach-through’ claim. If such claims were valid the patentee could hope to derive a royalty stream from some future blockbuster drug found using the screen. Reach-through claims have been granted, but only in rare instances and those cases appear to have ‘slipped through the net’.

The final chapter in the saga of the EU Biotech Directive has still to be written. Despite its acceptance by the European Parliament in 1998, only 4 of the 15 Member States had implemented the key provisions of the Directive by the deadline of 30 July 2000. Many of the others are still debating the issue and Holland has tried (unsuccessfully) to challenge the legality of the Directive at the European Court of Justice![20]

## 9 ADVANCED MODULE

This section will not provide a comprehensive advanced course, but will clear up a few more complex points for the interested reader. If nothing else, I hope what follows will provide a little more insight into what a fascinating and complex subject patent law can be.

In the discussion on novelty, two points were relegated to this section. The first concerned the question of someone abusing your confidence and disclosing your invention before you had filed a patent application on it. What is the remedy in such circumstances?

I do not propose to discuss legal remedies outside the sphere of patent law. These would no doubt depend on the facts, for example, were papers relating to the invention stolen? But whatever the circumstances, patent law, at least in Europe, is clear. You can still file a patent application within 6 months of the abuse and the non-confidential disclosure will be ignored. However, it is important to note that within this 6 month period a *final* European patent application must be filed. A recent case has established that you cannot file a priority application within 6 months and a final European patent application 12 months thereafter.

Such a problem is rarely encountered, but another point raised earlier in relation to novelty is relatively common. I refer to the situation in which you believe you are the first to file a European patent application on an invention but find a competing application having an earlier priority date is published later.

These circumstances, while depressing, do not necessarily mean that all is lost. The rules in Europe state that because the conflicting application was not published at the time you filed, its contents are citable against your claims only in relation to novelty. The conflicting disclosure cannot be used to attack your claims on the grounds of lack of inventive step. In practice this means that often both parties can walk away with something, since there will rarely be total overlap of subject matter. If your application discloses a preferred compound and an earlier-filed conflicting application does not, you may still be able to establish an important position. The other party may succeed in obtaining a dominating generic claim, but you may still achieve the necessary leverage to conclude a cross-licensing arrangement.

The situation becomes even more complex if each party has several priority documents containing differing subject matter. For example, it can happen that your competitor, while having the earliest priority date, may not be entitled to that earlier date for key subject matter he has included only in a later application. If you got to the patent office before that later application was filed you might still win the novelty contest in relation to what is really important!

This discussion leads naturally to the law on priority entitlement. How much do you have to disclose in your earliest priority document to establish a date you can rely on for later refinements of the invention? The European Patent Convention is not very helpful on this point, stating simply that a right of priority exists in relation to the 'same invention'. Is 'same invention' to be interpreted liberally, as it was in a celebrated EPO case called 'Snackfood'!? Or should it be interpreted conservatively (with emphasis on the precise language used) as in some other decisions? Finally, an Enlarged Board of Appeal decision has laid the matter to rest: the conservative approach seems to have won the day and Snackfood has had its chips.[21] This decision is critically important, not just in relation to conflicting applications as discussed above, but also in relation to the relevance of publications occurring during the 12 month priority interval. To be sure of priority entitlement, the applicant must have clearly described the essential parameters of the invention in the priority document.

I should like to conclude this section with a caveat pertaining to over-broad claims. The validity (or otherwise) of such claims is a complex subject and is one which has challenged even the huge intellect of the judges in the House of Lords, notably in the Biogen case.[22] It is all the more important, since in many countries (including the UK) and in the European Patent Office, claims cannot formally be attacked in revocation or opposition proceedings solely on the ground that they are too broad. One way in which this difficulty has been circumvented is to argue that in order for the sufficiency requirement to be met (insufficiency **is** a ground for revocation; see Section 3) the patentee has to teach adequately how to perform the invention *over the full width of the claims*. If, because the claim is so broad, embodiments within it cannot be made or do not have the claimed pharmaceutical activity, then it is bad for insufficiency.

In the Biogen case, this concept was cleverly refined by Lord Hoffmann. He saw that there was no classical insufficiency attack (the directions given in the specification gave the promised result). But, more subtly, he ruled that there were ways one could envisage of performing the invention within the boundaries of the claim that owed nothing at all to the teaching of the patent, or the technical contribution the inventor had made.

In the more recent UK case of Monsanto *v* Merck,[23] decided by the UK High Court in 2000, an insufficiency argument was also successful in bringing down a broad claim. In that case, it was held on the evidence that not all the tens of thousands of compounds within the scope of the main claim had the promised technical effect. The claim was insufficient because the activity of the compounds as anti-inflammatory COX II inhibitors was an essential feature of the invention. Accordingly, it is clear that the medicinal chemist should be consulted in determining what constitutes a realistic and defensible claim. Extrapolate too far at your peril! Having said that, it is true that broad claims may be sustained, even if only one way is given of carrying out the invention, for a fundamental principle or for a pioneering invention in a new field. Whether broad 'mechanism of action' claims (see Section 2) fall into this category has yet to be tested by the courts.

## 10 SITES FOR SORE EYES

The Internet has transformed the world of patent searching. No more painstaking manual trawls through thick volumes in subterranean libraries, patent documents can be called up, free, on your screen at the touch of a button. If you would like further information about patents, I would strongly recommend you follow some of the links set out below.

**http://www.uspto.gov/** – the home page of the US Patent Office. Best of all is the page at: **http://www.uspto.gov/patft/index.html** which provides access to the full text of all US patents issued since January 1976. Historians can also find full-page images of each page of every US patent issued since 1790. For example: **http://patf.uspto.gov/netahtml/search-bool.html** using the Boolean search page with *Cohen* as one inventor in search term 1 and *Boyer* as another inventor in search term 2 led, in the space of just a few seconds, to links to all 3 famous US patents on 'biologically functional molecular chimeras', setting out the basic principles of gene cloning.

**http://www.european-patent-office.org/** – the European Patent Office's well-maintained web site from which the case law of the Boards of Appeal can be accessed. Note also the twirling EPO logo, intended to represent a stylised finger print! One of the best features of this EPO site is the esp@cenet patent searching facility located at: **http://ep.espacenet.com/** which can be accessed from the EPO home page in two clicks. The brilliant esp@cenet allows you to search for patent documents published by the European Patent Office, PCT applications, and patent specifications with an English abstract and title from more than 30 million documents world-wide. There is also an online European Patent Register at: **http://register.epoline.org/espacenet/ep/en/srch-reg.htm**.

**http://www.wipo.int/index.html.en** – the WIPO web site, essential for those with an interest in the PCT (see Section 5).

Obviously it would be impossible to give the reader anything like a comprehensive listing of all the patent information sites available on the web. But just a mention of one of two more favourites may be of interest. Some of these links may be transitory in nature, but at the time of writing: **http://colitz,com/site/index.htm#** is well worth bookmarking. This is a link to the offices of Michael J. Colitz, Jr. who maintains the 'Wacky Patent of the Month' (Trade Mark) web page and archives thereof. I particularly like the "Annunciator for the Supposed Dead" (see under October 1996), patented in 1891, when fear of being buried alive was apparently of concern! Some practitioners swear by another (subscription) search site, found at: **http://www.delphion.com/**. This also maintains a 'Gallery of Obscure Patents'.

**http://www.courtservice.gov.uk/judgments/judg_frame.htm** – for UK Court judgments such as the Monsanto *v* Merck case mentioned above. This may be accessed through the database at: **http://www.kuesterlaw.-com/** – through which nearly everything mentioned in this section and much else besides can be found. Created and maintained by Jeffrey R. Kuester, this is intended to be the most comprehensive resource on the Internet for technology law information and is reportedly the most linked-to intellectual property web site on the Internet.

All this is an enormous help, but an essential point should not be overlooked. Patent resources available on the Internet will never replace the role of the skilled patent searcher. These are the people employed by large organisations or specialist search firms who have the ability to dig up dominating patents, find equivalents in other countries, monitor status, payment of renewal fees and carry out numerous other tasks which are beyond the ken of the average attorney. Long may they continue to flourish!

## 11 CONCLUSION

*"Patent law, to the casual observer at least, had seemed for decades to be a ponderous subject with the dynamism and dash of a moribund sloth. But in the last fifteen years, and notably in the last five, the creature has descended from its steamy branch and evolved new traits of agility and diplomacy."*

So wrote Emma Johnson in *Nature Biotechnology* in 1996.[24] Six years on and the subject of patents has become even more prominent. Supplements on 'patenting life' help to sell newspapers.[25] Debates on the subject even make good television! We read about the controversy surrounding the patenting of business methods and the US lawsuit between Amazon.com and Barnes and Noble relating to the '1-Click' patent. Substantial 'bounties' are even offered for relevant prior art documents with which patents can be attacked and perhaps 'busted'![26]

It is hoped that the reader can now see why patenting sometimes has to be treated diplomatically and why, in order to embrace new technologies, patent law has had to become more agile. But above all it is hoped that the research scientist will understand that he or she has a crucial part to play in this exciting, patent-conscious age. With the huge amounts of money being devoted to R&D, protection of the company's IP rights has become even more critical.

In this chapter, I have tried to provide a snapshot of where patent law stands today, focussing particularly on the important part the research scientist has to play. There are huge areas that have not been covered. One is the enormous upheaval expected to hit us in Europe with the proposed introduction of a new system, the Community Patent Convention, and perhaps the setting up of a European Patents Court.

I have concentrated on the basic principles of the patent system but have also tried to illuminate some of the deeper reaches of the subject. That is a deliberate policy. In my experience, once scientists understand the framework properly they are much more willing to engage in a positive dialogue with their patent attorney, rather than treat him or her as someone who exists simply to block their scientific publications. However, please do not take my word for it. [Indeed, please do not take my word for anything: it should be made clear that nothing in this chapter is meant to constitute legal advice.[27]] Please, get to know your own patent attorneys and work closely with them. A really effective partnership between R&D and the IP department can create spectacular results, often seen years later to have created value for one's employer measurable in many millions of pounds. Such is the power of patents!

## 12 REFERENCES AND NOTES

1. *Medicinal Chemistry Principles and Practice*, F.D. King (ed.), The Royal Society of Chemistry, Cambridge, 1994, Chapter 10.
2. Approximately 70% of medicines do not earn a sufficient return to recover the investment (EFPIA figures, 1999).
3. An application for a drug marketing authorisation relying on cross-referral to the originator's full dossier is known as an abridged application. RE is a temporary prohibition from referral to the originator's data without the originator's consent. It does not restrict the independent development of identical or similar products but does

give protection against quick generic copying. RE is acquired as a result of the regulatory approval process and is particularly important in Europe because for many countries the exclusivity period is 10 years and sometimes extends beyond patent expiry. There are also RE periods in the US, Japan and many other countries.

4. This is known as the "Bolar" exemption.
5. Pfizer had patents on sildenafil citrate (Viagra) before its use for treating male erectile dysfunction (MED) was found. The case in question concerned European Patent 0702555 relating to that use. The decision by the UK court to revoke the patent was upheld by the Court of Appeal. At the time of writing, opposition proceedings involving 13 opponents are still in progress at the EPO.
6. In US practice, it seems that any DNA sequence must *per se* have been unpredictable and accordingly is non-obvious, whether or not it was obvious to set out to identify it. The European Patent Office view is rather more robust. If it is obvious to obtain the sequence, then the sequence itself, whatever it turns out to be, is generally also obvious.
7. For a discussion of the Gunning Fog Index see ref. 1, page 156.
8. See ref. 1, pages 156–157 for a discussion of the Catnic case. This approach has been used many times since, albeit somewhat refined. A critical example of claim interpretation emerged recently in the case of American Home Products *v* Novartis relating to the use of rapamycin for suppressing transplant rejection. The claim recited only rapamycin but was nevertheless held by the UK court of first instance to cover the use of Novartis' related analogue. However, the decision was reversed on Appeal.
9. It is possible to file in countries not on the original list before any publication occurs, but if the additional filings are made after the 12 month convention period, they will no longer be able to claim the priority date of the first filed application.
10. A PCT application may be filed in any language acceptable to a PCT 'Receiving Office', or in fact, if filed at the WIPO Office in Geneva, in any language. It may be necessary to supply a translation later, however, as PCT applications are published in only one of seven languages. For further information, see the PCT Applicants Guide or the WIPO web site link given in Section 10.
11. Currently it is necessary to pay only a maximum of five designation fees in order to designate all PCT Contracting States.
12. EPC countries are not identical to EU countries. There are currently 20 member states of the European Patent Convention (EPC), namely: Austria, Belgium, Cyprus, Denmark, Finland, France, Germany, Greece, Ireland, Italy, Liechtenstein, Luxembourg, Netherlands, Mon-

aco, Portugal, Spain, Sweden, Switzerland, Turkey and UK. Liechtenstein and Switzerland are covered in just one designation. 'Extension agreements' have been concluded with Albania, Latvia, Lithuania, Macedonia, Romania and Slovenia. Bulgaria, the Czech Republic, Estonia and Slovakia will join the EPO in July, 2002.

13. A PCT application can be made designating the EPO (all countries) as just a single designation (attracting a single designation fee). At the 30 month stage one would then enter the 'regional phase' before the EPO (paying appropriate EPO fees) and prosecute the application to grant as one would for a regular European Patent Application. Entry into the European regional phase can be delayed until 31 months from priority.
14. Michelle Paver, formerly of Simmons and Simmons, has published a best-selling novel "Without Charity". See also Daily Mail, 4 April 2001.
15. 'Least Developed' country members were originally given until January 2006 to meet their TRIPS obligations. It now appears that the deadline will be extended by 10 years.
16. The widely publicised South African litigation relating to generic AIDS drugs was finally settled in 2001, following intense public pressure. In a joint statement issued by the parties in which the South African Government confirmed that the South African legislation would be implemented in a TRIPS-compliant fashion the action by the pharmaceutical companies was withdrawn.
17. J. Barrett-Major, *J. Commercial Biotechnol.*, 1999, **6**, 43–52.
18. See US Patent Number 141072 on *Manufacture of Beer and Yeast* (L. Pasteur).
19. European Patent Number 0169672 relating to the oncomouse was opposed by numerous animal rights groups on ethical grounds. Under Article 53(a) of the European Patent Convention inventions 'shall not be granted in respect of inventions the publication or exploitation of which would be contrary to "order public" or morality ...'. In 2002, the patent was upheld in an amended form.
20. The opinion of the Advocate General, given in June 2001, was that the Dutch challenge should be rejected. The ECJ confirmed this.
21. Decision of the Enlarged Board of Appeal of the EPO; Case G2/98.
22. Biogen *v* Medeva, a case relating to hepatitis B vaccine. See [1997] Reports of Patent Cases, page 1 *ff.* (House of Lords).
23. Monsanto *v* Merck, Judgment of 4 February 2000 by Pumfrey, J., upheld in 2001 by the court of Appeal.
24. E. Johnson, *Nat. Biotechnol.*, 1996, **14**, 288–291.
25. Guardian 12 page special report, "Patenting Life", 15 November 2000. See also 'A Patent Lust for Life' by Jon Evans in *Chem. Br.*, January 2001, pp. 27–30.

26. See the BountyQuest web page at **http://www.bountyquest.com/bountylist/bountyguide.htm.**
27. Readers are urged to consult their own patent representative. This chapter does not constitute legal advice and the author cannot be held responsible for any action taken (or not taken) as a result of studying it. The views expressed herein are those of the author and do not necessarily coincide with those of GlaxoSmithKline.

CHAPTER 14

# An Introduction to Molecular Biology

RALPH RAPLEY AND ROBERT J. SLATER

## 1 INTRODUCTION

Molecular biology is a term used to describe the field of science that is concerned with studying the chemistry and physical structure of biological macromolecules. It is a relatively young discipline. It was not until 1944 that deoxyribonucleic acid was shown to be the carrier of hereditary information, and the structure of the molecule was not known until 1953 when the, now familiar, 'double helix' structure was first proposed by Francis Crick and James Watson in Cambridge. They interpreted data from X-ray crystallography gathered by Rosalind Franklin, who was working in the laboratory of Maurice Wilkins in London. Following these discoveries, it was possible to describe the mechanism of inheritance and genetic mutation at the molecular level and experiments in molecular genetics began to provide new concepts at a staggering rate. It is a remarkable achievement that, from a position of almost complete ignorance about the chemistry of gene action fifty years ago, we are now in a position to read the entire genetic code of an organism, understand the mechanism of gene expression and manipulate DNA by genetic engineering. For a more detailed account, see a current molecular biology text such as that by Brown[1] or Lewin.[2]

## 2 NUCLEIC ACIDS

Cells contain two forms of nucleic acids: deoxyribonucleic acid (DNA) and ribonucleic acid (RNA). They both have the same basic structure incorporating three types of structural units: a five carbon sugar (pentose), nitrogenous bases and phosphate groups. The nucleic acids are polymers of repeating, identical pentose units (deoxyribose in DNA and ribose in

RNA) linked by phosphate groups, there is a nitrogenous base attached to each pentose. Although there are five bases in total, each form of nucleic acid includes only four. Adenine (A), guanine (G) and cytosine (C) are common to both forms, but DNA contains thymine (T) and RNA contains uracil (U). A and G are purines; C, T and U are pyrimidines. A simple diagram to represent the structure of DNA is shown in Figure 1.

It is important to realise that nucleic acid polymers have polarity; that is their two ends, or termini, are not the same. At one end of the sugar phosphate backbone there is a free 5′-phosphate group, whereas the other end has a 3′-hydroxyl group. For reasons that will become apparent later, it is the convention to describe the sequence of bases, using the single letter abbreviations, in the 5′ to 3′ direction.

Nucleic acids can form secondary structures that are based on the ability of the bases to associate *via* relatively weak interactions called hydrogen bonds. DNA consists of two strands running in opposite directions (one 5′ to 3′ the other 3′ to 5′) held together by these hydrogen bonds, wound together to form the double-helix. The association between the bases of the individual strands follows strict rules, originally proposed by Chargaff in 1950. A pairs with T and G with C. The two strands of a stable DNA molecule are, therefore 'complementary'; the sequence of bases in one strand is determined by the sequence in the other. This explains the elegantly simple principle behind DNA's role as the store of genetic

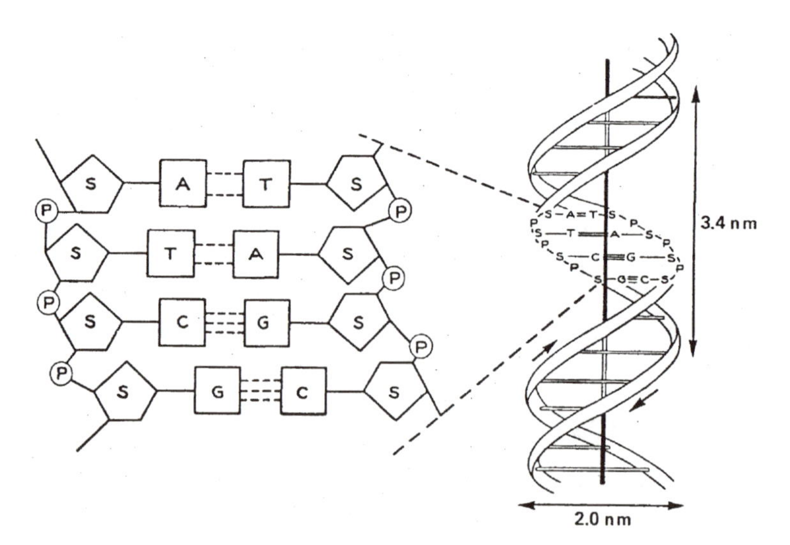


**Figure 1** *The Watson and Crick model of the double helical structure of DNA, A – adenine, C – cytosine, G – guanine, T – thymine, P – phosphate, S – sugar (deoxyribose)*

material; in Watson and Crick's words 'the structure suggests a copying mechanism'. If two complementary strands of DNA are separated, each strand can be used as a template for the formation of two identical DNA molecules (see DNA replication below). The process is semi-conservative, in that each daughter strand contains one of the original template strands.

In eukaryotic cells (*i.e.* yeasts, plants and animals), DNA molecules are organised into chromosomes (large DNA-protein complexes). Each chromosome contains one very long DNA molecule. Human cells contain 46 chromosomes (23 from each of our parents) and the total length of DNA in one set of chromosomes per cell is approximately one metre (the average size of human cells is 8 μm). The human genome (one set of chromosomes 1 to 22 and the sex chromosomes X and Y) contains approximately 3000 million base pairs. Although chromosomes of higher organisms consist of a length of DNA with ends (telomeres), a DNA molecule need not have any ends at all. The genetic material of bacteria (and chloroplasts and mitochondria) exists as circular molecules with super-helical forms where the double helix is coiled upon itself (referred to as 'supercoiling').

RNA is usually a single stranded polymer and is the genetic material of certain viruses (*e.g.* human immunodeficiency virus, HIV, thought to be the causative agent of AIDS). RNA is also found in all living cells and acts as an intermediate in gene expression. With the exception of some viral RNA molecules, RNA does not normally exist as a double strand of two polymers. The polymers can, however, fold back on themselves and form regions of secondary structure. Also a strand of RNA can base pair with a single strand of DNA. In this case, uracil in the RNA, hydrogen bonds with adenine in the DNA. RNA is classified by function into three main types: messenger RNA (mRNA), ribosomal RNA (rRNA) and transfer RNA (tRNA), discussed later.

## 3 PROTEINS

Proteins are made of one or more polypeptides: that is, long, complex polymers of 20 different units called amino acids. Polypeptides vary considerably in length and sequence of the 20 units. The number of permutations is therefore considerable and to all intents and purposes "infinite" within the context of biochemistry. Some proteins include other chemical structures such as sugar residues or porphoryins and they may be made up of several polypeptides. It is the great variety in the structure of proteins that provides the diversity of protein function such as: structural (*e.g.* collagen), hormone (*e.g.* insulin), catalyst (enzymes such as amylase to break down starch) and oxygen carrier (haemoglobin).

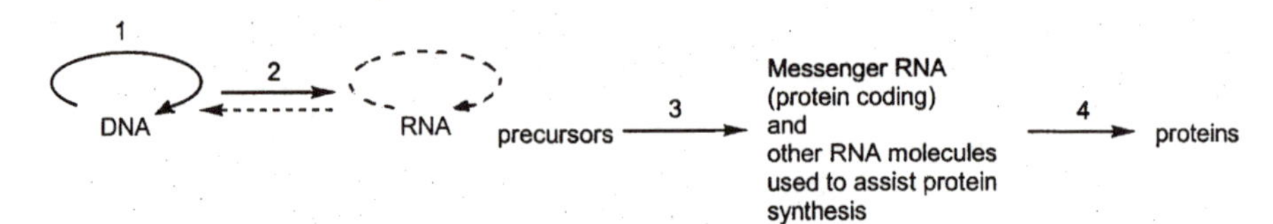


**Figure 2** *The flow of genetic information* 1. *DNA replication;* 2. *Transcription;* 3. *RNA processing;* 4. *Translation. (The dotted lines refer to activities concerned with the biology of certain RNA viruses)*

## 4 THE FLOW OF GENETIC INFORMATION

The store of genetic information held within the DNA of cells determines the structure and activity of living organisms. The DNA not only provides a template for more DNA (replication), but also directs the synthesis of protein molecules.

The process by which information is transferred from DNA to protein was first proposed by Francis Crick and is known as the central dogma of molecular biology (Figure 2).

The sequence of bases along the DNA molecule provides a template for the synthesis of complementary strands of RNA ('transcription') which, in turn, is used as a template for the synthesis of proteins ('translation'). The normal route of information flow is thus DNA to RNA to protein. A modification, however, is required in the case of some RNA viruses that use DNA as a replication intermediate. These are called retroviruses, the best known one being HIV.

## 5 DNA REPLICATION

Replication of DNA must occur to provide daughter cells with their complement of genetic information. The process is highly complex, involving many different enzymes and can only be summarised here. DNA replication commences by unwinding the DNA to form the templates for production of the new strands. DNA synthesis requires a primer, which is a short strand of RNA, complementary and hydrogen bonded to the template. An early step in DNA replication is the synthesis of these short RNA molecules by RNA primase.

DNA polymerases are the enzymes that replicate DNA and they can only synthesise the polymer in a $5'$ to $3'$ direction. The arrows in Figure 3 show the direction of DNA synthesis. The substrates for the reaction are nucleoside triphosphates of the four bases (that is a deoxypentose sugar, a base and three phosphates) termed dATP, dGTP, dCTP and dTTP. DNA

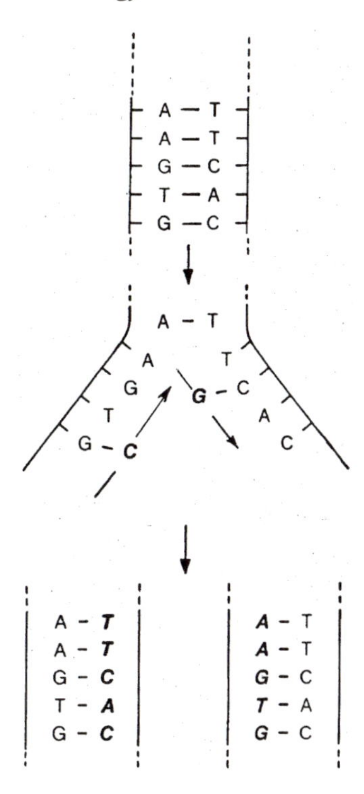


**Figure 3** *DNA replication*

polymerases require a primer because they 'proof read' the last base pair before adding the next nucleotide: this minimises mutation rate.

The base pairing rules ensure that the two daughter DNA molecules are replicas of the original template. Primer RNA sequences are excised and the gaps filled in by DNA polymerase. Because DNA can only be synthesised 5′ to 3′, one strand cannot be made continuously. It has to be made as a series of short fragments, synthesised away from the replication fork (Figure 3), called Okazaki fragments; these are eventually joined up by an enzyme called DNA ligase.

To facilitate replication there are enzymes to unwind DNA (helicase), proteins that keep the templates single stranded (DNA binding proteins) and enzymes to relieve three-dimensional tensions in the DNA molecule that result from it being unwound. The latter enzymes are called topoisomerases. They differ between animals and bacteria and are, therefore, a target for a group of antibiotics based on quinone.

DNA replication in animals and plants is instigated at more than one point (replication origin) along a DNA molecule. This is essential for replication of the DNA of higher organisms because of the length of the molecules involved.

## 6 TRANSCRIPTION

Transcription is the process by which DNA strands are copied into RNA, it is the first stage in gene expression. A whole chromosome is never transcribed into RNA. Instead, short regions of DNA (called "transcription units") are transcribed, starting at a sequence called a "promoter" and ending at a termination sequence. These regions of DNA contain the "genes". Transcription units in bacteria often include more than one gene (*i.e.* the units code for more than one protein) but in eukaryotes they include only one coding region for one polypeptide.

RNA synthesis is catalysed by RNA polymerase. Synthesis is in the 5′ to 3′ direction and the substrates for the reaction are the four nucleoside triphosphates, ATP, GTP, CTP and UTP. Transcription is a continuous process and produces an RNA molecule complementary to one of the DNA strands (Figure 4). This strand is referred to as the template strand (sometimes called antisense or non-coding strand). The RNA product has the same sequence (but for U in place of T) as the non-transcribed strand (sometimes called the sense or coding strand). Unlike DNA synthesis no primer is required.

Gene expression is regulated principally at the point of transcription. Cells do not transcribe all their genes at the same level all the time, and multi-cellular organisms do not express every gene in every cell. Indeed,

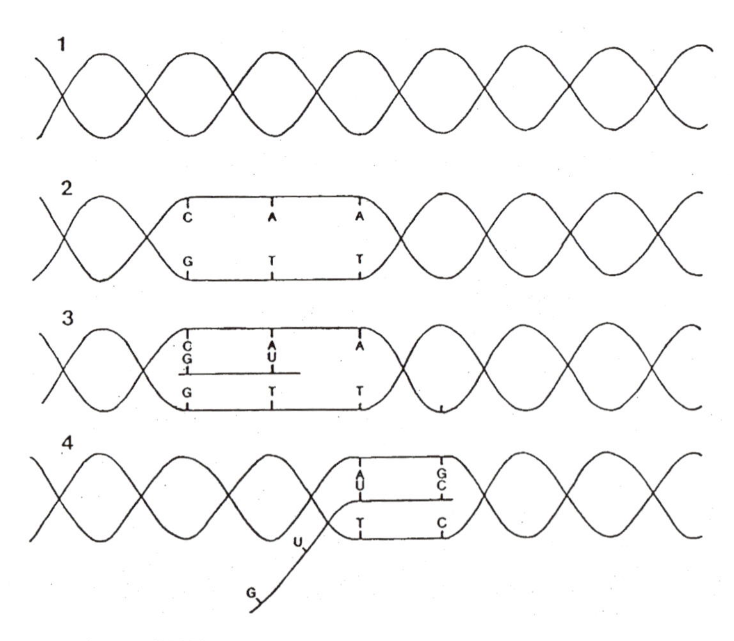


**Figure 4** *Synthesis of RNA*

this is the basis of cell differentiation. Different promoters have different affinities for RNA polymerase (thus affecting the rate of initiation of transcription). Also, there are DNA binding proteins that either assist the binding of RNA polymerase (activator proteins, often called transcription factors) or inhibit the binding (repressors). These proteins of course have their own genes (called regulatory genes). The study of these is crucial to our understanding of the control of gene expression, cell development, differentiation and the aetiology of many diseases.

(1) DNA double helix prior to RNA synthesis
(2) DNA begins to unwind
(3) RNA is synthesised 5′ to 3′ using one chain of the DNA as the template (read 3′ to 5′) for complementary base pairing.
(4) The DNA chain unwinds further and RNA synthesis continues. The part of the DNA which has been used as a template (*i.e.* the "gene" or "transcription unit") rewinds.

## 7 RNA PROCESSING

Following transcription, RNA molecules are, in many cases, modified to produce the mature, biologically active structure. In bacteria, RNA processing is largely confined to rRNA and tRNA, but in eukaryotes all forms of RNA are processed.

In all organisms, the initial products of rRNA synthesis are trimmed with nucleases, methylation occurs and double-stranded structures form. Transfer molecules are significantly processed. Many of the bases are chemically modified and additional nucleotides are added to the 3′ end.

The messenger RNA of eukaryotes is processed to a very great extent. The most significant step is the removal of introns, or intervening sequences, which are non-coding regions within the gene and initial transcripts. Removal of introns requires ribonuclease and ligation activities and is referred to as splicing. Apart from splicing, the mRNA molecules are processed by having additional nucleotides covalently attached to the 5′ and 3′ ends, referred to as capping (addition of a methylated G) and tailing (addition of polyA) respectively. These processes do not occur in bacteria; a fact that has implications for the exploitation of molecular biology in genetic engineering and is discussed in more detail in later.

## 8 PROTEIN SYNTHESIS

The specific three-dimensional structure created by a defined amino acid sequence is the basis of protein versatility and individual specificity. It is the sequence of bases along the coding regions of DNA that determines the amino acid sequence of a protein. The various steps in protein synthesis are described below.

## Activation of Amino Acids

A prerequisite for protein synthesis is the activation of the twenty amino acids. The activated intermediates are amino acid esters in which the carboxyl group of an amino acid is linked to the 2′ or 3′ OH group at the 3′ terminus of a tRNA molecule. There is at least one specific tRNA for each amino acid and the activation is catalysed by specific aminoacyl-tRNA synthetases, using adenosine triphosphate (ATP) as the energy

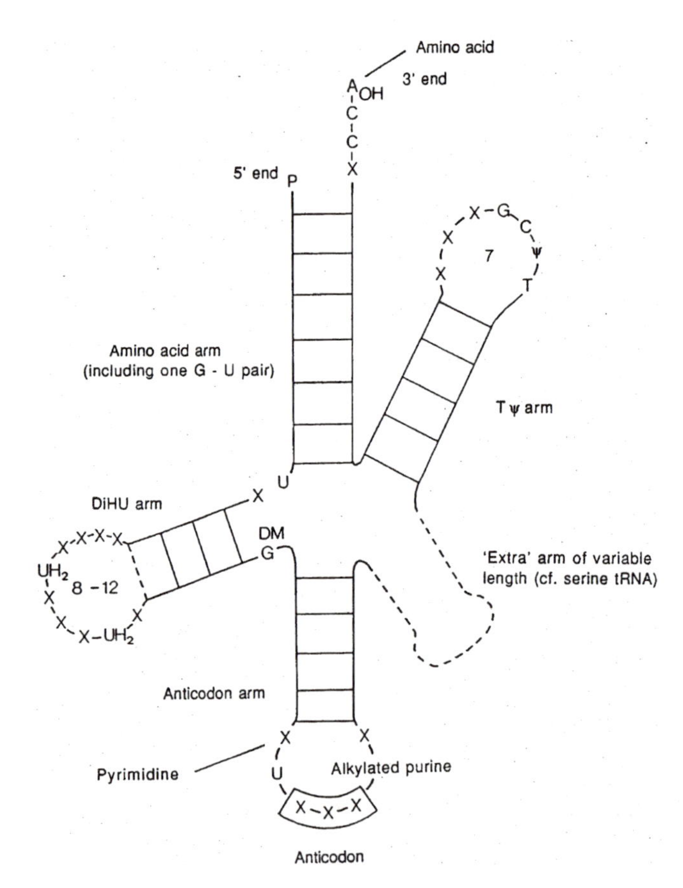


**Figure 5** *A generalised diagram of the characteristic clover-leaf structure of tRNA. Base pairs are indicated by lines across the structure, constant nucleotides are shown as such, and variable represented by X. Numbers indicate how many nucleotides form the loop*

source. Transfer RNA molecules have distinct structural features (Figure 5), the most significant of which is the anticodon loop which contains three nucleotides specific to the particular tRNA and amino acid. The anticodon is responsible for the insertion of the correct amino acids during polypeptide formation.

## Translation

The site of protein synthesis is the ribosome. There are many thousands of ribosomes in each cell and each is made up of three rRNA molecules and more than fifty proteins. The ribosomes can be divided into two subunits referred to as 50 S and 30 S in bacteria and 60 S and 40 S in eukaryotes.

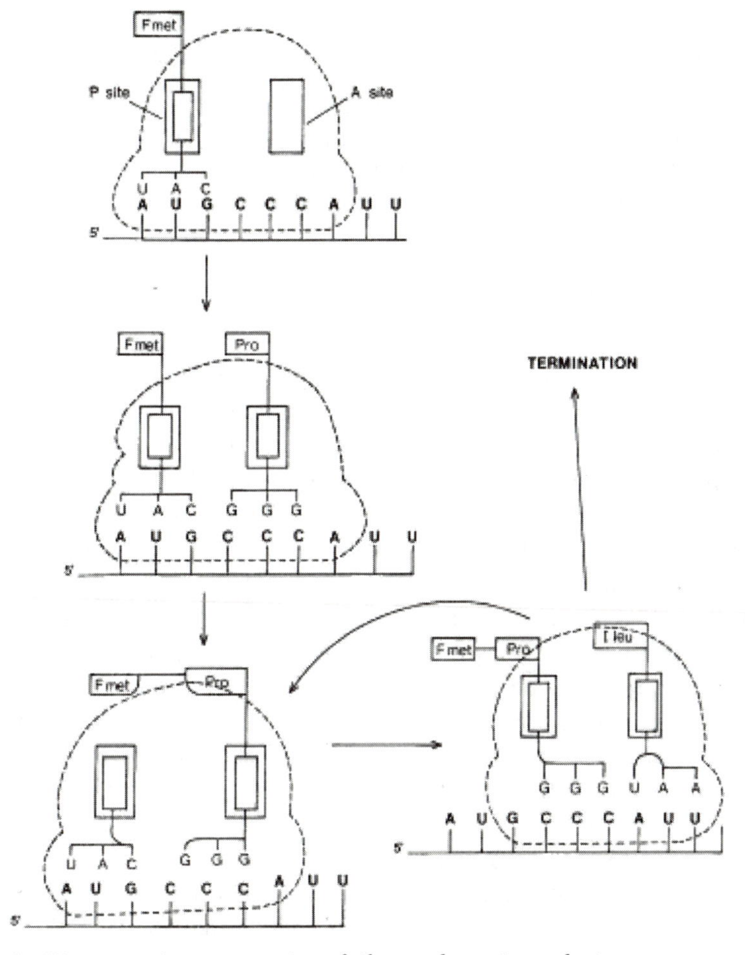


**Figure 6** *Diagrammatic representation of ribosomal protein synthesis*

The figures refer to the extent of sedimentation of the particles in a density gradient ultracentrifugation analysis. The intact ribosomes are 70 S in bacteria and 80 S in eukaryotes.

The mRNA carries the sequence of nucleotides that determines the sequence of amino acids in a particular protein. A complex forms between the two subunits of a ribosome, a sequence near the 5′ end of the mRNA and an aminoacyl tRNA molecule carrying a methionine amino acid. Each amino acid in the polypeptide is determined by three bases on the mRNA, these are referred to as triplet codons. Base pairing occurs between the codons on the mRNA and the anticodon loop of the aminoacyl tRNA molecules. There are two binding sites (the P and A sites) for tRNA molecules on the large subunit of the ribosome. Thus two amino acids are brought together, determined by the base sequence on the mRNA, where they can be condensed to form a peptide bond by catalytic activity in the ribosome (Figure 6).

The ribosome moves down the mRNA, in the 5′ to 3′ direction, until protein synthesis is complete, the energy being supplied by the hydrolysis of phosphate groups from two guanosine triphosphate molecules for each amino acid added. Many additional protein factors have been characterised, which are necessary for protein analysis.

## 9 THE GENETIC CODE

The link between the sequence of bases in DNA and amino acids in proteins is referred to as the genetic code (Figure 7). It is essentially universal, the exception being a small difference in mitochondria. The code is based on triplets, it is continuous within mRNA and is non-overlapping (*i.e.* ratherlikeasentencemadeofwordsofthreeletterswherethere isnopunctuationorgaps). The need for triplets of bases in DNA for coding amino acids is simple: $4 \times 4 \times 4$ combinations, 64, is required to code for the 20 different amino acids found in proteins. Single 4, or doublet, $4 \times 4$, codons would be insufficient. Sixty-four codons, however, appears too many but you can see from Figure 7 that all combinations are used. It is easy to see how the code evolved this way, because it ensures that if there is a mutation (*i.e.* an accidental change in the base sequence in DNA occurring during DNA replication) a protein will still be made from the gene, albeit with a different amino acid sequence. If only 20 of the triplets were used the consequence of mutation would be far more serious and such theoretical species would rapidly become extinct.

Three codons do not code for amino acids but act as stop signals and one codon, A U G, is the universal start codon, coding for methionine. A given codon only designates one amino acid, but there is more than one

2nd letter

| 1st letter | U | C | A | G | 3rd letter |
|---|---|---|---|---|---|
| U | UUU Phe | UCU Ser | UAU Tyr | UGU Cys | U |
| | UUC Phe | UCC Ser | UAC Tyr | UGC Cys | C |
| | UUA Leu | UCA Ser | UAA stop | UGA stop | A |
| | UUG Leu | UCG Ser | UAG stop | UGG Trp | G |
| C | CUU Leu | CCU Pro | CAU His | CGU Arg | U |
| | CUC Leu | CCC Pro | CAC His | CGC Arg | C |
| | CUA Leu | CCA Pro | CAA Gln | CGA Arg | A |
| | CUG Leu | CCG Pro | CAG Gln | CGG Arg | G |
| A | AUU Ileu | ACU Thr | AAU Asn | AGU Ser | U |
| | AUC Ileu | ACC Thr | AAC Asn | AGC Ser | C |
| | AUA Ileu | ACA Thr | AAA Lys | AGA Arg | A |
| | AUG Met | ACG Thr | AAG Lys | AGG Arg | G |
| G | GUU Val | GCU Ala | GAU Asp | GGU Gly | U |
| | GUC Val | GCC Ala | GAC Asp | GGC Gly | C |
| | GUA Val | GCA Ala | GAA Glu | GGA Gly | A |
| | GUG Val | GCG Ala | GAG Glu | GGG Gly | G |

| | | |
|---|---|---|
| Ala – Alanine | Arg – Arginine | Asn – Asparagine |
| Asp – Aspartic acid | Cys – Cysteine | Glu – Glutamic acid |
| Gly – Glycine | His – Histidine | Ileu – Isoleucine |
| Leu – Leucine | Lys - Lysine | Pro – Proline |
| Met – Methionine | Phe – Phenylalanine | Trp – Tryptophan |
| Ser – Serine | Thr – Threonine | Gln – Glutamine |
| Tyr – Tyrosine | Val – Valine | |

**Figure 7** *The genetic code with a list of the amino acids and their abbreviations*

codon for most amino acids, thus the code is said to be degenerate. Finally, the code is only ever read in the 5′ to 3′ direction on mRNA and it determines the sequence of amino acids from the amino to carboxy termini of polypeptides (Figure 8).

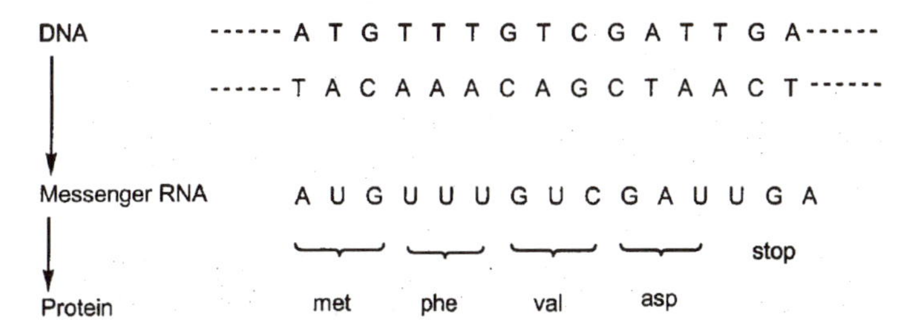


**Figure 8** *Reading the genetic code: an imaginary gene coding for a protein of just four amino acids. (In practice genes are considerably longer than this)*

## 10 POST TRANSLATIONAL MODIFICATION

These are specific to the protein concerned and can be significant. Modifications include: removal of the terminal methionine, formation of disulfide bonds, glycosylation, formation of multi-subunit complexes, and combination with metal ions.

## 11 THE CONTROL OF TRANSCRIPTION AND TRANSLATION

In bacteria transcription and translation are coupled. That is, protein synthesis begins before mRNA synthesis is complete. The principle level of control, therefore, is at transcription. There are proteins that can interact with the DNA to stop or stimulate transcription of specific genes. One such protein is the *lac* repressor in *Escherichia coli* which binds to a sequence of DNA with an intriguing palindromic structure and prevents the transcription of genes responsible for the catabolism of lactose.

Transcriptional control also occurs in eukaryotes; for example, the stimulation of transcription by steroid hormones *via* intracellular receptor molecules that bind to specific regions of DNA in the 'upstream' region (*i.e.* 5′ to the promoter) of certain genes. It is convenient to think of genes as having two parts: a coding region and a control region. Most gene control sequences are upstream of the coding region. They are sites for the binding of transcription factors, many of which may be required to activate a gene. A complication is that in eukaryotes the DNA is complexed with proteins (histones and other proteins) to form chromosomes. Control factors, therefore, have to interact with this complex to initiate transcription.

Translational control is perhaps more significant in eukaryotes than in bacteria because of the presence of a nuclear envelope. Hemin (an iron porphyrin), for example, is required in red blood cells before mRNA (predominantly coding for haemoglobin) can be translated. Generally, it is thought that cells regulate their overall level of translation rather than specifically regulating translation of specific mRNA, but there are exceptions.

## 12 GENOMICS

Discovering the structure of DNA gave us immediate insight into the nature of our genetic but the ability to chemically sequence the order of bases along a stretch of DNA, and 'read' our genes, gave us another quantum leap in understanding. We now live in a world where we can sequence the DNA of whole genomes (a 'genome' is all the DNA found in a cell of an organism and is the complete genetic code for that organism). For example we now know the entire sequence for *Escherichia coli*, (a simple gut bacterium with just over 4000 genes), *Saccharomyces cerevisae* (yeast, with approximately 6000 genes) *Caenorhabdidtis elegans*, (a worm, which is a model organism for studying developmental processes, about 19000 genes) and *Drosophila melanogaster*, (the fruit fly and mainstay of eukaryotic genetics, about 13000 genes). We can compare sequences with each other and identify homologous genes within genomes (paralogs) and between genomes (orthologs). This is valuable if we have say identified the function of a gene in yeast and find the ortholog in humans.

Of greatest significance, however, is the sequence of the human genome. On June 26th 2000, the preliminary working draft of the human genome project was announced (see http://www.sanger.ac.uk) This is a major milestone in human endeavour. At the time of the announcement, the assembly covered 97% of the human genome, 85% of the DNA had been sequenced, and very accurate information had been obtained for 24% of the genome including the entire sequence of chromosomes 21 and 22 (the two smallest human autosomes). This working draft is useful for most biomedical research. The target to reach 99.99% accuracy for the entire genome by 2003 is ahead of schedule.

The sequencing has produced a lot of information on genetic variation at the level of single nucleotide polymorphisms (SNPs, *i.e.* single base differences between individuals). SNPs are a relatively simple measure of genetic variation. They can be used to link regions of the genome with inheritance of a disorder or susceptibility to a disorder. Ideally we would like the entire DNA sequence of everyone within a population that suffer from a particular disease but this is currently impractical so SNPs are a short cut.

By June 2000 we knew virtually the entire sequence of Human Chromosomes 21 and 22 and this information provided us with a lot of information about the human genome.[3] Chromosome 21, the smallest human autosome (that is chromosomes excluding the sex chromosomes X and Y) contains 225 genes: 127 known genes, 98 predicted genes and 59 pseudogenes (*i.e.* non-functional genes or gene-like sequences). In comparison, chromosome 22 has 545 genes. At the time of writing, 20 disease-related loci had been mapped to chromosome 21. Forty percent of the chromosome is made up of interspersed, non-coding DNA repeats. Sixty-eight genes are anonymous transcription units with no similarity to known proteins so 41% of the genes on chromosome 21 are of completely unknown function. Twenty-two genes are larger than 100 kb, the average size of genes on chromosome 21 is 39 kb. Importantly, all known genes previously mapped to chromosome 21 can be identified by *in silico* methods (*i.e.* identifying genes by reading the DNA sequence in a computer database without necessarily knowing anything about the function of the genes being investigated). This provides significant confidence in *in silico* methods and is rather like a 'positive control' for the technique. Mutations in 14 known genes on chromosome 21 have been identified as causes of monogenic disorders including a form of Alzheimer's disease and a form of epilepsy. Loss of a part of one of the two copies of chromosome 21 in some individuals is associated with certain cancers. Interestingly, Down's syndrome individuals (who have three copies of chromosome 21) show a lower frequency of these cancers and the region is, therefore, a candidate locus for a tumour supressor gene. This relatively simple observation illustrates the power of whole genome sequencing and analysis understood by the general term 'genomics'.

As it happens, chromosome 21 is gene poor, whereas chromosome 22 is gene rich. If they are taken together as representative, then it suggests that the human genome has approximately 40000 genes. This is many fewer than was previously thought. However, it may turn out that either one of the sequenced chromosomes is an exception, in which case the estimate will have to be revised.

Having the complete sequence of human chromosomes will completely change the approach taken to identify disease-related genes. Disease genes will be found by analysing mutations in candidate regions of the genome (*e.g.* by looking for deletions or SNPs) and linking them with phenotype (the visible result of gene expression such as disease symptoms).

## 13 NUCLEIC ACID ANALYSIS AND RECOMBINANT DNA TECHNOLOGY

### Nucleic Acid Extraction Techniques

The use of DNA for analysis or manipulation usually requires a certain amount of isolation and purification. DNA is recovered from cells by the gentlest possible method of cell rupture to prevent the DNA from fragmenting by mechanical shearing.[4] This is usually in the presence of EDTA which chelate the $Mg^{2+}$ ions needed for enzymes that degrade DNA termed DNase. Ideally, cell walls, if present, should be digested enzymatically (*e.g.* lysozyme treatment of bacteria), and the cell membrane should be solubilised using detergent. After release of nucleic acids from the cells, RNA can be removed by treatment with ribonuclease (RNase), which has been heat treated to inactivate any DNase contaminants and protein can be removed by centrifugation. Solutions of DNA in a buffer containing

EDTA (to inactivate DNases) can be stored at 4 °C for at least a month. DNA solutions can be stored frozen although repeated freezing and thawing tends to damage long DNA molecules by shearing.

The methods used for RNA isolation are very similar to those described above for DNA. However, RNA molecules are relatively short, and therefore less easily damaged by shearing, so cell disruption can be rather more vigorous.[5] However, RNA is very vulnerable to digestion by RNases which are present endogenously in various concentrations in certain cell types and exogenously on fingers. DNase treatment can be used to remove DNA and RNA can be precipitated by ethanol. One reagent in particular, which is commonly used in RNA extraction, is guanadinium thiocyanate, which is both a strong inhibitor of RNase and a protein denaturant. It is possible to check the integrity of an RNA extract by analysing it by agarose gel electrophoresis.

In many cases it is desirable to isolate eukaryotic mRNA which constitutes only 2–5% of cellular RNA from a mixture of total RNA molecules. This may be carried out by affinity chromatography on oligo(dT)-cellulose columns. Nucleic acid species may also be subfractionated by more physical means, such as electrophoretic or chromatographic separations based on differences in nucleic acid fragment sizes or physicochemical characteristics.

## Electrophoresis of Nucleic Acids

In order to analyse DNA by size, electrophoresis in agarose (>100 base pairs) or polyacrylamide (high resolution and short DNA molecules) gels is usually undertaken. Electrophoresis may be used analytically or preparatively, and can be qualitative or quantitative.[6] Large fragments of DNA such as chromosomes may also be separated by a modification of electrophoresis termed pulsed field gel electrophoresis (PFGE).[7] The easiest and most widely applicable method is electrophoresis in horizontal agarose gels as indicated in Figure 9. This is followed by staining of the DNA with the dye ethidium bromide. This dye binds to DNA by insertion between stacked base pairs, termed intercalation, and exhibits a strong orange/red fluorescence when illuminated with ultraviolet light.

In general 'mini-gels' are used to check the purity and intactness of a DNA preparation or to assess the extent of a enzymatic reaction during, for example, the steps involved in the cloning of DNA. In recent years, a number of acrylic gels have been developed, which may be used as an alternative to agarose and polyacrylamide.

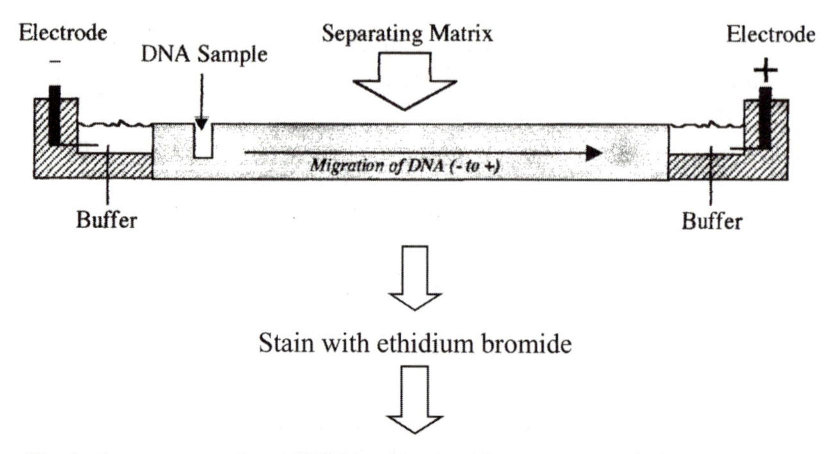


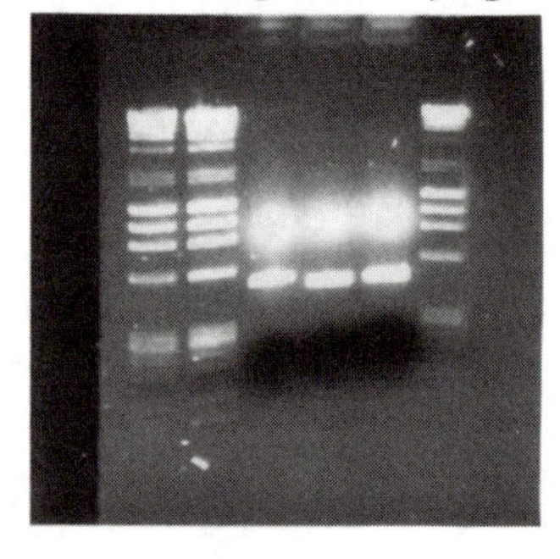

**Figure 9** *Electrophoresis in horizontal agarose gels*

## Restriction Mapping of DNA Fragments

Restriction mapping involves the size analysis of restriction fragments produced by several restriction enzymes individually and in combination. Comparison of the lengths of fragments obtained allows their relative positions within the DNA fragment to be deduced. Any mutation which creates, destroys or moves the recognition sequence for a restriction enzyme leads to a restriction fragment polymorphism (RFLP).[8] An RFLP can be detected by examining the profile of restriction fragments generated during digestion. Routine RFLP analysis of genomic DNA samples generally also involves hybridisation with labelled gene probes to detect a specific gene fragment. The first useful RFLP was described for the detection of sickle cell anaemia. In this case, a difference in the pattern of digestion with the restriction endonuclease *HhaI* could be identified between DNA samples from normal individuals and patients with the disease. This polymorphism was later shown to be the result of a single base substitution in the gene for $\beta$-globin which changed a codon, GAG specific for the amino acid glutamine to GUG which encoded valine.

## Nucleic Acid Blotting and Hybridisation

Electrophoresis of DNA restriction fragments provides no indication as to the presence of a specific, desired fragment among the complex sample. This can be achieved by transferring the DNA from the intact gel onto a piece of nitrocellulose or nylon membrane placed in contact with it. This provides a more permanent record of the sample since DNA begins to diffuse out of a gel that is left for a few hours.[9] This transfer is named a Southern blot after its inventor Ed Southern. Transfer of the DNA from the gel to the membrane allows the membrane to be treated with a labelled DNA gene probe. This single stranded DNA probe will hybridise under the right conditions to complementary single stranded DNA fragments immobilised onto the membrane.

The conditions of hybridisation are critical for this process to take place effectively and is usually referred to as the stringency of the hybridisation. Two of the most important components are the temperature and the salt concentration. The steps involved in Southern blotting are indicated in Figure 10. It is also possible to analyse DNA from different species or organisms by blotting the DNA and then using a gene probe representing a protein or enzyme from one of the organisms. In this way it is possible to search for related genes in different species. This technique is generally termed zoo blotting.

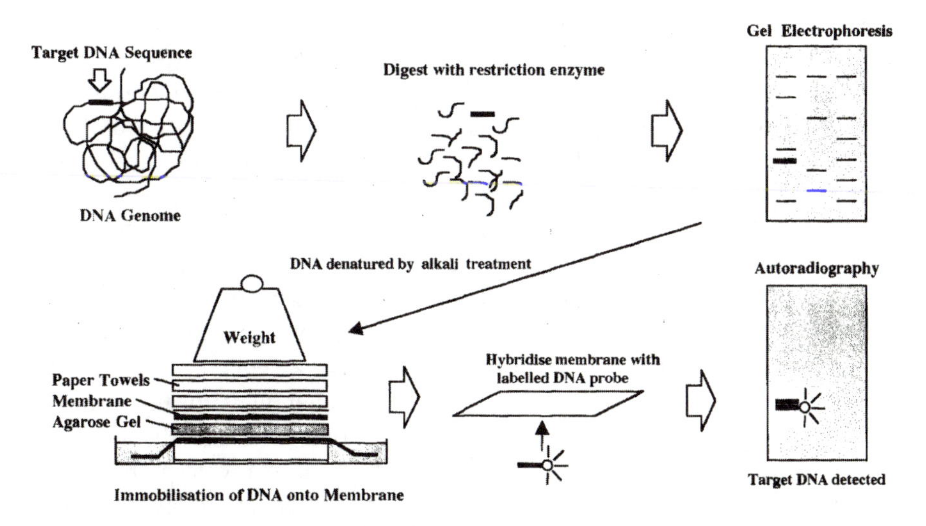


**Figure 10** *Steps involved in Southern blotting*

A similar process of nucleic acid blotting can be used to transfer RNA, separated by gel electrophoresis, onto membranes similar to that used in Southern blotting. This process, termed northern blotting, allows the identification of specific mRNA sequences of a defined length by hybridisation to a labelled gene probe.[10]

## Production of Gene Probes

The availability of a gene probe is essential in many molecular biology techniques, yet in many cases is one of the most difficult steps. The information needed to produce a gene probe may come from many sources but with the development and sophistication of genetic databases this is usually one of the first stages.[11] There are a number of genetic databases throughout the world and it is possible to search these over the internet and identify particular sequences relating to a specific gene or protein. In some cases it is possible to use related proteins from the same gene family to gain information on the most useful DNA sequence. Similar proteins or DNA sequences but from different species may also provide a starting point with which to produce a so called heterologous gene probe. Although in some cases probes are already produced and cloned it is possible, armed with a DNA sequence from a DNA database, to chemically synthesise a single stranded oligonucleotide probe. This is usually undertaken by computer-controlled gene synthesisers which link dNTPs together based on a desired sequence.

Where little DNA information is available to prepare a gene probe, it is possible in some cases to use the knowledge gained from analysis of the corresponding protein. Thus it is possible to isolate and purify proteins and sequence part of the N-terminal end of the protein. From our knowledge of the genetic code, it is possible to predict the various DNA sequences that could code for the protein and then synthesise the appropriate oligonucleotide sequences chemically. Due to the degeneracy of the genetic code, most amino-acids are coded for by more than one codon, therefore there will be more than one possible nucleotide sequence which could code for a given polypeptide. The longer the polypeptide, the greater the number of possible oligonucleotides which must be synthesised. Fortunately, there is no need to synthesise a sequence longer than about 20 bases, since this should hybridise efficiently with any complementary sequences, and should be specific for one gene. Ideally, a section of the protein should be chosen which contains as many tryptophan and methionine residues as possible, since these have unique codons and there will therefore be fewer possible base sequences which could code for that part of the protein. The synthetic oligonucleotides can then be used as probes in a number of molecular biology methods.

## DNA Gene Probe Labelling

An essential feature of a gene probe is that it can be visualised by some means. There are two main ways of labelling gene probes, traditionally this has been carried out using radioactive labels, but gaining in popularity are non-radioactive labels.[12,13] Perhaps the most used radioactive label is phosphorous-32 ($^{32}P$), although for certain techniques sulfur-35 ($^{35}S$) and tritium ($^{3}H$) are also used.

Non-radioactive labels are increasingly being used to label DNA gene probes as recent developments have led to similar sensitivities, which, when combined with their improved safety, have led to their greater acceptance. The labelling systems are either termed direct or indirect. Direct labelling allows an enzyme reporter such as alkaline phosphatase to be coupled directly to the DNA. However, indirect labelling is at present more popular. This relies on the incorporation of a nucleotide which has a label attached, such as biotin, flourescein and digoxygenin. These molecules are covalently linked to nucleotides using a carbon spacer arm of 7, 14 or 21 atoms. Specific binding proteins may then be used as a bridge between the nucleotide and a reporter protein, such as an enzyme. For example, biotin incorporated into a DNA fragment is recognised with a very high affinity by the protein streptavidin. This may either be coupled or conjugated to a reporter enzyme molecule such as alkaline phosphatase or horse radish peroxidase (HRP). Alternatively, labels such as digoxygenin incorporated into DNA sequences may be detected by monoclonal antibodies, again conjugated to reporter molecules including alkaline phosphatase.

The simplest form of labelling DNA is by 5′ or 3′ end labelling. Other forms of labelling of DNA are random primer and nick translation labelling, these methods are summarised in Table 1.

## The Polymerase Chain Reaction

Polymerase chain reaction, or PCR, is frequently one of the first techniques used when analysing DNA and has opened up the analysis of cellular and molecular processes to those outside the field of molecular biology.[14] The PCR is used to amplify a precise fragment of DNA from a complex mixture of starting material, usually termed the template DNA, and in many cases requires little DNA purification. It does require the knowledge of some DNA sequence information which flanks the fragment of DNA to

**Table 1** *Methods of labelling DNA*

| *Labelling method* | *Enzyme* | *Probe type* | *Specific activity* |
|---|---|---|---|
| 5′ end labelling | Alkaline phosphatase<br>Polynucleotide Kinase | DNA | Low |
| 3′ end labelling | Terminal Transferase | DNA | Low |
| Nick translation | DNaseI<br>DNA PolymeraseI | DNA | High |
| Random hexamer | DNA PolymeraseI | DNA | High |
| PCR | *Taq* DNA Polymerase | DNA | High |
| Riboprobes (cRNA) | RNA PolymeraseI | RNA | High |

be amplified (target DNA). From this information, two oligonucleotide primers may be chemically synthesised, each complementary to a stretch of DNA to the 3′ side of the target DNA, one oligonucleotide for each of the two DNA strands (see Figure 11). Further reagents required for the PCR include a DNA polymerase, each of the four nucleotide dNTP building blocks of DNA in equimolar amounts (50–200 μM) and a buffer appropriate for the enzyme. The PCR may be thought of as a technique analogous to the DNA replication process that takes place in cells since the outcome is the same, the generation of new complementary DNA stretches based upon the existing ones. It is also a technique that has replaced, in many cases, the traditional DNA cloning methods since it fulfils the same function, the production of large amounts of DNA from limited starting material. However, this is achieved in a fraction of the time needed to clone a DNA fragment. Although not without its drawbacks, the PCR is a remarkable development which is changing the approach of many scientists to the analysis of nucleic acids and continues to have a profound impact on core biosciences and biotechnology.[15]

*Elements Involved in the PCR.* The PCR consists of three defined sets of times and temperatures, termed steps; (i) denaturation, (ii) annealing and (iii) extension as seen in Figure 12. Each of these steps is repeated 30–40 times, termed cycles. In the first cycle, the double stranded template DNA is denatured and annealed. DNA synthesis proceeds from both 3′ ends of the primers until the new strands have been extended along and beyond the target DNA to be amplified. As the system is taken through successive cycles of denaturation, annealing and extension all the new strands will act as templates and so there will be an exponential increase in the amount of DNA produced. The net effect is to selectively amplify the target DNA and the primer regions flanking it.

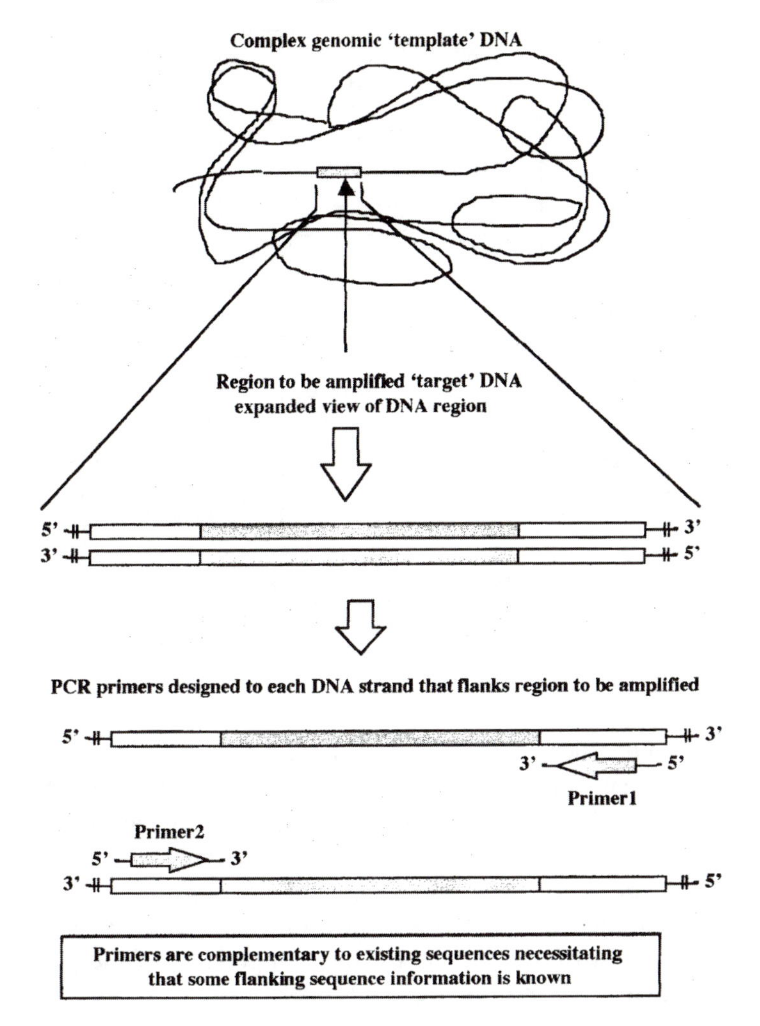


**Figure 11** *Design of two oligonucleotide primers complementary to a stretch of DNA to the 3′ side of the target DNA, one for each of the two DNA strands*

The widespread utility of the technique is also due to the ability to automate the reaction and as such many thermal cyclers have been produced in which it is possible to program in the temperatures and times for a particular PCR reaction.

*Primer Design in the PCR.* The key to the PCR lies in the design of the two oligonucleotide primers. These have to, not only be complementary to sequences flanking the target DNA, but must not be self-complementary or bind each other to form dimers since both prevent DNA amplification.[16] A number or software packages such as Oligo, Primer *etc.* have allowed the

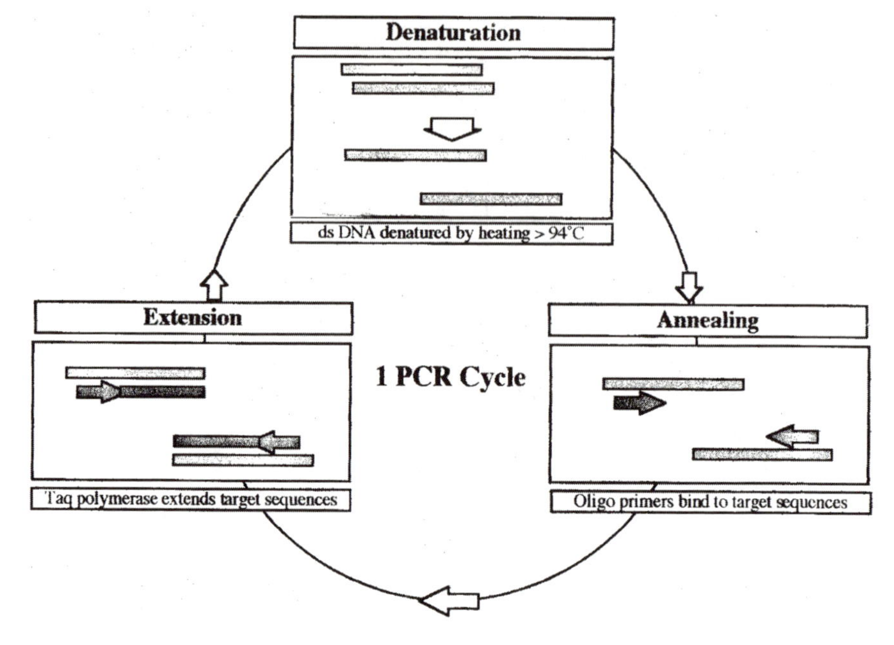


**Figure 12** *Elements involved in the PCR*

process of primer design to be less troublesome. It is also possible to include more than one set of primers in a PCR. This method, termed multiplex PCR, allows the amplification of more than one product in a single reaction tube and is especially useful in molecular diagnosis of clinical disorders.[17]

In general, primers are usually designed to be between 20 and 30 bases in length. It is best to balance the melting temperature of the primer pair and to have a GC content of between 40–60%.

*PCR Amplification Templates.* The PCR may be used to amplify DNA from a variety of sources or templates. It is also a highly sensitive technique and requires only one or two molecules for successful amplification. Unlike many manipulation methods used in current molecular biology, the PCR technique is sensitive enough to require very little template preparation. Indeed, the extraction of DNA from many prokaryotic and eukaryotic cells may involve a simple boiling step. However, the components of many DNA extraction techniques such as SDS, phenol, ethanol and proteinase K may adversely affect the PCR at certain concentrations.[18]

The PCR may also be used to amplify RNA, a process termed RT-PCR (reverse transcriptase-PCR, see Figure 13). Initially a reverse transcription reaction, which converts mRNA into cDNA, is carried out. In addition, the

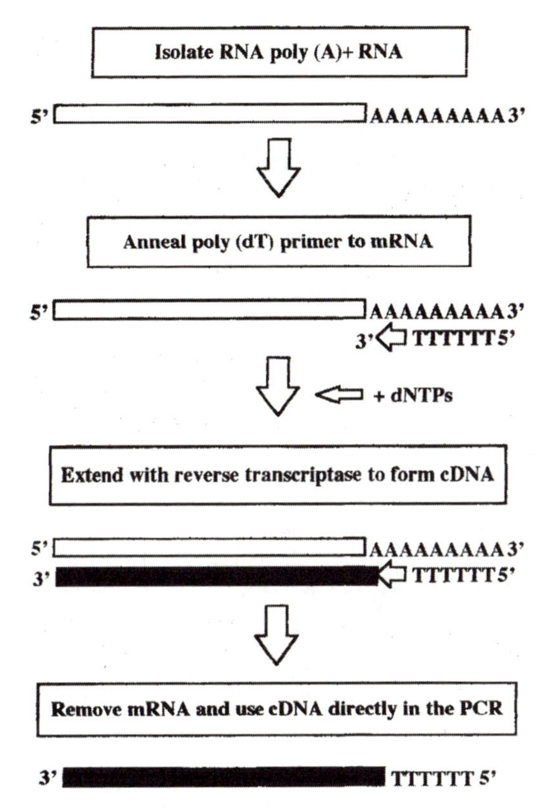


**Figure 13** *Reverse transcriptase-PCR (RT-PCR)*

PCR may be extended to determine relative amounts of a transcription product.[19] Many traditional methods in molecular biology have now been superseded by the PCR and the applications for the technique appear to be unlimited. Some of the uses to which the PCR has been put are summarised in Table 2.

*Applications of the PCR.* There are a number of molecular biology methods where the PCR has been used to great effect. The labelling of gene probes is one such area which has traditionally been undertaken by techniques such as nick translation. The nature of the PCR makes it an ideal method for gene probe production and labelling. PCR products may be labelled at the 5′ or 3′ end using the methods indicated earlier. This may be undertaken before the PCR by labelling the oligonucleotide primers or the resulting PCR product. However, since the PCR is essentially two primer extension reactions it allows the incorporation of nucleotides that have been labelled either radioactively, or with a non-radioactive label such as biotin. The advantage of the PCR as a gene probe

**Table 2** *Selected applications of the PCR*

| *Field of study* | *Applications* | *Specific uses* |
|---|---|---|
| DNA amplification | General molecular biology | Screening gene libraries |
| Production/labelling | Gene probe production | Use with blots/hybridisations |
| RT-PCR | RNA analysis | Active or latent viral infections |
| Scenes of crime | Forensic science | Analysis of DNA from blood |
| Microbial detection | Infection/disease monitoring | Strain typing/analysis RAPDs |
| Cycle sequencing | Sequence analysis | Rapid DNA sequencing possible |
| Referencing points in genome | Genome mapping studies | Sequence tagged sites (STS) |
| mRNA analysis | Gene discovery | Expressed sequence tags (EST) |
| Detection of known mutations | Genetic mutation analysis | Screening for cystic fibrosis |
| Detection of unknown mutations | Genetic mutation analysis | Gel based PCR methods, DGGE |
| Quantitative PCR | Quantification analysis | 5′ nuclease (TaqMan assay) |
| Production of novel proteins | Protein engineering | PCR mutagenesis |
| Retrospective studies | Molecular archaeology | Dinosaur DNA analysis |
| Sexing or cell mutation sites | Single cell analysis | Sex determination of unborn |
| Studies on frozen sections | *In situ* analysis | Localisation of DNA/RNA |

Abbreviations: RT, reverse transcriptase; RAPDs, rapid amplification polymorphic DNA; DGGE, denaturing gradient gel electrophoresis; STS, sequence tagged sites; EST, expressed sequence tags.

and labelling system is the fact that it offers great flexibility and may be rapidly produced.

A further important modification of the PCR is termed quantitative PCR.[20] This allows the PCR to be used as a means of identifying the initial concentrations of template DNA and is very useful for the measurement of, for example, a virus or a mRNA representing a protein expressed in abnormal amounts in a disease process. Early quantitative PCR methods involved the comparison of a standard or control DNA template amplified with separate primers at the same time as the specific target DNA. These types of quantitation rely on the reaction being exponential and so any factors affecting this may also affect the result. Other methods involve the incorporation of a radiolabel through the primers or nucleotides and their subsequent detection following purification of the PCR product.

An alternative automated method of great use is the 5′ exonuclease detection system or TaqMan assay. Here an oligonucleotide probe is labelled with a fluorescent reporter and quencher molecule at each end. When the primers bind to their target sequence, the 5′ exonuclease activity of Taq polymerase degrades and releases the reporter from the quencher. A signal is thus generated, which increases in direct proportion to the number of starting molecules. Thus a detection system is able to induce and detect fluorescence in real time as the PCR proceeds. This has important implications in, for example, the rapid detection of bacterial and viral sequences in clinical samples.[21]

## Recombinant DNA Technology and Gene Libraries

*Digesting Genomic DNA Molecules.* The isolation and purification of genomic DNA is the first step to many gene cloning experiments. DNA is then digested with restriction endonuclease enzymes.[22] These are the key to molecular cloning because of the specificity they have for particular DNA sequences. It is important to realise that every copy of a given DNA molecule from a specific organism will give the same set of fragments when digested with a particular enzyme. DNA from different organisms will, in general, give different sets of fragments when treated with the same enzyme. By digesting complex genomic DNA from an organism, it is possible to reproducibly divide its genome into a large number of small fragments, each approximately the size of a single gene. One group of enzymes cut straight across the DNA to give flush or blunt ends, whilst other restriction enzymes make staggered cuts, generating short, single-stranded projections at each end of the digested DNA. These ends, termed cohesive or sticky ends are complementary and are therefore able to base-pair with each other. In addition, the phosphate groups of the DNA are always retained the 5′ end.

*Ligating DNA Molecules.* The DNA products with cohesive or sticky ends may be joined to any other DNA fragments treated with the same restriction enzyme. Thus, when the two sets of DNA fragments are mixed, base pairing between complementary sticky ends will result in the annealing together of fragments originally derived from different starting DNA. In addition, there will also be pairing of fragments derived from the same starting DNA molecules, termed reannealing. However, present day cloning vectors have in built systems to overcome this. All the pairings between DNA fragments are transient because of the weakness of hydrogen bonding between the few bases in the sticky ends. However, they can be stabilised by use of an enzyme termed DNA ligase in a process termed ligation. This enzyme, usually isolated from bacteriophage T4 and termed T4 DNA ligase, forms a covalent bond between the 5′-phosphate at the end of one strand and the 3′-hydroxyl of the adjacent strand (Figure 14).

The ligation of different DNA molecules also reconstructs the site of cleavage. Therefore recombinant molecules, produced by ligation of sticky ends, can also be cleaved again at the junctions, using the same restriction

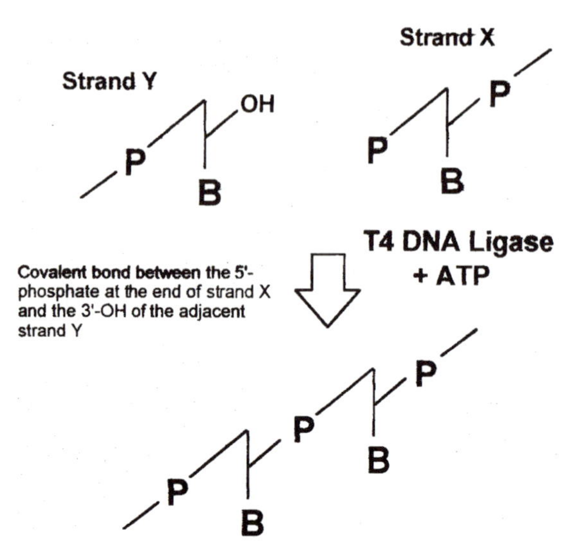


**Figure 14** *Ligation of DNA molecules using the enzyme T4 DNA ligase*

enzyme that was used initially to generate the fragments. In order to propagate digested DNA from an organism it is necessary to join or ligate that DNA with a specialised DNA carrier molecule termed a cloning vector. Thus each DNA fragment is inserted by ligation into the vector DNA molecule, which allows the whole recombined DNA to be replicated indefinitely within an appropriate host cell. In this way a DNA fragment can be cloned to provide sufficient material for further detailed analysis, or for further additional manipulation. Thus, all of the DNA extracted from an organism and digested with a restriction enzyme will result in a collection of clones. This collection of clones is known as a gene library.

*Considerations in Gene Library Preparation.* There are two general types of gene library. A genomic library, which consists of the total chromosomal DNA of an organism and a cDNA library, which represents the mRNA fraction from a cell or tissue at a specific point in time. If the ultimate aim is understanding the control of protein production for a particular gene or its architecture, then, genomic libraries must be used. However, if the goal is the production of new or modified proteins, or the determination of the tissue specific expression and timing patterns, cDNA libraries are more appropriate. Since the genome of an organism is fixed, chromosomal DNA may be isolated from almost any cell type in order to prepare a genomic library. In contrast, however, cDNA libraries only represent the mRNA being produced from a specific cell type at a particular time in the cells development. Thus, in the construction of cDNA libraries, it is important to consider carefully the cell or tissue type from which the mRNA is to be derived.

*Screening Gene Libraries.* Following the construction of a cDNA or genomic library, the next step requires the identification of the specific DNA fragment of interest. Clones containing the desired fragment need to be located from the library and in order to undertake this a number of techniques, mainly based on hybridisation, have been developed.

Colony hybridisation is a method frequently used used to identify a particular DNA fragment from a plasmid gene library (Figure 15).[23] A large number of clones are grown up to form colonies, replica plated onto a nylon membranes and allowed to grow. The colonies are then lysed and the liberated DNA is denatured and bound to the membranes so that the pattern of colonies is replaced by an identical pattern of bound DNA. The membranes are then incubated with a prehybridisation mix containing non-labelled, non-specific DNA such as salmon sperm DNA to block non specific sites. Denatured, labelled gene probe is then added and the binding detected by autoradiography of the membranes. By comparison of the patterns on the autoradiograph with the original plates of colonies, those which contain the desired gene (or part of it) can be identified and isolated for further analysis.

The polymerase chain reaction may also be applied to cDNA or genomic library screening where the libraries are constructed in vectors such as plasmids or bacteriophage. This is usually undertaken with primers that are designed to anneal to the vector rather than the foreign DNA insert. The size of an amplified product may be used to characterise the cloned DNA. Subsequently, restriction mapping is then usually undertaken. The main advantage of the PCR over traditional hybridisation-based screening is the rapidity of the technique, PCR screening may be undertaken in three to four hours, whereas it may be several days before detection by hybridisation is achieved. The PCR screening technique gives an indication of the size of the cloned inserts rather than the sequence of the insert; however, PCR primers that are specific for a foreign DNA insert may also be used. This allows a more rigorous characterisation of clones from cDNA and genomic libraries.

*Screening Expression cDNA Libraries.* In a number of cases, the protein for which the gene sequence is required may be partially characterised and it may, therefore, be possible to raise antibodies to that protein. This allows immunological screening to be undertaken rather than gene hybridisation.[24] Such antibodies are useful since they may be used as the probe if little or no nucleic acid sequence information is available. In such cases, it is possible to prepare a cDNA library in a specially adapted

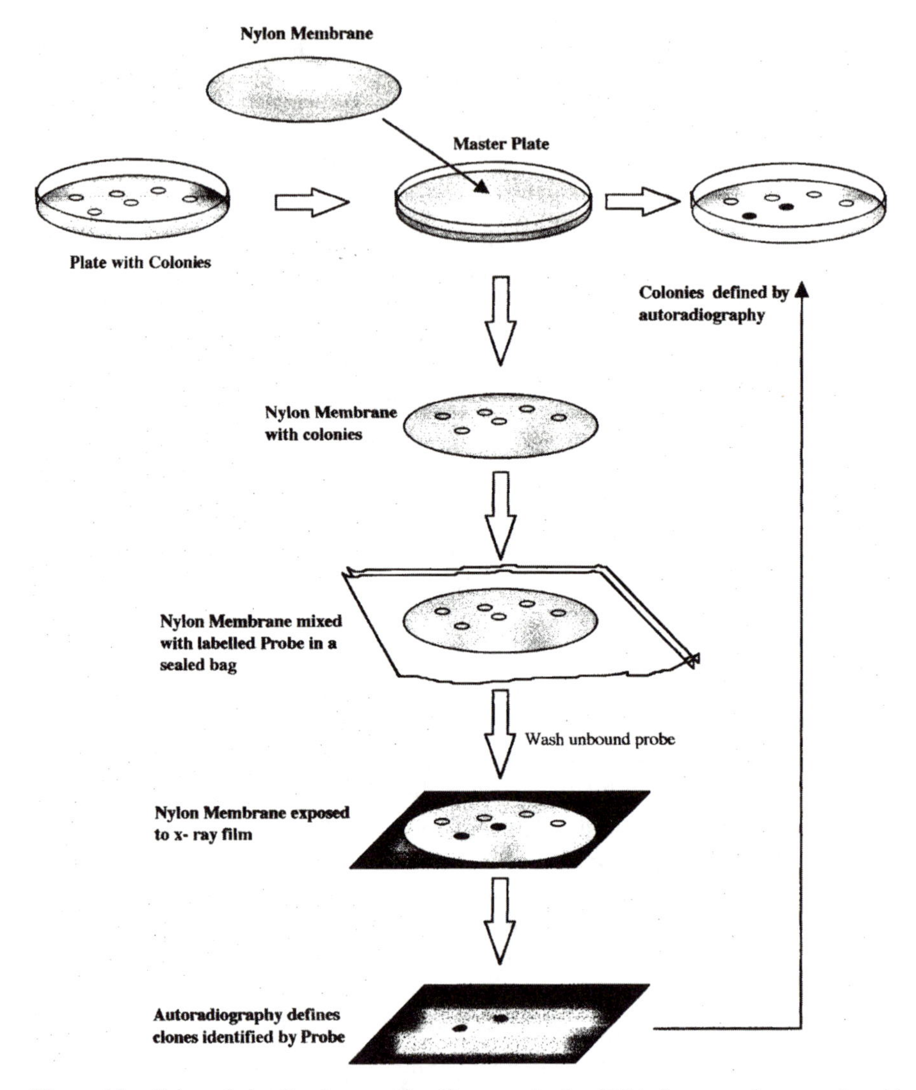


**Figure 15** *Colony hybridisation to identify a particular DNA fragment from a plasmid gene library*

vector termed an expression vector. These transcribe and translate any cDNA inserted into the vector. The protein is usually synthesised as a fusion with another protein such as β-galactosidase.

In some cases, expression vectors incorporate inducible promoters which may be activated by, for example, increasing the temperature allowing stringent control of expression of the cloned cDNA molecules. The cDNA library is plated out and nylon membrane filters prepared as for colony/plaque hybridisation. A solution containing the antibody to the desired protein is then added to the membrane. The membrane is then washed to remove any unbound protein and a further labelled antibody, which is directed to the first antibody, is applied. This allows visualisation of the plaque or colony that contains the cloned cDNA for that protein and this may then picked from the agar plate and pure preparations grown for further analysis.

*Nucleotide Sequencing of DNA.* The determination of the order or sequence of bases along a length of DNA is one of the central techniques in molecular biology. The precise usage of codons, information regarding mutations and polymorphisms and the identification of gene regulatory control sequences are also only possible by analysing DNA sequences. Two techniques have been developed for this, one based on an enzymatic method, frequently termed Sanger sequencing after its developer, and a chemical method, Maxam and Gilbert, named for the same reason. At present Sanger sequencing is by far the most popular method and many commercial kits are available for its use. However, there are certain occasions such as the sequencing of short oligonucleotides where the Maxam and Gilbert method is more appropriate.

One absolute requirement for Sanger sequencing is that the DNA to be sequenced is in a single stranded form.[25] The Sanger method is simple and elegant and mimics in many ways the natural ability of DNA polymerase to extend a growing nucleotide chain based on an existing template. Initially, the DNA to be sequenced is allowed to hybridise with an oligonucleotide primer, which is complementary to a sequence adjacent to the 3′ side of DNA within a vector such as M13 or in a PCR product. The oligonucleotide will then act as a primer for synthesis of a second strand of DNA, catalysed by DNA polymerase. Since the new strand is synthesised from its 5′ end, virtually the first DNA to be made will be complementary to the DNA to be sequenced. One of the deoxyribonucleoside triphosphates (dNTPs) which must be provided for DNA synthesis is radioactively labelled with $^{32}P$ or $^{35}S$ and so the newly synthesised strand will be labelled.

The reaction mixture is then divided into four aliquots, representing the four dNTPs A, C, G and T. In addition to all of the dNTPs being present in the A tube, an analogue of dATP is added (2′,3′-dideoxyadenosine triphosphate (ddATP) see Figure 16). This is similar to A but has no 3′ hydroxyl group and so will terminate the growing chain since a 5′ to 3′ phosphodiester linkage cannot be formed without a 3′-hydroxyl group. The situation for tube C is identical except that ddCTP is added. Similarly the G and T tubes contain ddGTP and ddTTP respectively.

Since the incorporation of ddNTP rather than dNTP is a random event,

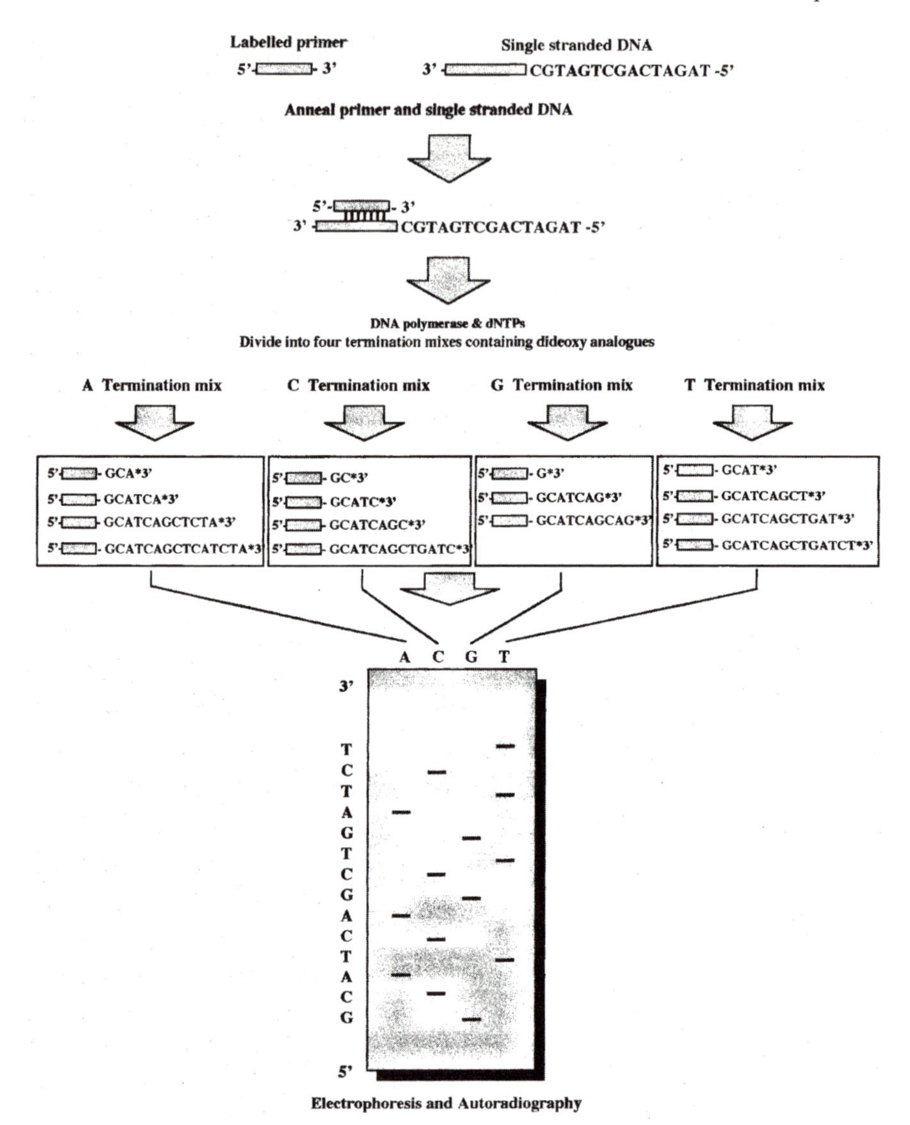


**Figure 16** *The Sanger method for nucleotide sequencing of DNA*

the reaction will produce new molecules varying widely in length, but all terminating at the same type of base. Thus four sets of DNA sequence are generated, each terminating at a different type of base, but all having a common 5′ end (the primer). The four labelled and chain terminated samples are then denatured by heating and loaded next to each other on a polyacrylamide gel for electrophoresis. The positions of radioactive DNA bands on the gel are then determined by autoradiography. Since every band in the track from the dideoxyadenosine triphosphate sample must contain molecules which terminate at adenine, and that those in the ddCTP terminate at cytosine, *etc.*, it is possible to read the sequence of the newly synthesised strand from the autoradiogram, provided that the gel can resolve differences in length equal to a single nucleotide. Under ideal conditions, sequences up to about 300 bases in length can be read from one gel.

*PCR Cycle Sequencing.* One of the most useful methods of sequencing PCR products is termed PCR cycle sequencing. This is not strictly a PCR since it involves linear amplification with a single primer. Approximately 20 cycles of denaturation, annealing and extension take place. Radio-labelled or fluorescent labelled dideoxynucleotides are then introduced in the final stages of the reaction to generate the chain terminated extension products. Automated direct PCR sequencing is increasingly being refined, allowing greater lengths of DNA to be analysed in one sequencing run and provides a very rapid means of analysing DNA sequences.

*Automated Fluorescent DNA Sequencing.* Recent advances in dye terminator chemistry have led to the development of automated sequencing methods that involve the use of dideoxynucleotides labelled with different fluorochromes. Thus the label is incorporated into the ddNTP and this is used to carry out the chain termination as in the standard reaction. The advantage of this modification is that, since a different label is incorporated with each ddNTP, it is unnecessary to perform four separate reactions. Therefore the four chain terminated products are run on the same track of a denaturing electrophoresis gel. Each product with their base specific dye is excited by a laser and the dye then emits light at its characteristic wavelength. A diffraction grating separates the emissions, which are detected by a charge coupled device (CCD) and the sequence interpreted by a personal computer. In addition to real time detection the lengths of sequence that may be analysed are in excess of 500 bp. Further improvements are likely to be made, not in the sequencing reactions themselves, but in the electrophoresis of the chain terminated products. Here capillary electrophoresis may be used where liquid polymers in thin capillary tubes would substantially decrease the electrophoresis run times. The consequence of automated sequencing and the incorporation of PCR cycle sequencing has substantially decreased the time needed to undertake sequencing projects. This has given rise to the use of banks of automated robotic sequencing systems in factory style units which are now in operation in various laboratories throughout the world, especially those undertaking work for the various genome sequencing projects.

*Maxam and Gilbert Sequencing.* The chemical cleavage method of DNA sequencing developed by Maxam and Gilbert is often used for sequencing small fragments of DNA such as oligonucleotides.[26] A radioactive label is added to to either the 3′ or the 5′ ends of a double-stranded DNA preparation. The strands are then separated by electrophoresis under denaturing conditions, and analysed separately. DNA, labelled at one end, is divided into four aliquots and each is treated with chemicals which act on specific bases by methylation or removal of the base. The conditions are selected so that, on average, each molecule is modified at only one position along its length. Every base in the DNA strand has therefore an equal chance of being modified. Following the modification reactions, the separate samples are cleaved by piperidine, which breaks phosphodiester bonds exclusively at the 5′ side of nucleotides whose base has been modified. The result is similar to that produced by the Sanger method. Each sample contains radiolabelled molecules of differing lengths, however all the labelled end in common. The other end is cut at the same type of base. Analysis of the reaction products by electrophoresis is as described for the Sanger method.

## 14 BIOINFORMATICS AND THE INTERNET

Increasingly molecular biology methods such as DNA amplification are being automated. Furthermore, sequencing technology has now reached such a level of sophistication that it is quite common for a large stretch of DNA to be sequenced and that sequence be manipulated or stored in a computer database.[27] This has given rise to a whole new area of molecular biology termed bioinformatics. A number of large sequence facilities are now fully automated and download sequences automatically to those databases from robotic work station servers. This increase in genetic information has luckily been matched by developments in computer hardware and software. There are now a large number of genetic databases which have sequence information representing a variety of organisms. The largest include GenBank at the National Institutes of Health (NIH) in the USA, EMBL at the European Bioinformatics Institute (EBI) at Cambridge, UK and the DNA database of Japan (DDBJ) at Mishima in Japan. There are also many other databases within which specialist DNA and protein sequences are stored, all of which may be accessed over the internet. A number of these important databases and internet resources are listed in Table 3.

It is possible, once a nucleotide sequence has been deduced, to search an existing database for similar, homologous, sequences and for generic gene or protein coding sequences. Thus it is possible to search for open reading

**Table 3** *DNA and protein sequence databases and internet sources*

| *Database or resource* | *URL (uniform resource locator)* |
|---|---|
| **General DNA sequence databases** | |
| EMBL: European genetic database | http://www/ebi/ac.uk |
| GenBank: US genetic database | http://ncbi.nlm/nih/gov |
| DDBJ: Japanese genetic database | http://ddbj.nig.ac.jp |
| **Protein sequence databases** | |
| Swiss-Prot: European protein sequences | http://expasy.hcuge.ch/sprot/sprot-top.html |
| TREMBL: European protein sequences | http://www.ebi.ac.uk/pub/databases/tembl |
| PIR: US protein information services | http://www-nbrf.georgetown.edu/pir |
| **Protein structure databases** | |
| PDB: Brookhaven protein database | http://www.pdp.bnl.gov |
| NRL-3D: Protein structure database | http://www.gdb.org/Dan/proteins/nrl3d.html |
| **Genome project databases** | |
| Human mapping: John Hopkins, USA | http://gdbwww.gdb.org |
| dbEST: cDNA and partial sequences | http://www.ncbi.nih.gov |
| Genethon: genetic maps based on repeat markers | http://www.genethon.fr |
| Whitehead Institute: (YAC & physical maps) | http://www-genome.wi.mit.edu |

frames, *e.g.* sequences beginning with a start codon (ATG) and continuing with a significant number of 'coding' triplets before a stop codon is reached. There are a number of other sequences that may be used to define coding sequences; these include ribosome binding sites, splice site junctions, poly A polymerase sequences and promoter sequences that lie outside the coding regions. It is now relatively straightforward to use sequence analysis software to search a new sequence for homology within a chosen database. Software programs such as BLAST and FASTA provide the means to search for sequences of homology, and allow such sequences to be aligned, allowing important clues of the potential structure and function of a given DNA sequence. In some cases it is also possible to generate a graphical three-dimensional model of a putative protein encoded by a DNA sequence by using existing sequence information. The atomic coordinates of protein structures generated from X-ray crystallography or nuclear magnetic resonance (NMR) data are also held in databases. The largest of these is the protein databank (PDB) held at Brookhaven in the US. It is possible, although difficult at present, to predict secondary protein structures from translated nucleotide sequences. Such predicted molecular models are very complex to produce, requiring sophisticated numerical processing; however, they do provide important insights into protein structure and function and are constantly being refined. Another exciting future possibility is that molecular modelling could be combined with virtual reality systems allowing real time interaction of proteins and ligands to be observed. Whatever the means of displaying modelled proteins, there is no doubt that even now they are extremely important in the rational design and modification of proteins and enzymes.

The main development in computing that has allowed the explosion in sequence analysis is the internet. This is a world-wide system that links numerous computers, local networks, research, commercial and government institutions and establishments, and all parts of the world-wide web (WWW). DNA databases and other nucleic acid sequence and protein analysis software may all be accessed over the internet given the relevant software and authority. This is now relatively straightforward with so-called web browsers which provide a user-friendly graphical interface for sequence manipulation. Consequently, the new expanding and exciting areas of bioscience research are those that analyse genome and cDNA sequence databases (genomics) and also their protein counterparts (proteomics). This is sometimes referred to as *in silico* research and there is no doubt that for basic and biotechnological research, it is as important to have internet and database access as it is to have equipment and reagents for laboratory molecular biology.

## 15 HUMAN GENOME MAPPING PROJECT

There is no doubt that the mapping and sequencing of the human genome is one of the most ambitious projects in current science. It will certainly bring new insights into gene function and gene regulation, provide a means of identifying DNA mutations or lesions found in current genetic disorders and point to new ways of potential therapy. It is also a multi-collaboration effort that has engaged many scientific research groups around the world and given rise to many scientific, technical, financial and ethical debates. One interesting issue is the sequencing of the whole genome in relation to the coding sequences. Much of the human genome appears to be non-coding and composed of repetitive sequences. Only a small portion of the genome appears to be encoding enzymes and proteins. Nevertheless, this still corresponds to approximately 40 000 genes and their mapping and sequencing is an exciting prospect. The study further aims to understand and possibly provide the eventual means of treating some of the 4000 genetic diseases, in addition to other diseases whose inheritance is multi-factorial. The diversity of the human genome is also an area of great interest and is currently under study. Despite some initial reservations, a partial sequence has already been completed; however, the equally difficult task of decoding and interpreting the complete sequence will no doubt still take many years.

## 16 REFERENCES


1. T.A. Brown, *Genomes*, 1999, Bios, Oxford.
2. B. Lewin, *Genes VII*, Wiley, New York, 2000.
3. Hattori *et al.*, *Nature*, 2000, **405**, 311.
4. J. Sambrook *et al.*, *Molecular Cloning: A Laboratory Manual*, 2nd edn., Cold Spring Harbour, NY, 1989.
5. P. Jones *et al.*, *RNA Isolation and Analysis*, Bios Scientific Publishers, Oxford, 1994.
6. P. Jones, *Gel Electrophoresis of Nucleic Acids*, Wiley, Chichester, 1995.
7. D.C. Schwartz and C.R. Cantor, *Cell*, 1984, **37**, 67.
8. E.K. Green, in *Molecular Biomethods Handbook*, R. Rapley and J.M. Walker (eds.), Humana Press, Totowa, NY, 1998.
9. E.M. Southern, *J. Mol. Biol.*, 1975, **98**, 503.
10. J.C. Alwine *et al.*, *Proc. Natl. Acad. Sci. USA*, 1977, **74**, 5350.
11. M. Aquino de Muro, in *Molecular Biomethods Handbook*, R. Rapley and J.M. Walker (eds.), Humana Press, Totowa, NY, 1998.
12. G.H. Keller and M.M. Manak (eds.), *DNA Probes*, 2nd edn., Stockton Press, NY, 1993.
13. T.P. McCreery and T.R. Barrette, in *Molecular Biomethods Handbook*, R. Rapley and J.M. Walker (eds.), Humana Press, Totowa, NY, 1998.
14. R.K. Saiki *et al.*, *Science*, 1985, **230**, 1350.
15. K.A. Eckert and T.A. Kunkel, *Nucleic Acids Res.*, 1990, **18**, 3739.
16. W. Rychlik, in *Nucleic Acids Protocols Handbook*, R. Rapley (ed.), Humana Press, Totowa, NY, 1999.
17. C.R. Newton and A. Graham, *PCR*, 2nd edn., Bios Scientific Publishers, Oxford, 1997.
18. J. Bickley and D. Hopkins, in *Analytical Molecular Biology*, G.C. Saunders and H.C. Parkes (eds.), The Royal Society of Chemistry, Cambridge, 1999.
19. Y.M. Dennis Lo, *Clinical Applications of PCR*, Humana Press, Totowa, NY, 1998.
20. S. Cheng *et al.*, *Nature (London)*, 1994, **369**, 684.
21. C.A. Heid *et al.*, *Genome Res.*, 1996, **6**, 986.
22. H.O. Smith and K.W. Wilcox, *J. Mol. Biol.*, 1970, **51**, 379.
23. U. Gubler and B.J. Hoffman, *Gene*, 1983, **25**, 263.

24. R.A. Young and R.W. Davis, *Proc. Natl. Acad. Sci. USA*, 1983, **80**, 1194.
25. F. Sanger *et al.*, *Proc. Natl. Acad. Sci. USA*, 1977, **74**, 546.
26. A.M. Maxam and W. Gilbert, *Proc. Natl. Acad. Sci. USA*, 1977, **74**, 560.
27. S. Misener and S.A. Krawetz, *Bioinformatics: Methods and Protocols*, Humana Press, Totowa, NJ, 1999.

CHAPTER 15

# Strategy and Tactics in Drug Discovery

FRANK D. KING

## 1 INTRODUCTION

Since the publication of the first edition, there have been dramatic changes in the strategies and tactics applied to drug discovery. This has come about for two reasons:

1. The realisation that the commercial demands for the industry require that most major pharmaceutical companies need to market 2–4 novel chemical entities (NCEs) per year, whereas the average was only 0.75![1]
2. The rapid advances in technologies, such as genomics, combinatorial chemistry and high throughput screening.[2]

The commercial demand requires a spectacular improvement in research productivity. This has been the major driver for increasing the size of compound collections, the implementation of high throughput screening and the development of combinatorial chemistry techniques to improve the number and quality of lead molecules to accelerate the lead optimisation process. In addition, an ever increasing number of assays are being introduced into the lead optimisation phase in an attempt to reduce the failure rate in development. The genomics revolution has also opened up incredible opportunities for a huge number of novel targets, all of which are relatively unvalidated but some will almost certainly provide the novel treatments of the future, though how many and which ones is still unknown. All of this puts an incredible pressure on the pharmaceutical industry, as these novel technologies are not cheap, are high risk and, although they promise longer term gains, the industry still needs to maintain a good flow of new products using existing technologies, to maintain a flow of cash to support the more basic research.

In chemistry, the focus has been on developing rapid, parallel synthesis and combinatorial methods (HT chemistry) to produce many more compounds in a shorter period of time, with the hope of reducing the drug discovery cycle time. As we applied these methods to lead optimisation, it rapidly became clear that these techniques were frequently not the most appropriate and so their place within the drug discovery process is still evolving. However, HT chemistry has changed the way that many medicinal chemists think about lead optimisation.

In spite of these changes, the overall drug discovery process has basically remained the same, namely target identification and validation, lead identification, lead optimisation, development candidate identification and, finally, development of that compound. Where the differences lie are in the relative resources applied to each phase and tactics employed.

## 2 TARGET IDENTIFICATION AND VALIDATION

Historically, drug targets were identified from investigative pharmacology – looking for the effects of compounds on a biological function that related to a disease. When I started in the industry over 20 years ago, I was working on analogues of a marketed compound, metoclopramide, which stimulated gastric motility by an unknown mechanism. In this case, the primary screen was an *in vivo* measure of gastric motility in the rat. However, following the pioneering work from James Black on $\beta$-adrenergic and histamine $H_2$ receptors, the majority of approaches have been to identify modulators of the actions of specific proteins (receptors or enzymes), which from pharmacology or clinical efficacy are known to be involved in disease processes. For example, the identification of the mechanism of action of metoclopramide as activation of the 5-$HT_4$ receptor led to a rekindling of interest in that area, using affinity and efficacy at the 5-$HT_4$ receptor as the primary screen.

This pharmacological approach is still valid and many targets being worked on in the industry are derived from an understanding of the actions of existing drugs and/or biochemical pathways. In these cases, the commercial requirement would be to identify a differentiating factor, such as target selectivity, improved ADME or reduced side-effect profile, and this would form a key part of the justification. Such differentiation distinguishes the ‘me-too’ approach, which nowadays is considered unlikely to be commercially viable, from a ‘me-better’, which could be a potential block buster. Excellent examples of 'me-better' compounds are the anti-ulcer ranitidine, which satisfied an issue of cytochrome P450 inhibition associated with cimetidine, and fluconazole, which was the first orally bioavailable triazole antifungal agent. It is worth noting that it is frequently not the first marketed compound in a class that realises the full market potential. A good example of this is the angiotensin converting enzyme (ACE) inhibitor enalopril, which addressed key side-effect issues with the first-in-class, captopril.

From the human genome project, $\sim$40 000 human genes have been identified and only a small fraction of these have been investigated as potential targets. For example, estimates of the number of kinases range between 500 and 2000, yet only a small, but rapidly increasing, proportion have been investigated from a drug discovery perspective. However, the high number of potential targets raises major challenges. We can only work on a limited number – so how do we choose which ones, when all the information we have is a DNA sequence and gene location? High speed methods of tissue localisation have been developed, for example Taqman where the levels of mRNA are measured, but localisation alone does not validate a target, and can even be quite misleading. For example, the mRNA for the histamine $H_2$ receptor is high in the heart, but low in the stomach. So, using that information alone, would one have identified that receptor as a target for a gastric anti-secretory agent?

Clearly, with the publication of the human genome sequence there is a wealth of information to 'mine', but the challenges involved in translating that wealth of knowledge into drugs, as well as the cost and risks involved, should not be underestimated. In addition, many of the diseases currently not well treated are multifactorial, thus a single target modulator may not offer significant clinical efficacy. A future challenge facing the industry could be how to exploit such polypharmacy – combinations of drugs that each modulate single targets or designed single drugs which modulate multiple targets.

Although the medicinal chemist is not heavily involved at this early stage of the drug discovery process, it is important to be aware of where the targets are coming from and the strength of supporting data so that judicious decisions on resource allocation can be made. This is particularly important where the target validation comes down to identifying a tool compound for the pharmacologists to use to validate the target. Even that can be extremely resource intensive. With such a wealth of targets available, it may be possible to prioritise on the basis of targets with a good track record of success, for example 7-transmembrane receptors.

## 3 LEAD IDENTIFICATION

Traditionally a 'lead' came either from natural product screening, an established drug or pharmacological tool or from competitor patents or publications. For most diseases, natural product screening has recently had a poor track record when compared with other approaches, although for anti-infectives and oncology, it is still a potentially viable approach. However, to address this, the 'natural diversity' is being augmented by genetic modification of the organisms that produce the natural products and by altering the conditions under which they are grown. Using known drugs or competitor compounds is still a valid approach for lead identification, especially for 'me-betters' but there are commercial risks involved in this approach as it relies upon finding a novel, differentiated series of compounds, the benefits of which translate to the clinic.

Nowadays, most 'leads' are derived from the high throughput screening (HTS) of collections of compounds. Most large pharmaceutical companies have compound collections ranging between 0.5–2 million compounds. There are also commercially available compound collections with varying levels of diversity and uniqueness. Even with the advent of miniaturised HTS methodologies, large compound collections are very expensive and time consuming to test. To circumvent this, companies either test as mixtures or pre-select subsets of compounds to screen. This selection can be made on the basis of, for example, chemical structure, physico-chemical properties, classes of compounds with a historically high rate of success or use structural information of the target, so-called virtual screening. HTS technology has spawned numerous terms such as 'early hit', 'validated hit', 'lead' and 'validated lead', with various meanings in each company. Generally a 'hit' is a compound which has been identified from screening but with minimal validation, maybe even only affinity/activity data from a stock solution. For some screens there could be many thousands of hits requiring validation and therefore a triage process needs to be in place to reduce the number to manageable proportions.

Validation of the hit includes confirmation of activity from a solid (if available), structure confirmation, additional characterisation in bio-assays for selectivity, and screening of analogues. If the 'hit' is a combinatorial mixture, then the validation process would include deconvolution and preferably re-synthesis of the individual member as a single compound. If no analogues are available, either in house or commercially, then the production of a focused library is an option to be considered.

The medicinal chemist is normally intimately involved in this process, selecting compounds for confirmation. Once confirmed, a decision needs to be made over whether to initiate a hit-to-lead programme. Decision criteria applied include potency and selectivity, structural novelty, amenability to (rapid) synthesis, essential functionality with known liabilities and physicochemical properties. One can roughly classify hits into the following categories with the likely decision:

- High potency, high selectivity – definitely yes
- High potency, poor selectivity – probably yes unless chemically intractable
- Low potency, high selectivity – possibly yes if very good biological rationale
- Low potency, low selectivity – only if desperate, *i.e.* readily amenable to high throughput chemistry, for very well validated targets, of high priority and little competition.

The definition of high/low potency should be relative to the target potency required to provide a drug. For example, typically nM potency would be required for a GPCR antagonist, whereas only μM potency may be needed for an ion channel inhibitor.

The level of target validation, unmet medical need, competition and potential market size are also important considerations. Thus for a well validated target in a highly competitive environment, one may only work on high quality hits as speed is of the essence. On the other hand, if there were little known competition, for a well validated target a lower quality hit may be considered, whilst bearing in mind that it is likely to take longer before a development candidate would be identified. However, if the chemistry is amenable, it is at this stage that high throughput chemistry can make a big impact and rapid advances in potency and selectivity can be achieved, or alternatively if no improvements are found, a swift termination of the programme.

A good example of an excellent lead was from our HTS for the 5-$HT_6$ receptor. The arylsulfonamide **1** was identified as an antagonist with a p$K_i$ of 8.3 and was very selective over all other receptors tested, the closest being 50-fold selectivity over 5-$HT_{1A}$ and 5-$HT_{1D}$ This compound was an intermediate in the preparation of sulfonamide analogues of a series of amides reported to be mixed 5-$HT_{1B/1D}$ antagonists. As these sulfonamides had low 5-$HT_{1B/1D}$ potency, very few compounds had been prepared. However, additional analogues were rapidly prepared using commercially available arylsulfonyl chlorides and the 5-chloro-2-methylbenzothiophene aromatic nucleus was identified within 2 months of **1** being confirmed. Subsequent metabolism studies showed that the *N*-methyl group was metabolically labile, thus removal of that group gave SB-271046 **2**, a potent and selective 5-$HT_6$ receptor antagonist (p$K_i$ 8.9) with good pharmacokinetic properties.[3]

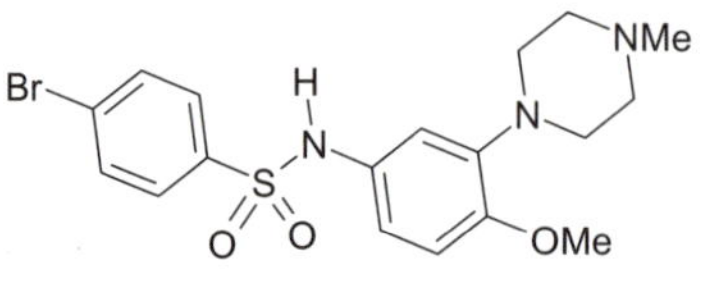

**1** 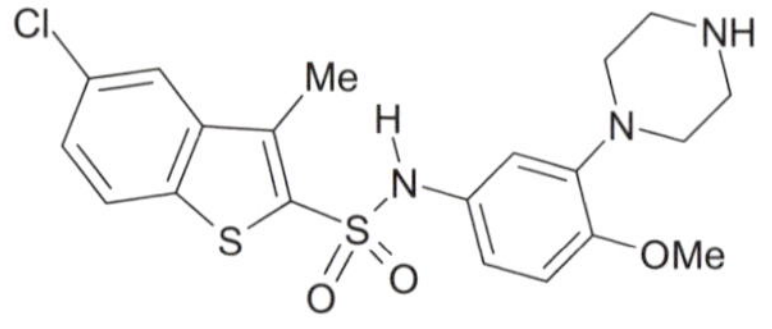 **2** SB-271046

Once a "hit" has been validated, it becomes a "lead" and normally decisions need to be made over resources required for lead optimisation. This is a crucial decision point and care needs to be taken as over-optimism can commit a large resource to a project which, once started, can prove difficult to stop.

## 4 LEAD OPTIMISATION

This is the time when costs and resource increase dramatically, and forward planning is essential. Major considerations are the screening cascade, the capacity of the screen and cycle times. A typical screening cascade and its relationship to the iterative cycles is shown in Figure 1 and covered in more detail in Chapter 5. The primary screens are the first assays that the newly synthesised compound is tested in. These assays should be rapid and robust, typically binding displacement or simple enzyme inhibitor assays.

Normally a key selectivity assay is included at an early stage in the cascade, either a closely related protein or one that the lead compounds are known to interact with. Other rapid generic assays, such as cytochrome P450 inhibition and *in silico* physicochemical calculations, can be included to aid in the design of better compounds. Potency and selectivity filter criteria are set for progression down the screening cascade, potency being close to that required to give a reasonable dose for *in vivo* activity and selectivity being that which would be unlikely to cause unacceptable side-effects. Typically, for a GPCR antagonist this could be 10 nM affinity and 30-fold selectivity. In the early stages of a programme, one is usually exploring the diversity to find the 'activity/inactivity boundaries', so relatively few compounds may be expected to pass these filters. As one focuses on the activity, the numbers progressing down the screening cascade should increase.

Secondary assays tend to be lower throughput, more time intensive functional tissue or cell-based assays, *in vitro* metabolism studies and

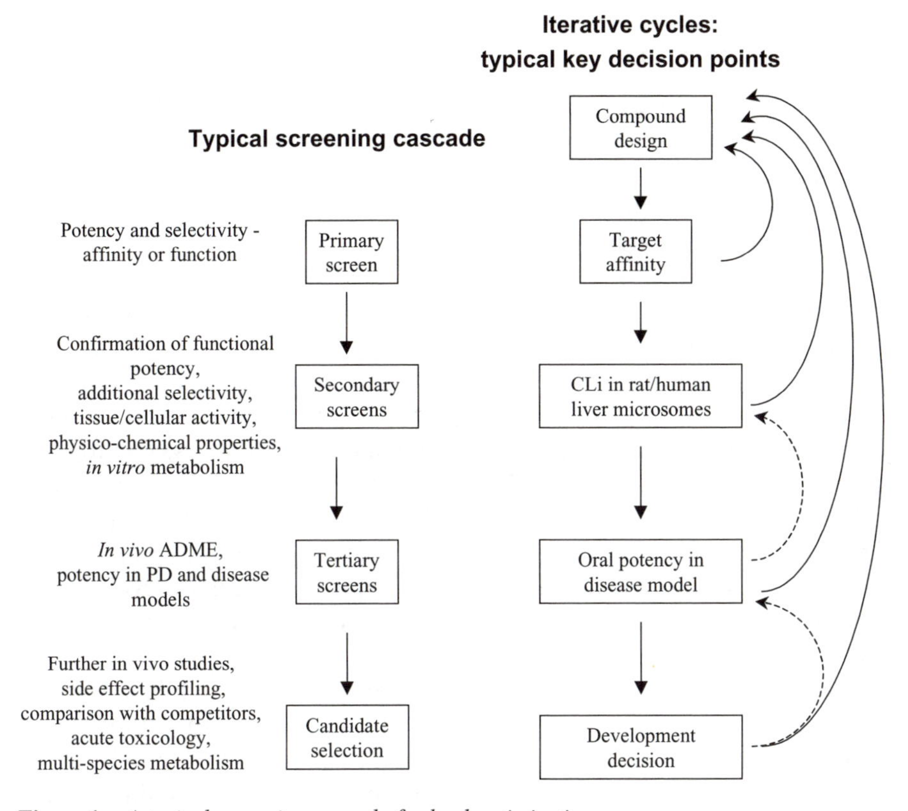


**Figure 1** *A typical screening cascade for lead optimisation*

preliminary *in vivo* assays. However, it is essential that these assays will be able to cope with the anticipated increasing number of compounds passing primary screening criteria as the work matures. Tertiary assays, again, tend to be even lower throughput, such as *in vivo* ADME and animal disease models. Ideally pharmacodynamic (PD) and pharmacokinetic (PK) studies are run in parallel to establish a PK/PD relationship. If the target is not well validated, compounds which are not ideal can be progressed to these tertiary assays as so-called 'tool compounds'. Demonstration of activity in the disease model with these tool compounds then gives confidence to continue to identify compounds more likely to be successfully developed.

This process of sequentially addressing the issues of potency, selectivity, activity in secondary and tertiary assays, and finally satisfying candidate selection criteria is the major reason why drug discovery often takes so long. Multiple chemical series need to be optimised, frequently the optimised compounds from each series do not satisfy all of the required properties and rapidly finding the series which is successful is the real challenge. It is reminiscent of finding your way through a maze.

## 5 DEVELOPMENT CANDIDATE

The decision to progress a compound into development is the successful outcome of a lead optimisation programme and is the result of a compound successfully passing all of the hurdles set within the screening cascade. The requirements that a development candidate has to meet varies from company to company but there has been a tendency for the number to increase, with more and more development functions getting involved in the candidate selection process. This has been driven by the need to reduce failure rates in development, which, historically, is around 50–75% prior to first-time-in-human studies. Many of the reasons for failure are due to poor pharmacokinetcs, cytochrome P450 interactions, poor physicochemical properties and toxicity. Most companies, nowadays, require data on these as a part of the candidate selection package and in some even a full 28 day toxicology study is required. Adverse data may not necessarily prevent the compound going forward, but it adds to the risk of failure and may result in additional studies being required in early development. Increasingly these properties are also being incorporated into hit selection and lead optimisation phases.

As programmes mature towards a development candidate, greater emphasis is placed upon targeting the drug to the site of action. Thus, factors such as bioavailability by the proposed route of administration, duration of action, distribution to the site of action, *e.g.* crossing the so-called *blood–brain barrier*, all become important factors for optimisation. As a part of this process, formulation issues also need to be considered. Thus, if an intravenous product is required, a level of solubility, often $>1$ mg ml$^{-1}$ (pH range 8–4.5) is required so that the dose can be given in an acceptable volume. For oral products, solubility can be much lower but solubilities lower than 10 µg ml$^{-1}$ (pH range 8–2) are likely to be problematic and cause delays in development as formulation issues are addressed. Salt and crystal forms are also crucial considerations, particularly for compounds with low solubility where rate of dissolution can be important.

Because the compound will, hopefully, eventually be marketed, an adequate patent position needs to have been established. This is discussed in greater detail in Chapter 13, but normally the compound should be novel and covered as a specific example within a patent filing. Because of timings, the patent has normally not been granted at the time of the development decision but there should be no prior art that would prevent marketing. However, in some cases a licence may be needed if there is overlap with another patent and this forms part of the risk assessment in the decision process.

## 6 BACK-UP/FOLLOW-UP

Once the first candidate, the *front runner* has entered development, the search then continues to identify *back-up* or *follow-up* compounds. The definition of these varies between companies but generally a *back-up* will be a compound within the same generic structural class with broadly similar pharmacology. The purpose of a back-up is effectively an insurance policy, to replace the *front runner* if any compound-specific issue arises. The intention is normally that the *back-up* would progress through development until the *front runner* passes the key high risk studies. The development of the *back-up* may then be halted, or, if there are alternative therapeutic uses, continued. In contrast, the purpose of a *follow-up* is to address issues identified with the *front runner* and *back-up*, with the intention of fully developing the compound to market. For this, the follow-up must be sufficiently differentiated over the front runner for it to be commercially viable. A totally different chemical series is normally required, often with an alternative pharmacological and/or pharmacokinetic profile. For many companies, *back-up* and *follow-up* strategies are automatic. However, a more judicious approach to *back-up/follow-up* strategies may be called for with less a validated target, where the major risk of failure in development would be target-related.

As the *front runner* progresses through development, new issues can arise which may need addressing with a new compound. These are often toxicities unrelated to the pharmacological mode of action. Rapid feedback of these findings to discovery research is vital so that a potential replacement compound can be identified that overcomes these issues. Frequently this can occur some years after, which may necessitate re-starting the research project. It is, therefore, not surprising that it can take up to 20 years from the start of a programme to finally marketing a product.

## 7 OPTIMISING THE CHANCES FOR SUCCESS

The drug discovery process is expensive and high risk, with on average only about 1 in 10 000 compounds synthesised making it to the market. These historical data also include *me-too* and *me-better* drugs where there is a higher than average probability of success, so for a novel or 'difficult' target, the compound numbers could be very much higher. Therefore, many projects that we work on may never deliver development candidates, never mind drugs. Reasons for this are either that the biological hypothesis was wrong or that a compound could not be found (or does not exist!) that satisfies all the properties of potency, selectivity and 'developability' such as ADME and safety.

As we move into the post-genomics era, we are working on targets with less biological target validation and therefore the rate of failures due to insufficient activity in animal models and clinical trials is likely to increase. Increasing confidence in the target often requires a tool compound, the identification of which requires the early stages of the drug discovery process of lead identification and optimisation. In addition, we will be working on more classes of target that so far have not delivered drugs with ever increasing stringency of selectivity as more and more closely related targets are identified. As more and more criteria need to be satisfied, the probability of success will be further reduced. This is particularly true for diseases where current therapies already exist and so a much better product profile, either in terms of efficacy or tolerability, needs to be achieved. There is also still a widely held view that, with the theoretical diversity of compounds available, if we make enough compounds then we should be able to find development candidates as potential drugs for most of the targets. This belief puts the onus on target validation as the key for future success, as by increasing the size of compound banks coupled with high throughput chemistry we should be able to identify compounds for most validated targets. However, if this belief is wrong and that success will depend upon both biological target validation plus a better assessment of chemical tractability, then the onus falls upon the medicinal chemist to better understand chemical tractability and to assess probability of success based upon the structure of the target coupled with the physicochemical requirements needed for a compound to interact with that target. For example, the properties of the target protein may be such that to interact with that protein with sufficient affinity requires properties of the molecule that fall way outside the Lipinski 'rule of 5' parameters for an orally bioavailable drug.

An example of where we put this into practice within SB was where it was agreed that the interference of protein–protein interactions which involved large surface area interactions, could only met with large molecules. This would be unlikely to give orally bioavailable drugs, so the probability of success was unacceptably low. The decision was therefore made that HTS would only be done if a small binding epitope had been identified. Similarly we ran a number of efforts on the inhibition of RNA for antibacterial targets, but the HTS and subsequent lead optimisations only gave polyamines that were unsuitable as drugs, so again, the probability of success was very low. These approaches were therefore terminated. Only time will tell whether these decisions were right!

This probability concept can be visualised by Figure 2, which needs some explanation. The area of the graph covers 'molecular property space', the definition of which is as yet undefined but could, for example, be molecular weight and polar surface area. For the visualisation process this is two-dimensional but in reality could be multi-dimensional. For a particular binding site on a target there will be a defined area of property space which is required for potency. Assuming that binding site is unique, within that area there would be a smaller area required for selectivity. There will also be another area of space that defines the properties needed to satisfy the product profile required to make a drug, for example, oral bioavailability, brain penetration, duration of action and freedom from toxicity. It is reasonable to believe this latter requirement is a constant, as it is dependent upon a single species – human, but may differ between, for example, an oral or parentral product. If we can better define that area of space, we can focus on the property space within which our potential drug molecules should lie and Lipinski's 'rule of 5' goes some way towards that. The property space for potency is dependent upon the binding sites

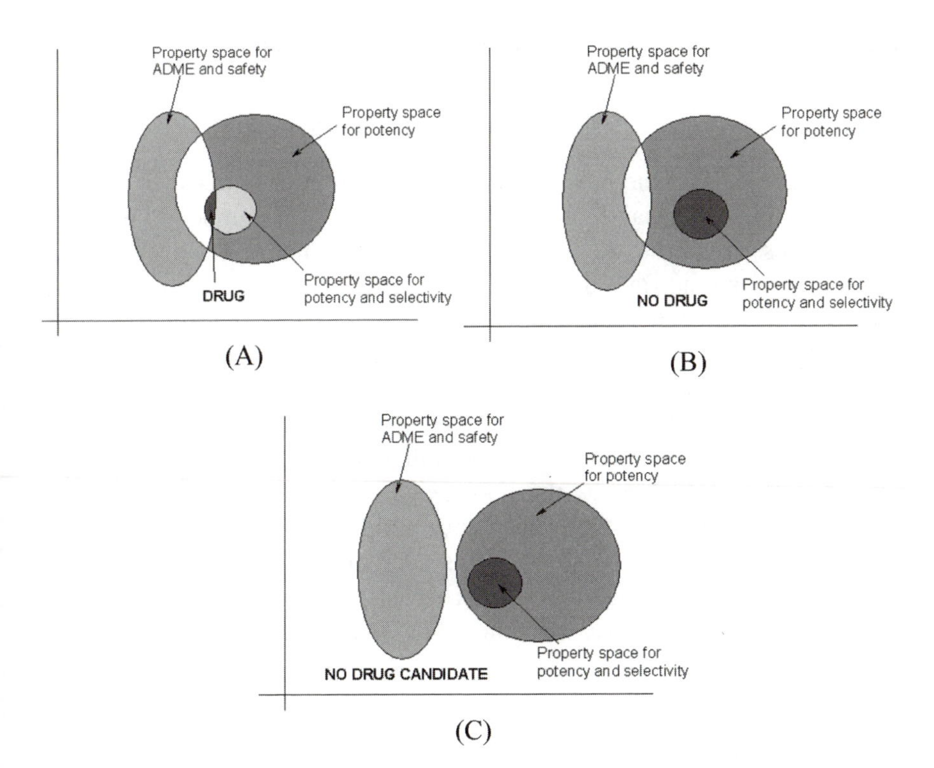


**Figure 2** *Venn diagrams illustrating the inter-relationship between property space requirements for different requirements for a drug that need to be met by a single molecule*

available for the target protein and that for selectivity, differentiation from all other binding sites on other proteins. The degree of overlap indicates the size of property space from which a drug-like molecule will be found, and, therefore, the probability that compounds with the right properties can be identified.

Using this concept, one can visualise at least three scenarios. The first (A) is where there is overlap for potency, selectivity and 'developability' such that a drug is feasible. The size of overlap could give an indication of the probability of finding single, or multiple chemical series. The second (B) is where potency and selectivity, and potency and 'developability' are mutually exclusive. In this case one may need to progress a non-selective compound if this is a viable proposition. However, in practice if selectivity is high up on the screening cascade, these compounds may be eliminated from further study and hence never be identified. The third (C) is where no overlap occurs, so one could spend many years identifying potent, selective compounds but never get a drug. This concept opens up the challenge that better predictions of these areas of property space from a combination of the developability properties and a knowledge of the target structure could be used to better prioritise targets and leads.

## 8 DECISION MAKING IN MEDICINAL CHEMISTRY

It can be argued that the most important decision in drug discovery is that which is made every day by medicinal chemists; what compound to make next. Success or failure of the company depends upon that decision. So how do we make the right decision? Strategically the two extremes to lead optimisation, and hence decision making, are: rational design (carefully designed single molecules using target structural information, pharmacophore identification and (Q)SAR) and random synthesis (make all possible analogues from readily available starting materials and trust to luck). In practice we do a combination of both, although the relative proportion of rational:random increases as the knowledge of SAR develops. The more random-based approaches have been covered in the chapter on combinatorial chemistry (Chapter 16) so the remainder of this section will concentrate on the more rational-based approaches, additional to that covered in the chapters on QSAR (Chapter 11) and computational chemistry (Chapter 12).

### Pharmacophore

In the early stages of any project, often little is known about what properties of the hit are required for potency at the target. At this stage one, therefore, needs to investigate as many parts of the molecule as possible. The intention is to identify a *pharmacophore,* that is the spacial orientation of the functional groups necessary for activity. Two representative examples of pharmacophores for 5-$HT_3$ receptor antagonists are shown in Figure 3.[4]

It is in these early stages of hit/lead optimisation that high throughput chemistry techniques (see Chapter 16) are probably of most value for, say, optimising an aromatic substituent. Other approaches are more resource intensive but equally valuable, such as conformational restriction to define the active conformation. However, it is important to note that a single change can affect more than one property. Thus, in the dopamine $D_2$ receptor antagonist area, dramatic increases in potency and selectivity were gained by going from an ethylenediamine **3** to an aminopiperidine **4** to an aminotropane **5**. However, each change also alters conformational freedom, substituent orientation, lipophilicity as well as the basicity of the amine, any of which could contribute to the improved potency.

ArCONH~~N(Et)~CH$_2$Ph < ArCONH–(piperidine)N–CH$_2$Ph < ArCONH····(tropane)N–CH$_2$Ph

**3** **4** **5**

Similarly, for simple *N*-methylation of an amide, changes in activity could be due to conformational effects, *cis*/*trans* isomer ratio, removal of an N–H hydrogen bond donor or simply steric hindrance (Figure 4). It is therefore important to build up a comprehensive picture of interdependent activities to come to the right conclusions.

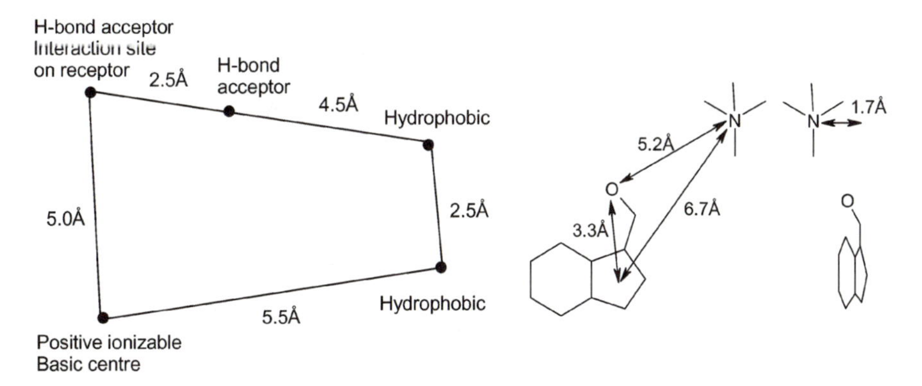


**Figure 3** *Two pharmacophores for 5-$HT_3$ receptor anatagonists*

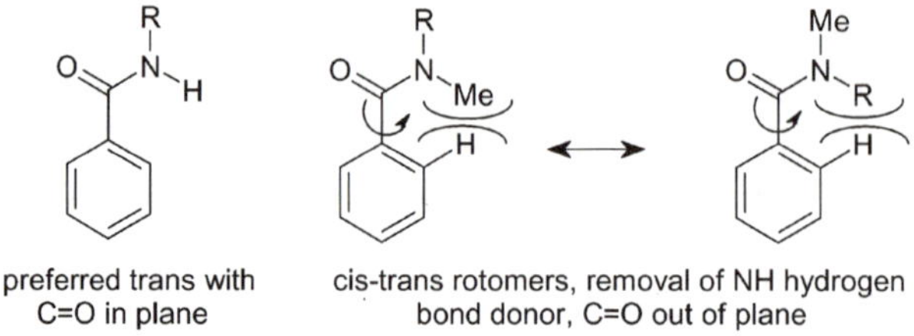

preferred trans with C=O in plane

cis-trans rotomers, removal of NH hydrogen bond donor, C=O out of plane

**Figure 4** *Multiple effects of simply N-methylating an amide*

Once a working model of a pharmacophore has been determined, by application of one or more of the standard techniques described in the next few sections, a number of closely related series may be identified. In an ideal world with limitless resources, all series may be worked on to identify the best substituents. However, often resource and time are limited so some choice needs to be made as to which of the series is worked on. The temptation is to work on the most potent series, but occasionally useful information can be obtained more rapidly by working on chemically more amenable series, then applying the SAR knowledge to the less chemically tractable series. Two related examples where we found this particularly useful were in the 5-$HT_3$ and the 5-$HT_{2C}$ antagonist areas. In the 5-$HT_3$ antagonist area, the 6,5-heterocyclic amides **7** were the lead series, but based upon our pharmacophoric hypothesis, the *o*-methoxyphenyl ureas **6** were found to be equally active.[5] Although they had poorer oral bioavailability, this relatively simple synthesis allowed us to explore rapidly substitution around the aromatic ring. Similarly in the 5-$HT_{2C}$ antagonist area, the most potent series was the indolines **9**, but the number of commercially available and readily synthesised indolines was limited.

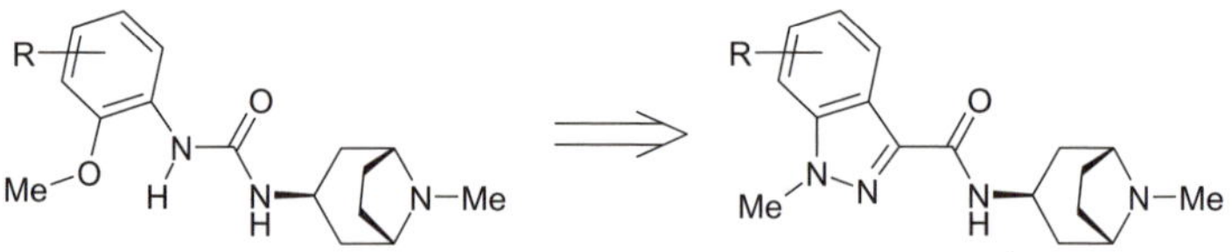

**6** Optimise R

**7** Optimised R

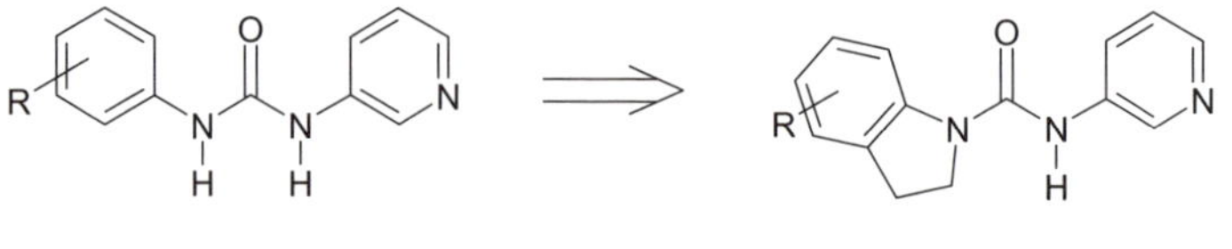

**8** Optimise R

**9** Optimised R

However, by preparing a series of simpler, but less potent, phenyl ureas **8** we were able to explore rapidly the SAR for substitution around the phenyl ring, then apply this knowledge to the synthesis of specific indolines (see Chapter 17).

## Bioisosteres

The definition of bioisosterism is the identification of two different, interchangeable functionalities that retain biological activity. From a chemical viewpoint these groups can be quite different, but clearly the target recognises them both equally well. Tables of classical bioisoteres have been documented in the first edition of this book.[6] Bioisosteric replacements can offer advantages in potency, metabolic stability, reduced cytochrome P450 interactions and improved physicochemical properties.

The concept of bioisoterism was applied before many of the structure-based design principles were recognised. Now that we have much more information on drug/target interactions we can define different kinds of bioisosterism as *pharmacophoric* and *template*.

*Pharmacophoric Bioisosterism.* This can be defined as two different functionalities which have similar specific interactions with the target protein. For example, a carboxylic acid may be replaced by a tetrazole or an acylsulfonamide, all of which have similar p$K_a$s. They can also act as H-bond acceptors as shown in Figure 5.

Other examples of pharmacophoric bioisosteres are amides/esters/heterocycles, where the H-bonding to the carbonyl or hetero atom is a key pharmacophoric element; phenyl/heterocycles where the aromaticity/lipophilicity is a key interaction; phenyl/cyclopentyl for lipophilicity alone and imidazole/amine for basicity. Some examples are shown in Table 1. However, a full list is beyond the scope of this book but one can refer to previously reported tables and, by using the knowledge of the kinds of interactions as above, determine which of the traditional bioisoteres could be applied.

*Template Bioisosterism.* Template bioisosterism can be defined as functional group replacements which do not have a specific interaction with

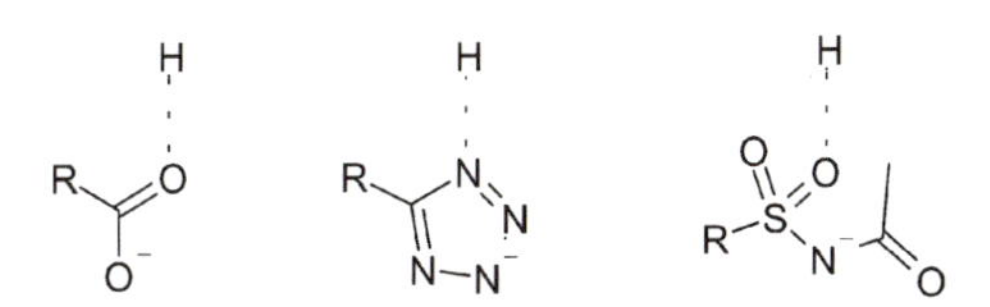

**Figure 5** *Pharmacophoric bioisosterism of carboxylate, tetrazolyl and acylsulfonamide anions*

**Table 1** *Pharmacophoric bioisosteres*

| *Active group* | *Biosisosteric replacement* |
|---|---|
| -COOH | $-PO(OH)_2$; $-SO_2NHR$; $-SO_2NHCOR$, -CONHCN; 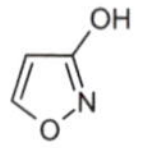 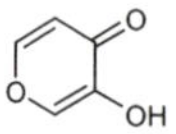 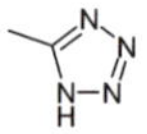 |
| -COOR | 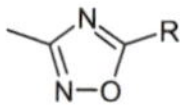 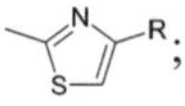 ; plus other heterocycles 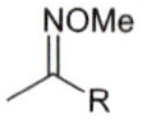 |
| -OH | -NHCOR; $NHSO_2R$; $CH_2OH$; NHCN; $CH(CN)_2$ |
| $-NR_2$ | $-C(=NR)NR_2$; $-NRC(=NR)NR_2$; $NR_3^+$; 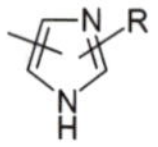 |
| Aryl | Phenyl; pyridyl; thienyl, furyl, pyrrolyl |
| Halogen | F; Cl; Br; I; $CF_3$; CN; $N(CN)_2$; $C(CN)_3$ |

**Table 2** *Examples of template bioisosteres*

| | |
|---|---|
| -CONH- | -CSNH-; $-CH_2NH-$, -NHCO-; >C=C<; $-CH_2S-$; $-CH_2O-$ |
| Phenyl | pyridine, pyrimidine, thiophene, furan, pyrrole |
| $-CH_2CH_2-$ | $-CH_2S-$; $-CH_2O-$; cyclohexane; cyclopentane; phenyl. |

the target, but act as equivalent spacer groups that similarly orientate the pharmacophoric elements. Examples from traditional biosisoteres are phenyl to pyridyl or other heterocycles for disubstitution orientation; amide bioisosteres such as thio-, amino- and oxy-methylene; disubstituted cycloalkanes for alkanes and simple O, N and S replacements for $CH_2$ in alkyl chains (Table 2).

Most of these examples are simple alternative frameworks that replace the original. However, it is also possible to make use of preferred conformations to hold the pharmacophoric elements in their required conformations. This can be exemplified by the previously mentioned use of *o*-methoxyanilides as bioisosteric replacements for 6,5-heterocyclic carbamates and esters as 5-$HT_3$ receptor antagonists (Figure 6).[4,7] Although there does not appear to be a formal, strong H-bond between the *o*-methoxy group and the NH of the urea, the system does seem to hold the phenyl ring, the carbonyl and the R substituent in the required planar orientation, as the des-methoxy compound was very much less potent.

## Conformational Restriction

Restriction of the possible conformations of a molecule is a standard method for improving the properties of a lead molecule. By so doing, all

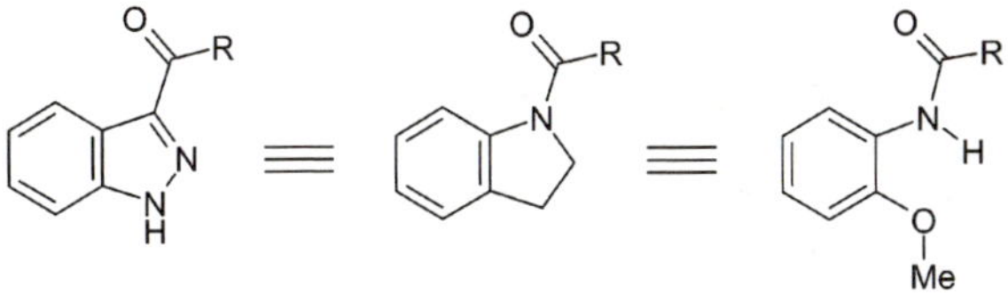

**Figure 6** *Bioisosteric replacement of a 6,5 heterocycle by an* o*-methoxyanilide*

of the key functionalities involved in binding to the target are retained, but their relative positions are fixed in the active conformation. In fact, virtually every drug that contains a cyclic 'template' would have less potent acyclic equivalents. However, the starting point for drug design is always with the acyclic 'parent'. For example, morphine **10** was a known natural product with analgesic properties, and it was only subsequently that it was found to be a conformationally restricted mimetic for the pentapeptide enkephalins **11**.

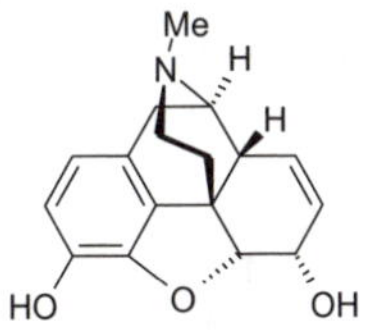

Tyr - Gly - Gly - Phe - Leu/Met

**10** Morphine

**11** Leu/Met Enkephalins

Nowadays, especially with the advent of 'diversity arrays' for screening, a hit obtained from high throughput screening is often relatively conformationally unrestricted, and therefore the design of conformationally restricted analogues can give rapid progress during the lead optimisation phase.

If the pharmacophore is well defined, then the form of conformational restriction can be predicted, otherwise trial and error must be used. Clearly the outcome of this work then defines the active conformation. As a simple example, for the 3-chlorobenzamide in Figure 7, two rotamers are possible for the *trans* amide conformation. By conformationally restricting the amide as, for example, a benzotriazinone, the more active rotamer can be identified. Of course this is only successful if the amide NH is not a key pharmacophoric element.

What one can expect to achieve from a conformational restriction approach are:

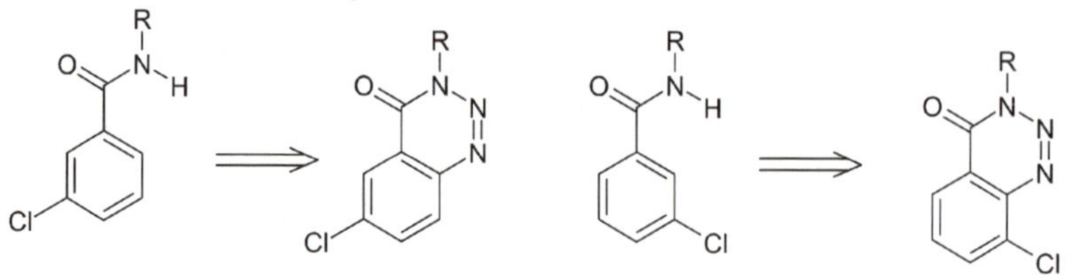

**Figure 7** *Conformational freezing of amide rotamers*

- improved affinity
- improved selectivity
- pharmacophore definition
- improved chemical/metabolic stability
- novelty.

*Improved Affinity.* From my own experience, the most spectacular increase in affinity achieved by conformational restriction was by the introduction of a bridging propylene into the piperidine ring of clebopride **12** to give the granatane BRL 26175 **13**. The piperidine already confers partial conformational restriction, but by this simple additional change an improvement in dopamine $D_2$ receptor affinity of 100-fold was achieved.[7] A similar conformational restriction was applied in the 5-$HT_3$ receptor antagonist area where addition of the bridging ethylene to the *N*-methylpiperidine **14**, but now as the axial isomer BRL 24682 **15**, improved 5-$HT_3$ receptor antagonist affinity by 1000-fold.[8] This dramatic improvement in affinity was believed to be due to the requirement of both an axial

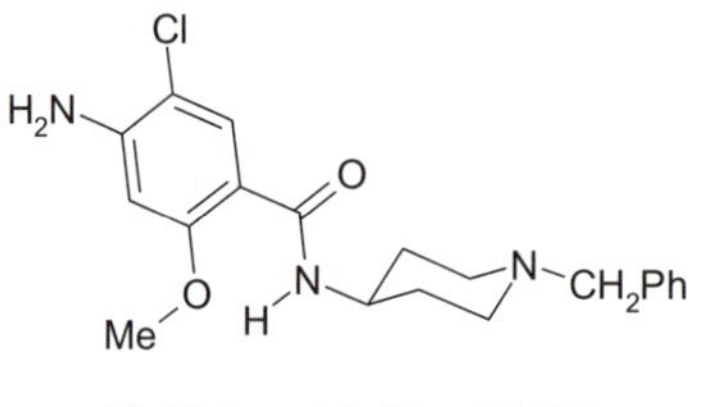

**12** Clebopride $D_2$ pKi 7.9

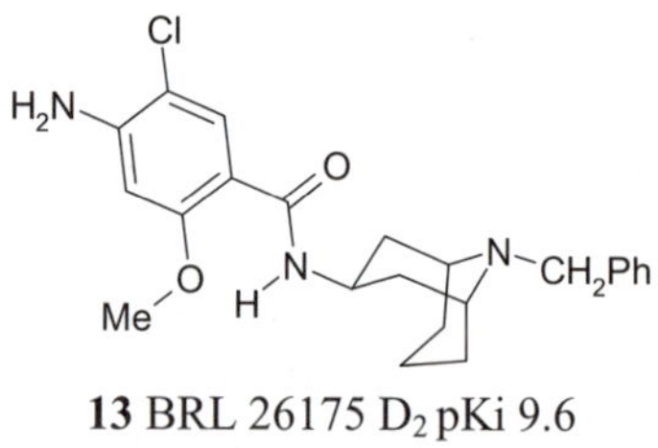

**13** BRL 26175 $D_2$ pKi 9.6

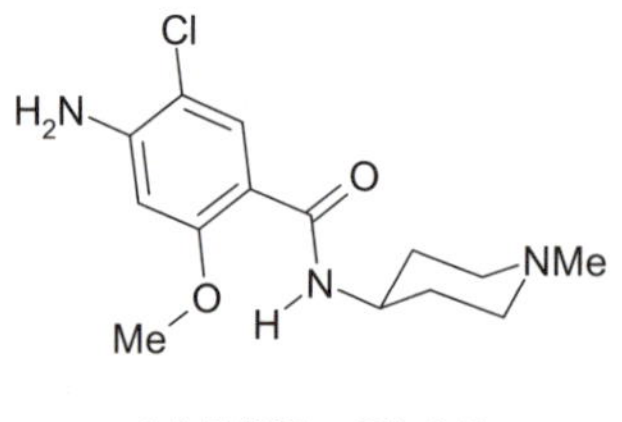

**14** 5-$HT_3$ pKi 6.5

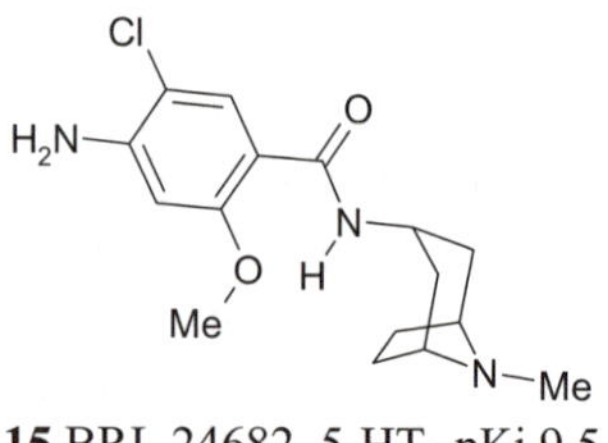

**15** BRL 24682 5-$HT_3$ pKi 9.5

amide and *N*-methyl group for the piperidine to bind to the 5-$HT_3$ receptor, both of which were energetically disfavoured. In contrast, for the tropane both the axial and equatorial *N*-methyl orientations are similarly favoured.

An excellent recent example of conformational restriction is in the development of RGD mimetics as antagonists of the fibrinogen receptor.[9] Although the interaction of fibrinogen with its receptor is a protein–protein interaction, the small tripeptide sequence of Arg-Gly-Asp **16** (RGD) in fibrinogen was identified as the key binding region. The key pharmacophoric elements for antagonists are: mimetics of the basic guanidine of the arginine – a spacer unit – and the terminal carboxylic acid of the aspartate. The simple pentapeptide Gly-Arg-Gly-Asp-Ser was found to be a weak inhibitor of platelet aggregation ($IC_{50}$ 25 μM). However, cyclisation to the peptide disulfide SKF 107260 **17** gave a highly potent antagonist ($IC_{50}$ 90 nM). Potency was also improved with NSL-95301 **18** ($IC_{50}$ 190 nM), which contains Arg bioisostere, the benzamidine, and the *gem*-dimethyl group, which restricts the freedom of rotation such that the distance between the N and C-termini in the low energy conformations falls within the range required for recognition at the fibrinogen receptor. Based on the structures of the cyclic peptides, non-peptide antagonists such as SB-214857 **19** were identified ($IC_{50}$ 28 nM) which also contain the less polar piperidine as a bioisosteric replacement for the guanidine.

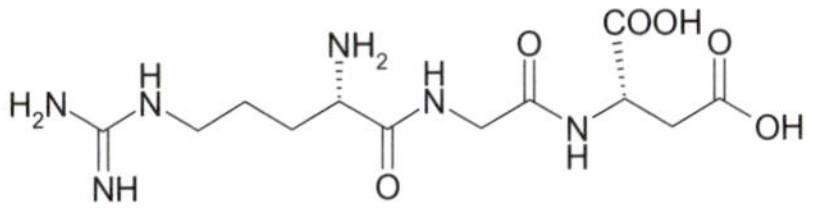

**16** Arg----------Gly-----Asp

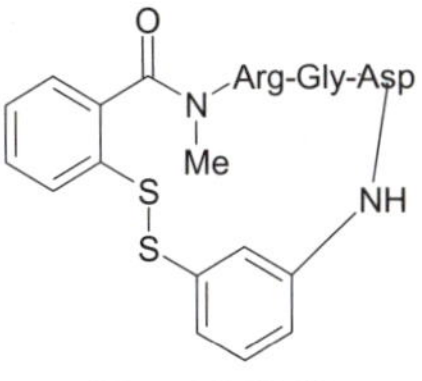

**17** SKF 107260

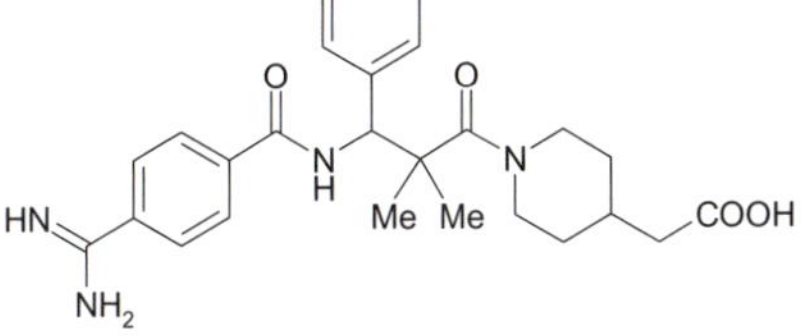

**18** NSL-95031

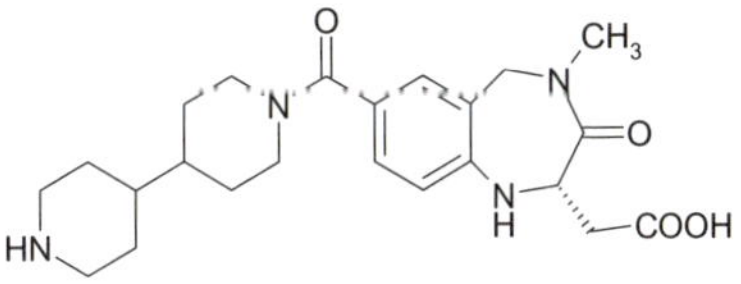

**19** SB-214857

*Improved Selectivity.* Provided that sufficient differences exist between the binding sites of two or more targets, conformational restriction would be expected to be an approach to gaining selectivity for one protein over others. Depending upon the nature of the active sites, selectivity increase can be achieved by either increasing potency at the desired target, or by reducing potency at the other sites. All of the examples previously discussed improved selectivity as a consequence of increased potency. However, a recent example of where conformational restriction afforded improvements in selectivity without increases in target potency is in the dopamine $D_4$ receptor antagonist area. Here the pyrazole **20** had good potency but suffered from ion channel liabilities, but by conformational restriction of the phenyl group, a series of antagonists, for example **21**, with lower ion channel liability was identified.[10]

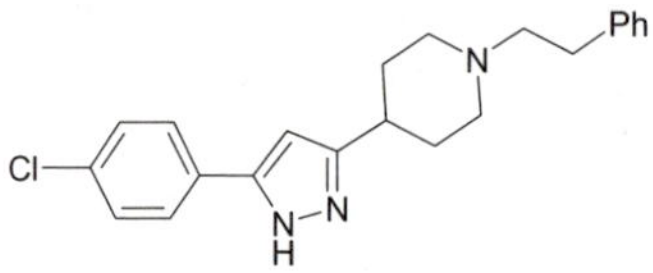

**20**
hD4 Ki 5.8nM;
Na channel
Ca channel Ki 1.4μM
$Ik_r$ channel $EC_{25}$ 0.4μM

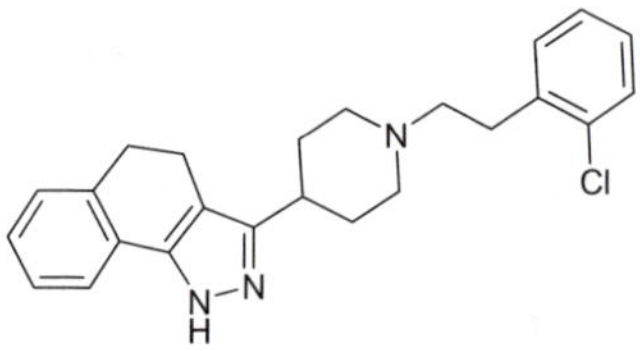

**21**
hD4 Ki 4.3nM
Ca channel 58% @ 10μM
$Ik_r$ channel $EC_{25}$ >30μM

In another example, selectivity for the endothelin A and B receptors was reversed by what was proposed to be a conformational restriction of the amide substituents (**22** *vs.* **23**).[11]

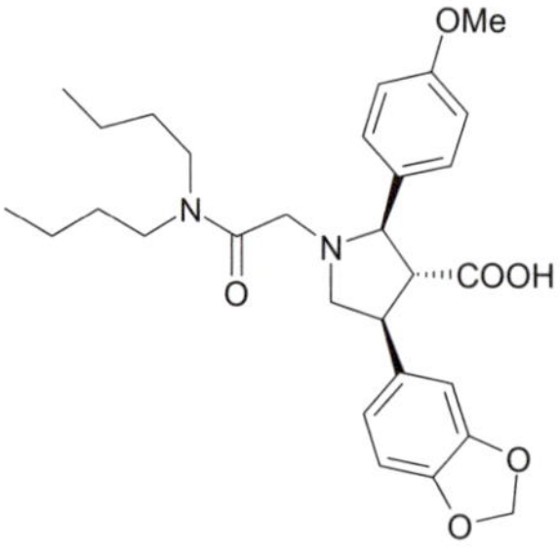

**22**
$hET_A$ $IC_{50}$ 0.08nM
$hET_B$ $IC_{50}$ 145nM

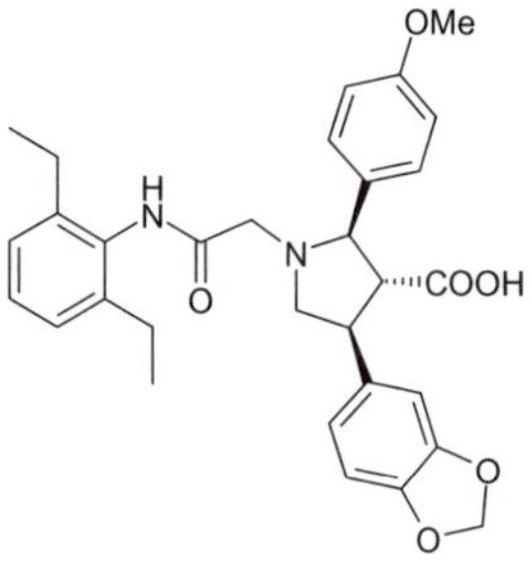

**23**
$hET_A$ $IC_{50}$ 89nM
$hET_B$ $IC_{50}$ 0.27nM

*Improved Chemical/Metabolic Stability.* Peptides in general find little use as drugs due to their metabolic instability, in particular due to cleavage by proteases, and to their poor physicochemical properties. Cyclisation forms a key part of the overall strategy for designing peptidic drugs to address these issues. In the previously described examples of fibrinogen receptor antagonists, the cyclised peptides were also more stable than the corresponding linear peptides. The resistance of cyclic amino acids, such as proline and piperidine-2-carboxylic acid, to human protease cleavage was used to produce SC-67655, a stabilised pentapeptide ligand for the DR4 peptide binding cleft of major histocompatibility complex molecules. Both the heptapeptide SC-64762 **24** and SC-67655 **25** had high affinity for the DR4 cleft, but only SC-67655 was stable in human serum.[12]

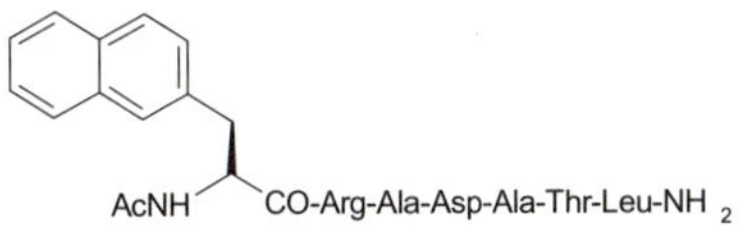

AcNH CO-Arg-Ala-Asp-Ala-Thr-Leu-NH $_2$

**24** SC-64762
97% degradation in human serum 1h @ 37 °C

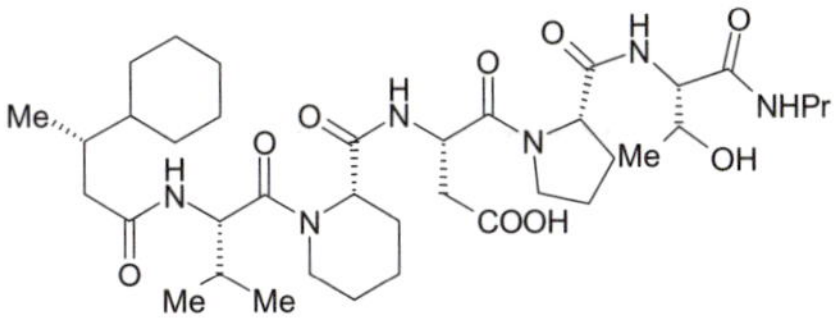

**25** SC-67655
7% degradation in human serum 1h @ 37 °C

## Pro-drugs

In many cases, required improvements in the *in vitro* potency and selectivity of lead molecules can be achieved relatively rapidly, often within one year of starting a chemical programme. Structure-based design, combinatorial (array) chemistry, bioisosteric replacements and conformational restriction all contribute to this achievement. However, it can then take many years to obtain a compound that has the desired level of oral activity and bioavailability. High throughput pharmacodynamic assays plus the increasing use of *in vitro* and *in vivo* metabolism studies all help in the identification of orally bioavailable compounds. However, there is often functionality which is an absolute requirement for potency but which limits oral bioavailability or tolerability. In many cases this can spell the end of the chemical programme. However, in some cases that essential functionality can be modified such that, although the parent compound is inactive, the modified functionality is converted *in vivo* into the active species, this is the *pro-drug* concept. This is essentially a chemical solution to the problem and in some areas has been highly successful. Thus, this approach has attracted the attention of medicinal chemists for many years. Reasons for adopting a pro-drug approach are for:

- improved physicochemical properties, *e.g.* solubility
- improved absorption and distribution
- improved drug targeting and reduced toxicity
- improved stability and/or prolonged release
- improved patient compliance, *e.g.* better taste.

Some examples of successful pro-drugs are given in Table 3.

Nabumatone was not a designed pro-drug but was found from an *in vivo* screening programme, where its anti-inflammatory activity was identified. On further study, it was found to be oxidatively transformed into an arylacetic acid, a classical non-steroidal anti-inflammatory related to

**Table 3** *Examples of successful pro-drugs*

| *Pro-drug* | *Active species* |
|---|---|
| 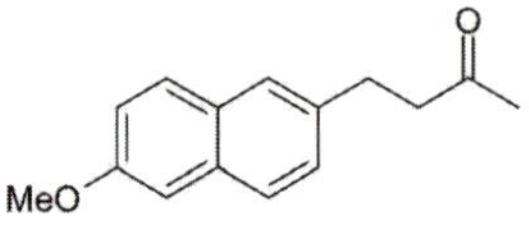 Nabumetone | 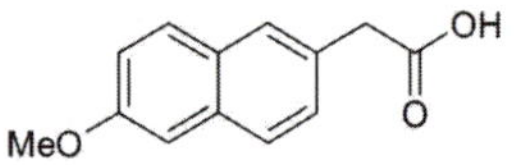 |
| 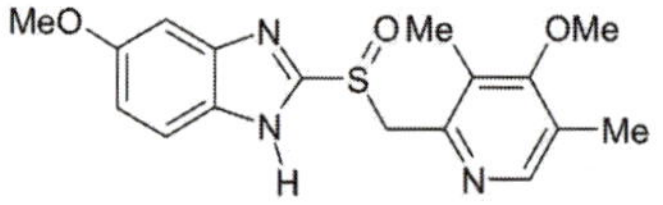 Omeprazole | 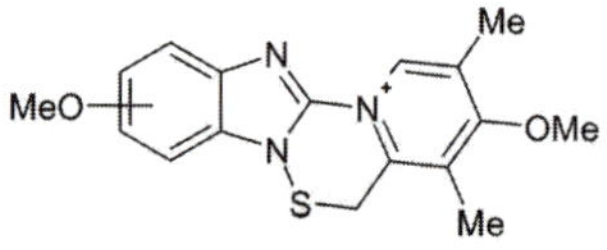 (or sulfenic acid) |
| 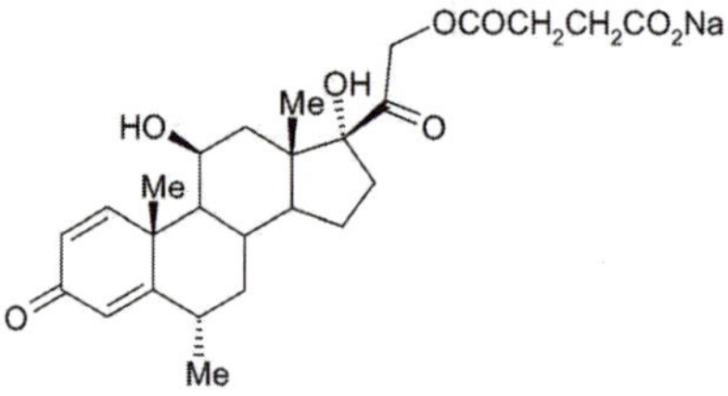 Methylprednisolone sodium succinate | 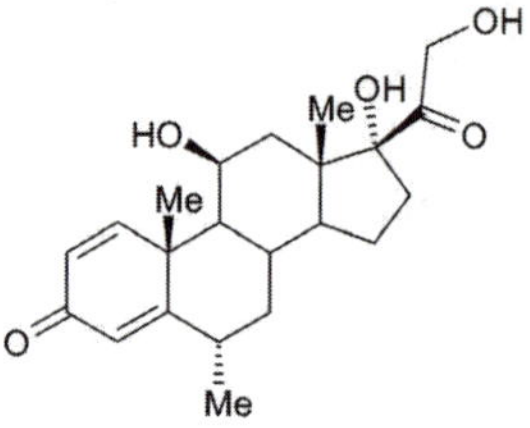 Methylprednisolone |
| 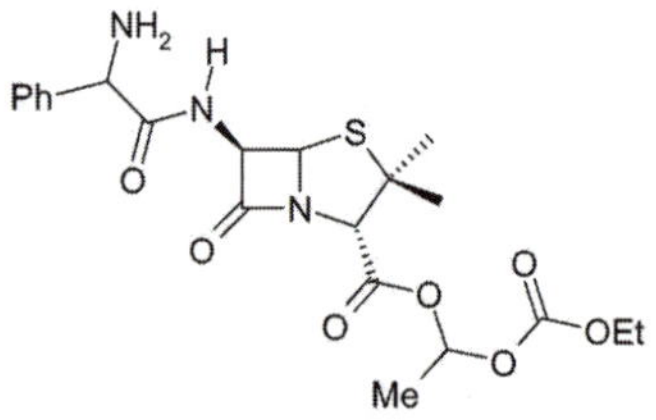 Bacampicillin | 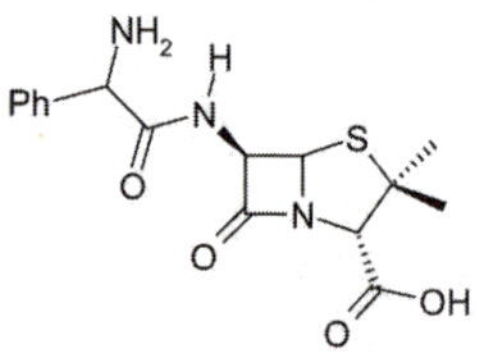 Ampicillin |
| 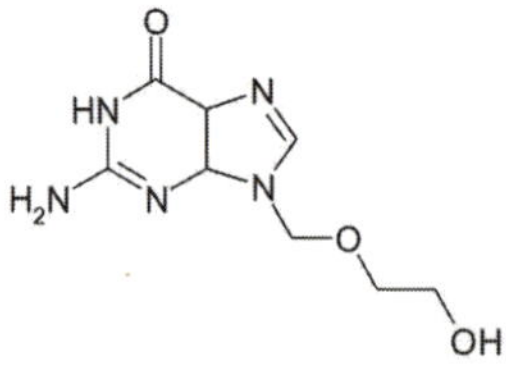 acyclovir | 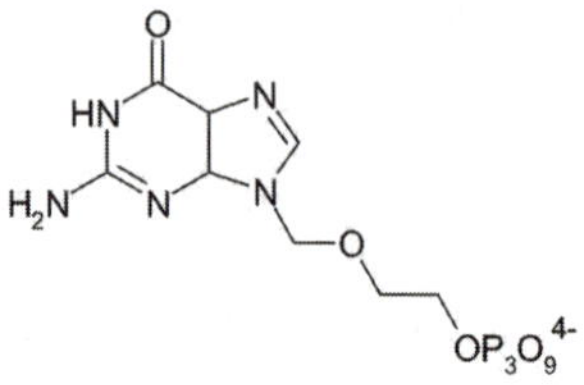 |

indomethacin. Nabumatone has a reduced propensity to cause gastric damage due, in part, to its Cox2 selectivity, but also due to it being absorbed as the inactive ketone, thus high levels of active species are not present in the gut.

Omeprazole is another example of a compound identified from *in vivo* experiments on compounds which reduce gastric acid secretion. The compound undergoes an acid-catalysed rearrangement to the active species, which is an irreversible inhibitor of the $H^+/K^+$-ATPase enzyme, the so-called proton pump, in the gastric parietal cells. Omeprazole gains its relative safety because it is only transformed into the active species in the acid-containing parietal cell. However, because of its acid instability it needs to be administered as an enteric coated capsule, which releases the compound into the duodenum and thus protects the compound from the acid in the stomach.

Methylprednisolone sodium succinate is an example of a *dipartate* pro-drug, in which a single chemical step converts the pro-drug into the active species and is the most common form of pro-drug design, shown schematically below.

$$\text{Carrier–drug} \rightarrow \text{drug}$$

The active drug, methylprednisolone, is poorly water soluble and therefore not suitable for aqueous injection. The succinate half-ester has the required water solubility and is the marketed form of the drug. Unfortunately, the ester has relatively low solution stability and is therefore sold as a freeze-dried solid and reconstituted in water immediately prior to use. This illustrates a fundamental problem with the pro-drug approach, that of getting the appropriate balance between being sufficiently stable to have an adequate shelf-life and being sufficiently unstable that it is rapidly converted *in vivo* into the active species.

Bacampicillin is an example of a *tripartate* or *double* pro-drug as shown schematically below:

$$\text{Carrier–linker–drug} \rightarrow \text{linker–drug} \rightarrow \text{drug}$$

The active species, ampicillin, is only about 40% absorbed on oral administration with significant amounts remaining in the GI tract, leading to destruction of the gut flora. In contrast, bacampicillin is almost totally absorbed and is transformed into ampicillin by a two-step process. The first step is the hydrolysis of the carbonate by esterases to form the unstable hydroxymethylester, the second step is a spontaneous decomposition

**Table 4** *Examples of the most common carriers for pro-drugs*

| | | *Result* |
|---|---|---|
| Drug-OH → Drug-O-X: esters and carbonates | | |
| X = | -(CO)R/Ar, -(CO)OR | Reduced polarity |
| | -(CO)$CH_2CH_2$COOR | |
| | -(CO)$CH_2NR_2$; -(CO)$CH_2SO_3H$; -$PO_3^{2-}$; | Increased water solubility |
| | -(CO)$CH_2CH_2CO_2^-$ | |
| $Drug-NH_2$ → Drug-NH-X: amides, carbamates, imines and aminomethyls | | |
| X = | -(CO)R/Ar; -(CO) OR/Ar; =CHAr, | Reduced polarity |
| | =NAr; -$CH_2$NHCOAr | |
| | -(CO)$CHRNH_2$; | Water solubility with Reduced basicity |
| Drug>C=O → Drug>C<X: imines, ketals and heterocycles | | |
| X = | =NR; =NOR; -$(OR)_2$; ; | Increased water solubility |
| Drug-COOH → Drug-COO-X: esters | | |
| X = | R; CHRO(CO)R/Ar/OR/$NR_2$; | Reduced polarity |
| | -$CH_2CONR_2$; -$CH_2CH_2$OH | |

to give ampicillin. The side-products of this reaction are ethanol, carbon dioxide and acetaldehyde, all of which are non-toxic.

It is the design of the carrier that provides the intellectual challenge to the medicinal chemist and numerous different carriers have been produced for different reasons. Table 4 summarises many of the most common carriers.

Although there are many examples of successful pro-drugs making the market, one cannot underestimate the added complexity and risk in development. Chemically labile carriers often require specialised formulations which take time and money to develop. Enzyme-mediated activation may suffer from interspecies variability which makes predictions from animals to man difficult. In addition, the cleavage products need to be non-toxic. Thus, in the modern day industry, where added development complexity and cost can lead to commercial failure, a pro-drug approach is applied when all other approaches fail.

## Soft Drugs

Whereas a pro-drug is pharmacologically inactive and is metabolically converted into the active form, a 'soft' drug is a term used to describe a drug which is also metabolically unstable, but is pharmacologically active but rapidly converted into inactive metabolites. A classic example of a 'soft' drug is that of procaine **26**, which was used as a local anaesthetic by dentists. Procaine is an ester and is rapidly metabolised by esterases to give the inactive acid and alcohol. Its action is therefore restricted to the site of injection and therefore does not show the systemic cardiovascular activity of the more stable amide analogue, procainamide **27**, which is used as an anti-arrhythmic.

The major advantage of soft drugs are reduced systemic side effects and toxicity, thus an improved therapeutic ratio. Clearly the 'soft' drug needs to get to the site of action intact (*cf.* local action of procaine) and have sufficient pharmacodynamic activity to maintain duration of action. Because of this the concept of soft drugs are mainly restricted to local applications, for example topical to the skin, (the antibacterial mupiricin), where the physicochemical properties retain the compound at the site of action (*cf.* procaine) or where the pharmacodynamics rely on peak levels rather than exposure. Clearly the distinction between a 'soft' drug and a rapidly metabolised drug that has found a specific use and where the boundary lies between a 'soft' and so-called 'hard' (more metabolically stable) drug are often mute points.

## Data Interpretation

An important consideration throughout the hit validation and lead optimisation process is that of data interpretation. Data is what drives medicinal chemistry, whether determined experimentally or *in silico*. It is from the interpretation of data that the medicinal chemist derives structure–activity relationships, derives binding hypotheses, attempts to solve the issues of potency, selectivity, metabolism and pharmacokinetics, and decides what to make next. It is therefore vital that medicinal chemistry is supported with the provision of high quality data. However, even with the very best experimental procedures, data, particularly from biological systems, is never precise. Data derived from *in silico* calculations can be even less precise, being orders of magnitude away from the true value. There are always errors, and hence uncertainties associated

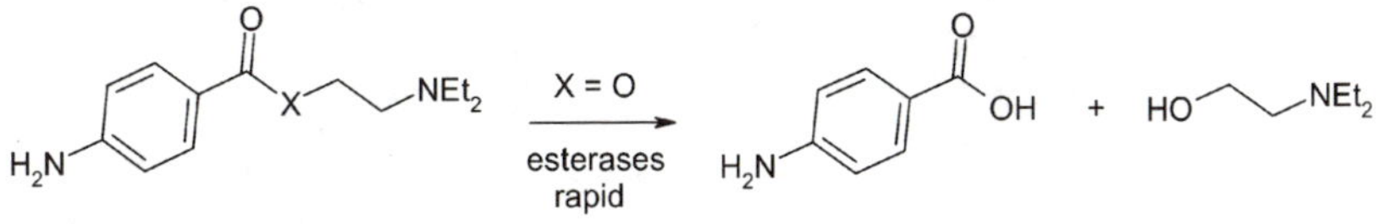

X = O; esterases, rapid

**26** X = O; procaine
**27** X = NH; procainamide

Inactive metabolites

**Figure 8** *Procaine as an example of a "soft drug"*

with the data. For primary screening this was alluded to in the chapter on biological evaluation (Chapter 5) but it is so important that it is worthy of further consideration.

Frequently one finds that the first compound from a series is the best, or that on re-test, the better compounds are always less potent. Often this can be explained by statistical outliers. This can be exemplified by Figure 9, which shows the results from a recombinant receptor ligand displacement binding assay ($y$-axis = p$K_i$) in which a standard compound has been tested over a number of months (96 separate assays). As one can see, the value for an individual result varies from a p$K_i$ of 7.1 to 8.1. This variability is also exemplified in Figure 3, Chapter 5 where, for 1251 assays, a typical range of individual values was from p$K_i$ 7.5 to 9.6! The implication of this is that if one were to make a combinatorial array of 1200 compounds, all with the same p$K_i$, then if only tested to $n = 1$, one member of the array would appear to have a p$K_i$ of 9.6 and one of 7.5 just through pure chance. Often in the heat of excitement the p$K_i$ of 9.6 would spark off the synthesis of a related array. It is therefore important to have some idea of variability and errors, and what these mean, so that appropriate judgement can be used before embarking upon additional syntheses.

Errors are normally defined as either a standard deviation (SD, $\sigma$) or standard error of the mean (SEM). The standard deviation is defined by Equation (1):

$$\mathrm{SD}(\sigma) = \sqrt{\frac{\sum(x - x')^2}{n}} \tag{1}$$

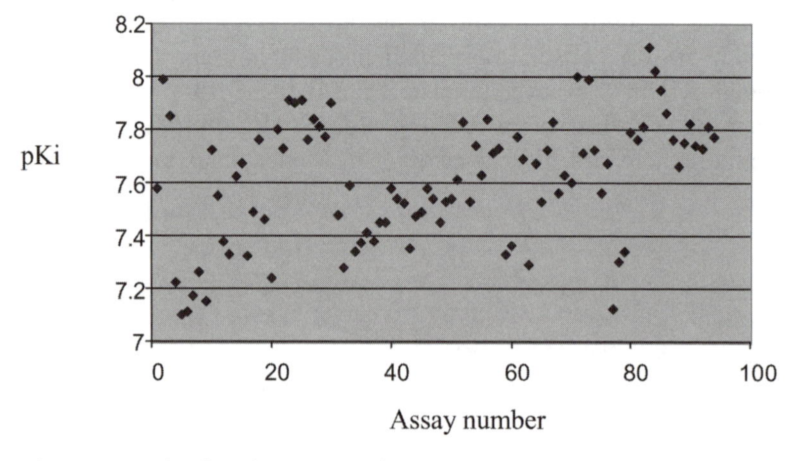


**Figure 9** *An example of single point variability of a ligand displacement assay*

where $x$ = measured values, $x'$ = mean value and $n$ = number of determinations. For a normal distribution, this value tells us that for ~65% of assays, a single determination of a $pK_i$ value will fall within the true $pK_i \pm SD$ and for 95% of the assays within $\pm 2 \times SD$. The SD for a compound included as a positive control in all assays is appropriate for assessing the variability within the whole population of assay determinations. Ideally the positive control should be structurally related to the chemical series being tested, so that this SD figure can be used to estimate the error for each individual compound tested.

The error normally quoted for a single compound is the SEM. The SEM is related to the SD by Equation (2) and is related to the number of determinations $n$ for that compound:

$$\text{SEM} = \frac{\text{SD}}{\sqrt{n}} \quad (2)$$

Thus the larger the $n$ number, the smaller the SEM and therefore the closer the mean $pK_i$ is to the true $pK_i$. Frequently the SD used for an individual compound is that derived from the individual results for that compound. However, a better assessment for comparing values for compounds over time is to use the SD for the assay derived from the positive control, as the $n$ number is often very high and therefore gives a better assessment of the variability of the assay. From Figure 9, this assay variability results in a standard deviation (SD) of 0.2. Thus for an $n = 2$ determination, the SEM would be $\pm 0.14$, for $n = 3$ $\pm 0.12$ and for $n = 4$ $\pm 0.10$. Normally we use $n = 3$ for our assays and thus a mean $pK_i$ of, say, 8.5 has a ~65% probability of being between 8.38 and 8.62 and a 95% probability of being between 8.26 and 8.74.

This error becomes particularly important when comparing the potency values of two compounds. Often a change in a molecule results in a large change in potency (normally a reduction!) which is obviously real. However, in optimising activity, the difference in measured potency between compounds can be small (say 0.5 log unit or less) and it is important to know whether any observed increase in potency is real. For this we need to consider the standard error of the difference (SED) in $pK_i$s, the $\Delta pK_i$ value. The SED defines the 65% confidence that a difference in $pK_i$ values is real, and $2 \times$ SED defines the 95% confidence (called the least significant difference, LSD, for 95% confidence). If both compounds are assayed to $n = 3$, then the SED for the $\Delta pK_i$ is calculated from Equation (3):

$$\text{SED for } \Delta pK_i = \sqrt{2\text{SEM}^2} \quad (3)$$

Thus, for the above example of $n = 3$, with SEM for the p$K_i$ of 0.12, the SED for the Δp$K_i$ is 0.17. This means that to be 95% sure that the difference in p$K_i$ between compound A and compound B is real, the Δp$K_i$ needs to be 2 times the SED, *i.e.* 0.34 or (equating to 2.2-fold difference in $K_i$) and for 65% confidence, 0.17 (or 1.5-fold). Ways of improving confidence are to increase the $n$ number, thus testing to $n = 6$ would reduce the SED to 0.07, giving a 95% LSD of 0.14 or 1.4-fold. The frequently applied alternative is to make the same change to a molecule from a second, closely related series to see if the improvement in potency is repeated.

Clearly this assessment of errors and confidence also has consequences for selectivity, where one is comparing activities in two or more different screens, and each of those screens will have its own SD. Thus, similar to comparing Δp$K_i$ values within an assay, errors on the Δp$K_i$ values between assays can be calculated from Equation (4):

$$\text{SED for } \Delta\text{p}K_i = \sqrt{(\text{SEM1})^2 + (\text{SEM2})^2} \tag{4}$$

where SEM1 and SEM2 are the SEM values for assays 1 and 2 based on the respective SDs of the assays. If the SD for both assays are the same, then Equations (3) and (4) are the same. Thus, assuming a SD of 0.2 in both assays, for an $n = 3$, similar to the above, the difference in potency at the two targets needs to be greater than 2.2-fold for the selectivity to be "statistically significant" (95% probability). In real numbers this means that if the potency at the primary target is 10 nM, then the potency at the second target needs to be >22 nM before we can say there is any real selectivity. For many years we had a standard 100-fold target for selectivity. So again, applying this to say 1 nM potency at the target of interest, to be sure we had 100-fold selectivity we had to have achieved >220 nM at the second target, so nearly always our compounds that had 100-fold selectivity on $n = 3$ testing gave lower selectivity on retest!

The fun really starts when one compares selectivities between different compounds – a 4-way matrix determination! This is important as often we are not only trying to improve potency, but also selectivity. For comparing a selectivity Δp$K_i$ for compound A and compound B, the SED on the ΔΔp$K_i$ is:

$$\text{SED for } \Delta\Delta\text{p}K_i = \sqrt{\text{SEMA}^2 + \text{SEMB}^2} \tag{5}$$

where SEMA and SEMB are the SEMs for the selectivity of compound A and B respectively. For two assays with a SD of 0.2, for an $n = 3$ the SED for the ΔΔp$K_i$ is therefore 0.24. Thus for 95% confidence that a difference in selectivity is real, there must be almost a 3-fold difference and for ~65% confidence, 1.7-fold. It is interesting to note that reducing the SD of the assays to, say 0.15 (a major technological challenge for some assays) only reduces the SED to 0.21. Putting this into practice, just by chance for every 20 compounds synthesised and tested, all with a true selectivity of 10-fold, one would have an apparent selectivity of 30-fold and six compounds would have >17-fold. Clearly this has major implications on interpretation of SAR.

All these numbers are only intended as guides; assumptions have been made about experimental design and the actual values are highly dependent upon the SD of the assays. Therefore, it is important for the chemist to know what the SD of the assays are, and to insist on as high an *n* number as is reasonably possible.

Errors are particularly important when interpreting *in vivo* data where the variability is much greater. *In vivo* data on a single compound, therefore, needs to be treated with great caution, even though there is apparently a good *statistically significant* effect, and one should always ask – does this result make sense? A recent example of this was in an *in vivo* experiment where a compound had an unusual pro-convulsant activity at high dose, all other compounds tested were anti-convulsant! Although the effect was small, because the individual values were very close, the statistical measure of this occurring by chance were 1 in 200. However, because it didn't make sense it was retested and the pro-convulsant activity was never again observed!

## Chemistry

Synthetic organic chemistry is beyond the scope of this book. However, there are some common principles, specific for medicinal chemistry which are worthy of mention.

The synthetic organic chemist will normally devise a route which gives the maximum yield of final product following route optimisation and full purification and characterisation of both intermediates and final products. The sole intention is to define beyond any doubt the chemistry, to provide definitive characterising data for novel compounds and to provide sufficient detail to define the synthetic pathway to the highest standard so that any chemist can repeat the procedures. However, for the medicinal chemist the objectives are different. The primary objective for the first synthesis of any compound is to find out what effect the modification made to the molecule has on activity in the primary and secondary *in vitro* assays. On average about 1000 compounds need to be prepared and tested for each development compound. Therefore, in simplistic terms the quicker the 1000 compounds are prepared, the quicker a development compound will be identified.

Speed is therefore of the essence, so it is important to get the compound prepared and tested as quickly as possible. Excess time spent on route optimisation, purification and characterisation of reaction products is counter-productive. Another reason for taking intelligent short-cuts is that, due to a combination of the variability in most primary assays, plus the logarithmic relationship between ligand concentration and biological effect, in most assays compounds that are only 70% pure cannot be differentiated from 98% pure material (provided that the impurities do not interfere with the assay, for example excessive amounts of TFA). Furthermore, especially in the early stages of a programme, the majority of compounds do not pass the initial *in vitro* assay criteria, so excessive time spent in synthesis should be avoided.

The concept of quick synthesis with minimal purification and characterisation is inherent in the basic concept of high throughput chemistry, but can also be applied to the iterative synthesis of individual compounds. However, for subsequent batches of material, the standards set should be very much higher. Definitive data is likely to be required for these compounds, they will probably be tested in assays that may be more sensitive to impurities (for example cytochrome P450 assays) or be tested *in vivo*, for which impurities could give rise to adverse events and misleading PK/PD data (*e.g.* bioavailability). Typically at this stage, purity standards are set at around 98% with no single impurity $>0.5\%$.

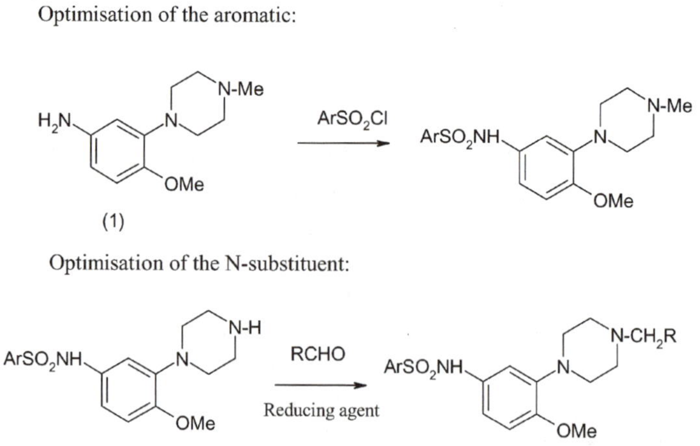

Optimisation of the aromatic:

Optimisation of the N-substituent:

**Figure 10** *A simple example of adjusting the chemistry for maximum diversity in the final reaction step*

Another important aspect for the medicinal chemist is in the design of the synthetic route. The desire is to devise a route which allows the maximum level of diversity to be introduced at a late stage in the synthetic sequence, preferably at the last step. This is so that the maximum number of analogues can be prepared with the minimum number of reactions. Again this has come to the forefront of thinking with combinatorial and array design, but it has always been an important concept in medicinal chemistry. A typical example of this is shown in Figure 10 from our 5-$HT_6$ receptor antagonist programme, where different routes were used investigating different parts of the molecule.

## 9 PATENTS

The subject of patents and processes of obtaining them are described in Chapter 13. Patents and patenting strategy have a major impact on the medicinal chemistry within a project. The strategy of patenting varies between companies, and can even vary between projects within a company. Thus, if you are working in a highly competitive area where the danger of a competitor patenting first is high, then an early patenting strategy would be wise. The downside to this is that you will be patenting with an incomplete knowledge of the SAR, that you often create a lot of your own prior art, and, if one eventually identifies a product from that patent, the time of exclusive use of that product can be reduced unless there is room to file a new patent case specifically to that product. In less competitive areas, a later patenting strategy may be preferable, when more is known about the SAR and there is a better defined level of interest in the compounds. The later patenting strategy has the advantage that competitors are less likely to be aware of your interest in the area and means that they cannot use the information disclosed within that patent to mount a me-too or me-better strategy. However, one does run the risk of being scooped by a competitor if, by chance, they identify the same chemical series.

The scope of the patent is usually agreed between the project chemists and the patent attorney, based upon what the patent attorney feels would be valid in the light of what the chemist believes would be active. Thus, for example it would be reasonable to include known, structural bioisosteres and substituents that one would expect would retain activity. However, the chemist must always bear in mind that during the 12 months between initial and foreign file, sufficient exemplification of the scope needs to be done to justify the scope of the claims. However, the chemical programme should not be driven by the need to exemplify the scope of the patents. The chemical programme should follow the activity and modify the claims appropriately.

## 10 CONCLUSION

Medicinal chemistry is a truly exciting subject. It combines the enjoyment of synthetic organic chemistry with the wider involvement of many other disciplines including physical chemistry, structural biology, biochemistry, pharmacology, toxicology, drug metabolism, pharmacokinetics and patents. However, because of this, medicinal chemistry can also be frustrating because success depends upon so many factors outside the control of the medicinal chemist. Even the worst chemist can have success in terms of a development candidate if all the other factors are right and a good lead is given. On the other hand, very good chemists can go through their whole careers without ever getting a development candidate. It is, therefore, not surprising that many chemists put success or failure down to luck. However, I have always believed that good medicinal chemists make their own luck and the purpose of this chapter is to help the good chemist become a successful medicinal chemist.

## 11 REFERENCES

1. J. Drews, *Drug Disc. Today*, 1997, **2**, 72.
2. See Chapters 14 and 16.
3. S.M. Bromidge *et al.*, *J. Med. Chem.*, 1999, **42**, 202.
4. L.M. Gaster and F.D. King, *Med. Res. Rev.*, 1997, **17**, 163.
5. J. Bermudez *et al.*, *J. Med. Chem.*, 1990, **33**, 1932.
6. See also C.W. Thornber, *Chem. Soc. Rev.*, 1979, **8**, 563.
7. M.C. Coldwell *et al.*, *BioOrg. Med. Chem. Lett.*, 1995, **5**, 39.
8. M. Langlois *et al.*, *BioOrg. Med. Chem. Lett.*, 1995, **5**, 795.
9. C.D. Eldred and B.D. Judkins, *Prog. Med. Chem.*, 1999, **36**, 29.
10. I. Collins *et al.*, *BioOrg. Med. Chem.*, 1998, **6**, 743.
11. T.W. von Geldern *et al.*, *J. Med. Chem.*, 1999, **42,** 3668.
12. G.J. Hanson *et al.*, *BioOrg. Med. Chem. Lett.*, 1996, **6**, 1931.

CHAPTER 16

# Combinatorial Chemistry: Tools for the Medicinal Chemist

MORAG A.M. EASSON AND DAVID C. REES

## 1 INTRODUCTION

Combinatorial chemistry (CC) can be viewed as a set of tools which allows large numbers of compounds to be synthesised simultaneously in the time taken to prepare only a handful of compounds by traditional synthetic strategies. The number of compounds produced far exceeds the number of chemical steps required to make them. A chosen set of building blocks are reacted together to make every available product and the collection of these products is referred to as a 'library' or an 'array'. These collections of compounds may be synthesised as mixtures or individuals using a variety of techniques. CC describes a broad range of techniques associated with automated, parallel or high speed synthesis. This chapter outlines some important aspects of the combinatorial 'toolbox' in the context of their application to medicinal chemistry research programmes.

Faced with the challenge of cutting the time taken to find lead candidates for drug discovery programmes by as much as possible, modern techniques for rapid synthesis can be extremely useful. Shortening the drug discovery process by one or two years is worth millions or even billions of pounds in sales to pharmaceutical companies. With the advent of high-speed bioassays, testing new compounds ceased to be the rate determining step in the drug discovery process. Modern high throughput assays and robotics are capable of screening more than ten thousand compounds per week. Providing suitable compounds for testing to satisfy this demand is therefore the rate-determining issue. Traditional sources of compounds for high throughput screening (HTS) include:

- Corporate collections of compounds – amassed from in-house research programmes, these may be biased towards certain compound classes
- natural products – often complex structures for subsequent optimisation
- commercial sources.

## 2 CONCEPTS IN COMBINATORIAL SYNTHESIS

### Compound Libraries and Arrays

Libraries of diverse compounds can be generated from selected building blocks, constructed in a highly organised and systematic manner. Consider the formation of a simple compound, 'AB', by the reaction of 'A' with 'B'. This example represents more traditional synthesis, where one reaction of two building blocks gives one product (Figure 1). Now, if there were ten different derivatives of A available to react with B, obviously ten different compounds, $A^1B$ to $A^{10}B$, could be formed. Similarly, if ten different derivatives of B were also selected for reaction then all possible combinations of these building blocks could be produced, giving one hundred possible compounds. Furthermore, every combination of $A^1$ to $A^n$ and $B^1$ to $B^n$ could be produced in an 'array' (Figure 1).

The roots of CC are firmly planted in peptide and oligonucleotide chemistry. For example, the sheer power of numbers can be illustrated by imagining the use of twenty natural amino acids in the construction of a decapeptide library. This library would consist of a massive $20^{10}$ possible combinations or 10 240 000 000 000 members. This is a colossal number of different compounds – several orders of magnitude more than the total number previously described in the entire history of *Chemical Abstracts*. However, in this combinatorial library the compounds are present as a **mixture** and not as individual, characterised structures.

The advent of combinatorial techniques heralded a renaissance in solid-phase organic chemistry (Figure 2) which was initially developed for peptide synthesis by Merrifield in 1963.

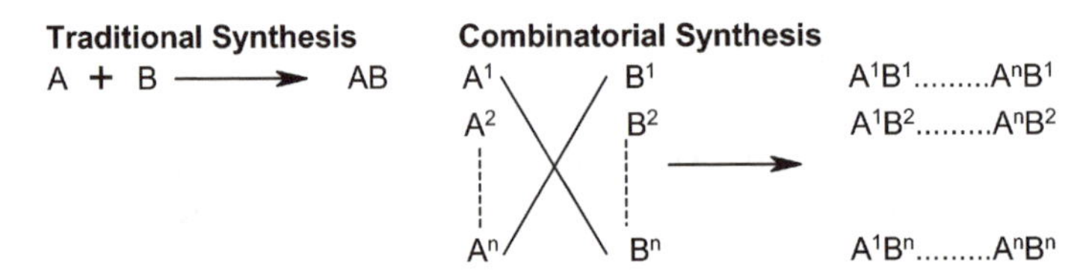


**Figure 1** *Schematic representation of traditional and combinatorial synthesis*

linker | excess A → linker—A | excess B → linker—AB | cleave → AB

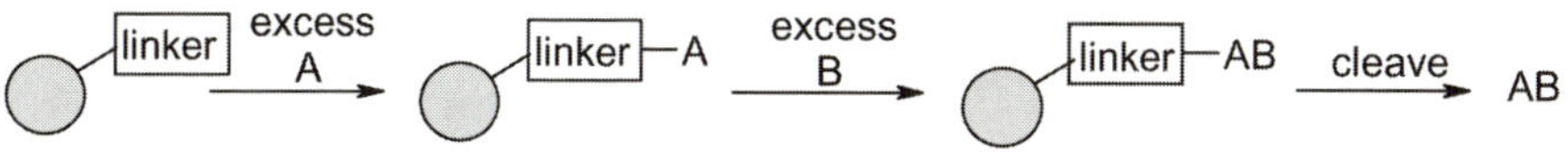

**Figure 2** *Solid-phase synthesis (solid-phase resin,* e.g. *polystyrene, indicated as filled circle)*

Solid-phase synthesis (SPS) initially found great utility in polypeptide and oligonucleotide chemistry because this type of synthesis consists of repetitive cycles of high yielding, room temperature coupling reactions accompanied by deprotection and subsequent coupling. The tedious washing steps and repetitive procedures could be automated on peptide synthesisers (robots).

Originally, SPS methods suffered from very limited chemistry as very few synthetic transformations had been adapted to the solid support, for example, peptide SPS is based on the treatment of an amine with an activated carboxylic acid to form an amide. However, the scope of SPS was an area of active, growing research during the 1990s and one of the most cited reports of early library synthesis was Ellman's solid-supported synthesis of 1,4-benzodiazepines in 1992. This represented the synthesis of a small 'drug-like' molecule rather than peptide synthesis and, consequently, stimulated a wealth of research in this area. Merrifield himself had envisaged as early as 1969 that 'a goldmine awaits discovery by organic chemists . . .'.

## Mix and Split Synthesis

Another driving force behind the increased empowerment of SPS was the 'mix and split' method. First reported in 1988 by Furka, in the context of peptide chemistry, this elegant but simple concept was subsequently pioneered by Houghten and Lam independently. An explosion of research and innovation in CC followed throughout the 1990s. Split–mix synthesis offers an efficient method for the simultaneous preparation of hundreds or thousands of related compounds, using only relatively few reaction steps.

A batch of resin is divided into equal portions in different reaction vessels, for example, three (Figure 3). Each portion of resin is treated with a different derivative of the first building block (A, B and C) as the first step. After washing to remove the excess reagents and by-products the beads are all pooled together in one pot and mixed thoroughly before being split into equal portions again for coupling to the next building block. This gives nine different combinations (AA, BA, CA, AB, BB . . . CC) in statistically equal amounts. Thus, the beads are split into portions, mixed together and re-split depending upon the number of

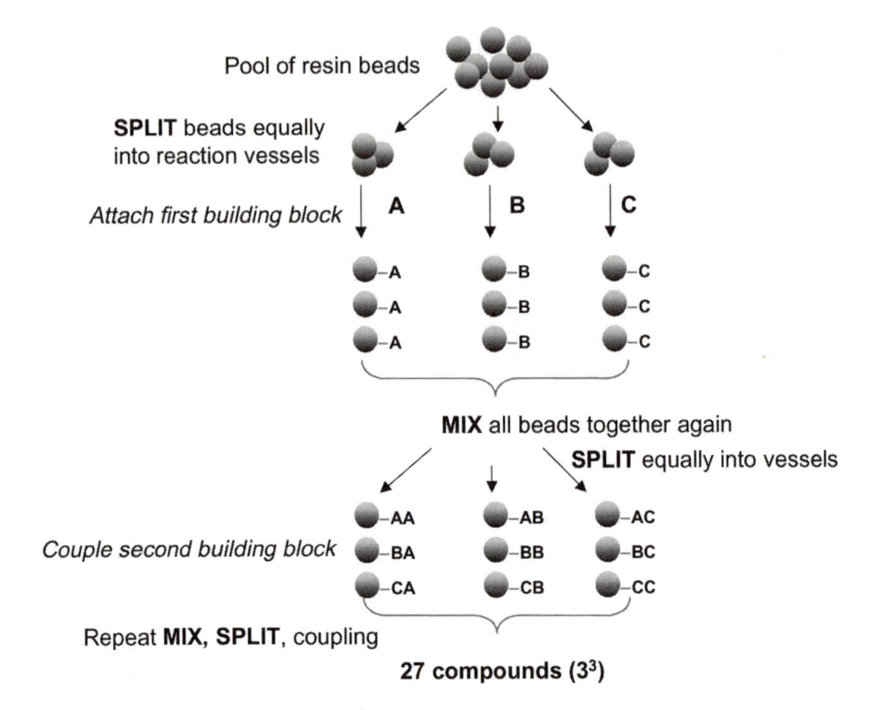


**Figure 3** *Schematic representation of the mix and split method*

different building blocks to be used. This governs the size of a library. For example, two rounds of coupling three building blocks, as above, gives nine compounds ($3^2 = 9$) while a third round of three building blocks give twenty-seven ($3^3 = 27$). Similarly, four rounds give eighty-one ($3^4 = 81$) and so on. In theory, some colossal numbers could be achieved, for example, twenty building blocks used in each of six rounds or synthetic steps would give sixty-four million compounds ($20^6 = 64\,000\,000$). In practice, many medicinal chemists working on drug discovery projects have chosen to make more libraries with just a few hundred or a few thousand members in each. This gives diversity to the overall compound collection due to the different structural templates and synthetic methods exploited for each class of compounds produced.

Finding biological activity from screening such library mixtures can be rapid, but 'deconvoluting' the identity of a single 'active' molecule provides the next challenge. Since all combinations of the building blocks are present, the hit molecule could be any one of these. Several approaches have been developed. One approach to identify the structure of an active hit is to perform further rounds of synthesis and screening, 'iterative resynthesis', where the synthetic history of the active pool is retraced. Another approach involves 'tagging' the resin bead to give the information on the structure. Elegant and sophisticated methods of tagging or encoding beads are currently in widespread use (for examples see Section 4).

## 3 IMPACT OF CC ON THE DRUG DISCOVERY PROCESS

The potential for generating large numbers of compounds in parallel fashion is now being tapped by many areas of science and technology, in particular, the pharmaceutical industry. CC impacts both lead discovery (hit finding and hit optimisation) as well as lead optimisation (LO) programmes (Figure 4). Potential applications are:

- Library synthesis for *de novo* lead generation (large and diverse libraries)
- Optimisation of lead compounds (smaller, focused libraries)
- Structure–activity relationship (SAR) studies of existing leads
- Expansion of patent scope.

Huge increases in numbers and novelty of biological targets are expected from genomics, therefore new molecules as well as traditional sources are required.

If little information is known about a biological target, screening hundreds of thousands of compounds present in compound collections and picking out those members which exhibit biological activity in a particular assay could provide a useful starting point for a drug discovery programme.

Planning library synthesis must be given careful consideration. *Which libraries should be produced? Which method of synthesis? Should compounds be produced as mixtures or singles? What purity and quantity?*

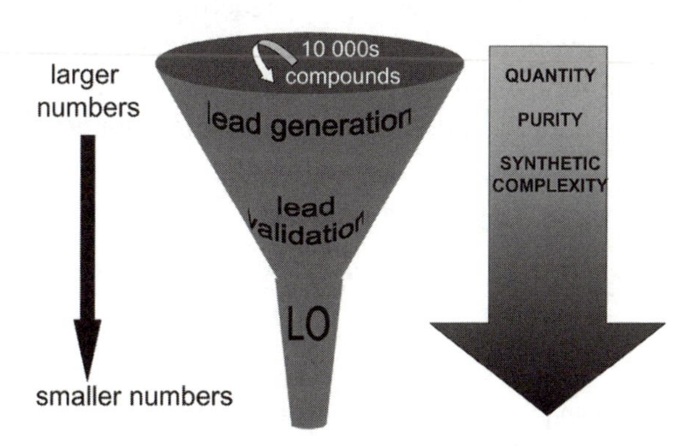


**Figure 4** *Combinatorial chemistry in the drug discovery process*

These choices are governed by the intended purpose of the library, whether it is designed primarily for lead finding or for focused optimisation work.

## 4 SOLID-PHASE SYNTHESIS OF 'DRUG-LIKE' MOLECULES

### Linkers and the Solid Support

In order to benefit CC in the pharmaceutical industry, SPS shifted from peptide routes and progressed to expedited synthesis of small organic 'drug-like' molecules. Synthetic methods are crucially important as the class of molecules of interest will be affected. For example, 'extra' substituents required for linking purposes such as $-CO_2H$ and $-CONH_2$ may be necessary for, or detrimental to, biological activity.

Solid-phase organic synthesis can take place on a variety of solid supports designed with shapes to give maximal surface area. The range includes polymeric resin beads, polyethylene pins, crowns, tubes, photolithographic chips, paper, glass and membranes. Far from being 'solid', a resin bead is a gel-type polymeric matrix swollen by solvent molecules to hold a solvated environment. Different solvents swell the resin by varying degrees. Bead sizes are in the order of 100 mesh (150 μm) to 400 mesh (45 μm) and the capacity of resin beads is given in millimoles of functionality per gram of resin (typically 0.25–2 mmol $g^{-1}$). The most commonly used resins are:

- Polystyrene cross-linked with 1–2% divinylbenzene
- TentaGel® – 80% poly(ethylene glycol) grafted to cross-linked polystyrene.

Polystyrene resin is physico-mechanically and thermally robust while TentaGel® is less so. However, the polyethylene glycol (PEG) spacer arms of the latter facilitate the projection of reacting groups into solution, allowing reactivity closer to solution phase chemistry. In addition to the selection of an appropriate solid support, the choice of solvent, reagent concentration, temperature and linker to the solid support are important factors.

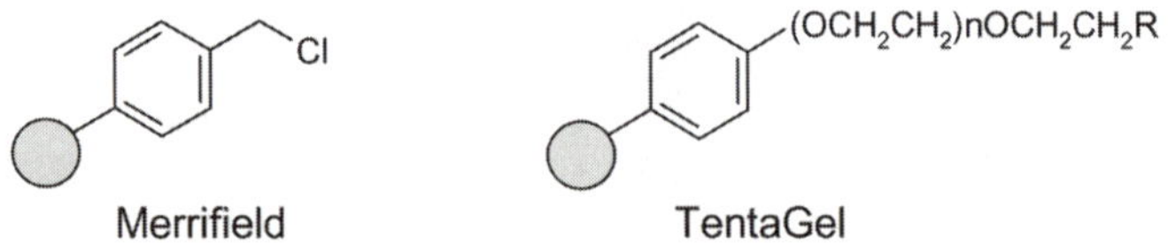

**Figure 5** *Resins for solid-phase synthesis (circle is polystyrene)*

Attachment of the molecule of interest to the solid support *via* a 'handle' or cleavable linker is of crucial importance. The linker can be likened to a protecting group in conventional solution-phase synthesis as similar principles apply. There are several considerations:

- The linker must withstand all conditions required in the synthetic sequence
- the linker must be labile to a suitable cleavage reagent
- the cleavage reagent itself should be readily removed from the cleaved products (particularly important for handling large numbers of products in combinatorial libraries)
- cleavage conditions should not be so harsh that the products are degraded
- selection of a *suitable* point of attachment to the molecule of interest is crucial
- choice of synthetic route may determine the choice of linker (and *vice versa*)
- attention should be given to the functional group obtained after cleavage.

Acid labile linkers are extremely common and cleavage reagents are easily removed due to their volatility. The largest number of linkers reported release carboxylic acids and amides upon cleavage, owing to roots in peptide synthesis. The antibiotic Ciproxin (Figure 6) is an early example of a widely used non-peptide drug synthesised by SPS. A common example of an amide-releasing linker is Rink amide resin (Figure 6) which has found utility for many classes of compounds, such as 1,4-dihydropyridines which are useful as calcium channel blockers.

Other main categories of cleavage/linkers include base labile, photolabile, 'traceless', cyclative cleavage, 'safety catch' and biolabile. While space here does not permit a comprehensive discussion worthy of the vast variety of linkers available, there are many excellent reviews of the area.[1]

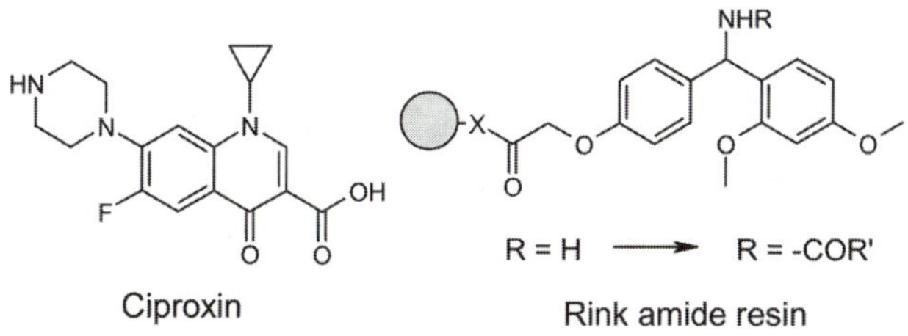

**Figure 6** *Ciproxin and acid-labile Rink amide resin*

## On-bead Monitoring

*'Doing organic chemistry on solid support has been likened to working with a blindfold on because of the limited analytical and purification techniques available relative to those available in solution.'*
Barry A. Bunin 1998

While rigorous confirmation of a chemical structure can be obtained by cleavage of a small sample of resin followed by analysis using normal techniques, some resin-bound intermediates may not be stable to the cleavage conditions required. Cleavage may also be more than one step, requiring several hours which may not be convenient to the chemist trying to monitor the path of a transformation in the middle of the synthetic route. Following the progress of SPS reactions (on-bead monitoring) can be achieved by a several techniques depending on the nature of the chemical structure of interest. For example, classical elemental analysis by combustion, or chemical colour tests such as a positive test (blue) with bromophenol blue, indicating the presence of a basic nitrogen. Cleavage of UV-active protecting groups when present (*e.g.* Fmoc) can provide quantitative information. Spectroscopic methods such as FTIR, magic angle spinning (MAS) and gel-phase NMR are becoming more and more widely used.

## Encoded Libraries

Identification of the structure of the most active library member (deconvolution) is a challenge which has produced several elegant solutions. Two popular techniques are chemical tagging (Figure 7) and radiofrequency tagging (Figure 8). Not being able to write a list of the building blocks upon each bead as simply as a marker pen on glassware, a code of chemical tags was devised by Still (1993) and subsequently developed commercially by Pharmacopeia.[2] A haloaromatic derivative (with varying length of hydrocarbon chain) is added to the polymeric matrix of the bead by rhodium-catalysed carbene insertion whenever a building block is added, allowing a record of each bead's chemical history to be built up. There is a limited set of tags to act as a binary code rather than one tag per library building block. The tagging molecules are released oxidatively, silylated and identified at subpicomolar levels by sensitive electron-capture GC by their retention times. This technology is compatible (orthogonal) with the chemistry of linking together the 'real' building blocks for synthetic construction, although it does involve extra chemical steps.

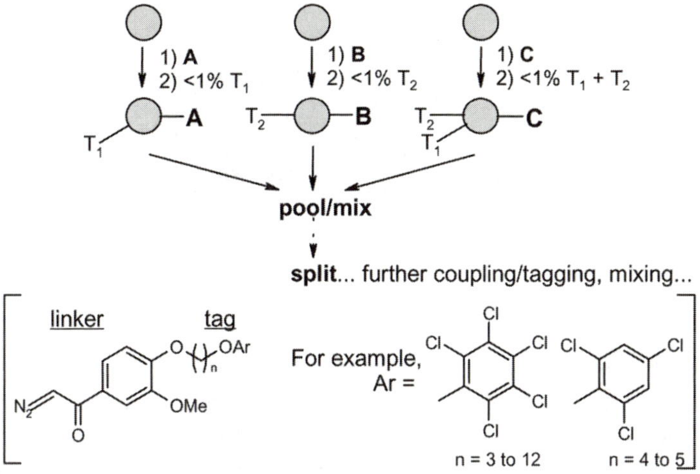

**Figure 7** *Tagging molecules (T) used for binary encoded solid-phase synthesis*

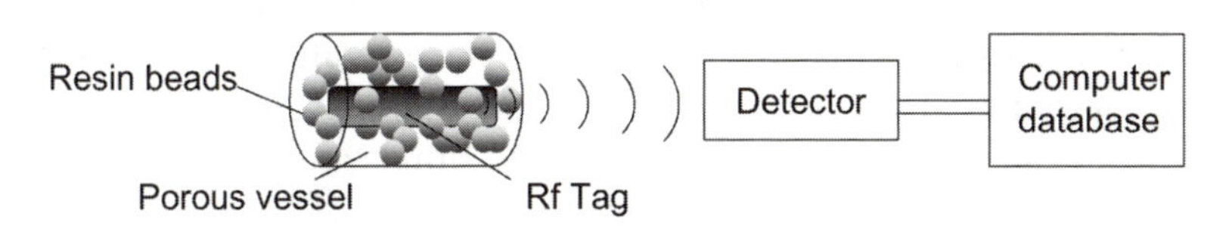


**Figure 8** *Schematic representation of RF tagging*

Radiofrequency (RF) tagging employs a microchip tag like a barcode to label each library member. Tags are robust, encapsulated in a glass casing, withstanding exposure to most chemicals and synthetic conditions. Each individual tag code is associated with the identity of its library member – the set of building blocks used – in a database on a computer and accompanies a particular portion of resin throughout its synthetic journey in a porous 'tea bag' vessel. This form of non-invasive encoded library synthesis has been applied to complex chemical structures. For example, a taxoid library (Figure 9) of 400 compounds was synthesised (Xiao 1997) as part of on-going attempts to overcome some limitations posed by the formulation and multiple drug resistance difficulties associated with the anticancer compound Taxol.

## Scope of Reactions and Structures on Solid Phase

It is now possible to carry out a vast range of types of reactions on solid support but it should be noted that only a few of these have actually been used for library synthesis.[3] In the early to mid-1990s many 'standard'

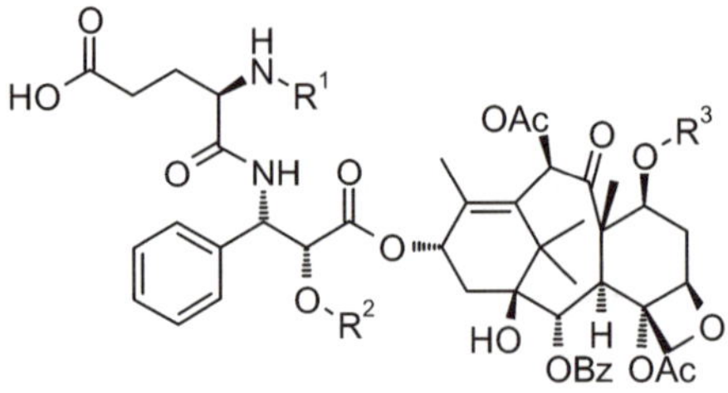

**Figure 9** *Taxoid library template elaborated on solid-support with RF tagging*

reactions were transposed from solution phase to solid phase, for example, oxidation, reduction, reductive amination, Mitsunobu reactions, phorphorylations, acylations (*N*- and *O*-), hydrolysis, condensations and cycloadditions. More reactions are continually being applied to solid phase, with recent examples including the Baylis–Hilmann reaction and ring closing metathesis.

The types of structural classes of libraries reported are similarly amplified. For example, heterocyclics alone incorporate structures such as pyrroles, prolines, indoles, $\beta$-carbolines, pyridines, isoquinolines, quinolones, $\beta$-lactams, furans, pyrans, sulfolenes, imidazoles, benzimidazoles, benzimidazolones, pyrazoles, piperazines, benzodiazepines, hydantoins, dihydropyrimidines and oxadiazoles.

### Singles *Versus* Mixtures

The current trend in the pharmaceutical industry is towards the preparation of single, 'discrete' compounds rather than mixtures (pools).

***Mixtures:***

- maximise screening throughput
- libraries require greater chemical development
- libraries require deconvolution or tagging

***Singles:***

- easier quality control
- increased reliability of biological data (less false positives)
- higher screening costs

## 5 SOLUTION-PHASE LIBRARY SYNTHESIS

### Strategies for Solution-phase Synthesis

Solution-phase library synthesis involves different challenges to SPS, although both methods require reaction conditions which are high yielding and of known scope. In general, it takes less time to establish the experimental conditions required for solution phase library synthesis than solid phase. With today's focus on libraries of smaller numbers and greater design, and advances in parallel purification strategies and equipment, solution-phase methods have become increasingly important in medicinal chemistry.[4] Solution phase library strategies range from the preparation of compound mixtures as 'indexed' libraries, to arrays of discrete compounds using elegant purification approaches involving resin-capture, soluble polymers (liquid-phase combinatorial synthesis) or related dendrimer-supported techniques. Furthermore, the development and popularisation of techniques referred to as 'support bound reagents and scavengers' (see below) is playing a substantial role in the synthesis of solution phase libraries.

## Parallel Solution-phase Libraries

Parallel solution-phase chemistry is useful for rapid synthesis of smaller optimisation libraries. For example, Pfizer identified an aminothiazole (Figure 10) as an antiviral agent against herpes simplex virus (HSV-1) from in-house screening. This hit was optimised by parallel synthesis of a series of libraries around three different parts of the structure. High-yielding substitutions and palladium catalysed couplings were used to generate more than 400 analogues in one week. Repetitive aspirate and dispense solution procedures were carried out by a liquid handling robot

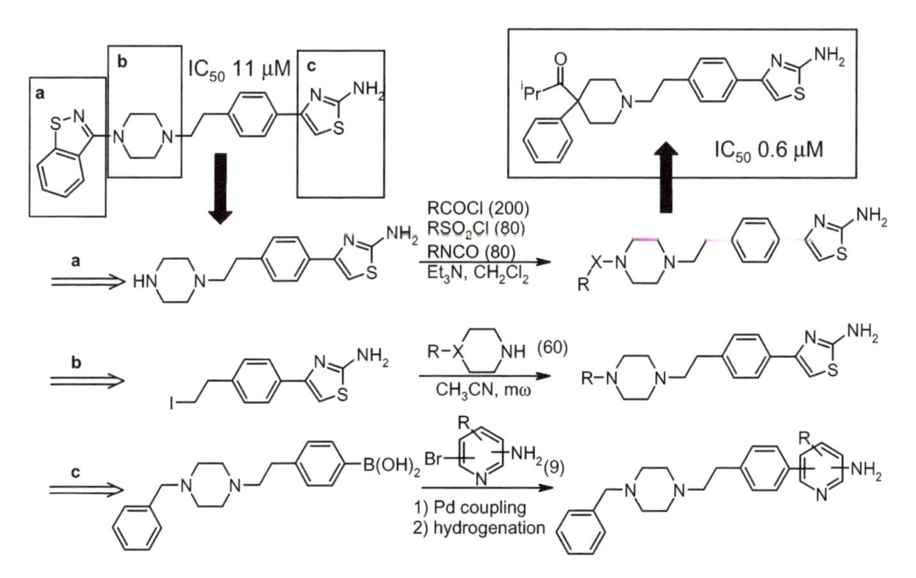


**Figure 10** *Solution-phase libraries for rapid HSV-1 hit optimisation*[5]

and all crude products were checked by thin-layer chromatography and mass spectrometry prior to testing. Screening of the crude library (micromolar concentrations) allowed rapid identification of biologically active compounds for resynthesis, purification and testing. At least one compound was identified ($IC_{50} = 0.6$ μM) which was 18 times more potent in the HSV-1 plaque reduction assay than the initial hit ($IC_{50} = 11$ μM).

## Support-bound Reagents and Scavengers

Polymeric supports can be applied to solution-phase chemistry to clean up the products by immobilising reagents, catalysts and by-products. Resin 'scavengers' are solid-supported nucleophiles and electrophiles for work-up and purification of a variety of different reactions, such as primary and secondary amine alkylations and acylations. Actual synthesis of products occurs in solution phase and is followed by the selective removal of unreacted excess starting material from solution in a "quenching" step. These impurities become covalently bonded to the solid-support and simple filtration allows separation of reagents and starting materials from the desired products (Figure 11). Evaporation of solvents gives products of high purity in a method that lends itself to parallel processing and automation.

Examples of common scavengers and their applications are given in Figure 12. Higher loading resins (high proportion of functionality per gram of resin) are finding increased utility due to their ability to sequester more reactants per gram of resin.[6]

Support-bound reagents work by a similar principle – unwanted by-products and excess reagents are immobilised and easily removed. Now that the potential is being realised, a wealth of reagents is being immobilised and exploited. Examples commonly used include support-bound bases such as diisopropyl-ethylamine and *N*-methylmorpholine. Syntheses employing support-bound reagents and scavengers are becoming increasingly ambitious and many successful multi-step processes have been reported. Synthetic routes can combine several support-bound reagents and scavengers, for example, in the parallel synthesis of acyl-aminopiperidines[7] (Figure 13).

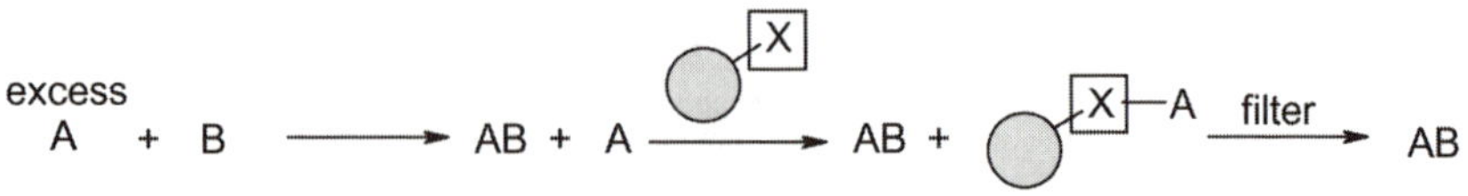

excess A + B ⟶ AB + A ⟶ AB + [resin]–X–A ⟶(filter) AB

**Figure 11** *Schematic representation of solid-supported 'scavengers'*

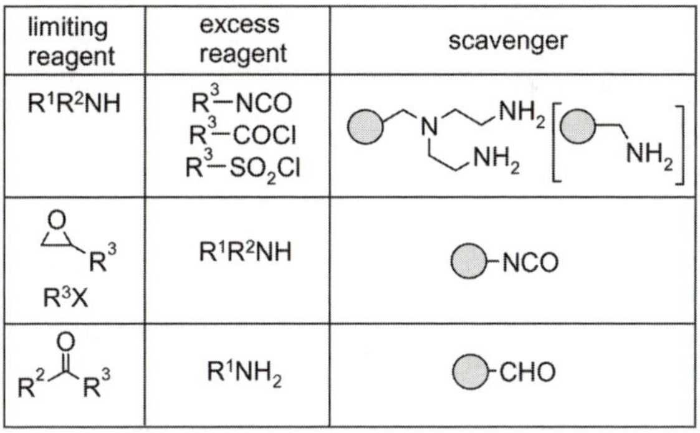

| limiting reagent | excess reagent | scavenger |
|---|---|---|
| $R^1R^2NH$ | $R^3$–NCO<br>$R^3$–COCl<br>$R^3$–$SO_2Cl$ | N, $NH_2$, $NH_2$ [$NH_2$] |
| O, $R^3$<br>$R^3X$ | $R^1R^2NH$ | –NCO |
| O, $R^2$, $R^3$ | $R^1NH_2$ | –CHO |

**Figure 12** *Examples of useful resin 'scavengers'*

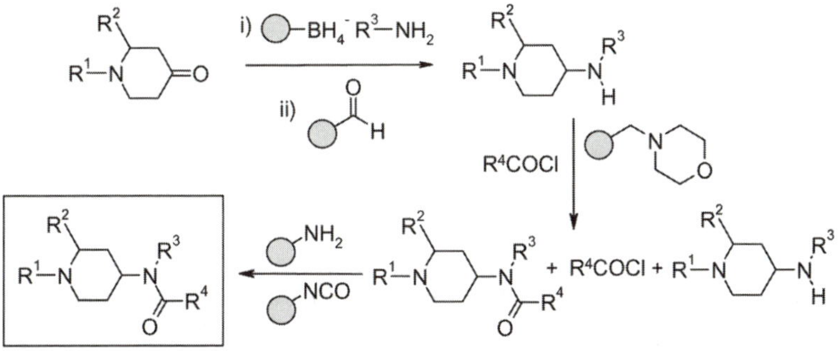

**Figure 13** *Synthesis of an acyl-aminopiperidine library*[7]

In the context of polymer-assisted solution-phase library synthesis, ion-exchange resins should also be mentioned as they are finding increasing utility as scavengers for rapid purification of solution-phase libraries. In addition, they are relatively cheap when compared to commercially available polymeric scavengers.

## Multi-component Condensations

In contrast to libraries constructed using linear, multi-step synthesis, multi-component condensations (MCCs) are essentially one-pot reactions, where three or more building blocks combine to form a new product that contains sections of all reactants. While linear multi-step construction requires a stepwise approach of reaction followed by resin-washing, re-treatment with new reagents, the MCC is a single process occurring in a single vessel and therefore suited to the construction of arrays in microtitre plate format. Although the Ugi reaction has been known for many years, it has recently

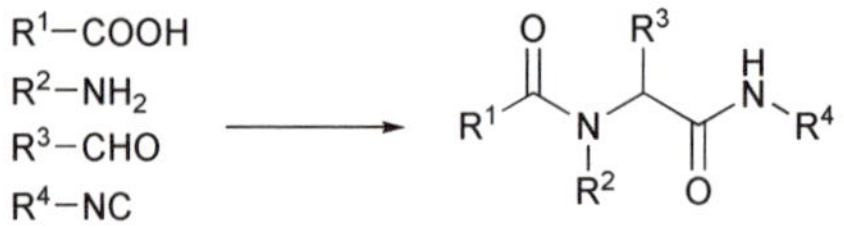

**Figure 14** *Ugi MCC*

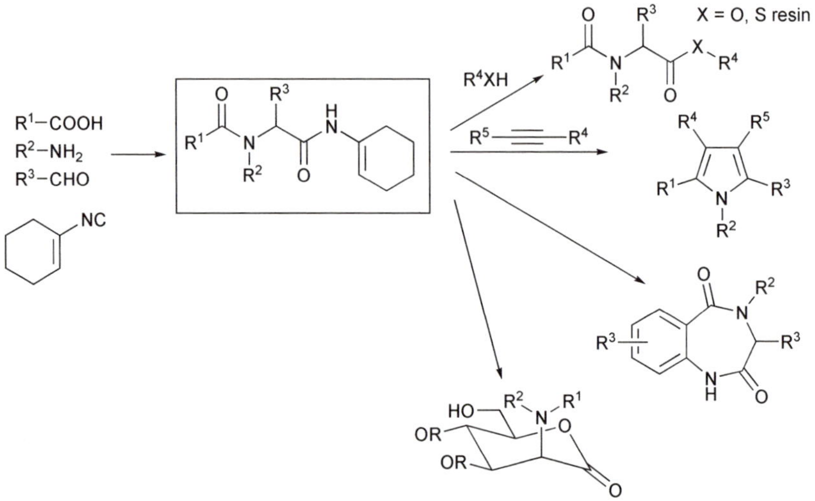

**Figure 15** *Ugi reaction with convertible isocyanide: conversion into other templates*[8]

been applied to both solution-phase and SPS strategies for the synthesis of small-molecule libraries (Figure 14).

Post-condensation modifications of products provide another source of library templates. Due to the commercially availability of only a limited number isocyanides, a convertible isocyanide, 1-isocyanohexene, was exploited by Armstrong[8] for accessing further templates (Figure 15). Besides the Ugi reaction, examples of MCCs include the Hantzsch, Strecker, Biginelli, Mannich, Passerini, and Pauson–Khand reactions. Solid-phase approaches have also been combined with MCC chemistry in resin capture strategies.

## 6 SOLID PHASE *VERSUS* SOLUTION PHASE

It is apparent that solid-phase and solution-phase techniques have different strengths and limitations (Table 1). They are complementary and should be used in an integrated manner. Generally, the investment of time spent developing a route on solid phase is recouped by generation of larger numbers in a library and relatively short time required to generate these members.

**Table 1** *Comparison of solid* vs. *solution phase chemistry*

| *Solid phase* | *Solution phase* |
|---|---|
| *Pros:* | |
| • Synthesis of very large numbers of compounds possible *via* mix and split technique<br>• Products generated are physically separate<br>• Reactions driven to completion by large excess of reagents<br>• Facile purification as excess reagents and by-products removed by simple washing/filtration<br>• Reactive sites on solid support are distinct and aid reactions requiring 'high dilution' conditions, *e.g.* intramolecular cyclisations<br>• Relatively long synthetic sequences possible<br>• Lends itself to automation | • Known chemistry, most reactions/reagents studied in solution<br>• Large excess reagents not generally used<br>• No extra steps, loading or cleavage from resin are necessary<br>• Reaction conditions applicable to wider range of substituents and steric effects generally overcome more easily in solution by varying conditions<br>• Large quantities of material possible |
| *Cons:* | |
| • Requires a SPS route with high yielding steps to be established experimentally<br>• Relative added expense of solid supports, large quantities of washing solvents and equipment<br>• More difficult to monitor reactions on solid-phase<br>• Requires a linker and a cleavage step<br>• Usually small scale ($<$50 mg final product) | • Conditions must be chosen to reduce by-products (can use solid-phase reagents)<br>• Time-consuming purifications necessary although automated purification equipment (*i.e.* HPLC) available<br>• Set-up of parallel handling/automation requires more planning/input |

## 7 LABORATORY AUTOMATION AND EQUIPMENT

### Revolution at the Bench

The modern chemistry laboratory has undergone significant changes from the mid-1990s to the present day: probably more so than in the whole of the preceding century (Figure 16).

With the rapid increase in chemistries in routine use for CC, there has also been a surge of innovation in commercial equipment and automation for synthesis, purification and analysis. These instruments aim to increase the quality and throughput of products. Today's automated systems aim for speed, safety and reduction of the cost per compound synthesised. Continued innovation for laboratory instrumentation and equipment has been consumer-driven as 'bottlenecks' have arisen. Suppliers of laboratory

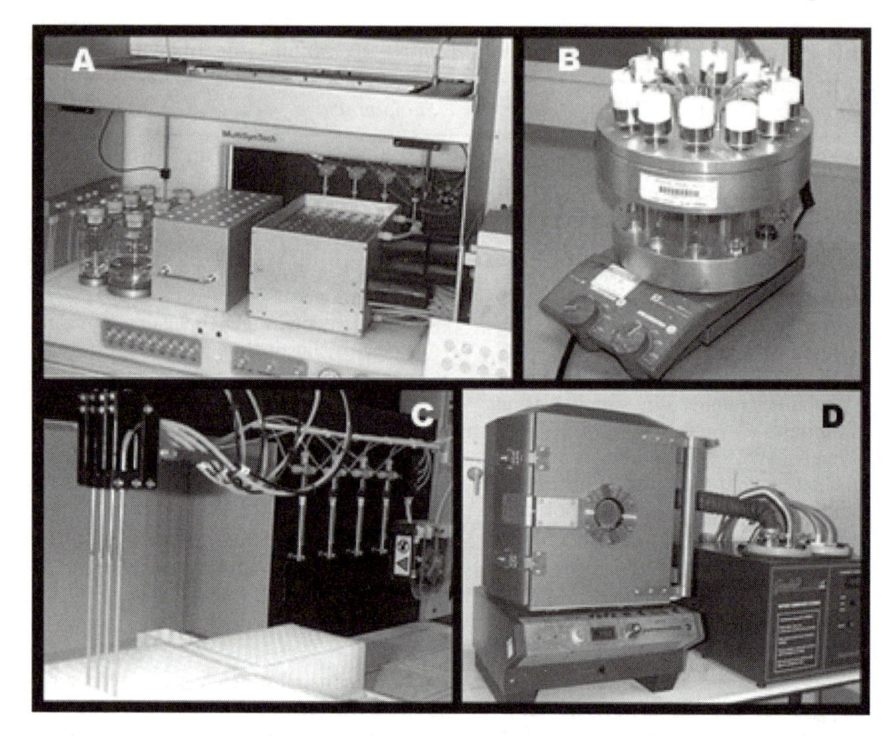


**Figure 16** (a) *Automated synthesiser,* (b) *parallel synthesis block,* (c) *liquid handling robot,* (d) *vacuum centrifuge*

equipment have been swift to exploit new ideas and satisfy the ever-demanding chemists, in some cases producing partnerships between engineering and pharmaceutical companies.[9]

## Synthesis

Today's laboratory automation, like SPS, has its roots in peptide chemistry as peptide synthesisers were developed to carry out the necessary repetitive deprotection and coupling procedures. Liquid handling is fundamental to chemistry automation and requires resistance to corrosive reagents. Modern synthesis robots have several key purposes:

- precision handling of small volumes of building blocks and reagents
- increased throughput by working 'round the clock' and reduction in human errors
- repetitive processes (tedious to the human chemist) are carried out
- allows the chemist freedom to pursue other work requiring more intellectual input

- advantages of SPS exploited – work-up consisting of washing and filtrations.

A typical **automated synthesiser** consists of a reaction block where reaction vessels are fixed (typically up to 96 vessels), reagent block/vessels and robotic arm(s) linked to common solvents (system fluids). Systems are capable of mixing, heating (some under reflux) and cooling, while maintained under inert atmosphere, and facilitate collection of products cleaved from solid-phase. A typical commercially available automated synthesiser is computer based and is pictured (Figure 16(a)). Generally, software requires programming simple line commands of processes such as dispensing building blocks, mixing or heating, which make up each step in the synthetic sequence. Limitations of robotic synthesisers may be specific to the particular chemistry. Solid reagents and building blocks must be completely dissolved and be stable in solution for the required time-frame prior to reaction. Instruments are geared to the production of discrete compounds, most using compatible, interchangeable blocks, with automated liquid handling. Indeed, a modular approach, where several simple device units are linked together, allows flexibility and increases throughput.

A variety of equipment for **semi-automated**, parallel **manual synthesis** and **microwave-assisted synthesis** is finding increasing utility for high speed synthesis. Many reaction blocks in various formats (*e.g.* $2 \times 5$ or $8 \times 12$ matrix) are available which can heat standard sized glass tubes with agitation of reactants by multiple stirring (*e.g.* Figure 16(b)) or orbital shaking. Other manual blocks (up to 96 positions) for solution-phase or SPS offer options of 'inert conditions', reflux or filtration. The latter can also be adapted to solid-phase extraction (SPE) for purification, if desired. There are also small, simple systems based upon a microtitre plate with filter plate. In general, manual workstations are finding widespread acceptability among chemists to complement automated systems, as they are easy to use without investment of time programming them or skills training to use them, and they are also relatively cheap compared with fully automated robotic synthesisers.

A **liquid handling robot** is another essential instrument for processing large numbers of compounds (Figure 16(c)). Its robotic arm, with 4 or 8 probes (needles), dispenses to and from every format, accelerating many different protocols including dispensing reagents for solution-phase library synthesis, reformatting samples or generating multiple copies of finished products for screening. Driven by 'user-friendly' software, other options for sample handling, such as barcode scanning and sample tube weighing, may be interfaced and associated with library data through spreadsheets. A

**weighing robot** (with robotic arm and gravimetric balance) can also help in this capacity when calculating amounts of finished samples present and subsequent concentrations required for screening. Removal of solvents from multiple samples, either in tubes or microtitre plates, is carried out by parallel evaporation or lyophilisation. A **vacuum centrifuge** (or several) is commonplace in medicinal chemistry laboratories today and can efficiently cope with removing even 'stubborn' solvents like DMSO (Figure 16(d)).

### Purification

Similar to synthesis equipment, instruments for purification range from automated to manual parallel systems:

- preparative LC-MS (mass-selective purification) – autosampler, software allows sample tracking with MS combined with UV/ELS (evaporative light scattering) detection (96 well plate format or tube racks)
- semi-prep HPLC – software allows sample tracking with UV/ELS detection (96 well plate format or tube racks)
- parallel flash column chromatography – including normal phase, reversed phase, ion-exchange – for larger scale products, based on column chromatography
- SPE – several automated instruments available, differing in throughput, based on individual SPE cartridges or sorbant-filled microtitre plates
- liquid–liquid extraction – by phase-boundary detection or by volume (dead reckoning)
- 'lollipops' for parallel extraction of (80) frozen aqueous samples from the organic phase.

### Analysis

Mass spectral analysis is the mainstay of high throughput analytical methods, combining sufficient sensitivity with speed. Systems combining HPLC (automated sample preparation and delivery systems) and MS are rapidly becoming the instruments of choice. LC-MS systems allow direct correlation of purity with identified mass ion. This circumvents the problem of separate HPLC and MS analysis where the major product detected by the former may not necessarily be the most intense mass ion detected by the latter and *vice versa*. Other sophisticated means of autosampling directly from microtitre plates are now possible include NMR analysis. Computer software now allows more rapid data handling of microtitre plates of samples where desirable parameters such as purity thresholds can be set. Results may be displayed immediately as a colour-coded well map (green/red for 'yes/no') or processed manually. Data associated with each compound, from library design and construction to analysis, is tracked by spreadsheets or similar computational means.

## 8 BIOLOGICAL ACTIVITY FROM COMPOUND LIBRARIES

From the perspective of a medicinal chemist, the test that CC must pass in order for it to become widely regarded as a useful 'tool', rather than just an emerging 'technique', is the identification and optimisation of biologically active compounds. This section describes the progress that has been achieved in this respect.

From the perspective of the pharmaceutical industry, the ultimate test for CC is its application to the discovery of commercially successful drugs. This is tricky to quantify at present partly due to the long time scale of drug development (>10 years) and partly due to the understandable requirement for corporate confidentiality. Whilst literature reviews provide an overview of the status of the diversity of chemical series and biological properties that have been studied, they almost certainly represent only a small fraction of the research that is still locked away inside confidential corporate files. There have been several anecdotal reports of small molecules identified from a 'combinatorial chemistry' approach in preclinical development in large established pharmaceutical companies and in the newer 'biotech' sector but there are very few refereed publications in the chemical literature that can be cited. Eli Lilly, Trega Biosciences, Corvas International and Ontogen Corporation are reported to have compounds that progressed to clinical trials from a library approach.

### Lead Generation Compound Libraries

Many of the early compound libraries available for screening for new lead structures were based upon the 'mix and split' peptide libraries (described earlier in this chapter). Since then, medicinal chemists have gained experience that has led to many improvements. The process that is used today in most drug discovery laboratories has many common features. This process and some of the questions raised are summarised in the schematic outline in Figure 17. The overall aim is to identify a chemical lead with *in vitro* activity in a particular screening assay that is amenable to further chemical and biological optimisation. The ability to synthesise and screen huge numbers of compounds has led to a corresponding requirement for

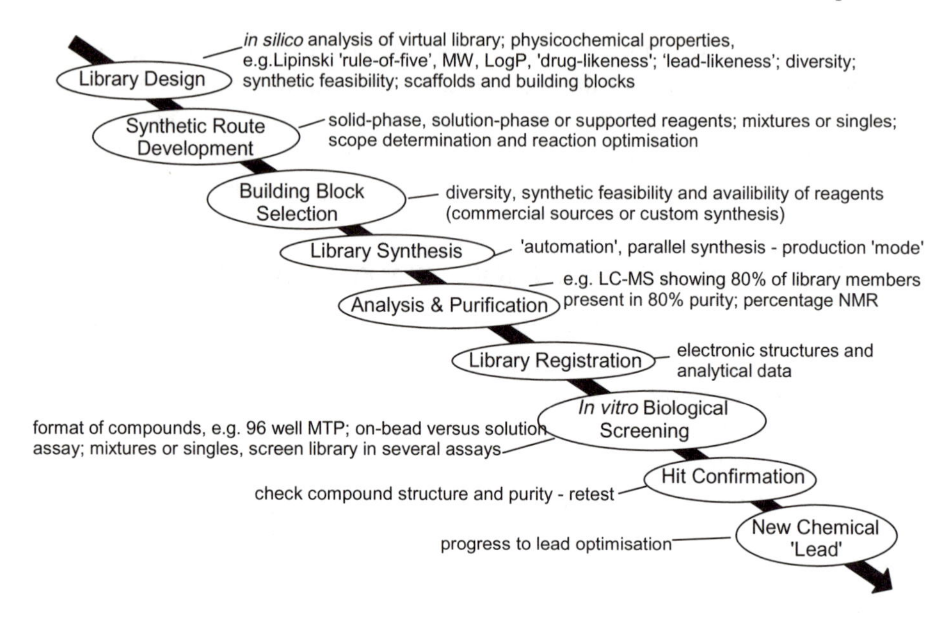


**Figure 17** *Schematic outline of lead identification using compound libraries*

increasing the capacity of analysis and purification and IT activities associated with, for example, data handling, compound tracking and library design.

General description of 'Lead Identification' libraries:

- target structures are selected according to feasibility of chemistry for library synthesis
- diverse sets of building blocks are selected
- each library screened in several different assays – aim for a 'hit' with only moderate *in vitro* activity
- integrate with molecular modelling for 'diversity' analysis and calculation of general 'drug-like' or 'lead-like' properties; trend is to use *in silico* analysis of virtual library prior to synthesis and to design 'smarter' libraries rather than just larger libraries
- typically obtain low quantity ($<10$ mg) and purity (70%) and partial characterisation
- fully automated or semi-automated synthesis techniques allowing thousands of compounds per chemist per year.

## Lead Optimisation Compound Libraries

During lead optimisation, the aim is to synthesise analogues of an existing lead structure to elucidate the structure activity relationships (SAR) and to improve biological activity *in vitro* (*e.g.* affinity, efficacy, selectivity) and *in vivo* (*e.g.* behavioural pharmacology, metabolism, kinetics). The challenge for the medicinal chemist is to speed up the traditional 'synthesis and testing' cycle, *i.e.* to decide which particular structures to synthesise for testing in a specific biological assay in order to deduce how the physicochemical properties correlate with biological activity.

General description of 'Lead Optimisation' libraries:

- the target structures are closely related to the biologically active 'hit'
- similar building blocks are used (rather than a very diverse set)
- improve compounds to optimise specific biological activity *e.g.* find better pharmacokinetic profile
- integrate with molecular modelling, *e.g.* for pharmacophore, QSAR and structure-based design
- typically aim for traditional quantities (*e.g.* >50 mg), purity (*e.g.* >90%) and characterisation of final compound
- parallel or semi-automated solid phase or solution phase techniques with resin bound scavengers allowing, *e.g.* a hundred compounds per chemist per year
- provides many structures for patent exemplification.

### Examples of Library Structures Demonstrating Biological Activity

Compound libraries are being used to identify new lead series with *in vitro* binding affinity for biological targets and to optimize *in vivo* activity and structure activity relationships of existing leads.

A clear and comprehensive overview of the chemical structures and biological activity of published compound libraries is available in 'graphical abstract' format in three literature review articles.[10] The structures shown below (Figure 18) and the primary literature references in which they appear are selected from one of the above review articles (Dolle, 1999).

## 9 CONCLUSIONS

Combinatorial chemistry refers to a box of tools available to the medicinal chemist rather than one defined technique. The range of synthetic techniques spans solid-phase to solution-phase and resin bound reagents, mixtures to singles, automated to manual and everything part way in between.

The impact of CC is not only on the number of compounds synthesised and analysed. There have been corresponding developments in IT for

- **protease inhibitors**: e.g. diketopiperazines **1** as collagenase inhibitors (metalloprotease), statine analogues **2** as cathepsin D or plasmepsin-II inhibitors (aspartic acid protease), and several series of thrombin inhibitors **3** (serine protease).

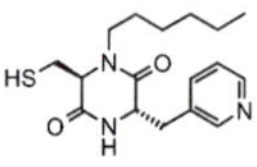

**1** Collagenase-1 inhibitor $IC_{50} = 2\ \mu M$

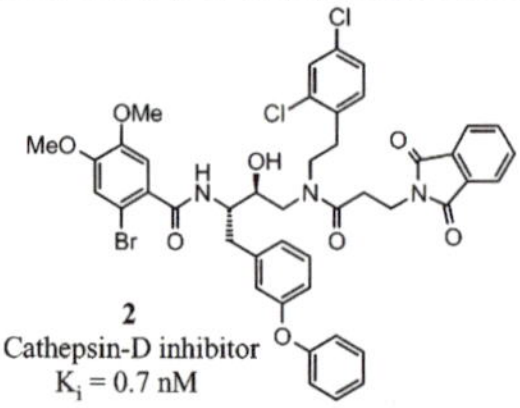

**2** Cathepsin-D inhibitor $K_i = 0.7$ nM

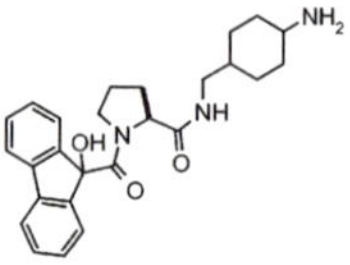

**3** Thrombin inhibitor $K_i = 1.5$ nM

- **non-proteolytic enzyme inhibitors**: e.g. phenothiazines as cyclooxygenase-2 inhibitors **4**, hydroxamic acid derivatives as phosphodiesterase PDE-4 inhibitors **5**, and pyrimidones as reverse transcriptase inhibitors **6**.

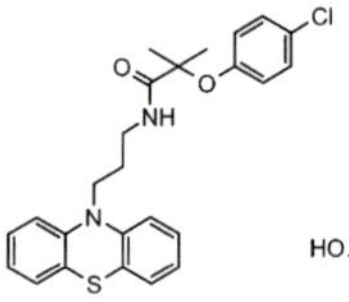

**4** Cyclooxygenase-2 inhibitor $IC_{50} = 1.3\ \mu M$

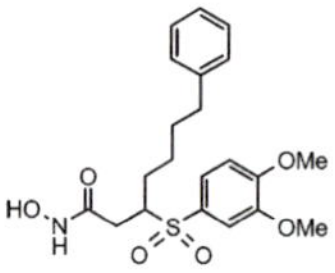

**5** Phosphodiesterase-4 inhibitor 1.0 nM

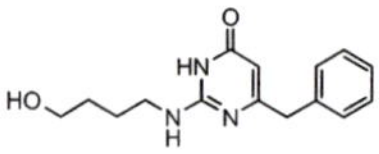

**6** HIV-1 reverse transcriptase inhibitor $K_i = 75\ \mu M$

- **G-protein coupled receptor (GPCR) ligands**: e.g. tetrapeptides or aryl piperazines **7** as kappa opioid agonists and antagonists, 'peptidomimetics' as somatostatin receptor subtype selective agonists and antagonists **8**.

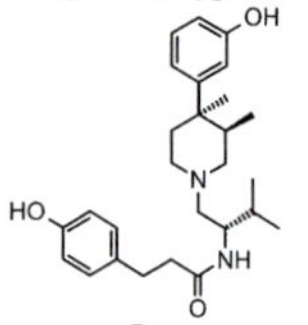

**7** Kappa opioid antagonist $K_i$ 6.9 nM

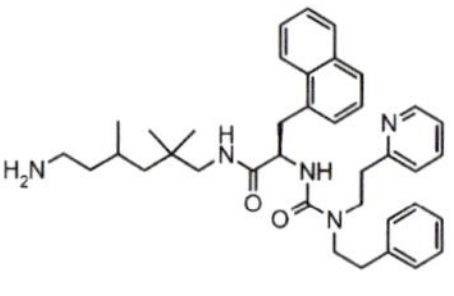

**8** Somatostatin-1 $K_i = 1.4$ nM

- **Modulators of non-G-protein coupled receptors**: e.g. dihydropyridines **9** as calcium channel modulators, polyamines **10** as putrescine uptake inhibitors, peptidomimetics with SH3 domain interaction activity, N-aryl tyrosine derivatives **11** as nuclear receptor peroxisosome proliferator-activated receptor gamma (PPARγ) modulators.

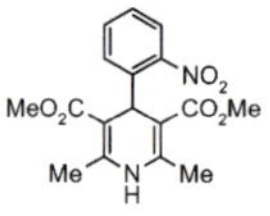

**9** Calcium channel blocker $IC_{50} = 12$ nM

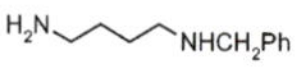

**10** Putrescine uptake inhibitor $IC_{50} = 14\ \mu M$

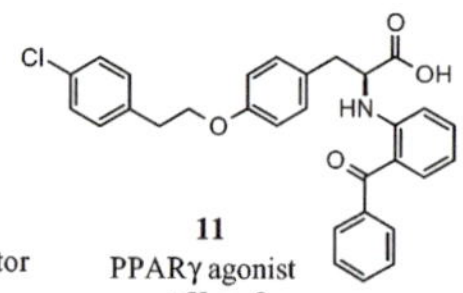

**11** PPARγ agonist $pK_i = 8$

- **Cytotoxic and Antimicrobial activity**: e.g. taxol analogues with microtubule activity, pyridinopolyamine derivatives **12** with antimicrobial activity against *S. pyrogenes*, quinolones and triazines **13** with activity against *S. aureus*.

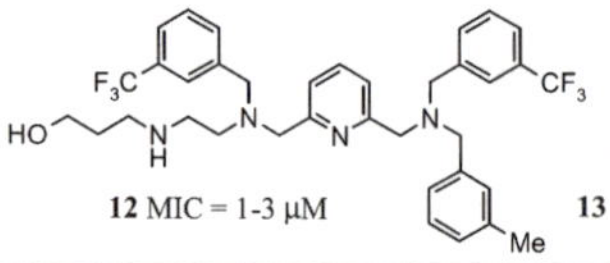

**12** MIC = 1-3 μM

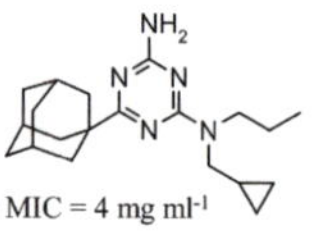

**13** MIC = 4 mg $ml^{-1}$

**Figure 18** *Examples of library structures demonstrating biological activity*

compound registration, *in silico* screening and informatics. In the scientific literature CC is recognised with its own journals (*e.g. Molecular Diversity* and the American Chemical Society's *Journal of Combinatorial Chemistry*), and on the commercial side there are several new and highly successful companies which specialise in producing compound libraries, equipment, consumables or software.

Medicinal chemists have applied these techniques to both lead identification and lead optimisation. In the former, there is a drive for numbers although design and molecular modelling play a role in measuring diversity and drug likeness; in the latter, there remains a preference for single, pure compounds of known structure. The integration of these activities has established CC as a useful tool for drug discovery which complements the previously established 'traditional' techniques. Indeed, if this chapter could make only one point, it would be that for 'combinatorial chemistry' to be fully utilised by medicinal chemists, it is essential that it is *integrated* with existing established techniques. As the boundary between these two approaches becomes increasingly blurred it becomes increasingly desirable for all medicinal chemists to appreciate and apply both.

## 10 REFERENCES

1. I.W. James, *Tetrahedron*, 1999, **55**, 4855.
2. J.J. Baldwin *et al.*, *J. Am. Chem. Soc.*, 1995, **117**, 5588.
3. S. Booth *et al.*, *Tetrahedron*, 1998, **54**, 15385; B.A. Bunin, *The Combinatorial Index*, Academic Press, San Diego, 1998; D. Obrecht and J.M. Villalgordo, *Solid-Supported Combinatorial and Parallel Synthesis of Small-Molecular-Weight Compound Libraries*, *Tetrahedron Organic Chemistry Series*, J.E. Baldwin and R.M. Williams (Series eds.), Pergamon, Elsevier Science, Oxford, 1998; *Solid-Phase Organic Synthesis*, K. Burgess (ed.), John Wiley and Sons Inc., New York, 2000.
4. N.K. Terrett, *Combinatorial Chemistry*, Oxford University Press, Oxford, 1998.
5. C.N. Selway and N.K. Terrett, *Bioorg. Med. Chem.*, 1996, **4**, 645.
6. S.W. Kaldor *et al.*, *Tetrahedron. Lett.*, 1996, **40**, 7193; D.L. Flynn *et al.*, *Curr. Opin. Drug Discovery Dev.*, 1998, **1**, 41.
7. M.W. Creswell *et al.*, *Tetrahedron*, 1998, **54**, 3983.
8. R.W. Armstrong *et al.*, *J. Am. Chem. Soc.*, 1996, **118**, 2574.
9. N.W. Hird, *Drug Discovery Today*, 1999, **4**, 265.
10. R.E. Dolle, *Mol. Diversity*, 1998, **3**, 199; R.E. Dolle and K.H. Nelson, Jr., *J. Comb. Chem.*, 1999, **1**, 235; C.D. Floyd *et al.*, *Prog. Med. Chem.*, F.D. King and A.W. Oxford (eds.), 1999, **36**, 91.

CHAPTER 17

# The Identification of Selective 5-HT$_{2C}$ Receptor Antagonists: A New Approach to the Treatment of Depression and Anxiety

STEVEN M. BROMIDGE

## 1 INTRODUCTION TO DEPRESSION

Although depression barely existed as a medical condition 50 years ago, according to the World Heath Organisation's 'Global Burden of Disease Study' it will become the world's second most debilitating disease by 2020, surpassed only by cardiovascular diseases.[1] Already, 3% of the population are suffering from major depression at any given time and more than 15% are expected to be afflicted at some stage in their lives. Major depression is overwhelming, incapacitating and very different in severity from everyday unhappiness.[2] It can also be fatal with approximately 15% of severely depressed patients committing suicide and two-thirds contemplating suicide. In addition to the devastating impact on individuals and their families, the illness is also a huge burden to society at large. In the USA the annual cost of depression has been estimated at $44 billion of which medical treatment accounts for $12 billion with most of the remainder a result of the lower productivity of affected workers.

The first generation of so called "tricyclic" antidepressants such as imipramine and amitryptiline are plagued by serious side-effects in addition to the risk of suicide by overdose. Over the last decade, the "selective serotonin reuptake inhibitors" (SSRIs), such as fluoroxetine and paroxetine, have, to a large extent, superceded the tricyclics as they are more effective and have reduced toxicity. However, the SSRIs have their own particular pattern of adverse reactions and new agents with less side-effects which will act more rapidly and effectively are still required.

**Figure 1** *Structures of 5-HT and mCPP*

## 2 RATIONALE FOR 5-HT$_{2C}$ ANTAGONISTS IN DEPRESSION

5-Hydroxytryptamine (5-HT or serotonin, Figure 1) has many diverse actions throughout the body and is an important neurotransmitter in the central nervous system. Multiple discrete receptor subtypes mediate these different actions and, so far, progress in receptor pharmacology and molecular biological techniques has led to the identification of 14 distinct 5-HT receptors belonging to 7 superfamilies (5-HT$_1$ to 5-HT$_7$).[3] With the exception of the 5-HT$_3$ receptor, which is a ligand-gated cation channel, all of these are G-protein coupled receptors (GPCRs). Abnormalities in 5-HT levels are known to be involved in a variety of disorders such as anxiety, depression and migraine.[4] Therefore, drugs which either stimulate (agonists) or block (antagonists) 5-HT receptors have been useful for treating these diseases. However, many of these drugs act at multiple receptors, which cause a variety of undesirable side-effects. Consequently, many pharmaceutical companies have focused on identifying agents that are selective for particular 5-HT receptor subtypes. Several lines of evidence suggest that selective antagonists of the 5-HT$_{2C}$ receptor may be useful in the treatment of anxiety and depression. For instance, the moderately selective 5-HT$_{2C}$ agonist *meta*-chlorophenylpiperazine (mCPP), which is a metabolite of the antidepressant trazodone, has been found to cause anxiogenisis in both animals and humans.[5]

## 3 INITIAL LEAD: IDENTIFICATION OF SB-200646

The chemical programme began in 1989 with the specific aim of identifying orally active 5-HT$_{2C}$ antagonists with selectivity over other receptors, in particular the very closely related 5-HT$_{2A}$ receptor. To aid us in this goal, separate 5-HT$_{2A}$, 5-HT$_{2B}$ and 5-HT$_{2C}$ receptor radioligand binding assays were established, initially a combination of rat cloned and native tissue receptor binding assays but later human cloned receptors expressed in immortalised cells. *In vivo* activity was assessed in rats by measuring the ability of compounds to block the decrease in normal

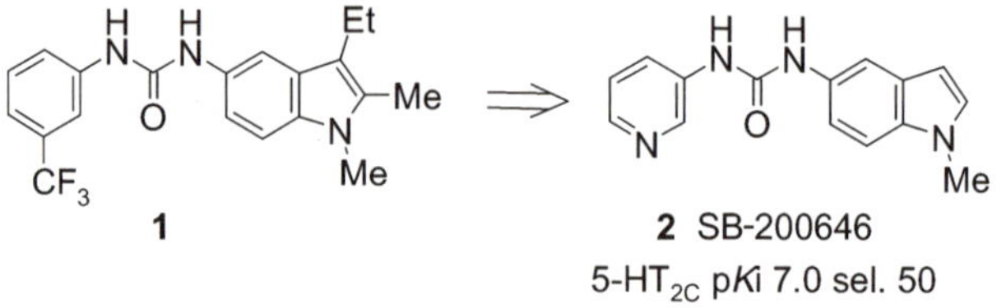

**Figure 2** *Discovery of 2*

exploratory behaviour (hypolocomotion) caused by administration of the 5-$HT_{2C}$ agonist mCPP, activity expressed as an $ID_{50}$, the dose required to reverse by 50% the effect of a set dose of mCPP.

Our initial lead was the 3-(trifluoromethyl)phenyl urea **1** which had been reported by the group at Lilly to block the 5-HT-induced contractions in isolated rat stomach fundus with about 100-fold greater potency than in a 5-$HT_{2A}$ functional preparation of rat jugular vein.[6] At that time the rat stomach fundus was thought to be a model of 5-$HT_{2C}$ activity (then classified as the 5-$HT_{1C}$ receptor); however, this model is now known to be 5-$HT_{2B}$ mediated. On synthesis, **1** proved to be too insoluble to test in our receptor binding assays. To increase aqueous solubility and to explore structure activity relationships (SAR), a range of indole analogues was synthesised in which the phenyl was replaced with pyridyl. From this investigation it was confirmed that the 3-pyridyl and 5-indolyl substitution were optimal. The best compound was **2**, SB-200646, which had moderate affinity for the 5-$HT_{2C}$ receptor (p$K_i$ 7) and 50-fold selectivity over 5-$HT_{2A}$ in the binding assays (Figure 2).[7] Encouragingly, **2** also had *in vivo* activity in the rat hypolocomotion model with an $ID_{50}$ of 20 mg kg$^{-1}$ when dosed orally.

## 4 CONFORMATIONAL RESTRICTION: IDENTIFICATION OF SB-206553

A combination of molecular modelling and NMR studies indicated several possible conformations for **2** resulting from rotation around the four C–N single bonds of the urea. In order to define the active conformation, we investigated conformational restraint by cyclisation onto either the 4- or 6-position of the indole ring (Figure 3). Replacement of one of the polar urea NH groups was also expected to have a beneficial effect on the ability of the compounds to cross the blood–brain barrier and therefore to increase brain penetration. The 'angular' five-membered constrained indoline **3** had about the same level of affinity (p$K_i$ 7.3) for the 5-$HT_{2C}$ receptor as the parent urea **2**. In contrast, the corresponding linear isomer **4**, SB-206553,

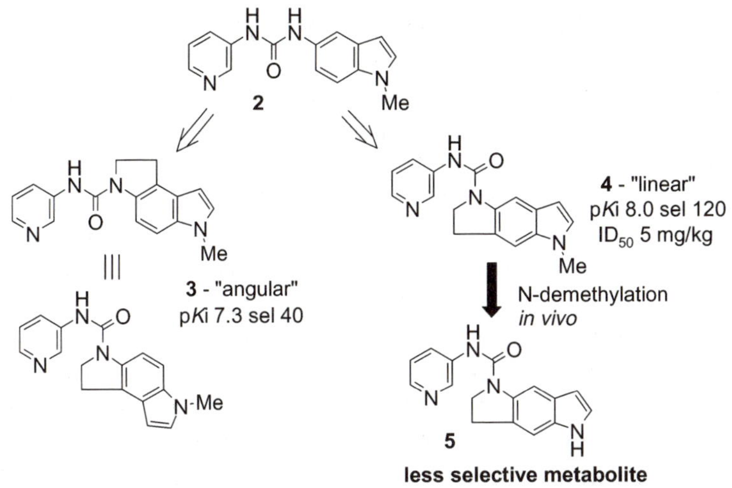

**Figure 3** *Conformational restriction to give 4*

gave a near 10-fold increase in affinity (p$K_i$ 8.0) with more than 100-fold selectivity over the 5-HT$_{2A}$ receptor.[8] In addition, **4** was confirmed to be a competitive antagonist in a functional model of receptor activation (phospholipase C) and was more potent *in vivo* than **2** with an ID$_{50}$ of 5.5 mg kg$^{-1}$ in the rat hypolocomotion model. Although described as cyclisation onto the 4- and 6-position of the indole, **3** and **4** can also be regarded as positional isomers with respect to the fused *N*-methylpyrrole ring.

Compound **4** was also shown to have anxiolytic activity in animal models and was free of the unwanted side-effects which are seen with existing therapies, such as sedation, rebound anxiety on withdrawal of treatment and interaction with alcohol and other drugs. These exciting results provided strong support for our original hypothesis that 5-HT$_{2C}$ antagonists would be useful anxiolytic agents. Unfortunately, the indole *N*-methyl group of SB-206553 was metabolically cleaved to produce the corresponding NH-indole **5** (Figure 3), which retained 5-HT$_{2C}$ receptor potency but had poor selectivity over other receptors.[9] The development of SB-206553 was therefore terminated and strategies to overcome this issue were initiated.

## 5 MOLECULAR AND RECEPTOR MODELLING STUDIES

In order to assist the strategy of finding a replacement for the *N*-methylindole, we investigated possible binding modes within a model of the 5-HT$_{2C}$ receptor. A 3-dimensional computational model of the 5-HT$_{2C}$

receptor was built from its primary sequence by analogy to the then known structure of bacteriorhodopsin. Possible ways that our active molecules could dock into the 5-$HT_{2C}$ receptor model were investigated in order to aid the definition of the active conformation and binding properties. Standard agonists and antagonists for 5-HT receptors have a strongly basic nitrogen that binds to a highly conserved aspartate residue on the third transmembrane helix (TM3).[11] However, the pyridyl nitrogen of these compounds is only weakly basic, so an alternative binding mode in which the carbonyl group of the urea group, essential for activity, forms hydrogen bonding interactions with two serine residues on TM3 was proposed. One of these serine residues (Ser-312) is unique to 5-$HT_2$ receptors. Energy minimisations then placed the two aromatic groups into two well-defined hydrophobic pockets in the receptor model (Figure 4).

In a complementary approach, we applied small molecule modelling to overlap both active and inactive compounds using the urea group as the common overlapping feature. The space occupied by each molecule was then analysed and compared with its affinity for both the 5-$HT_{2C}$ and 5-$HT_{2A}$ receptors. This allowed us to define an allowed volume at the 5-$HT_{2C}$ receptor (Figure 5; shown in light grey), which should be similar to the size and shape of the binding pocket. We were also able to define a small part of this 5-$HT_{2C}$ allowed volume, near the *N*-methyl group, that was disallowed for 5-$HT_{2A}$ receptor activity (Figure 5; shown in dark grey). Thus, compounds with substituents that interact with this crucial region maintained 5-$HT_{2C}$ affinity but had reduced 5-$HT_{2A}$ affinity and increased selectivity. This 'active analogue approach' is based on the assumption

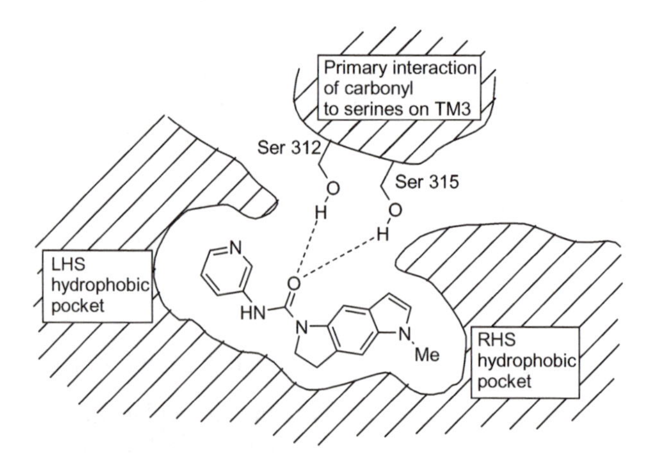


**Figure 4** *Schematic representation of **4** docked into a model of the 5-$HT_{2C}$ receptor*

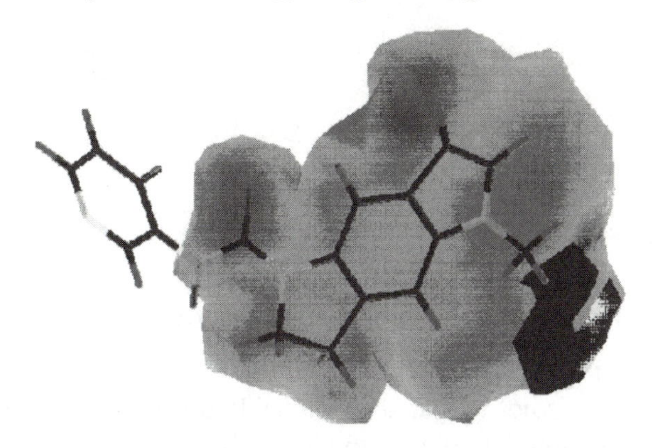

**Figure 5** *5-HT$_{2C}$ allowed volume (light grey) which includes a 5-HT$_{2C}$ disallowed volume (dark grey) viewed around the structure of **4***

that weakly active compounds have substituents in sterically disallowed areas of the relevant binding site.[13] Encouragingly, these results agreed with the receptor modelling which suggested that the RHS hydrophobic binding pocket in the 5-HT$_{2A}$ receptor was smaller than the corresponding pocket in the 5-HT$_{2C}$ receptor due to key amino acid differences in the respective protein sequences. We were able to use this combination of small molecule and receptor protein modelling in order to guide the design of further improved 5-HT$_{2C}$ receptor antagonists.

## 6 BIOISOSTERIC REPLACEMENT OF THE *N*-METHYLINDOLE

The replacement of the *N*-methylindoline by alternative *N*-alkyl groups and other heterocycles either lost potency and/or selectivity, or offered no metabolic stability. As an alternative, the modelling suggested that the fused pyrrole group was simply occupying space which could be similarly occupied by two *ortho*-groups. In order to explore the required properties of the substituents, rather than preparing analogues of **4**, analogues of the simpler **2** were prepared. This was because many anilines are commercially available to enable the rapid synthesis and evaluation of a large number of compounds, the aim being to identify the more favourable substituents which could then prepared as the equivalently substituted indolines (Figure 6).[14] Over 100 biaryl ureas were rapidly prepared by parallel reaction incorporating a diverse range of substituted anilines. As a consequence of the relatively free rotation about the *N*-aryl bond of the phenyl urea, the 3- and 5-positions are equivalent. However, for the equivalent indolines they are non-equivalent and therefore both the 4- and 6-isomers were initially prepared.

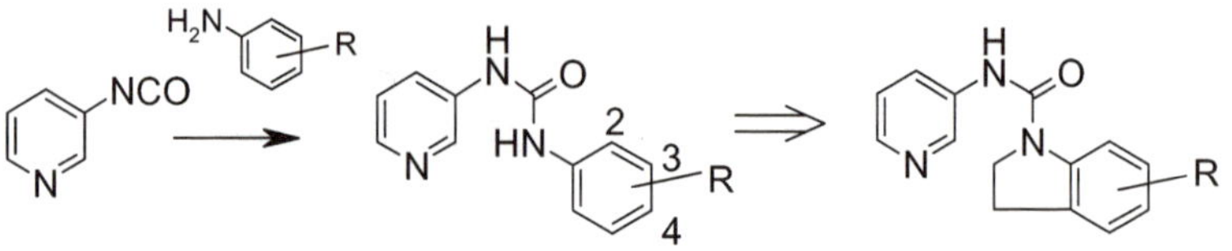

Figure 6 *Identification of metabolically stable isosteres of* **4**

From these studies the following conclusions were drawn:

- 2-substitution is not tolerated, presumably due to conformational twist
- 3,4-disubstitution is preferred to monosubstitution
- a 3-electron withdrawing group (6- in the indoline) gives higher potency
- a small lipophilic 4-substituent gives high potency, consistent with allowed volume map
- the optimised substitution is 3-$CF_3$-4-SMe.

Using this approach, phenyl ureas with improved 5-$HT_{2C}$ affinity and selectivity over **2** were rapidly identified and also key insights into QSAR gained. For example, within a series of 3-chloro-4-substituted analogues we found a good correlation between 5-$HT_{2C}$ affinity and the lipophilicity ($\pi$) of the 4-substituent (Table 1; Figure 7).

The preferred substituents in the phenyl urea series were then prepared as the equivalently substituted indolines. Initially, a small selection of

**Table 1** *The 5-$HT_{2C}$ affinities and selectivities of biaryl ureas*

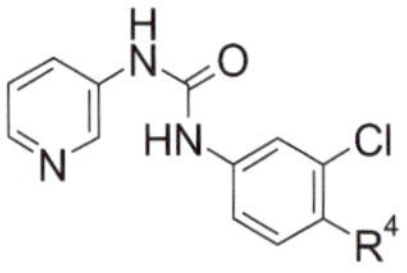

| $R^4$ | $\pi$ of $R^4$ | 5-$HT_{2C}$ ($pK_i$) | Selectivity 5-$HT_{2C/2A}$ |
|---|---|---|---|
| $SO_2Me$ | −1.63 | 5.0 | – |
| $CONH_2$ | −1.49 | 5.1 | – |
| $CH_2OH$ | −1.03 | 5.5 | – |
| CN | −0.57 | 5.9 | – |
| COMe | −0.55 | 6.6 | 25 |
| COOH | −0.32 | 7.1 | 80 |
| OMe | −0.02 | 6.7 | 30 |
| Me | 0.56 | 7.8 | 40 |
| SMe | 0.61 | 7.5 | 120 |
| Cl | 0.71 | 7.5 | 50 |

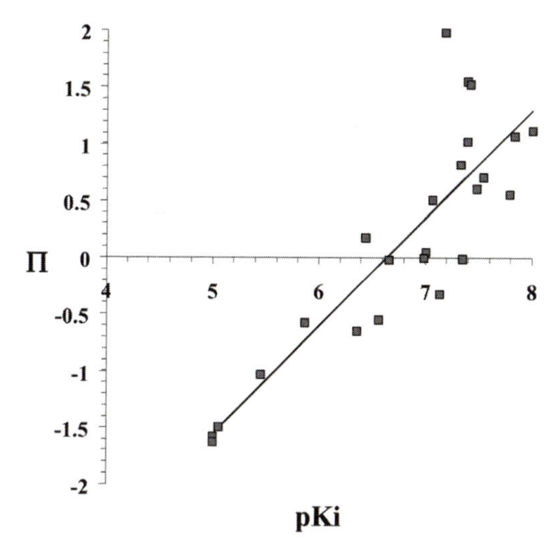


**Figure 7** *Correlation of lipophilicity at C-4 with pK$_i$*

substituted indolines predicted to give a range of activities was prepared in order to validate the SAR correlation between the two series. A plot of the 5-HT$_{2C}$ affinities of corresponding compounds from the two series (Figure 8) shows that, within the error of the primary binding assay, a good correlation existed with an average increase in p$K_i$ of ~0.5 for the indolines. Especially outstanding were several high affinity and selective 5-thioalkyl-6-trifluoromethylindolines (Table 2).

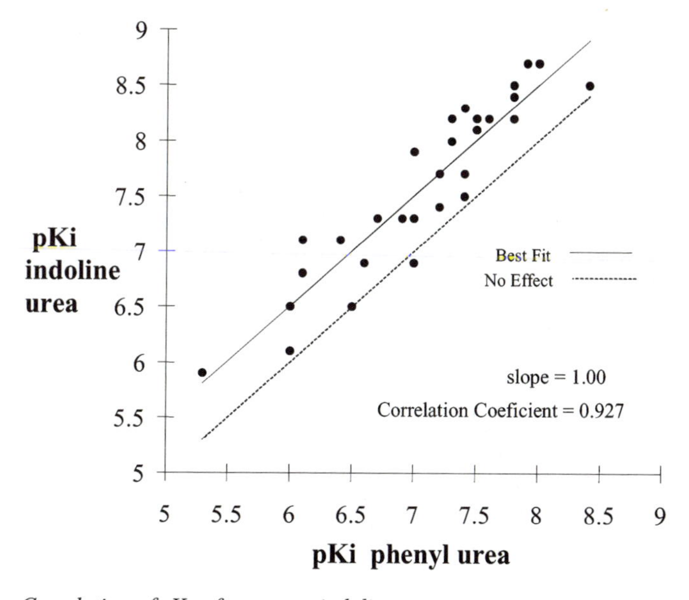


**Figure 8** *Correlation of* pK$_i$ *of ureas* vs. *indolines*

**Table 2** *Activity of 5-thioalkyl-6-$CF_3$ indolines*

| $R^5$ | 5-$HT_{2C}$ p$K_i$ | Selectivity 5-$HT_{2C/2A}$ | $ID_{50}$ vs mCPP ($mg\,kg^{-1}$ p.o.) |
|---|---|---|---|
| OMe | 8.0 | 120 | 0.8 |
| SMe[a] | 8.6 | 160 | 1.5 |
| SEt | 8.5 | 1000 | 4.4 |
| S$^n$Pr | 8.2 | >1000 | <5 |

[a] **6** (SB-221284).

Probing the key 5-$HT_{2C}$ allowed/5-$HT_{2A}$ disallowed region, identified from the modelling studies, by increasing the size of the 5-thioalkyl substituent gave an increase in selectivity but reduced oral activity in the rat hypolocomotion model. The thiomethyl analogue **6**, SB-221284 was 4-fold more potent at the 5-$HT_{2C}$ receptor than **4** with similar selectivity over the 5-$HT_{2A}$ receptor. In addition, **6** had improved oral activity in the rat hypolocomotion model and good activity in animal models of anxiety. Based on its favourable overall profile **6** was selected for pre-clinical development.

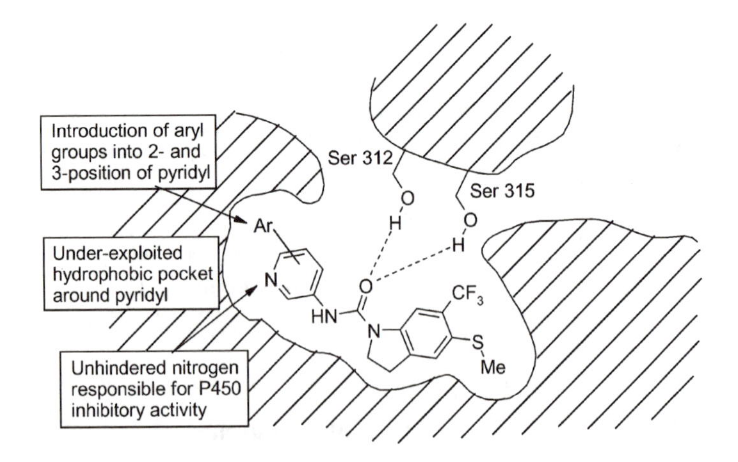


**Figure 9** *Evolution of (aryl-pyridylcarbamoyl) indolines from* **6** *SB-221284*

## 7 IDENTIFICATION OF BIARYLCARBAMOYLINDOLINES

Unfortunately, SB-221284 and related analogues were found to be potent inhibitors of a number of important human cytochrome P450 enzymes *in vitro* (see later, Table 4).[15] The P450 superfamily of enzymes all contain a porphyrin-haem catalytic centre and are pivotal in the oxidative metabolism and clearance of the majority of drugs.[16] Therefore, the development of SB-221284 was terminated and alternative candidates free from P450 inhibitory liability were sought. All compounds of interest were now screened against the five major human P450 enzymes to assess their potential to interact with the metabolism of other drugs.

SAR around the P450 inhibition indicated that the pyridyl group was the key pharmacophoric element, presumably through an interaction with the iron of the haem. Attention was therefore focused on alternatives to the simple 3-substituted pyridine. A number of substituted pyridyl analogues of **6**, in particular *ortho-* to the nitrogen, had greatly reduced P450 inhibitory activity and yet this increased steric bulk was tolerated in the binding pocket of the target receptor and 5-$HT_{2C}$ affinity was maintained. To guide this work, ligand docking of SB-221284 into our model of the 5-$HT_{2C}$ receptor showed that the previously identified hydrophobic pocket that the 3-pyridyl ring occupied was in fact quite deep, suggesting that introducing further aromatic substituents could increase binding interactions and also maintain the reduced P450 inhibitory potency (Figure 9).

The substitution of the pyridyl ring with a variety of aryl groups in combination with the previously optimised 5-methylthio- and 5-methoxy-6-trifluoromethylindoline[1] gave compounds such as **7**, which had improved affinity for the 5-$HT_{2C}$ receptor with much reduced inhibitory activity at the problem P450 1A2 enzyme (Table 3). Unfortunately, **7** was less than 100-fold selective over 5-$HT_{2A}$ receptors but the introduction of a 4-methyl substituent into the terminal pyridyl ring **8**, SB-234985, gave an increased 5-$HT_{2C}$ affinity and 250-fold selectivity over 5-$HT_{2A}$. We believe that the 4-methyl substituent increases the torsion angle between the pyridyl rings resulting in an orthogonal conformation which favours binding at the 5-$HT_{2C}$ receptor but not at the 5-$HT_{2A}$ receptor. Although SB-234985 had the highest 5-$HT_{2C}$ affinity we had yet seen (p$K_i$ 9.4), it showed only modest activity in the rat hypolocomotion model and too high a P450 inhibitory activity, particularly against the 1A2 enzyme (Table 3). It seemed likely that the relatively unhindered central pyridyl ring was responsible for this remaining P450 activity. Therefore, **9** was prepared in which the terminal 3-pyridyl group was moved from the 5- to the 6-position to increase the steric crowding of the central pyridine nitrogen.

**Table 3** *Activity of bisarylcarbamoylindolines*

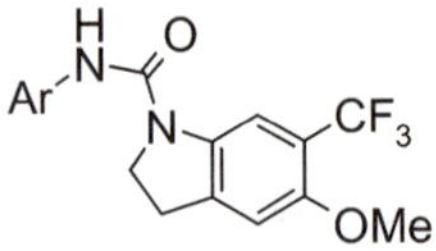

| *Cmpd.* | *Ar* | $pK_i$ $5\text{-}HT_{2C}$ | *5-HT selectivity* 2C/2A | 2C/2B | $ID_{50}$ *(po)* vs *mCPP* | $ID_{50}$ vs *CYP1A2* (μM) |
|---|---|---|---|---|---|---|
| 7 | 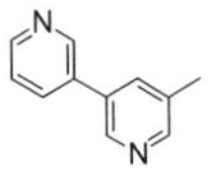 | 9.1 | 80 | 12 | – | ~5 |
| 8 | 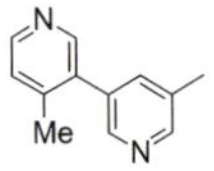 Me | 9.4 | 250 | 15 | 5 mg kg$^{-1}$ | 0.2 |
| 9 | 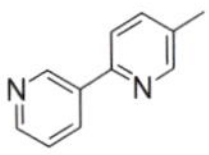 | 8.3 | >2000 | 2 | Inactive | >100 |
| 10 | 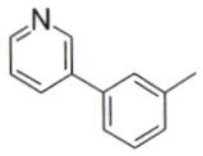 | 9.0 | 200 | 8 | 0.6 mg kg$^{-1}$ | 4 |
| 11[a] | 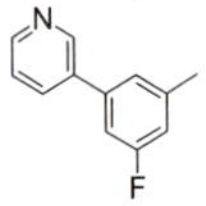 F | 9.0 | 130 | 10 | 0.7 mg kg$^{-1}$ | 28 |

[a] SB-228357.

Although this successfully reduced the P450 1A2 activity, it also resulted in reduced 5-HT$_{2C}$ affinity and abolished *in vivo* activity in the hypolocomotion model.

A more successful approach was to replace the central 3-amidopyridine with phenyl. The resultant compound **10** maintained excellent 5-HT$_{2C}$ affinity and selectivity over 5-HT$_{2A}$ with much reduced P450 activity and potent activity in the rat hypolocomotion model. However, time course studies revealed that the duration of activity of **10** in this model was less than 1 h following oral dosing at a dose of 3 times the ID$_{50}$, our biological measure of duration of action.

Hypothesising that this short duration of action of **10** may be due to metabolism of the relatively electron rich phenyl ring, a series of analogues was prepared containing additional substituents in this ring to block this metabolism. The best of these compounds was **11**, SB-228357, which had a similar profile to **10** but with an improved P450 profile (Table 4) and an excellent 6 h duration of action in the hypolocomotion model. SB-228357 also had good activity in animal models of anxiety and

**Table 4** *The inhibitory activity of selected lead compounds against major human cytochrome P450 enzymes*

| | $IC_{50}$ (μM) | | | | |
|---|---|---|---|---|---|
| *Cmpd.* | *1A2* | *2C9* | *2C19* | *2D6* | *3A4* |
| **6**[a] | 0.013 | >100 | >100 | 0.11 | >100 |
| **11** | 28 | 82 | >100 | 21 | 9 |
| **14** | >100 | 23 | >100 | >100 | >100 |

[a] SB-221284.

like previous compounds was found to have no side-effects such as sedation, withdrawal and ethanol interaction. This compound was selected for clinical development but, unfortunately, the very low aqueous solubility subsequently gave rise to formulation difficulties that halted its development. With the advantage of hindsight, it was apparent that, by using a pharmacodynamic assay without pharmacokinetic support, we were selecting for low solubility compounds which had good, but prolonged oral absorption.

## 8 BISPYRIDYL ETHERS: IDENTIFICATION OF SB-243213

Although the compounds described above had excellent 5-HT$_{2C}$ affinity and selectivity over the 5-HT$_{2A}$ receptor, they had very little selectivity over the 5-HT$_{2B}$ receptor. Further receptor–ligand modelling work suggested that SB-228357 was still not fully exploiting the LHS hydrophobic pocket within the 5-HT$_{2C}$ receptor and that introducing a linker group between the biaryl rings might more optimally occupy this binding pocket. As part of this work, the bispyridyl ether **12** was prepared and found to have excellent 5-HT$_{2C}$ affinity and almost 100-fold selectivity over 5-HT$_{2A}$ receptors.[17]

Encouraged by this result, a number of analogues, including the 2-methyl substituted analogue **13**, were prepared. This compound had increased 5-HT$_{2C}$ affinity and, not only over 1000-fold selectivity over 5-HT$_{2A}$ receptors, but also selectivity over the 5-HT$_{2B}$ receptor (80-fold) (Table 5). This increase in selectivity is again believed to be due to the steric bulk of the 2-methyl substituent favouring an orthogonal conformation of the pyridyl rings. Both **12** and **13** were found to have clean P450 profiles and also to have good oral activity in the rat hypolocomotion model. However, despite the excellent oral activity of these compounds,

**Table 5** *Activity of bispyridyl ether-carbamoylindolines*

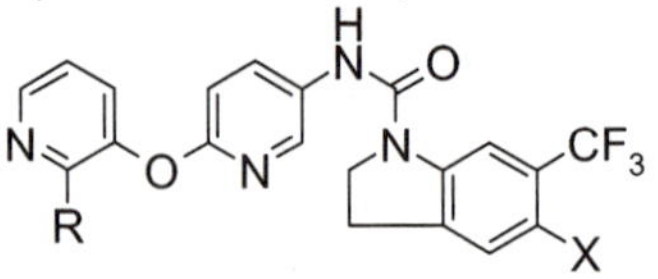

| *Cmpd.* | *R* | *X* | $pK_i$ *5-HT*$_{2C}$ | *5-HT selectivity* *2C/2A* | *2C/2B* | *ID*$_{50}$ *(po)* vs. *mCPP* (mg kg$^{-1}$) | *CYP1A2* (μM) |
|---|---|---|---|---|---|---|---|
| **12** | H | OMe | 8.9 | 80 | 16 | 0.7 | >100 |
| **13** | Me | OMe | 9.2 | 1300 | 80 | 2.8 | >100 |
| **14**[a] | Me | Me | 9.0 | 160 | 100 | 0.7 | >100 |

[a] SB-243213.

there was evidence that the 5-methoxyindoline substituent was metabolically labile. In previous series, the corresponding methyl analogues had shown reduced selectivity over 5-HT$_{2A}$, but for this series it was believed that a reduction of the 1300-fold selectively with **13** could be tolerated. In fact, replacing the 5-methoxy substituent with a methyl to give **14**, SB-243213, which maintained both 5-HT$_{2C}$ affinity and selectivity over 5-HT$_{2A}$ and 5-HT$_{2B}$. In addition, SB-243213 showed low P450 inhibitory activity and potent oral activity (ID$_{50}$ of 1 mg kg$^{-1}$) in the rat hypolocomotion model with a long duration of action.

On further cross-screening, SB-243213 was found to have >100-fold selectivity over more than 50 other receptor, ion-channel and enzyme binding sites. Like previous compounds it also had good activity in animal models of anxiety with no side-effects. Important for the ease of development, SB-243213 also showed considerably improved aqueous solubility (2-orders of magnitude) compared to SB-228357. Based on its excellent overall profile, SB-243213 was selected for progression and is currently in Phase 1 clinical trials in humans.

## 9 SYNTHESIS OF SB-243213

SB-243213 was prepared by the sequence of reactions shown in Figure 10.[17] Treating the sodium salt of 3-hydroxy-2-methylpyridine with 2-chloro-5-nitropyridine gave the nitrobispyridyl ether in 95% yield which was then reduced to the corresponding amine using tin chloride or catalytic hydrogenation. The aminobispyridylether was then coupled with 5-methyl-6-(trifluoromethyl)indoline *via* the phenyl carbamate to give SB-243213 in excellent overall yield.

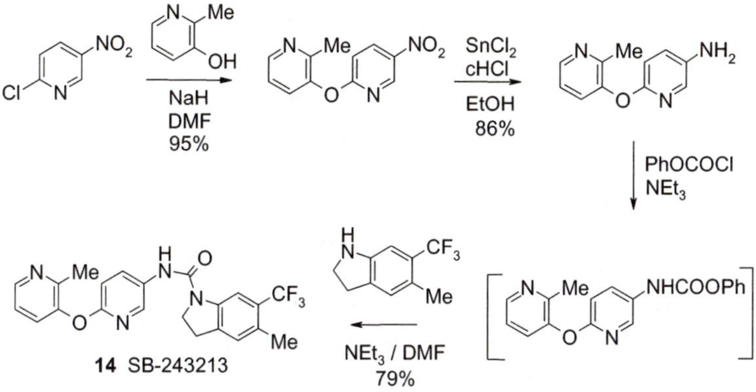

**Figure 10** *Synthesis of **14**, SB-243213*

## 10 SUMMARY

In summary, a series of selective 5-HT$_{2C}$ antagonists has been identified from the initial lead **2**. The first of these, **4** was metabolically dealkylated to give an active and non-selective metabolite. Efforts to overcome this by bioisosteric replacement of an indole led to **6**. However, this compound was found to be a potent cytochrome P450 inhibitor and so a series of biaryl analogues, such as SB-228357, was developed which circumvented this liability. This work also led to the evolution of bispyridyl ethers, in particular SB-243213, which have additional selectivity over the closely related 5-HT$_{2B}$ receptor. These compounds have potent oral anxiolytic activity in animal models with no side-effects such as sedation, withdrawal effects and ethanol interactions. SB-243213 is currently in Phase I clinical development for the treatment of depression and anxiety.

This case history summarises the efforts of many dedicated SmithKline Beecham scientists over a considerable number of years. As well as highlighting a number of important drug discovery and development issues, I hope it gives a flavour of the excitements and all too frequent disappointments that go hand in hand during research efforts of this nature. It also illustrates how very often identifying a pre-clinical development candidate is only a first step to producing a successful drug and frequently it is necessary to further refine the properties of subsequent candidates based on emerging pre-clinical and clinical data.

## 11 REFERENCES

1. C.J. Murray *et al.*, in 'The global burden of disease in 1990: summary results, sensitivity analysis and future directions', *Bull. World Health Organization*, 1994, **72**, 495.

2. L. Wolpert in *Malignant Sadness, The Anatomy of Depression*, Faber and Faber, London, 1999.
3. D. Hoyer *et al.*, *Pharmacol. Rev.*, 1994, **46**, 157.
4. T.P. Blackburn, in *Advances in Neuropharmacology*, F.C. Rose (ed.), Smith-Gordon and Nishimura, 1993, p. 51.
5. R.S. Kahn and S. Wetzler, *Biol. Psychiat.*, 1991, **30**, 1139.
6. P. Fludzinski *et al.*, *J. Med. Chem.*, 1986, **29**, 2415.
7. I.T. Forbes *et al.*, *J. Med. Chem.*, 1993, **36**, 1104.
8. I.T. Forbes *et al.*, *J. Med. Chem.*, 1995, **38**, 2524.
9. S.M. Bromidge *et al.*, *J. Med. Chem.*, 1998, **41**, 1598.
10. D. Julius *et al.*, *Science*, 1988, **241**, 558.
11. H. Luecke *et al.*, *J. Mol. Biol.*, 1999, **291**, 899.
12. K. Kristiansen and S.G. Dahl, *Eur. J. Pharmacol.*, 1991, **40**, 8.
13. G.R. Marshall *et al.*, in *Computer-Assisted Drug Design*, E.C. Olsen and R.E. Christofferson (eds.), ACS, Washington, DC, 1979, p. 205.
14. S.M. Bromidge *et al.*, *Bioorg. Med. Chem.*, 1999, **7**, 2767.
15. S.M. Bromidge *et al.*, *J. Med. Chem.*, 1997, **40**, 3494.
16. A. Smith *et al.*, *Drug Discovery Today*, 1997, **2**, 406.
17. S.M. Bromidge *et al.*, *J. Med. Chem.*, 2000, **43**, 1123.

CHAPTER 18

# The Identification of the HIV Protease Inhibitor Saquinavir

FRANK D. KING

## 1 INTRODUCTION

The human immunodeficiency virus (HIV) was first identified in 1985 as the probable cause of the acquired immunodeficiency syndrome (AIDS) and by 1995 over 2 million people had been infected, the majority of whom were dying.[1] A catastrophic disaster was awaiting on the horizon and in the late 1980s there was a tremendous drive to identify treatments for this terrible disease. Despite recent adverse press, the pharmaceutical industry can feel proud that it has managed to halt the progression of the disease, at least in the western world, and treatment will soon be extended to third world countries where the disaster has continued to spread unchecked for almost two decades.

HIV is a highly mutable lentivirus, a retrovirus whose genome is encoded in a singled-stranded RNA.[2] HIV infects T-lymphocytes and other cells by binding to the CD4 receptor and a second, so-called co-receptor on the cell surface. Following fusion, the RNA undergoes reverse transcription in the cytoplasm and the proviral DNA thus formed migrates to the nucleus and is integrated into the host cell genome by the viral enzyme integrase. Upon cell activation, transcription of the unregulated proviral DNA produces more genomic RNA, which contains the genetic material for new virus particles and also acts as a template for the synthesis of viral proteins necessary for the generation of infective viral particles. A key protein is HIV protease, which cleaves the *gag* ($Pr55^{gag}$) and *gag-pol* ($Pr160^{gag\text{-}pol}$) polyproteins to create the structural proteins and enzymes needed to produce the new viral particles.

HIV protease was shown to be an aspartyl protease from structural and inhibitor studies, and both frame shift mutations and mutation of the aspartic acid residues in the active site led to accumulation of non-infective, immature virus particles. Thus, inhibition of HIV protease became an attractive molecular target for the treatment of HIV infection. In addition, HIV protease is unusual in that it cleaves Tyr-Pro and Phe-Pro sequences in the *gag* and *gag-pol* polyproteins. Mammalian endopeptidases are incapable of cleaving amide linkages N-terminal to proline, thus holding out the promise of inhibitors with high selectivity.

The approach to inhibitor design adopted by the group at Roche,[3] and others,[1] was based upon the transition state (TS) mimetic approach used successfully for other aspartyl proteases, for example renin. Possible TS mimetics for inclusion in inhibitor molecules include statins, reduced amides, hydroxyethylenes, dihydroxyethylenes, difluoroketones, phosphinic acids, hydroxyethylamines and hydroxymethylcarbonyls (Figure 1).

Based upon the structure of the Ar-Pro cleavage site, the Roche group selected to investigate two mimetics that contain a proline-like moiety in the $P_1'$ site, namely the reduced amides and hydroxyethylamine mimetics (Figure 2). However, the reduced amides showed relatively poor potency and so the group concentrated on the hydroxyethylamine mimetics.

statin

reduced amide

hydroxyethylene

dihydroxyethylene

difluoroketone

phosphinic acid

hydroxyethylamine

hydroxymethylcarbonyl

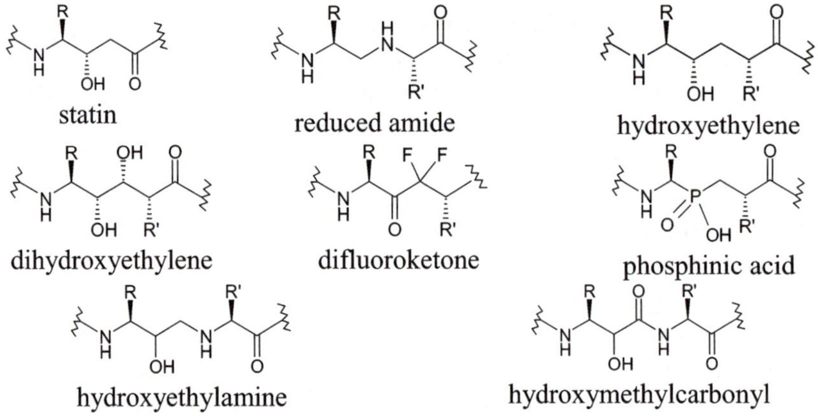

**Figure 1** *Transition state mimetics for aspartyl proteases*

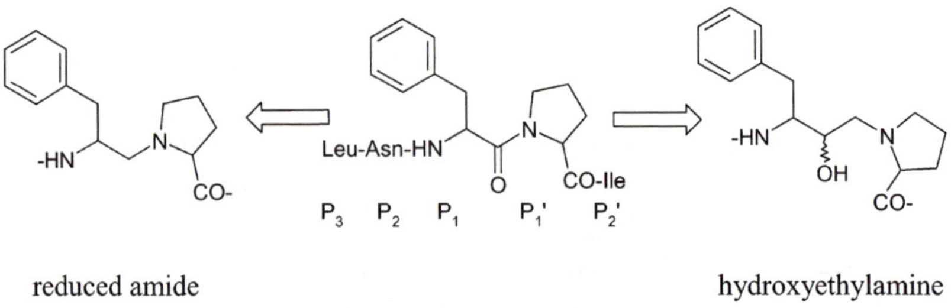

Leu-Asn-HN CO-Ile

$P_3$ $P_2$ $P_1$ $P_1'$ $P_2'$

reduced amide

hydroxyethylamine

**Figure 2** *Structures of HIV substrate and potential inhibitors*

## 2 PRIMARY ASSAY

The primary assay used cloned, expressed and part-purified HIV-1 protease. The preferred substrate was the protected heptapeptide succinyl-Val-Ser-Gln-Asn-Phe-Pro-Ile-isobutylamide. Formation of the cleavage product, H-Pro-Ile-isobutylamide was quantified by spectrophotometric analysis of the blue-coloured adduct formed by reaction of the terminal proline with isatin. The potency of the inhibitors was expressed as an $IC_{50}$, the concentration of inhibitor required to inhibit the enzyme by 50%.

## 3 INHIBITOR DESIGN

### Identification of the Minimum Inhibitor Sequence

The simple protected dipeptide **1**, the basic template for analogue synthesis, had weak, but encouraging potency of 6.5 μM (Table 1).

**Table 1** *The identification of minimum inhibitor sequence*

| *Compound no* | *Stereochemistry at -CHOH-* | *Structure* | $IC_{50}$ (nM) *HIV-1* |
|---|---|---|---|
| **1** | *R*[a] | 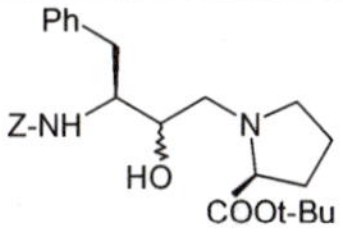 Ph; Z-NH; HO; N; COOt-Bu | 6500 |
| **2** | *R* | Ph; O; ZNH; N H; HO; N; CONH₂; COOt-Bu | 140 |
| **3** | *S* | | 300 |
| **4** | *R*[a] | 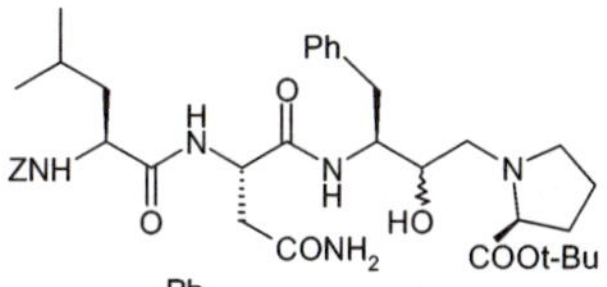 Ph; O; H N; ZNH; O; N H; HO; N; CONH₂; COOt-Bu | 600 |
| **5** | *R*[a] | 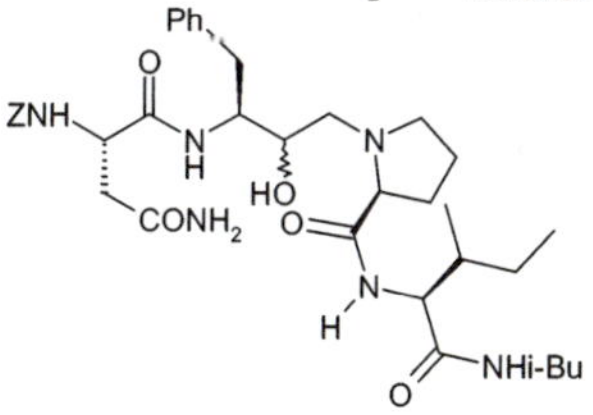 Ph; O; ZNH; N H; HO; N; CONH₂; O; N H; NHi-Bu; O | 130 |
| **6** | *R* | 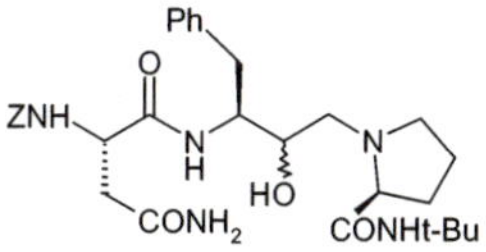 Ph; O; ZNH; N H; HO; N; CONH₂; CONHt-Bu | 210 |

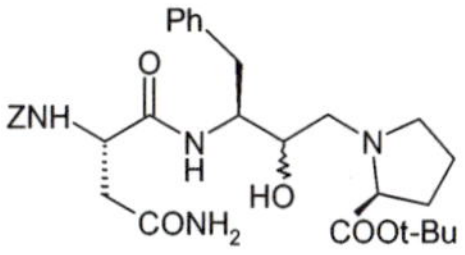

[a] More active diastereomer, probably *R*.

However, addition of the P2 asparagine **2** gave a 40-fold increase in potency, down to 140 nM for the *R* enantiomer. In contrast to the longer peptide inhibitors, such as JG-365, for which the *S* enantiomer was more potent, for this simpler series the *S* enantiomer **3** was less potent. Further elongation by addition of the P3 leucine **4** offered no advantage and **4** was, in fact, less potent than the tripeptide **2**. Similarly, C-terminal extension by addition of the isoleucyl-isobutylamide **5** gave no advantage but did demonstrate that the more metabolically stable C-terminal amide could be tolerated. This was confirmed with the tripeptide tert-butylamide **6** which retained most of the potency of **2**.

## Optimisation of the N-terminus

Over 50 N-terminal analogues were prepared and only a small selection are presented in Table 2. Replacement of the Z group of **2** by acetyl **7** or BOC **8** both resulted in reduced affinity. However, both the 3-phenyl-propionyl **9** and cinnamyl **10** derivatives retained good potency. Confor-

**Table 2** *The optimisation of N-terminus*

| *Compound no* | *Structure R =* | 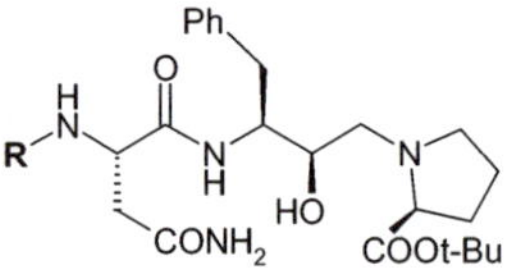 | *IC*$_{50}$ (nM) *HIV-1* |
|---|---|---|---|
| **2** | 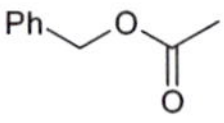 (Z) | | 140 |
| **7** | 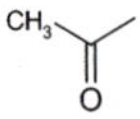 | | 8600 |
| **8** | 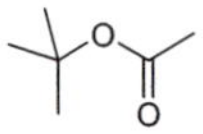 (BOC) | | 8000 |
| **9** | 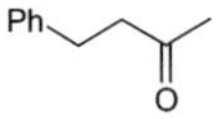 | | 240 |
| **10** | 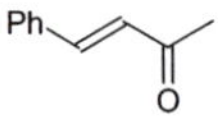 | | 240 |
| **11** | 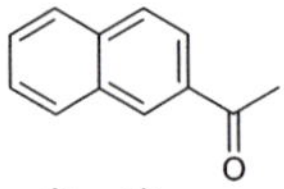 | | 46 |
| **12** | 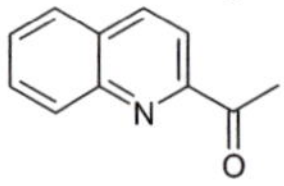 | | 23 |

mational restriction of the cinnamyl as the 2-naphthyl **11** gave an encouraging improvement in affinity which was further enhanced by the now optimised 2-quinolinyl **12**. From an investigation of the $P_1$ and $P_2$ substituents, benzyl and acetamide were found to be optimal, although both methylthiomethyl and cyanomethyl were as potent as the carboxamidomethyl at $P_2$.

## Optimisation of the Proline

A similar systematic investigation of the proline was undertaken with even larger enhancements in potency. Using the amide **6** as the comparator, the slightly larger thiazolidine **13** was found to be an amazing 25-fold more potent than **6**. Larger rings were therefore investigated but both the piperidine **14** and azepine **15** were less potent, which suggested a tight restriction on the directionality of the amide. Structural analysis and molecular modelling studies suggested that neither the proline nor the thiazolidine completely filled the P1′ pocket and therefore larger bicyclic

**Table 3** *Proline optimisation*

| *Compound no* | *Stereo-chemistry at -CHOH-* | *Structure R =* (ZNH–CH(CH$_2$CONH$_2$)–CO–NH–CH(CH$_2$Ph)–CH(OH)–CH$_2$–R) | 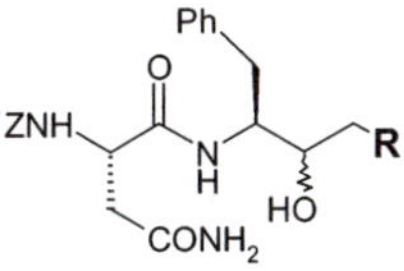 $IC_{50}$ (nM) *HIV-1* |
|---|---|---|---|
| **6** | *R* | 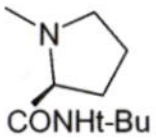 N CONHt-Bu | 210 |
| **13** | *R* | 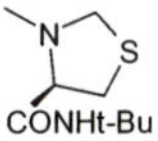 N S CONHt-Bu | 8.4 |
| **14** | *R* | 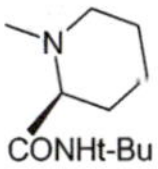 N CONHt-Bu | 18 |
| **15** | *R* | 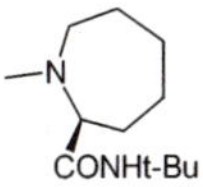 N CONHt-Bu | 92 |
| **16** | *R* | 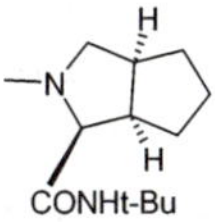 H N H CONHt-Bu | 5.6 |
| **17** | *R* | 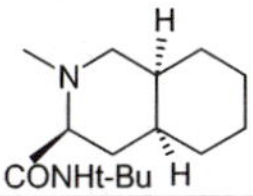 H N CONHt-Bu H | 2.7 |
| **18** | *S* | | ≫100 |

systems were investigated. Both the 5,5 **16** and the 6,6 **17** compounds were more potent, with the decahydroisoquinoline **16** now achieving low single figure nM potency. Interestingly, the *S* enantiomer **18** was found to be essentially inactive.

The optimised N-terminal 2-isoquinolinyl substituent was now incorporated into these more potent molecules and, very satisfyingly, the increases in potency were found to be additive. Thus the 2-isoquinolinyl piperidine **19** had an $IC_{50}$ of 2 nM potency and, when combined with the more potent decahydroisoquinoline, gave **21**, which had a potency below the limit of detection of the assay, but on further investigation was found to have a $K_i$ of 0.12 nM. For both these analogues, again the *S* enantiomers **20** and **22** were markedly less potent. It was also found that extension of the C-terminus resulted in a loss of activity. The potency of **21** (Ro 31-8959) prompted further biological studies to assess its potential as a development candidate.

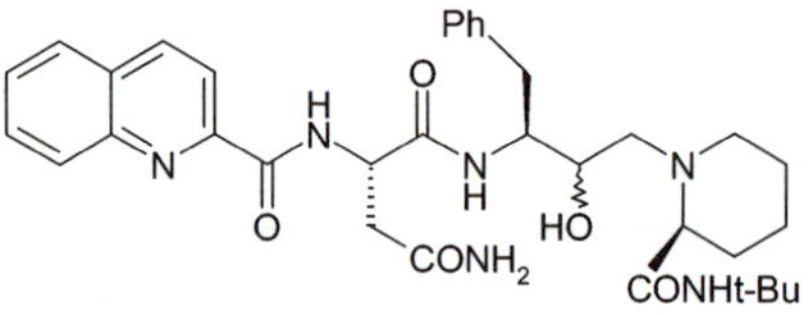

**19** R-isomer HIV-1 $IC_{50}$ 2 nM

**20** S-isomer HIV-1 $IC_{50}$ 470 nM

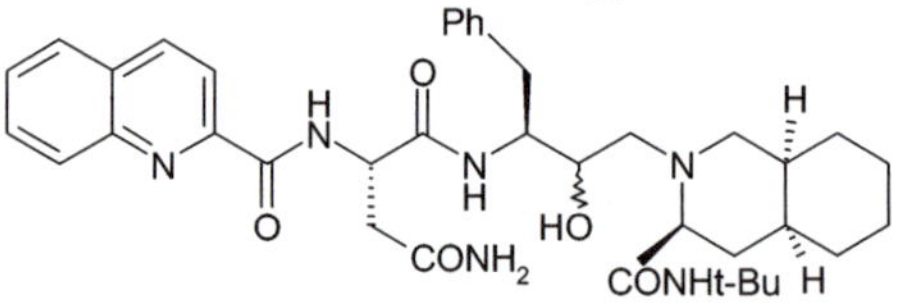

**21** Ro 31-8959 (saquinavir) R-isomer HIV-1 $IC_{50}$ <0.4 nM (Ki 0.12 nM)

**22** S-isomer HIV-1 $IC_{50}$ 620 nM

Ro 31-8959 inhibited both HIV-1 and HIV-2 with $IC_{50}$ values of <1 nM and did not inhibit the related human aspartyl proteases renin, pepsin, gastricin or cathepsins D and E at concentrations up to 10 μM. Ro 31-8959 also had low, but significant, oral bioavailability of 3% in the rat and 11% in the marmoset.[4] However, despite the low oral bioavailability, an oral dose of 10 mg kg$^{-1}$ in the rat gave a plasma concentration above the antiviral $EC_{50}$ for more than 6 h. Ro 31-8959 showed excellent activity in a range of cell-based models of both acute and chronic infections and also showed synergy with other HIV agents, such as nucleoside reverse transcriptase inhibitors. Ro 31-8959 entered clinical trials in 1991 and was the first HIV protease inhibitor to be marketed in 1995.

Subsequently, five other HIV protease inhibitors have reached the market: ritonavir **23** (Abbott),[5] indinavir **24** (Merck),[6] nelfinavir **25** (Agouron),[7] amprenavir **26** (GSK/vertex)[8] and lopinavir **27**/ritonavir combination (Abbott).[9]

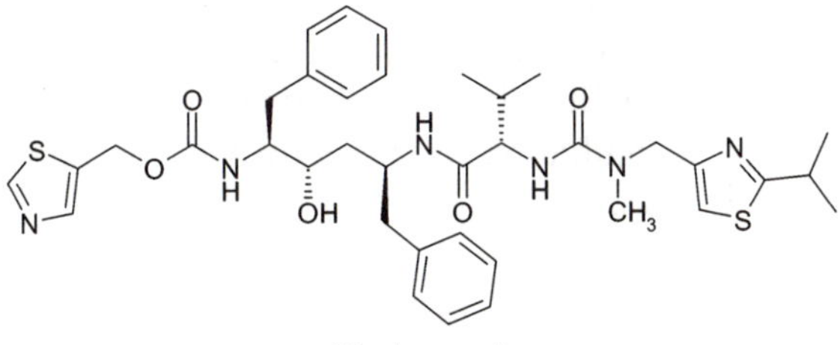

**23** ritonavir

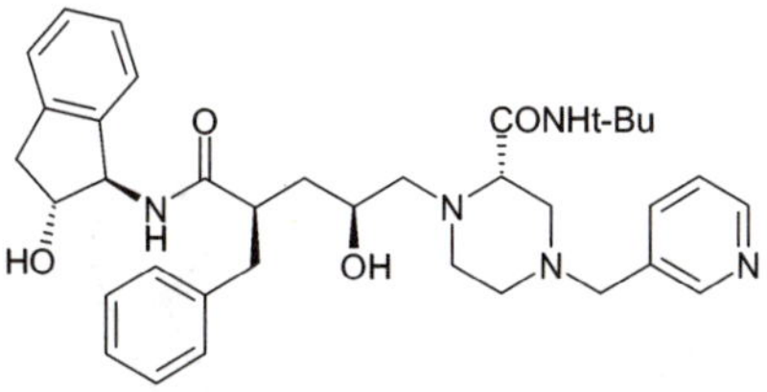

**24** indinavir

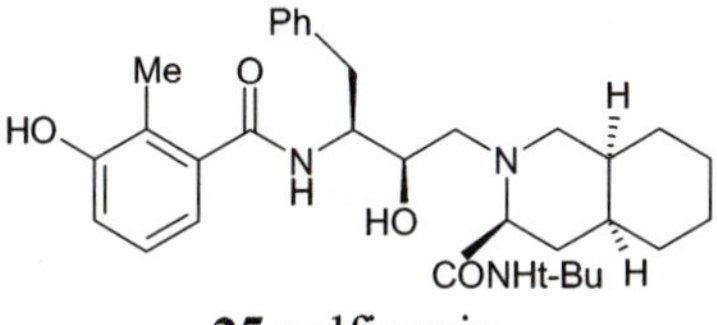

**25** nelfinavir

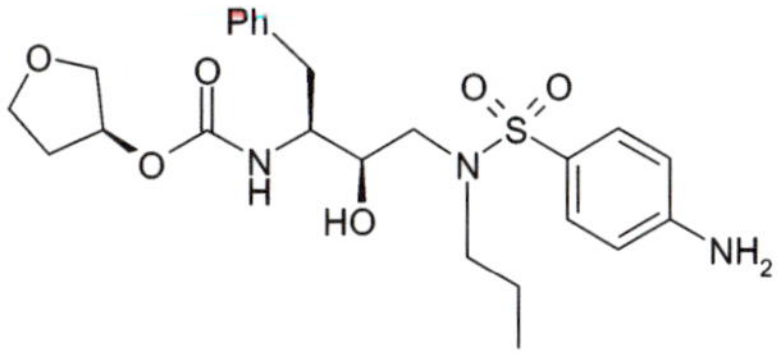

**26** amprenavir

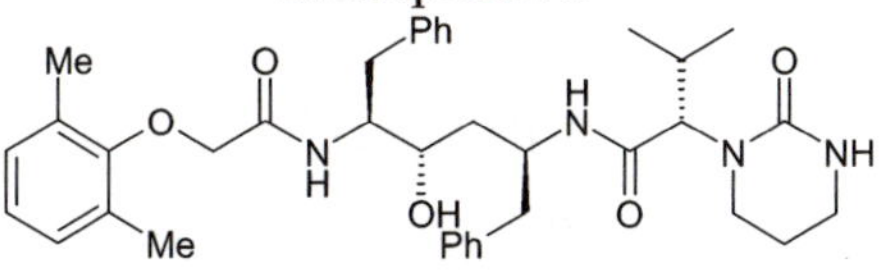

**27** lopinavir

## X-ray Structures[10]

In many of the potency optimisation studies, molecular modelling played an important role in the identification of the optimum substituents. The X-ray structure of HIV protease co-crystallised with **21** has been published (Figure 3). It shows that as expected **21** binds in an extended conformation and the *R*-hydroxy group is located between the catalytic aspartic acids. The *S,S,S* stereochemistry of the decahydroisoquinoline ring is optimal for activity since the decahydroisoqinoline moiety occupies nearly all of the $P_1'$ sub-site and the *tert*-butyl group fits tightly into the $P_2'$ pocket with all of the methyl groups completely buried. Compared to other HIV protease inhibitors, the amide is displaced by 1.8 Å which precludes its extension into the $P_3'$ pocket. Of particular interest in the case of the decahydroisoquinolines, is the directionality of the tert-butylamide, which is maintained as in the proline-based inhibitors, by adopting the opposite configuration of the ring nitrogen (Figure 4).[11]

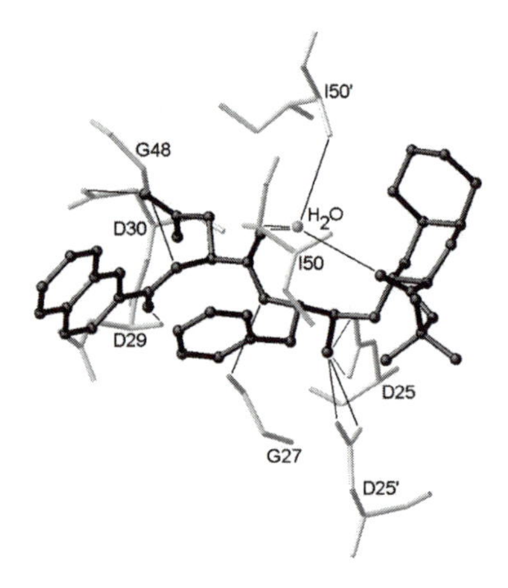


**Figure 3** *X-ray structure of saquinavir* ***21*** *bound into the active site of HIV protease*

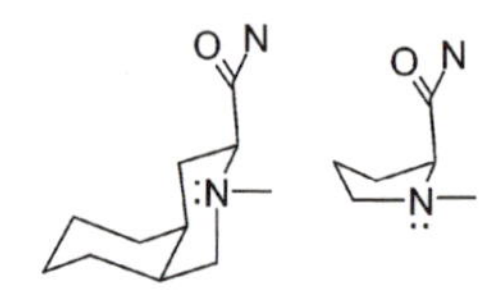


**Figure 4** *Differing bound conformations of the proline and decahydroisoquinoline*

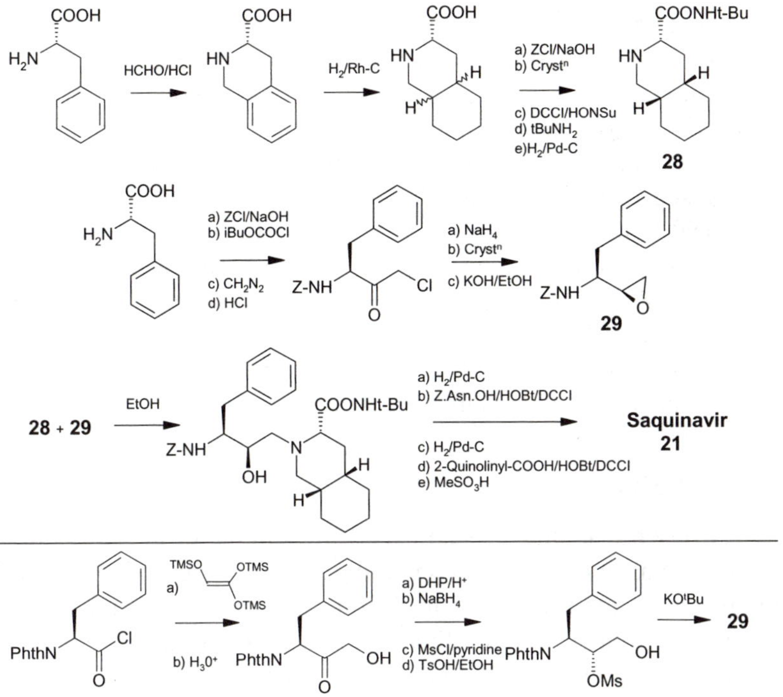

**Figure 5** *Synthesis of saquinavir **21***

## 4 SYNTHESIS OF SAQUINAVIR[12]

The research synthesis of saquinavir is shown in Figure 5. The epoxide **29** was a key intermediate for the synthesis of the N-terminal, $P_2$ and $P_1'$ analogues. Although this route was satisfactory for small scale synthesis, alternative routes were required for the kilogram quantities, in particular to avoid the large scale use of diazomethane. Five alternative syntheses of the epoxide **29** were investigated and the one selected for the large scale supply of material for clinical trials was the tris(trimethylsiloxy)ethene route (Figure 5).

## 5 CLINICAL DATA[13]

An oral suspension of saquinavir was rapidly absorbed in healthy volunteers with a $t_{max}$ of ~0.8 h but this was delayed as the hard capsule formulation (1–2 h) or following a meal (~4 h). Although 30% of the hard capsule dose is absorbed, the bioavailability was only 4% due to high presystemic metabolism in the gut and liver. More recently a soft gelatine capsule has been developed which increases bioavailability by ~3 times.

Saquinavir is metabolised (>90%) in the gut and liver by the cytochrome P450 enzyme 3A4 to species hydroxylated in the decahydroisoquinoline and *tert*-butyl groups, which have low enzyme inhibitory and antiviral activity. Metabolism is inhibited by 3A4 inhibitors (*e.g.* ritonavir) and increased by inducers (*e.g.* rifampin). Grapefruit juice, which contains 3A4 inhibitors, substantially increases bioavailability.

In clinical trials, saquinavir dosed 3 times daily at 600 mg showed good efficacy, increasing CD4 counts by 36 cells $mm^{-3}$ and a maximum fall in viral load of 80%. CD4 and viral load returned to near baseline values after 16 weeks. Due to improved efficacy and reduced incidence of resistance, saquinavir is used as part of a triple therapy regimen with nucleoside analogues, zidovudine and zalcitabine.

## 6 CONCLUSION

The discovery of saquinavir is an excellent example of the structure-based design of an enzyme inhibitor, combining previous knowledge gained from inhibitor design for related aspartyl proteases with a judicious choice of transition-state mimetic. Even though the animal PK data were less than encouraging, saquinavir performed well in the clinic and so justified the bold decision to progress the compound, which was rewarded fully when it was marketed in the USA in December of 1995.

## 7 REFERENCES

1. J.A. Martin, S. Redshaw and G.J. Thomas, *Prog. Med. Chem.*, 1995, **32**, 239.
2. A.G. Tomasselli and R.L. Heinrikson, *Biochim. Biophys. Acta*, 2000, **1477**, 189.
3. N.A. Roberts *et al.*, *Science*, 1990, **248**, 358.
4. J.A. Martin, *Drugs Future*, 1991, **16**, 210.
5. A. Graul and J. Castener, *Drugs Future*, 1996, **21**, 700.
6. *Therapeutic Drugs*, C. Dollery (ed.), 2nd edition, Churchill Livingston, Edinburgh, 1999, I36.
7. A. Bardsley-Elliot and G.L. Plosker, *Drugs*, 2000, **59**, 581.
8. H.B. Fung *et al.*, *Clin. Therapeut.*, 2000, **22**, 549.
9. E.M. Mangum and K.K. Graham, *Pharmacotherapy*, 2001, **21**, 1352.
10. H.M. Berman *et al.*, *Nucleic Acids Res.*, 2000, **28**, 235.
11. A. Krohn *et al.*, *J. Med. Chem.*, 1991, **34**, 3340.
12. K.E.B. Parkes *et al.*, *J. Org. Chem.*, 1994, **59**, 3656.
13. *Therapeutic Drugs*, C. Dollery (ed.), 2nd edition, Churchill Livingston, Edinburgh, 1999, S8.

CHAPTER 19

# Discovery of Vioxx (Rofecoxib)

FRANK D. KING

## 1 INTRODUCTION

Acetylsalicylic acid **1** (aspirin) is arguably the most successful drug ever discovered. It was originally extracted from willow bark and its synthesis by Hoffman in 1899 heralded the start of medicinal chemistry in the pharmaceutical industry. Its main use was for the treatment of pain and inflammation and it represents the first of a class of agents termed NSAIDs, non-steroidal anti-inflammatory drugs. Other examples of this class are the arylacetic acids such as indomethacin **2** and diclofenac **3**, which for many years have been the mainstay treatment for arthritis. Unfortunately, these drugs suffer from potentially life-threatening gastro-intestinal side effects, in particular gastric ulceration. In fact, the standard *in vivo* model used in our projects to find gastric anti-ulcer agents was the blockade of indomethacin-induced ulceration! Therefore, for many years a major target for the industry was to discover novel agents that have the beneficial effects of the NSAIDs but without the GI side-effects.

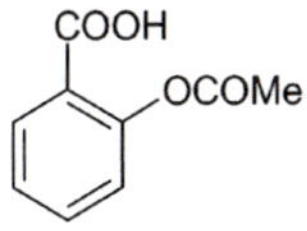

**1** Aspirin 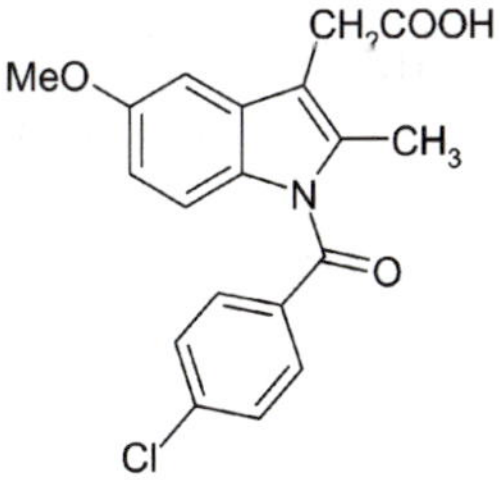    **2** indomethicin 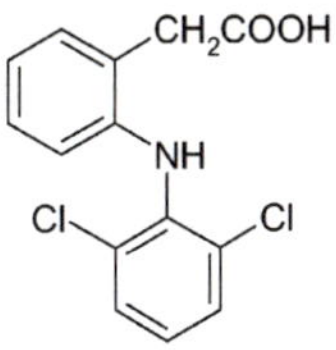    **3** diclofenac

The mechanism of action of the NSAIDs was elucidated in the early 1970s by Vane, who showed that aspirin inhibited the formation of prostaglandins *via* inhibition of the enzyme prostaglandin $H_2$ synthase (PGHS), more often referred to as cyclooxygenase (COX) (Figure 1).[1] COX converts arachidonic acid into prostaglandin $G_2$ ($PGG_2$) as a part of the arachidonic acid pathway to the synthesis of a wide range of prostaglandins, including prostacyclin and prostaglandin $E_2$, both of which have a role in the cytoprotection of the GI tract. Prostaglandins are also released in inflammation and are involved in nociception, so inhibition of synthesis results in anti-inflammatory and analgesic activity, but also reduces protection of the GI tract from, for example the acid released in the stomach.

The breakthrough came when it was discovered in the late 1980s that there were two forms of the COX enzyme, termed COX-1 and COX-2.[2,3] COX-1 was the originally identified PGHS purified from sheep seminal vesicles and is constitutively expressed throughout a wide range of tissues, including the GI tract. It is regarded as the "housekeeping" enzyme responsible for maintaining homeostasis. In contrast, COX-2 is not so widely expressed but is upregulated in inflammation, being induced by pro-inflammatory mediators such as TNF (tumour necrosis factor) and

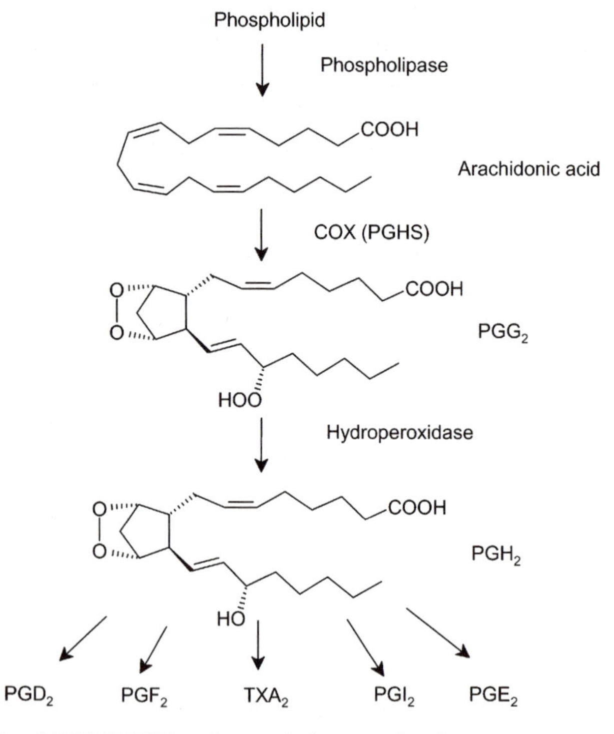


**Figure 1** *Role of COX (PGHS) in the arachidonic acid pathway*

IL-1 (interleukin-1). Thus, the possibility now existed that a selective COX-2 inhibitor may retain the analgesic and anti-inflammatory activity of the NSAIDs, but without the GI side-effects.

X-ray structures of both isoforms of the enzyme have now been determined.[4] The enzyme is dimeric and consists of three domains; an N-terminal beta sheet (for maintaining the dimer), a helical amphipathic region (for anchoring the enzyme into the cell membrane) and a lipophilic domain, which contains the active site and is where the NSAIDs bind. Although the sequence homology between COX-1 and COX-2 is relatively low (63%), in the active site there is only one amino acid different, an isoleucine in COX-1 replaced by a valine in COX-2 at residue 523. This single change has a dramatic effect on the size, shape and flexibility of the binding site. However, as often the case, none of this was known when the selective COX-2 inhibitors were discovered.

## 2 LEAD MOLECULES

Evidence that selectivity for COX-2 over COX-1 could be achieved first came from the studies on DuP 697 **4** identified in the early 1990s by the group at DuPont from an extensive investigation of 1,2-diaryl heterocycles related to the NSAID phenylbutazone **5**.[5] DuP 697 was found to retain the anti-inflammatory activity but with much reduced GI side-effects in animal models and subsequently was found to be a selective COX-2 inhibitor. Unfortunately, it had an extremely long plasma half-life in man (12 days!), believed to be due to enterohepatic recirculation. This long half-life was not predicted from animal ADME studies where the 5-bromo substituent was subject to displacement by sulfur containing groups.

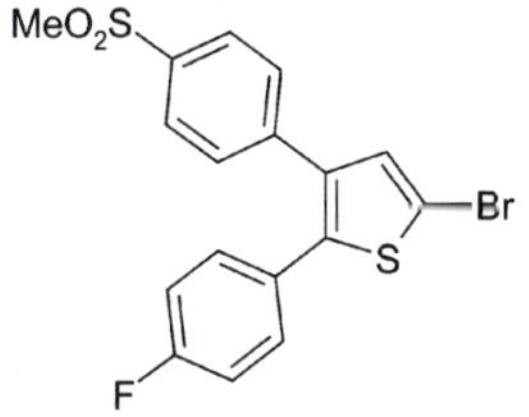

**4** DuP 697 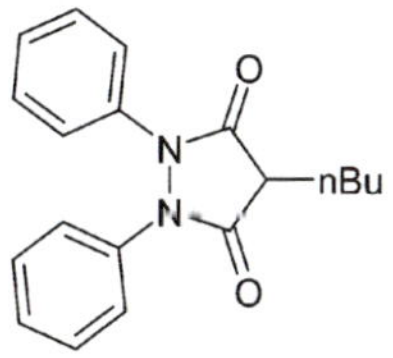 **5** phenylbutazone

An extensive investigation of the SAR of **4** led to fundamental principles for the 1,2-diaryl heterocyclic inhibitors. For the 2-aryl substituent, both a 4-sulfone and 4-sulfonamide gave potent and selective COX-2 inhibitors. The sulfonamides were generally more potent, had improved oral absorption but lower selectivity over COX-1. The 1-phenyl group can tolerate a wider range of substituents, though *meta*-substitution generally gave better selectivity and a *para*-substitutent resulted in lower selectivity. The aryl group can also be replaced by O-iPr or cyclohexyl.

In the thiophene ring, the 5-bromo substituent was found not to be necessary. Also for COX-2 inhibitory activity, the two aryl groups need to be in adjacent positions on the thiophene ring, thus the isomeric 3,4-diaryl thiophenes were also selective COX-2 inhibitors. Subsequently, a number of other cyclic templates have been identified including 4-, 5- and 6-membered rings and even simple 1,2-diarylethylenes.[6]

## 3 IDENTIFICATION OF ROFECOXIB (MK-966)[7]

### Assays

For enzyme inhibition studies, the group at Merck Frosst originally used osteosarcoma cells for the expression of COX-2 and undifferentiated U937 cells for COX-1. The group had previously used these cell lines for their leukotriene biosynthesis and phospholipase $A_2$ ($PLA_2$) inhibitor programmes. The measure was the inhibition of the conversion of arachidonic acid into $PGE_2$. Because the enzyme kinetics of the inhibitor molecules differed at the two enzymes (COX-1, rapidly reversible, competitive; COX-2, time-dependent, slowly reversible, Figure 2), the selectivity values for individual compounds were highly assay dependent, particularly on the arachidonic acid concentration. A whole blood assay was therefore developed as a secondary screen for a better predictor of *in vivo* selectivity. For COX-1, the measure was the inhibition of release of platelet thromboxane $B_2$ ($TXB_2$) following clotting of the blood and for COX-2, the measure was the $PGE_2$ levels following induction of COX-2 protein synthesis using bacterial LPS. In this latter assay, indomethacin was found to be 2.5 times more potent at COX-1 ($IC_{50}$ 200 nM) than COX-2 ($IC_{50}$ 500 nM).

**COX-1**

$$E + I \underset{}{\overset{K_i}{\rightleftharpoons}} EI$$

- Rapid, reversible, competitive inhibition
- $IC_{50}$ values independent of pre-incubation period (unless very short)

**COX-2**

$$E + I \overset{K_i}{\rightleftharpoons} EI \underset{K_{off}}{\overset{K_{on}}{\rightleftharpoons}} EI^*$$

- Initial phase rapid, reversible competitive inhibition
- Second phase slower, tight binding, pseudo-irreversible
- $IC_{50}$ values dependent on pre-incubation period

**Figure 2** *Comparative enzyme kinetics for inhibitors at COX-1 and COX-2*

## Medicinal Chemistry

The starting point for the Merck Frosst group was to address the lack of water solubility with DuP 697 and related methylsulfonylphenyl compounds, which was believed to limit the oral bioavialability. It was believed that if the thiophene could be replaced by a furanone **6**, this could be delivered using the highly soluble sodium salt of the ring opened hydroxyacid **7** as a pro-drug. It was satisfying for them to find that one of the furanone isomers was a potent and selective COX-2 inhibitor. Interestingly, the SAR for the regional isomer is similar to that found for the bromo-3,4-diarylthiophenes **8** from the DuPont work. As often happens in medicinal chemistry, the original idea did not work in that attempts to form the sodium salt by ring opening the furanone resulted in destruction of the molecule. However, luckily the furanones themselves were found to have excellent oral bioavailability, far more than previously seen with the thiophene class.

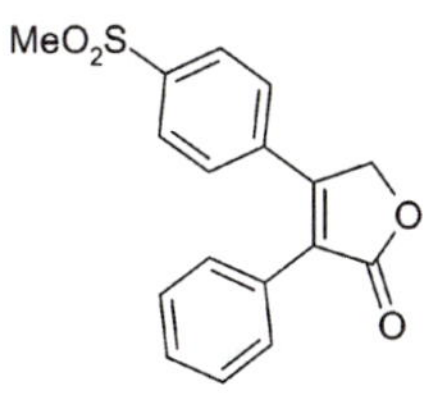

$MeO_2S$ OH COONa S Br F O O

**6** MK-966 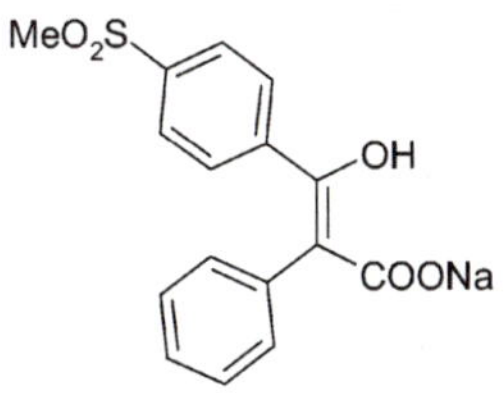 **7** 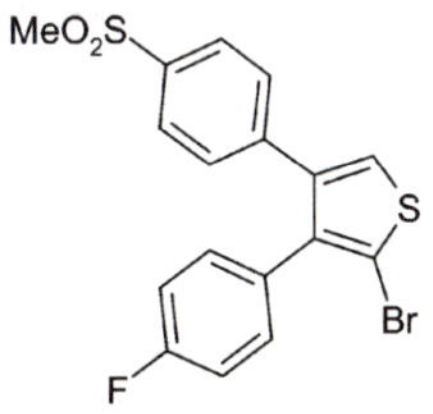 **8**

The SAR for the furanones was found to be similar for the thiophenes and related 5-membered heterocycles (Table 1):

- The *para*-sulfonamides retained COX-2 potency but were less selective over COX-1
- a *para*-substituent also retained COX-2 potency but also increased COX-1 potency
- a *meta*-substituent reduced COX-2 potency but retained high selectivity
- combined *meta*- and *para*-substitution also retained COX-2 potency and selectivity, though selectivity was reduced in the whole blood assay.

From a large number of compounds prepared, the unsubstituted phenyl analogue **6** (MK-966, rofecoxib) proved to have the best overall profile and was selected for development. In a number of *in vitro* COX-2 cellular

**Table 1** *SAR of the 3,4-diarylfuranones*

| R | X | Whole cells $IC_{50}$ μM | | Human whole blood $IC_{50}$ μM | |
|---|---|---|---|---|---|
| | | COX-2 | Selectivity | COX-2 | Selectivity |
| phenyl | $-SO_2Me$ | 0.02 | >750 | 0.5 | 38 |
| 4-fluorophenyl | $-SO_2Me$ | 0.01 | 470 | 0.6 | 17 |
| 3-fluorophenyl | $-SO_2Me$ | 0.02 | >2500 | 1.8 | 47 |
| 3,4-difluorophenyl | $-SO_2Me$ | 0.03 | >1600 | 0.9 | 14 |
| 4-fluorophenyl | $-SO_2NH_2$ | nd | nd | 0.8 | 7 |
| Indomethacin | | 0.03 | 0.7 | 0.4 | 0.5 |

**Table 2** *COX-1 and COX-2 activities of* ***6*** *in different* in vitro *assays*

| Assay: COX-2 | $IC_{50}$ (μM) |
|---|---|
| $PGE_2$ production by osteosarcoma cells | 0.026 |
| $PGE_2$ production by CHO cells | 0.018 |
| $PGE_2$ production by LPS-induced human mononuclear cells | 0.045 |
| *Assay: COX-1* | $IC_{50}$ (μM) |
| $PGE_2$ production by U937 cells | >50 |
| $PGE_2$ production by U937 cells (low substrate) | 2 |
| $PGE_2$ production by CHO cells | >15 |
| $PGE_2$ production by human kidney microsomes | 14 |

assays it had $IC_{50}$s of between 18–45 nM. In contrast in the COX-1 cellular assays its $IC_{50}$s ranged between 2–>50 μM (Table 2).

## 4 *IN VIVO* ACTIVITY OF ROFECOXIB

In *in vivo* anti-inflammatory and analgesia animal models, rofecoxib showed similar activity to indomethacin, further demonstrating that a

**Table 3** *Activities of indomethacin and rofecoxib in animal models of inflammation and pain ($ED_{50}$ mg $kg^{-1}$ p.o.)*

| *Animal model* | *Indomethacin* | *Rofecoxib* |
|---|---|---|
| Adjuvant arthritis | 0.2 | 0.7 (bid) |
| Rat paw oedema | 2.0 | 1.5 |
| Rat pyresis | 1.1 | 0.2 |
| Rat paw hyperalgesia | 1.5 | 1.0 |

highly selective COX-2 inhibitor would be as effective as NSAIDs in the clinic (Table 3). Rofecoxib was also found to have a lower propensity to cause gastric damage in a number of animal models. Thus, in a model using $^{51}Cr$-labelled red blood cells to detect intestinal damage in both rats and squirrel monkeys, rofecoxib at 100 mg $kg^{-1}$ b.i.d. for 5 days showed no leakage of the $^{51}Cr$ marker. In contrast, both indomethacin and diclofenac, at 10 mg $kg^{-1}$ caused a significant increase of the $^{51}Cr$ marker in the GI tract. Similarly, in a 14-day toxicology study in rats (300 mg $kg^{-1}$ p.o.), rofecoxib did not produce any gastric lesions, in contrast to indomethacin (single dose, 3 mg $kg^{-1}$) which clearly did.

## 5 CLINICAL RESULTS[8]

In a placebo controlled, pharmacokinetic study in healthy men, rofecoxib showed an elimination $t_{1/2}$ of 9.9–17.5 h with non-linear pharmacokinetics. At doses of 25, 100, 250 and 375 mg once daily for 12 days, COX-2 inhibition of 67, 96, 92 and 96% respectively was observed in a whole blood assay with no inhibition of COX-1. Adverse events were rare and transient. In comparative trials in patients with osteoarthritis, rofecoxib (25 mg $day^{-1}$) showed similar analgesic efficacy to celecoxib (200 mg $day^{-1}$), paracetamol (1000 mg $day^{-1}$), ibuprofen (2400 mg $day^{-1}$), diclofenac (150 mg $day^{-1}$) and naproxen (1000 mg $day^{-1}$). In patients with rheumatoid arthritis, rofecoxib (50 mg once daily) also showed similar efficacy to naproxen (500 mg twice daily) after a median of 9 months therapy. Good analgesic efficacy was also seen in clinical trials for postsurgical dental pain, postoperative surgical pain and primary dysmenorrhoea. The incidence of withdrawal due to GI adverse events was also lower for rofecoxib than for traditional NSAIDs, with a lower incidence of gastric perforations, ulcerations and bleeding.

## 6 CONCLUSION

The identification and development of rofecoxib (Vioxx) is a classic example of the initial identification of the mechanism of action of highly successful drugs, the NSAIDs. Subsequently, sub-types of the COX enzyme were identified, which suggested the possibility that a selective COX-2 inhibitor could retain the analgesic activity of the NSAIDs but without the GI side effects. Merck were not the first to identify selective COX-2 inhibitors and there was intense competition in the area. However, they identified a solubility issue with the known inhibitors and adopted a pro-drug strategy to overcome that issue. Although this approach was not successful, it did lead the team to investigate the diarylfuranones, which had sufficiently good properties in their own right to be considered for development. It then became a race to the market and Vioxx was marketed in 1999 and by the year 2000 it was the 12th largest selling drug with sales of over $2B (Appendix 1).

## 7 REFERENCES

1. J.R. Vane, *Nat. New Biol.*, 1971, **231**, 232.
2. A. Raz *et al.*, *J. Biol. Chem.*, 1988, **253**, 3022.
3. J.Y. Fu *et al.*, *J. Biol. Chem.*, 1990, **265**, 16737.
4. C. Luong *et al.*, *Nat. Struct. Biol.*, 1996, **3**, 927.
5. K. Gans *et al.*, *J. Pharmacol. Exp. Ther.*, 1990, **254**, 180.
6. J.J. Talley, *Prog. Med. Chem.*, 1999, **36**, 201.
7. P. Prasit, in *Therapeutic Roles of Selective COX-2 Inhibitors*, J.R. Vane and R.M. Botting (eds.), William Harvey Press, London, 2001, p. 60.
8. A.J. Matheson and D.P. Figgitt, *Drugs*, 2001, **61**, 833.

CHAPTER 20

# $NK_1$ Receptor Antagonists

CHRIS SWAIN

## 1 INTRODUCTION

A number of extensive reviews have been published which cover much of the current literature on receptor distribution, *in vitro* and *in vivo* pharmacology, potential clinical utilities of neurokinin (NK) receptor antagonists and the patent literature.[1–5] The neurokinin receptors are currently divided into three main populations. The endogenous agonists are a family of peptides that share the common C-terminal sequence 'Phe-X-Gly-Leu-Met-$NH_2$'. The preferred mammalian neurokinin agonist for each of the receptors is substance P (SP) for $NK_1$, neurokinin A (NKA) for $NK_2$ and neurokinin B (NKB) for $NK_3$. However, each of the agonists can act as a full agonist at each of the receptors and it is possible that under physiological conditions the actions of each agonist might be mediated by more than one receptor type.

Mammalian Tachykinins

| | |
|---|---|
| Substance P | Arg-Pro-Lys-Pro-Gln-Gln-*Phe* Phe-*Gly-Leu-Met* –$NH_2$ |
| Neurokinin A | His-Lys-Thr-Asp-Ser-*Phe*-Val-*Gly-Leu-Met* –$NH_2$ |
| Neurokinin B | Asp-Met-His-Asp-Phe-*Phe*-Val-*Gly-Leu-Met* –$NH_2$ |

All three human receptors have been cloned and expressed in cell lines, facilitating rapid screening of novel agents for receptor selectivity.[6] The cloning of the human receptor was of particular importance, because there are considerable species differences with respect to antagonist affinities.

Whilst the possible role of $NK_1$ antagonists in a variety of clinical indications has been proposed, we were particularly interested in possible

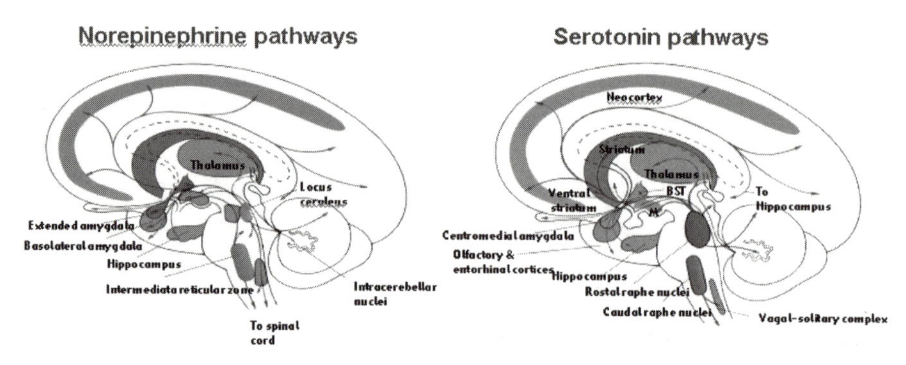


**Figure 1** *Brain distribution of the $NK_1$ receptor*

CNS indications. These include emesis, for which there is ample historical evidence supporting this as a possible indication, in particular SP found in brainstem emetic nuclei and the finding that local application of SP caused retching in ferrets.

Perhaps more interestingly, there is widespread distribution of $NK_1$ receptors in the brain (Figure 1: dark areas) and thus the potential for interactions with monoamine systems implicated in mental illness.

## 2 MEDICINAL CHEMISTRY PROGRAMME

Whilst we had an interest in the neurokinins for a number of years, two factors resulted in the initiation of a full-blown medicinal chemistry programme in the early 1990s. The first was the publication of the sequence of the $NK_1$ receptor that allowed the cloning of the human $NK_1$, $NK_2$ and $NK_3$ receptors, and transfection into stable cell lines. This greatly simplified the issues of species differences and the fact that native tissue often contained multiple receptor subtypes. The second factor was the publication by Pfizer of the first non-peptide $NK_1$ antagonist CP-96345, **1**.[2] CP-96345 has excellent affinity (Figure 2) but very poor oral availability.

NH OMe N Ph Ph

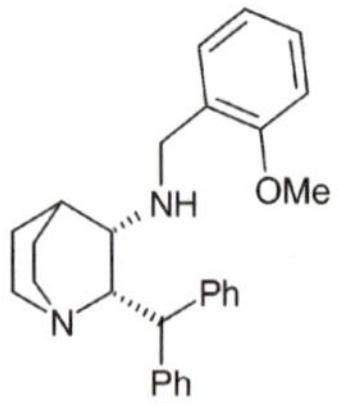

**1**

CP-96345

$hNK_1$ $IC_{50}$ 0.6 nM

Poor oral availability

Cardiovascular side-effects

(Possibly due to $Ca^{2+}$ channel blockade)

**Figure 2** *Structure and activity of CP-96345*

**Table 1** *Structure and activities of quinuclidines*

| *X* | $IC_{50}$ (μM) | *X* | $IC_{50}$ (μM) |
|---|---|---|---|
| –$NHCH_2$-Ph | 0.15 | –OCO-Ph | >1 |
| –$NHCH_2CH_2$-Ph | 0.70 | –OCONH-Ph | >1 |
| –NHCO-Ph | >1 | –$OCH_2$-Ph | 0.11 |
| –NHCOO-Ph | >1 | –$OCH_2$-3,5-$(CF_3)_2C_6H_3$ **2** | 0.002 |
| –NHCONH-Ph | >1 | | |
| –NHCSNH-Ph | >1 | | |

In addition, CP-96345 elicits cardiovascular side-effects, possibly due to L-type calcium channel blockade.

Based on analysis of the structure of **1**, it was felt that the 2-methoxybenzyl amine was a potential metabolic liability and the initial target was to identify substituents with improved metabolic stability, whilst maintaining high affinity. Molecular modelling studies, carried out in our laboratory, suggest that in the global energy minimum conformation there existed an intramolecular hydrogen-bonding interaction between the methoxy oxygen and the N–H of the benzylamine in **1** which may play an important role in stabilising the receptor-bound conformation. The hydrogen-bonding interaction is replaced by a destabilising lone pair–lone pair repulsion in the corresponding *o*-methoxybenzyl ether, resulting in quite different low energy conformations. Thus, in order to obtain a meaningful evaluation of the contribution of the linking heteroatom to binding it was essential to include the unsubstituted analogues in any comparative study (Table 1). This work lead to the identification of the 3,5-bis(trifluoromethyl)benzylethers **2**, with excellent affinity and oral availability.

## Reducing Calcium Channel Activity

These compounds still suffered from significant calcium channel affinity. A survey of the literature suggested that the ion channel activity might be dependent on the basicity of the quinuclidine nitrogen. Since modulation of the $pK_a$ in the quinuclidine series would represent a significant synthetic challenge, it was decided to investigate the effects of $pK_a$ in the corresponding acyclic series derived by excising much of the bicyclic framework. The simple primary amine **3** has modest affinity for the $NK_1$

**Table 2** *Structure and activities of substituted ethanolamines as basic pharmacophore*

| | $R^1$ | $R^2$ | $hNK_1$ (nM) | $Ca^{2+}$ (nM) |
|---|---|---|---|---|
| 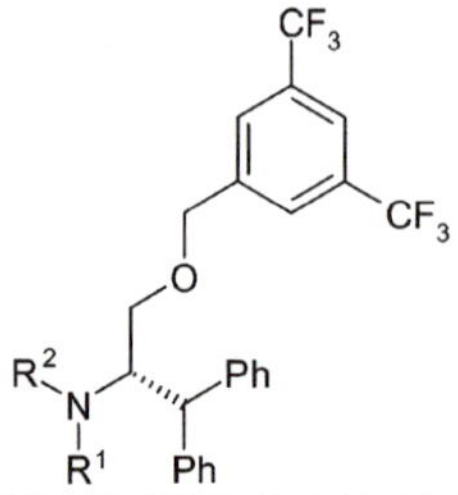 **1** | CP-96345 | | 0.6 | 240 |
| **3** | H | H | 10 | 190 |
| | Me | Me | 2.5 | 980 |
| | H | $CH_2$cPr | 100 | 357 |
| **4** | H | $CH_2CONH_2$ | 0.8 | 1700 |

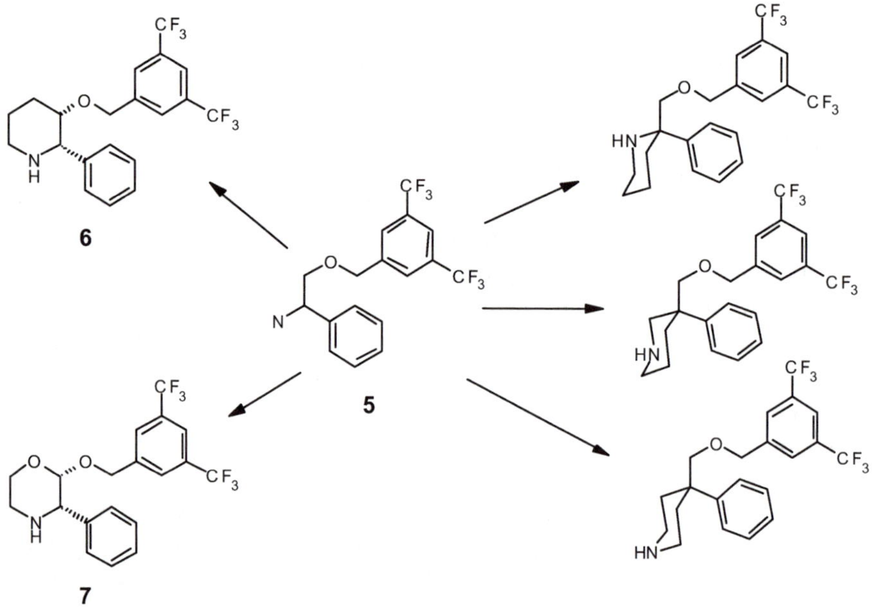

**Figure 3** *Alternative conformational restraints derived from the basic pharmacophore* **5**

receptor (Table 2) and significant activity for the L-type calcium channel. Alkyl substitution does not improve selectivity; however, the introduction of electron-withdrawing groups, **4**, reduces calcium channel binding and increases $hNK_1$ affinity.

The molecular modelling studies in conjunction with mutagenesis work had also indicated that one of the phenyl rings of the benzhydryl group merely served as a conformational anchor and it was shown that it was possible to replace the benzhydryl with a single phenyl ring. The substituted phenyl glycinol **5** represents the minimum pharmacophore for which a number of alternative conformational restraints were subsequently explored.[5] Whilst it was possible to design high affinity ligands in each series, of particular note are the piperidines **6** and morpholines **7** (Figure 3).

## Improving the Duration of Action

Whilst the carboxamide **4** served to demonstrate modulation of calcium channel binding, it was recognised that this moiety would be a significant metabolic liability and was thus replaced by a series of bioisosteric amide replacements (Table 3), in particular 5-membered heterocyclic rings, triazole **8** and triazolinone **9**. A range of polar electron-withdrawing heterocycles gave excellent affinity and outstanding selectivity over calcium channel binding; interestingly, introduction of the weakly basic imidazole **10** resulted in modest calcium channel affinity.

The presence of a 3,5-bis(trifluoromethyl)benzyl ether appears to play a significant role in enhancing the *in vivo* activity. *In vitro* metabolism studies using rat liver microsomes identified the presence of 3,5-bis-(trifluoromethyl)benzoic acid as a significant metabolite, presumably arising *via* oxidation at the benzylic position, highlighted in Figure 4.

In an effort to improve the duration of action of this class of $NK_1$ antagonists, we sought to block this site of metabolism by the introduction of a methyl substituent. Introduction of the alpha methyl affords two diastereoisomers; in the case of the 3,5-bis(trifluoromethyl) substitution, the *R*-diastereoisomer maintains excellent affinity whilst the *S*-isomer has

**Table 3** *Structure and activities of morpholines*

| | *No.* | $R^1$ | $hNK_1$ (nM) | $Ca^{2+}$ (μM) |
|---|---|---|---|---|
| 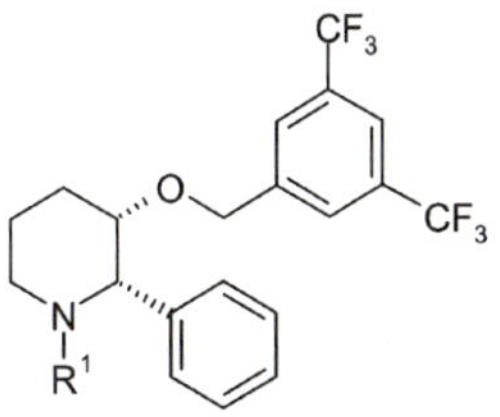 | | 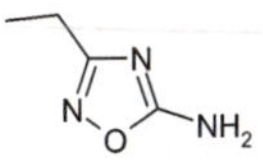 | 0.45 | >30 |
| | **8** | 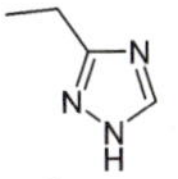 | 0.19 | >100 |
| | **9** | 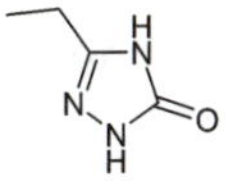 | 0.1 | >10 |
| | **10** | 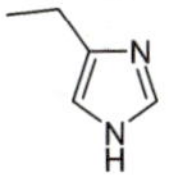 | 0.7 | 2.3 |

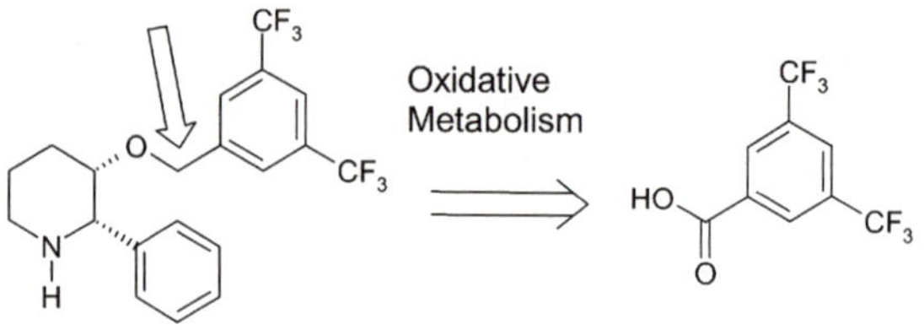

**Figure 4** *Site of metabolism of the 3,5-bis(trifluoromethyl)benzyl ether 7*

a 100-fold reduction in affinity (Table 4). The compounds were evaluated *in vivo* by their ability to antagonise the extravasation induced by the vannilloid sensorotoxin, resiniferatoxin, one hour after administration of the test drug or at longer treatment times. The extent of plasma protein extravasation was determined spectrophotometrically by using Evans Blue dye as a plasma marker. The bis(trifluoromethyl)benzyl ether **8** was a potent, dose dependent antagonist after oral administration ($ID_{50}$ 0.34 mg kg$^{-1}$ p.o.) but displayed modest duration (55% inhibition 8 h after 1 mg kg$^{-1}$ p.o.) and no evidence of activity at 24 h. Introduction of the alpha methyl **11** maintained the oral potency at the 1 h and gave a significant improvement at the 8 h time point. Replacement of the triazole with the triazolinone **12** gave a modest improvement in the oral activity (0.026 mg kg$^{-1}$) and a greatly improved duration (66% inhibition 24 h after 1 mg kg$^{-1}$ p.o.).

**Table 4** *Summary of* in vitro *and* in vivo *studies ($ID_{50}$ or % inhib. @ mg kg$^{-1}$ p.o.)*

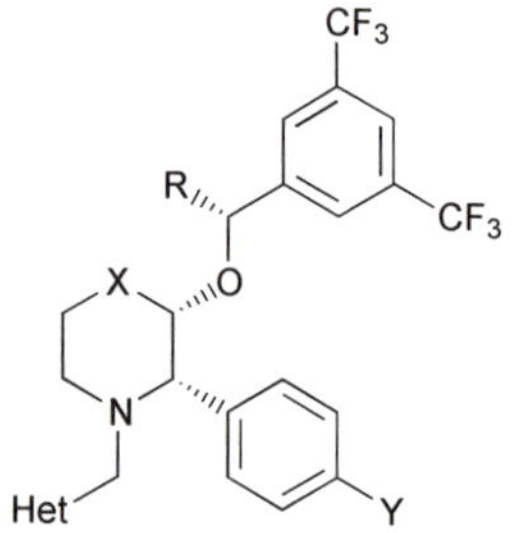

| *No.* | *X* | *Y* | *R* | *Heterocycle* | $IC_{50}$ (nM) | $ID_{50}$ at 1 h (nM) | *Inhibition at 8 h* | *Inhibition at 24 h* |
|---|---|---|---|---|---|---|---|---|
| **8** | $CH_2$ | H | H | Triazole | 0.18 | 0.034 | 55% @ 1 | 0% @ 1 |
| **11** | $CH_2$ | H | Me | Triazole | 0.16 | 0.06 | 78% @ 1 | 12% @ 1 |
| **12** | $CH_2$ | H | Me | Triazolinone | 0.16 | 0.026 | 97% @ 1 | 66% @ 1 |
| **13** | O | F | Me | Triazolinone | 0.09 | 0.008 | 100% @ 1 | 0.55 |

We were delighted to find that the corresponding morpholines also displayed a desirable profile. In addition, introduction of a 4-fluoro also afforded an improvement in duration of action that proved to be additive with the influence of the alpha methyl substitution **13** to afford the clinical candidate MK-869.

## Non-CNS Penetrant Compounds

At the same time as the medicinal chemistry programme was trying to identify novel $NK_1$ antagonists, we also tried to definitively address the issue of whether CNS penetration was essential for the target indications.

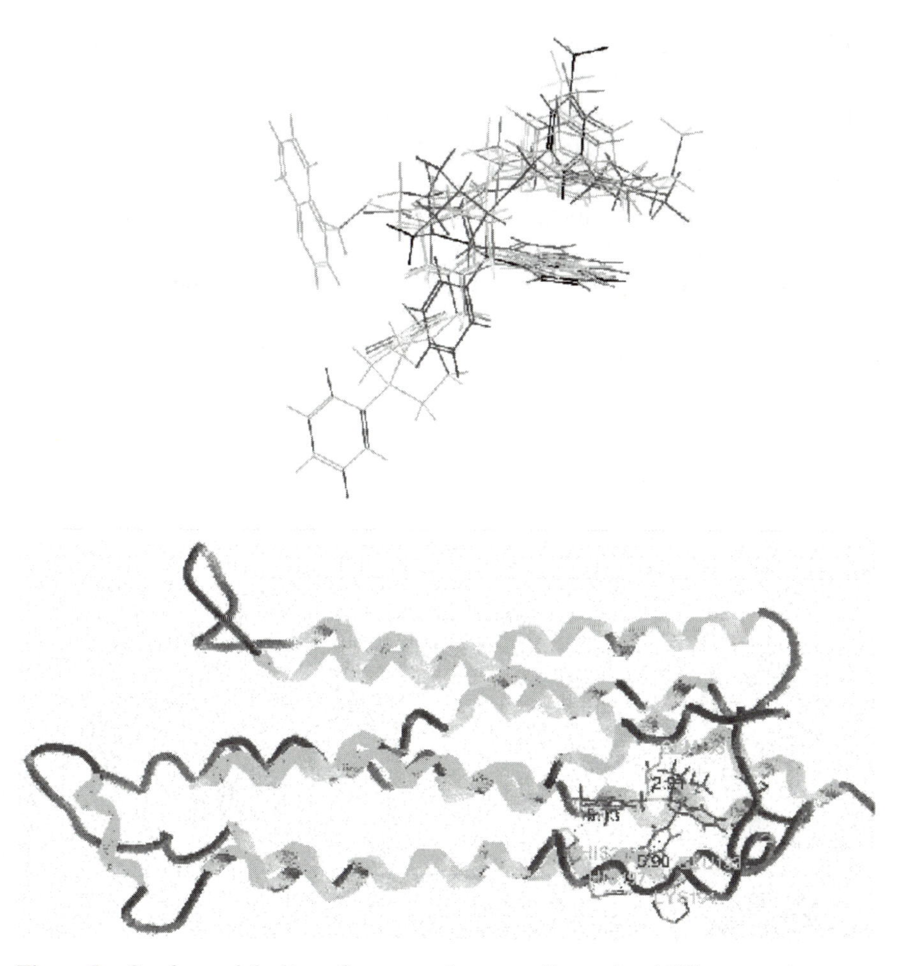

**Figure 5** *Overlay and docking of a range of structurally unrelated $NK_1$ antagonists*

To do so, a non-CNS penetrant compound was designed and compared with a CNS penetrant compound in a number of preclinical models. Using the pharmacophore developed from modelling and mutagenesis studies, a range of structurally unrelated $NK_1$ antagonists were overlaid and then docked with a model of the $hNK_1$ receptor (Figure 5).

Whilst the receptor model has insufficient resolution to identify new intramolecular interactions, it was useful to identify regions of the molecules that appeared to be directed towards the extracellular space. As a result of this work the quaternary ammonium species **14** was identified. In peripherally mediated assays both the penetrant and non-penetrant compounds had equivalent activity (Figure 6).

In the chemotherapy induced emesis assay (CIE), the non-brain penetrant compound was ineffective, confirming that CNS penetration was essential for activity (Figure 7).

Whilst CIE confirmed the need for CNS penetration it was far from ideal as an assay to evaluate the relative brain penetration of a range of compounds. In gerbils, central infusion of substance P agonists, such as GR73632, elicits a vigorous and readily quantifiable rhythmic drumming or tapping of the hind feet, which can be inhibited by systemic administration of brain-penetrant substance P receptor antagonists. Thus, inhibition of $NK_1$ agonist-induced foot tapping in this species provides an *in vivo* functional assay for the CNS penetration of antagonists, and this enabled us to identify optimal research tools with which to investigate the role of substance P in the brain.

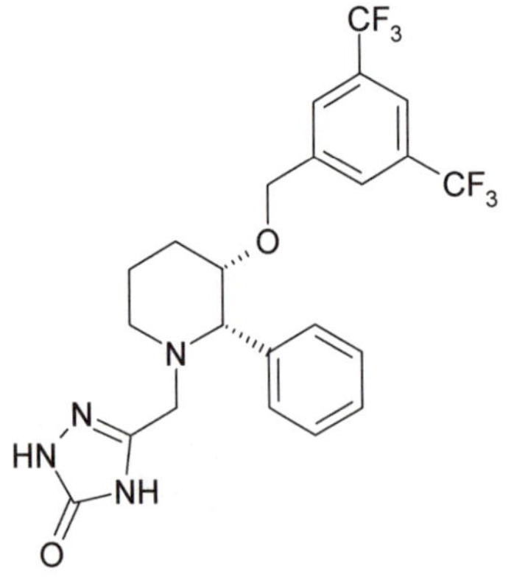

**10** L-741671

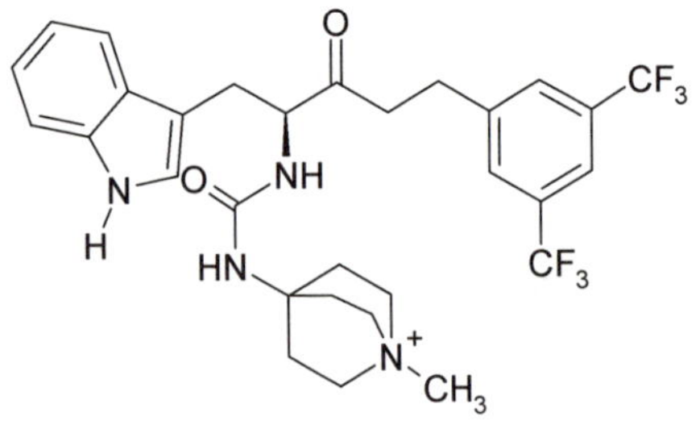

**12**

**Figure 6** *Structures of the brain penetrant and non-penetrant $NK_1$ antagonists*

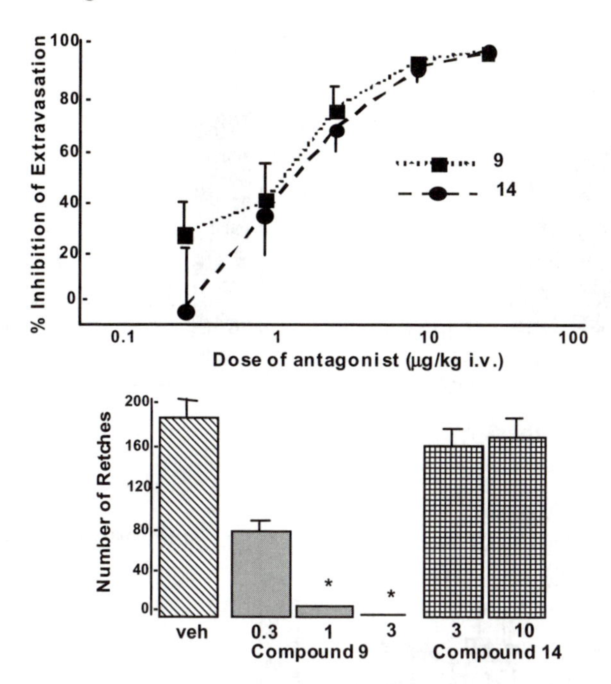


**Figure 7** *Activities of **10** and **11** in the extravasation and anti-emetic assays*

## 3 PROFILE OF MK-869: CLINICAL CANDIDATE

MK-869 (**13**) is a high affinity $NK_1$ antagonist with excellent CNS penetration and duration (Table 5). It is a potent anti-emetic in acute CIE studies in the ferret after both i.v. and p.o. administration (Figure 8). Perhaps more importantly, in studies to investigate the 'delayed' response MK-869 was also active. The emetic response to cisplatin in the ferret is biphasic, with a second bout of emesis some 36–72 h after the initial

**Table 5** *Structure and activities of MK-869 (13)*

|  |  |
|---|---|
| 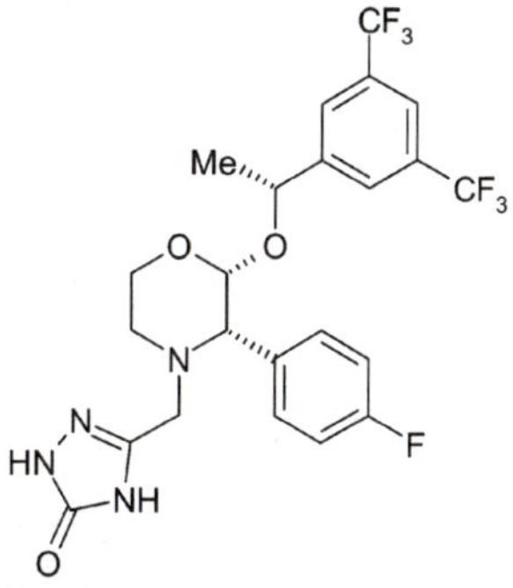 hNK1 $IC_{50}$ | 0.09nM |
| SYVAL $ID_{50}$ (po 1 h) | 0.008 mg kg$^{-1}$ |
| SYVAL $ID_{90}$ (po 24 h) | 1.8 mg kg$^{-1}$ |
| foot-tapping $ID_{50}$ (iv 1 h) | 0.33 mg kg$^{-1}$ |
| foot-tapping $ID_{50}$ (iv 24 h) | 0.36 mg kg$^{-1}$ |

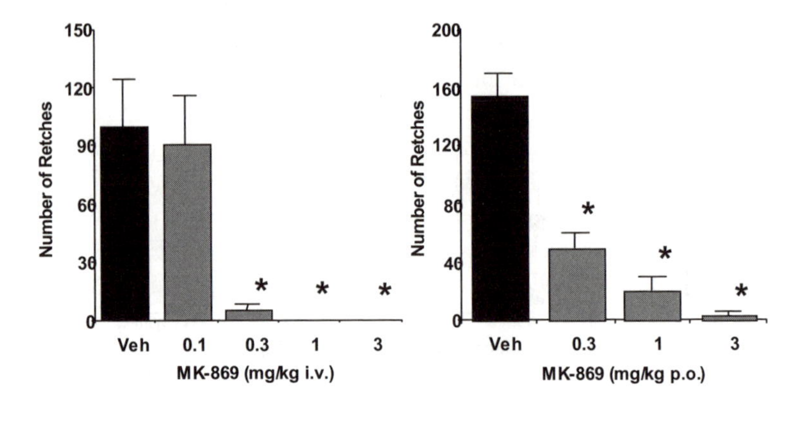


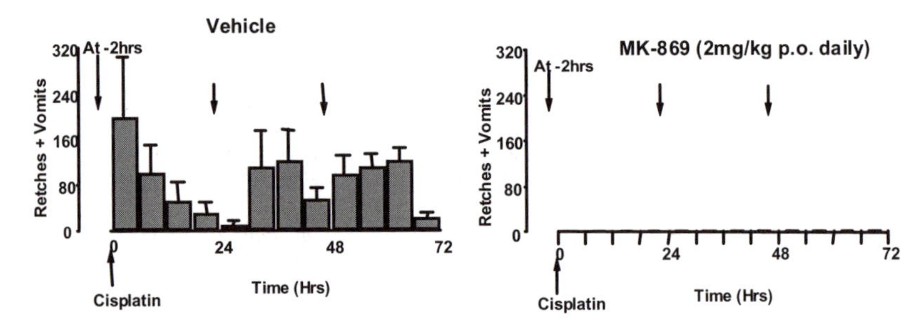


**Figure 8** *Activity of MK-869 in anti-emetic model*

insult, this response is largely refractory to treatment with 5-$HT_3$ antagonists. This biphasic response in ferret mimics that which is seen in the clinic. MK-869 given once a day (2 mg $kg^{-1}$ p.o.) completely blocks both the acute phase and the delayed phase emetic response.[9]

## 4 CLINICAL RESULTS

### Emesis

Patients undergoing cancer chemotherapy consistently report that vomiting and nausea significantly affect their quality of life and may cause them to refuse therapy. Cisplatin causes profound vomiting and nausea within 24 h of administration with a second "delayed" phase some 36–72 h after administration. Existing treatments consist of 5-$HT_3$ antagonists such as Ondansetron or Granisetron, in combination with dexamethasone.

The initial study compared L-758298, a water-soluble phosphate prodrug of MK-869, with the 5-$HT_3$ antagonist Ondansetron. Whilst both compounds gave similar protection in the acute phase (Table 6) there was a dramatic improvement in protection with the $NK_1$ antagonist in the delayed phase (Table 7). These data led to a study to evaluate a combination of a $NK_1$ antagonist, 5-$HT_3$ antagonist and dexamethasone. The results of the trial (Table 8) show that the combination of MK-869, the 5-$HT_3$ antagonist Granisetron together with dexamethasone provided significantly better protection than the combination of Granisetron and dexamethasone alone. The improved response being particularly apparent during the delayed phase.

## Pain

The preclinical data supporting a role for Substance P in pain is extensive, receptor distribution studies have shown localisation in dorsal horn, in

**Table 6** *Efficacy in cisplatin-induced acute emesis*

| *Treatment* | *Total* (N) | *CR*(%) | *MR*(%) | *F*(%) |
|---|---|---|---|---|
| L-758298 (60 or 100 mg) | 30 | 11 (36.7) | 6 (20) | 13 (43.3) |
| Ondansetron (32 mg) | 23 | 12 (52.2) | 3 (13) | 8 (34.7) |

No statistical significance in differences between groups. CR = Complete Responder (0 emetic episodes). MR = Major Responder (1–2 emetic episodes). F = Failure (3 or more emetic episodes).

**Table 7** *Efficacy in cisplatin-induced delayed emesis*

| *Treatment* | *Total* (N) | *CR*(%) | *MR*(%) | *F*(%) |
|---|---|---|---|---|
| L-758298 (60 or 100 mg) | 28 | 20 (71)[a] | 3 (1) | 5 (18) |
| Ondansetron (32 mg) | 21 | 6 (29) | 4 (19) | 11 (52) |

[a] $p = 0.004$. CR = Complete Responder (0 emetic episodes); MR = Major Responder (1–2 emetic episodes); F = Failure (3 or more emetic episodes).

**Table 8** *Efficacy in cisplatin-induced emesis – combinations*

| *Treatment* | *Total* (N) | *Acute phase CR*(%) | *Delayed phase CR*(%) |
|---|---|---|---|
| G/D/MK → MK | 52 | 92 | 82 |
| G/D/MK → PCB | 54 | 94 | 82 |
| G/D/PCB → PCB | 75 | 75 | 35 |

G = granisetron 10 mg $kg^{-1}$ I.V. D = dexamethasone 20 mg p.o. MK = MK-869 400 mg Day 1; 300 mg QD Day 2–5. CR = Complete Responder (0 emetic episodes).

addition, a number of studies have implicated NKs in spinal cord pain transmission. However, in three clinical trials (dental pain, neuralgia and migraine) MK-869 was shown to be ineffective for the treatment of pain.

## Psychiatric Indications

The treatment of psychiatric illness presents a number of significant challenges: the pathophysiology is poorly understood, the illness cannot be reproduced in animals and it can be difficult to measure efficacy in the clinic. Indeed only ~50% of clinical trials are deemed successful, in part due to significant placebo response. The majority of existing drugs have been discovered by accident rather than by design, with more recent compounds acting *via* established mechanisms with some selectivity changes.

In guinea-pigs, central infusion of substance P agonists causes locomotor activation accompanied by pronounced and long-lasting audible vocalisations. This observation was of particular interest because psychotropic drugs that alleviate symptoms of anxiety and depression in humans are known to inhibit stress-induced vocalisations in many mammalian species. In guinea-pigs, vocalisations elicited by intracerebroventricular (icv) infusion of GR73632 (0.1 nmol) were stereoselectively abolished by pretreatment with an $NK_1$ antagonist. GR73632-induced vocalisations were also markedly attenuated by acute pretreatment with the antidepressant drugs but not by the anxiolytics. These findings show that clinically used antidepressant drugs were able to block the behavioural effects of central substance P receptor stimulation.

Substance P antagonists also stereoselectively inhibit vocalisations evoked in guinea-pig pups by transient maternal separation and is comparable with clinically used antidepressant and anxiolytic drugs.

MK-869 was chosen to test the concept that it would be an effective anti-depressant clinically because of its high affinity, selectivity, brain penetrance, duration, and oral bioavailability that permitted a once daily oral dosing regimen. MK-869 was well tolerated in human volunteer studies at 300 mg, a dose for which pharmacokinetic data predicted 90% blockade of central substance P receptors.

The safety and tolerability of MK-869 were generally similar to placebo. The most common clinical adverse experiences (AEs) observed in patients receiving MK-869 were headache (32%), somnolence (20%), nausea (18%), and fatigue (14%); these were generally mild and transient. Nausea, which occurred in 29% of patients on paroxetine, was the chief AE causing discontinuation of treatment with paroxetine. Notably, the incidence of sexual dysfunction in patients receiving paroxetine (a problem observed with other serotonin reuptake inhibitors) was 26%, significantly greater than with MK-869 (3%) or placebo (4%).

The primary efficacy outcome measure was the 21-item Hamilton depression (HAM-D21) total score; secondary measures included the Hamilton anxiety (HAM-A) total score and the Clinical Global Impressions severity scale (CGI-S). The principal outcome was a difference in mean change from baseline to week 6 between MK-869 and placebo in total HAM-D21 score, suggesting the effect of MK-869 was similar to that of paroxetine.

**Acknowledgements**

The results described in this article are obviously the result of a large multi-disciplinary effort and I would like to thank all those involved.

## 5 REFERENCES

1. C.J. Swain and N.M.J. Rupniak, *Annu. Rep. Med. Chem.*, 1999, **34**, 51.
2. R.M. Snider *et al.*, *Science*, 1991, **251**, 435.
3. Z. Gao and N.P. Peet, *Curr. Med. Chem.*, 1999, **6**, 375.
4. T. Ladduwahetty *et al.*, *J. Med. Chem.*, 1996, **39**, 2907.
5. C.J. Swain *et al.*, *BioOrg. Med. Chem. Lett.*, 1997, **7**, 2959.
6. J.J. Hale, *J. Med. Chem.*, 1998, **41**, 4607.
7. M. Rudolph *et al.*, *New England J. Med.*, 1999, **340**, 190.
8. M.S. Kramer *et al.*, *Science*, 1998, **281**, 1640.
9. F.D. Tattersall *et al.*, *Neuropharmacology*, 2000, **39**, 652.

## Appendix 1: RANKING OF KEY ETHICAL DRUG PRODUCTS IN 2000 US$ SALES VALUE

| *Rank* | *Product* | *Generic* | *Product category* | *Mechanism* | *Company* | *$m* |
|---|---|---|---|---|---|---|
| 1 | Losec/Prilosec | omeprazole | proton pump inhibitor | $H^+/K^+$ ATPase inhibitor | AstraZeneca | 6260 |
| 2 | Zocor | simvastatin | hypolipidaemic | HMG CoA reductase inhibitor | Merck & Co. | 5280 |
| 3 | Lipitor | atarvastatin | hypolipidaemic | HMG CoA reductase inhibitor | Pfizer | 5031 |
| 4 | Norvasc | amlopidine | calcium antagonist | L-type inhibitor | Pfizer | 3362 |
| 5 | Claritin | loratadine | anti-histamine | $H_1$ antagonist | Schering-Plough | 3011 |
| 6 | Prevacid | lansoprazole | proton pump inhibitor | $H^+/K^+$ ATPase inhibitor | TAP | 2739 |
| 7 | Procrit/Eprex | epoetin alfa | anti-anaemic | human recombinant erythropoietin | Johnson & Johnson | 2709 |
| 8 | Celebrex | celecoxib | Cox-2 inhibitor | Cox-2 inhibitor | Pharmacia Corporation | 2614 |
| 9 | Prozac | fluoxetine | anti-depressant | SSRI | Eli Lilly | 2574 |
| 10 | Zyprexa | olanzepine | anti-psychotic | GPCR antagonist | Eli Lilly | 2350 |
| 11 | Seroxat/Paxil | paroxetine | anti-depressant | SSRI | GlaxoSmithKline | 2348 |
| 12 | Vioxx | rofecoxib | Cox-2 inhibitor | Cox-2 inhibitor | Merck & Co. | 2160 |
| 13 | Zoloft | sertraline | anti-depressant | SSRI | Pfizer | 2140 |
| 14 | Epogen | epoetin alfa | anti-anaemic | human recombinant erythropoietin | Amgen | 1963 |
| 15 | Glucophage | metformin | anti-diabetic | insulin sensitiser | Bristol-Myers Squibb | 1892 |
| 16 | Premarin | | hormonal product | | American Home Products | 1870 |
| 17 | Augmentin | amoxicillin/ clavulanic acid | penicillin antibiotic | $\beta$-lactam antibiotic + $\beta$-lactamase inhibitor | GlaxoSmithKline | 1847 |
| 18 | Pravachol | pravastatin | hypolipidaemic | HMG CoA reductase inhibitor | Bristol-Myers Squibb | 1817 |
| 19 | Vasotec | enalapril | ACE Inhibitor | ACE Inhibitor | Merck & Co. | 1790 |
| 20 | Cozaar/Hyzaar | losartan | angiotensin II antagonist | angiotensin II antagonist | Merck & Co. | 1715 |
| 21 | Insulins | – | anti-diabetic | – | Novo Nordisk | 1647 |
| 22 | Ciprobay | ciprofloxacin | quinolone antibiotic | DNA gyrase inhibitor | Bayer | 1638 |

| | | | | | | |
|---|---|---|---|---|---|---|
| 23 | Risperdal | risperidone | anti-psychotic | 5-$HT_2$/$D_2$ antagonist | Johnson & Johnson | 1603 |
| 24 | Taxol | paclitaxel | anti-cancer | microtubule stabiliser | Bristol-Myers Squibb | 1592 |
| 25 | Mevalotin | pravastatin | hypolipidaemic | HMG CoA reductase inhibitor | Sankyo | 1532 |
| 26 | Zithromax | azithromycin | macrolide antibiotic | protein synthesis inhibition | Pfizer | 1382 |
| 27 | Intron A/Rebetron | interferon $\alpha$-2b | cytokine/+ anti-viral | | Schering-Plough | 1360 |
| 28 | Viagra | sildenafil | phosphodiesterase inhibitor | PDEV inhibitor | Pfizer | 1344 |
| 29 | Neurontin | gabapentin | anti-convulsant | GABA releaser | Pfizer | 1334 |
| 30 | Flixotide | fluticasone | respiratory prophylactic | glucocorticoid | GlaxoSmithKline | 1333 |
| 31 | Fosamax | alendronic acid | bisphosphonate | calcium regulating agent | Merck & Co. | 1275 |
| 32 | Biaxin/Klaricid | clarithromycin | macrolide antibiotic | protein synthesis inhibition | Abbott | 1241 |
| 33 | Neupogen | fligrastim | colony stimulating factor | G-CSF | Amgen | 1224 |
| 34 | Sandimmun/Neoral | cyclosporine | immunosuppressant | T lymphocyte inhibitor | Novartis | 1214 |
| 35 | Zestril | lisinopril | ACE inhibitor | ACE inhibitor | AstraZeneca | 1188 |
| 36 | Effexor | venlafaxine | anti-depressant | SSRI | American Home Products | 1159 |
| 37 | Humulin | human insulin | anti-diabetic | | Eli Lilly | 1114 |
| 38 | Prinivil | lisinopril | ACE inhibitor | ACE inhibitor | Merck & Co. | 1075 |
| 39 | Allegra/Telfast | fexofenadine | anti-histamine | $H_1$ antagonist | Aventis | 1070 |
| 40 | Imigran | sumatriptan | anti-migraine | 5-$HT_{1B/D}$ agonist | GlaxoSmithKline | 1068 |
| 41 | Adalat | nifedipine | calcium antagonist | L-type inhibitor | BAYER | 1060 |
| 42 | Diflucan | fluconazole | anti-fungal | sterol 14-$\alpha$-demethylase inhibitor | Pfizer | 1014 |
| 43 | Rocephin | ceftriaxone | cephalosporin antibiotic | $\beta$-lactam antibiotic | Hoffmann-La Roche | 1012 |
| 44 | O.C.brands | | oral contraceptive | | Johnson & Johnson | 956 |
| 45 | Clexane/Lovenox | enoxaparin | low-m. wt. heparin | platelet aggregation inhibitor | Aventis | 956 |
| 46 | Serevent | salmeterol | $\beta_2$ agonist | $\beta_2$ agonist | GlaxoSmithKline | 942 |
| 47 | Levaquin | levofloxacin | quinolone antibiotic | DNA gyrase inhibitor | Johnson & Johnson | 941 |
| 48 | Plavix | clopidogrel | platelet antiaggregant | platelet aggregation inhibitor | Bristol-Myers Squibb | 903 |
| 49 | Zantac | ranitidine | $H_2$-antagonist | $H_2$-antagonist | GlaxoSmithKline | 871 |
| 50 | Vaccines | | vaccine | | Aventis | 867 |

# Appendix 2: SUMMARY OF RECEPTOR PROPERTIES

| | Agonists | Antagonists | Radioligands | Coupling | Structure |
|---|---|---|---|---|---|
| Acetylcholine receptors (muscarinic) *Pharmacol. Toxicol.*, 1996, **78**, 59 | | | | | |
| $M_1$ | sabcomeline; xanomeline | MT7; pirenzepine | [$^3$H]NMS [$^3$H]QNB | $G_{q/11}$ | h460 P11229 |
| $M_2$ | – | Tripitramine AFDX384 | [$^3$H]NMS [$^3$H]QNB | $G_{i/o}$ | h466 P08172 |
| $M_3$ | – | 4-DAMP darifenacin | [$^3$H]NMS [$^3$H]QNB | $G_{q/11}$ | h590 P20309 |
| $M_4$ | – | MT3 4-DAMP | [$^3$H]NMS [$^3$H]QNB | $G_{i/o}$ | h479 P08173 |
| $M_5$ | – | 4-DAMP darifenacin | [$^3$H]NMS [$^3$H]QNB | $G_{q/11}$ | h532 P08192 |
| Adenosine receptors Review: *Pharmacol. Rev.*, 1998, **50**, 413 | | | | | |
| $A_1$ | CPA; CCPA | DPCPX | [$^3$H]CCPA [$^3$H]DPCPX | $G_{i/o}$ | h326 P30542 |
| $A_{2A}$ | CGS 21680; HE-NECA | ZM241385; SCH58261 | [$^3$H]CGS21680 [$^3$H]ZM241385 | $G_s$ | h332 P29275 |
| $A_{2B}$ | – | MRS1754 | [$^3$H]MRS1754 | $G_s$ | h332 P29275 |
| $A_3$ | IB-MECA; 2-Cl-IB-MECA | MRS1220 VUF5574 | [$^{125}$I]AB-MECA | $G_{i/o}$ | h318 P33765 |
| Adrenoreceptors ($\alpha_1$-adrenoceptors) Review: *Trends Pharmacol. Sci.*, 1999, **20**, 94 | | | | | |
| $\alpha_{1A}$ | A61603 | KMD3213; SNAP5089 | [$^3$H]prazosin; [$^{125}$I]HEAT | $G_{q/11}$ | h466 P35348 |
| $\alpha_{1B}$ | – | AH1111OA | [$^3$H]prazosin; [$^{125}$I]HEAT | $G_{q/11}$ | h519 P35368 |
| $\alpha_{1D}$ | – | BMY7378; SKF105854 | [$^3$H]prazosin; [$^{125}$I]HEAT | $G_{q/11}$ | h572 P25100 |

| | | | | | |
|---|---|---|---|---|---|
| $\alpha_{2A}$ | oxymetazoline; guanfacine | BRL44408 | [$^{3}$H]RX821002; [$^{3}$H]UK14304 | $G_{i/o}$ | h450 P08913 |
| $\alpha_{2B}$ | – | imiloxan | [$^{3}$H]RX821002; [$^{3}$H]UK14304 | $G_{i/o}$ | h450 P18089 |
| $\alpha_{2C}$ | – | – | [$^{3}$H]RX821002; [$^{3}$H]UK14304 | $G_{i/o}$ | h461 P18825 |
| $\beta_1$ | denopamine; xameterol; | CGP20712A; betaxol; atenolol | [$^{125}$I]ICYP; [$^{3}$H]CGP12177 | $G_s$ | h477 P08588 |
| $\beta_2$ | salmeterol; formoterol | ICI118551 | [$^{125}$I]ICYP; [$^{3}$H]CGP12177 | $G_s$ | h413 P07550 |
| $\beta_3$ | BRL37344; CGP12177 | SR59230A; L748328 | [$^{125}$I]ICYP; [$^{3}$H]CGP12177 | $G_s/G_{i/o}$ | h408 P139415 |
| Angiotensin receptors Review: *Clin. Sci.*, 2001, **100**, 481 | | | | | |
| $AT_1$ | L162017 | EXP3174; SKF108566; losartan | [$^{3}$H]A81988; [$^{3}$H]L158809 | $G_{q/11}/G_{i/o}$ | h359 P30556 |
| $AT_2$ | [$p$-$NH_2$-Phe$^{6}$]-Ang II | PD123319: PD123177 | [$^{3}$H]CGP42112 | $G_{i\alpha2,3}$ | h363 P50052 |
| Atrial natriuretic peptide receptors *J. Mol. Biol.*, 2001, **100**, 481 | | | | | |
| $ANP_A$ | ANP(R3D, G9T, R11SM12L, R14S, G16R) | [Asu$^{7,23}$]-$\beta$-$ANP_{7\text{-}28}$ | [$^{125}$I]ANP | – | h1061 P16066 |
| $ANP_B$ | – | mAb3G12 | [$^{125}$I]CNP | | h1047 P20594 |
| Bombesin receptors *Ann. Med.*, 2000, **32**, 819 | | | | | |
| BB1 | NMB | PD165929 | [$^{125}$I]BH-NMB | $G_{q/11}$ | h390 P28336 |
| BB2 | GRP | kuwanon H | [$^{125}$I]GRP | $G_{q/11}$ | h384 P30550 |
| BRS-3 | – | – | [$^{125}$I][Tyr$^{6}$, $\beta$Ala$^{11}$, Phe$^{13}$, Nle$^{14}$]bombesin$_{6\text{–}14}$ | $G_{q/11}$ | h399 P32247 |

(*continued overleaf*)

**Appendix 2: (*continued*)**

| | *Agonists* | *Antagonists* | *Radioligands* | *Coupling* | *Structure* |
|---|---|---|---|---|---|
| Bradykinin receptors: *Curr. Pharmaceut. Design*, 2001, **7**, 135 | | | | | |
| $B_1$ | Lys[des-Arg$^9$]BK | B9958; R715 | [$^3$H]Lys[des-Arg$^9$]BK | $G_{q/11}$ | h353 P46663 |
| $B_2$ | [Hyp$^3$, Tyr(Me)$^8$]BK | HOE140; FR173657 | [$^3$H]NPC17731 | $G_{q/11}$ | h364 P30411 |
| Cannabinoid receptors: *Progress Neurobiol.*, 2001, **25**, 91 | | | | | |
| $CB_1$ | CP55244; HU210 | SR141716; LY320135 | [$^3$H]HU243; [$^3$H]SR141716 | $G_{i/o}$ ($G_s$) | h472 P21554 |
| $CB_2$ | HU308; L759633 | SR144528; AM630 | [$^3$H]HU243; [$^3$H]CP55940 | $G_{i/o}$ | h360 P34972 |
| Chemokine receptors: *Receptors Channels*, 2001, **7**, 417 | | | | | |
| CCR1 | MIP-1$\alpha$ RANTES | – | [$^{125}$I]RANTES; [$^{125}$I] MIP-1$\alpha$ | $G_{i/o}$ | h355 P32246 |
| CCR2 | MCP-1; MCP-3 | – | [$^{125}$I]MCP-1; [$^{125}$I]MCP-3 | $G_{i/o}$ | h374 P41597 |
| CCR3 | eotaxin; eotaxin-2 | mAb7B11; RS163883230 | [$^{125}$I]RANTES; [$^{125}$I]eotaxin | $G_{i/o}$ | h355 P51677 |
| CCR4 | TARC; MDC | – | [$^{125}$I]TARC | $G_{i/o}$ | h360 P51679 |
| CCR5 | MIP-1$\beta$ | TAK779; mAb2D7 | [$^{125}$I]RANTES; [$^{125}$I]MIP-1$\beta$ | $G_{i/o}$ | h352 P51681 |
| CCR6 | LARC; HBD2 | – | [$^{125}$I]LARC | $G_{i/o}$ | h374 P51684 |
| CCR7 | ELC; SLC | – | [$^{125}$I]ELC; | $G_{i/o}$ | h378 P32248 |
| CCR8 | I-309 | MC148R | [$^{125}$I]I-309 | $G_{i/o}$ | h355 P51685 |
| CCR9 | TECK | – | [$^{125}$I]TECK | $G_{i/o}$ | h357 P51686 |
| CCR10 | Eskine | – | – | $G_{i/o}$ | h354 AAA64593 |
| CCR11 | MCP-4 | – | – | $G_{i/o}$ | h350 AF193507 |
| CXCR1 | IL-8 | mAb5A12; | [$^{125}$I]IL-8 | $G_{i/o}$ | h350 P25024 |

| | | | | | |
|---|---|---|---|---|---|
| CXCR2 | GRO$\alpha$, $\beta$, $\gamma$; NAP-2; ENA78 | SB225002; mAb6C6 | [$^{125}$I]IL-8; [$^{125}$I]GRO$\alpha$ [$^{125}$I]NAP-2 | $G_{i/o}$ | h360 P25025 |
| CXCR3 | IP10; MIG; I-TAC | – | [$^{125}$I]IP-10 | $G_{i/o}$ | h368 P49682 |
| CXCR4 | SDF-1$\alpha$; $\beta$ | ALX40-4C; T22; AMD3100 | [$^{125}$I]SDF-1 | $G_{i/o}$ | h352 P30991 |
| CXCR5 | BLC; BCA-1 | – | – | $G_{i/o}$ | h372 P32302 |
| $CX_3CR1$ | fractalkine | – | [$^{125}$I]$CX_3C$-76 | $G_i$ | h360 |
| Chemotactic peptide receptors: *Immunol. Rev.*, 2000, **177**, 185 | | | | | |
| C3a | C3a | – | [$^{125}$I]C3a | $G_{i/o}/G_z$ | h482 Q16581 |
| C5a | C5a | AcPhe-[OPro-D-ChaTrpArg] | [$^{125}$I]C5a | $G_{i/o}/G_z/G_{\alpha 16}$ | h350 P21730 |
| fMLP | – | BOC-PLPLP | [$^{3}$H]fMLP | $G_{i/o}/G_z$ | h350 P21462 |
| Cholecystokinin receptors: *Curr. Med. Chem.*, 1999, **6**, 433 | | | | | |
| $CCK_1$ | A71623; JMV180; GW5823 | Devazepide; T0632; SR27897 | [$^{3}$H]devazepide | $G_{q/11}/G_s$ | h428 P32238 |
| $CCK_2$ | gastrin; BC264; RB400 | YM022; L740093; GV150013; RP73870 | [$^{3}$H]gastrin; [$^{3}$H]PD140376 | $G_{q/11}$ | h447 P32239 |
| Corticotropin-releasing factor receptors: *J. Med. Chem.*, 2000, **43**, 1641 | | | | | |
| $CRF_1$ | CRF; sauvagine; urotensin 1 | CP154526; NB127914; antalarmin | [$^{3}$H]urocortin | $G_s$ | h415 P34998 |
| $CRF_2$ | CRF; sauvagine; urotensin 1 | $\alpha$-helical CRF (9-41) | [$^{3}$H]urocortin | $G_s$ | h411 Q13324 |
| Dopamine receptors: *Ann. Rev. Pharmacol. Toxicol.*, 1999, **39**, 313 | | | | | |
| D1 | R(+)SKF81279 R(+)SKF38393 | SCH23390; SKF83566 | [$^{3}$H]SCH23390 | $G_s$ | h446 P21728 |
| D2 | (+)PHNO; lisuride | Raclopride; haloperidol | [$^{3}$H]raclopride; [$^{125}$I]iodosulpiride | $G_{i/o}$ | h443 P14416 |
| D3 | PD128097 | SB-277011 | [$^{3}$H]PD128097 | $G_{i/o}$ | h400 P35462 |
| D4 | PD168077 | L745870; U101958 | [$^{3}$H]NGD941; [$^{125}$I]L750667 | $G_{i/o}$ | h387 P21917 |

(*continued overleaf*)

**Appendix 2: (*continued*)**

| | *Agonists* | *Antagonists* | *Radioligands* | *Coupling* | *Structure* |
|---|---|---|---|---|---|
| D5 | R(+)SKF38393 | SCH23390 | [$^{3}$H]SCH23390 | $G_s$ | h477 P21918 |
| Endothelin receptors Review: *Pharm. Biotechnol.*, 1998, **11**, 113 | | | | | |
| $ET_A$ | ET-1; ET-2 | A127722; LU135252; SB234551 | [$^{3}$H]S0139 [$^{125}$I]PD164333 | $G_{q/11}$ | h427 P25101 |
| $ET_B$ | IRL1620; BQ3020 | A192621; BQ788 | [$^{125}$I]IRL1620; [$^{125}$I]BQ3020 | $G_{q/11}$ | h442 P24530 |
| GABA receptors: *Cell. Mol. Life Sci.*, 2000, **57**, 635 | | | | | |
| $GABA_B$ | L-baclofen; CGP35024 | CGP64213; SCH50911 | [$^{3}$H]L-baclofen; [$^{3}$H]CGP64213 | $G_{i/o}$ | h961 AJ012185 h941 AJ012188 |
| Galanin receptors: *Trends Pharmacol. Sci.*, 2000, **21**, 109 | | | | | |
| GAL1 | galanin | – | – | $G_{i/o}$ | h349 P42711 |
| gal2 | GALP | galanin-2-29 | – | $G_{i/o}/G_{q/11}$ | h387 O43603 |
| gal3 | – | – | – | $G_{i/o}$ | h368 O60755 |
| Glutamate (metabotropic) receptors: *Annu. Rep. Med. Chem.*, 2000, **35**, 1 | | | | | |
| $mglu_1$ | DHPG; 3HPG | LY393675 | – | $G_{q/11}$ | h1194 Q13255 |
| $mglu_2$ | LY389795 | LY341495 | – | $G_{i/o}$ | h872 Q14416 |
| $mglu_3$ | LY389795 | LY341495 | – | $G_{i/o}$ | h877 Q14832 |
| $mglu_4$ | L-AP4, LSOP | MAP4 | – | $G_{i/o}$ | h912 Q14833 |
| $mglu_5$ | CHPG; DHPG | SIB1893 | – | $G_{q/11}$ | h1212 P41594 |
| $mglu_6$ | homo-AMPA; 1-benzyl-APDC | MAP4; THPG | – | $G_{i/o}$ | h877 O15303 |
| $mglu_7$ | LSOP; L-AP4 | – | – | $G_{i/o}$ | h915 Q14831 |
| $mglu_8$ | LSOP; L-AP4 | MPPG | – | $G_{i/o}$ | h908 O00222 |

| | | | | | |
|---|---|---|---|---|---|
| Glycoprotein receptors: *Annu. Rev. Physiol.*, 1998, **60**, 461 | | | | | |
| FSH | FSH | – | [$^{125}$I]FSH | $G_s$ | h695 P23945 |
| LSH | LH; CG | – | [$^{125}$I]LH; [$^{125}$I]CG | $G_s/G_{q/11}/G_i$ | h699 P22888 |
| TSH | TSH | – | [$^{125}$I]TSH | all | h764 P16473 |
| Histamine receptors: *Trends Pharmacol. Sci*, 2000, **21**, 11; *Pharmacol. Rev.*, 1997, **49**, 253 | | | | | |
| $H_1$ | 2-(3-F-Ph)-histamine | triprolidine; mepyramine | [$^{3}$H]mepyramine; [$^{125}$I]iodbolpyramine | $G_{q/11}$ | h487 P35367 |
| $H_2$ | dimaprit; impromidine | tiotidine; ranitidine | [$^{3}$H]tiotidine; [$^{125}$I]iodoaminopotentidine | $G_s$ | h359 P20521 |
| $H_3$ | R-$\alpha$-methyl-histamine; imetit | clobenpropit; iodophenpropit; thioperamide | [$^{3}$H] R-$\alpha$-methyl-histamine; [$^{125}$I]iodoproxyfan | $G_{i/o}$ | h445 AF140538 |
| 5-Hydroxytryptamine (serotonin) receptors: *Annu. Rep. Med. Chem.*, 2000, **35**, 11 | | | | | |
| 5-$HT_{1A}$ | 8-OH-DPAT | WAY100635 | [$^{3}$H] WAY100635; [$^{3}$H] 8-OH-DPAT | $G_{i/o}$ | h421 P8908 |
| 5-$HT_{1B}$ | L694247 | SB236057 | [$^{3}$H]sumatriptan | $G_{i/o}$ | h390 P28222 |
| 5-$HT_{1D}$ | PNU109291 | BRL15572 | [$^{3}$H]sumatriptan | $G_{i/o}$ | h377 P28221 |
| 5-$HT_{1E}$ | – | – | [$^{3}$H]5-HT | $G_{i/o}$ | h365 P28566 |
| 5-$HT_{1F}$ | LY334370 | – | [$^{3}$H]LY334370 | $G_{i/o}$ | h366 P30939 |
| 5-$HT_{2A}$ | $\alpha$-Me-5-HT | ketanserin; MDL100907 | [$^{3}$H] ketanserin | $G_{q/11}$ | h471 P28223 |
| 5-$HT_{2B}$ | $\alpha$-Me-5-HT | SB204741 RS-127445 | [$^{3}$H]5-HT | $G_{q/11}$ | h481 P41595 |
| 5-$HT_{2C}$ | Ro600175 | SB242084 | [$^{3}$H]mesulergine | $G_{q/11}$ | h458 P28335 |
| 5-$HT_4$ | RS67506; ML10302 | SB204070; RS100235 | [$^{3}$H]GR113808; [$^{125}$I]SB207710 | $G_s$ | h387 Y09756 |
| 5-$ht_{5A}$ | – | – | [$^{3}$H]5-CT; [$^{125}$I]LSD | – | h357 P47898 |
| 5-$ht_{5B}$ | – | – | [$^{3}$H]5-CT; [$^{125}$I]LSD | – | – |

(*continued overleaf*)

**Appendix 2: (*continued*)**

| | *Agonists* | *Antagonists* | *Radioligands* | *Coupling* | *Structure* |
|---|---|---|---|---|---|
| $5\text{-HT}_6$ | – | Ro630536: SB271046 | [$^3$H]5-CT; [$^{125}$I]LSD | $G_s$ | h440 P50406 |
| $5\text{-HT}_7$ | – | SB258719 | [$^3$H]5-CT; [$^{125}$I]LSD | $G_s$ | h445 P34969 |
| Leukotriene receptors: *Pharmacol. Res.*, 1999, **40**, 3 | | | | | |
| BLT | $LTB_4$ | SB209247; SC53228 | [$^3$H]$LTB_4$; [$^3$H]CGS23131 | $G_{q/11}/G_{i/o}$ | h352 Q15722 |
| $CysLT_1$ | $LTD_4$ | ICI204219; MK476 | [$^3$H] $LTD_4$; [$^3$H]ICI198615 | $G_{q/11}$ | h337 AAD42778 |
| $CysLT_2$ | $LTC_4$ | – | – | $G_{q/11}$ | – |
| Lysophospholipid receptors: *Crit. Rev. Neurobiol.*, 1999, **13**, 151 | | | | | |
| edg1 | S1P | – | [$^{32}$P]S1P | $G_{i/\alpha 1,3}$ | h381 P21453 |
| edg2 | LPA | – | [$^3$H]LPA | $G_{i/o}$ | h364 Q92633 |
| edg3 | S1P | – | [$^{32}$P]S1P | $G_q/G_{i/o}$ | h378 Q99500 |
| edg4 | LPA | – | [$^3$H]LPA | $G_q/G_{i/o}$ | h382 AFO11466 |
| edg5 | S1P | – | [$^{32}$P]S1P | $G_q$ | h353 AF034780 |
| Melanocortin receptors: *Cell. Mol. Life Sci.*, 2001, **58**, 434 | | | | | |
| $MC_1$ | $\alpha$-MSH | – | [$^{125}$I]NDP-MSH | $G_s$ | h317 Q01726 |
| $MC_2$ | ACTH | – | [$^{125}$I]$ACTH_{1-24}$ | $G_s$ | h297 Q01718 |
| $MC_3$ | $\gamma$-MSH; $\beta$-MSH | – | [$^{125}$I]NDP-MSH | $G_s$ | h360 P41968 |
| $MC_4$ | $\beta$-MSH | HS014 | [$^{125}$I]NDP-MSH | $G_s$ | h332 P32245 |
| $MC_5$ | $\alpha$-MSH | – | [$^{125}$I]NDP-MSH | $G_s$ | h325 P33032 |
| Melatonin receptors: *Physiol. Rev.*, 1998, **78**, 687 | | | | | |
| $MT_1$ | $MEL_{1A}$; $ML_{1A}$ | – | [$^3$H]MLT | $G_{i/o}$ | h362 P49286 |
| $MT_2$ | $MEL_{1B}$; $ML_{1B}$ | K185; DH97 | [$^3$H]MLT | $G_{i/o}$ | h362 P49286 |
| $MT_3$ | N-Ac-5-HT; 5MCA-NAT | Prazosin | [$^{125}$I]2-I-MLT | – | – |

| | | | | | |
|---|---|---|---|---|---|
| Neuropeptide Y receptors: *Annu. Rep. Med. Chem.*, 1999, **34**, 31 | | | | | |
| $Y_1$ | [Pro[34]]NPY; [Pro[34]]PYY | BIBO3304; BIBP3226 | [[125]I][Leu[31], Pro[34]]NPY [[3]H] BIBP3226 | $G_{i/o}$ | h384 P25929 |
| $Y_2$ | $NPY_{13-36}$ | BIIE0246 | [[125]I]$PYY_{3-36}$ | $G_{i/o}$ | h381 P49146 |
| $Y_4$ | PP | – | [[125]I]PP | $G_{i/o}$ | h375 P50391 |
| $Y_5$ | $PYY_{3-36}$ | CGP71683 | – | $G_{i/o}$ | h445 Q15761 |
| $y_6$ | NPY; PYY | – | – | $G_{i/o}$ | h290 Y59431 |
| Neurotensin receptors: *Pharmacol. Ther.*, 1998, **79**, 89 | | | | | |
| NTS1 | JMV449 | SR48629 | [[3]H]SR48629; [[125]I]neurotensin | $G_{q/11}$ | h418 P30989 |
| nts2 | Levocobastine | – | [[125]I]neurotensin | $G_{q/11}$ | h410 Y10148 |
| Opioid and opioid-like receptors: *Biopolymers*, 2000, **55**, 334 | | | | | |
| OP1 ($\delta$) | DPDPE, DSBULET | Naltrindole NNDT | [[3]H]DPDPE [[3]H]TIPP4 | $G_{i/o}$ | h372 P41143 |
| OP2 ($\kappa$) | U69593; CI977 | Nor-binaltorphine | [[3]H]U69593 | $G_{i/o}$ | h380 P41145 |
| OP3 ($\mu$) | endomorphin-1; DAMGO | CTOP | [[3]H]DAMGO; [[3]H]PL017 | $G_{i/o}$ | h400 P35372 |
| OP4 | N/OFQ | J113397 | [[3]H]N/OFQ | $G_{i/o}$ | h370 P41146 |
| Purinergic P2Y receptors: *Prog. Med. Chem.*, 2000, **38**, 115 | | | | | |
| $P2Y_1$ | 2-MeSADP | MRS2197 | [[35]S]ADP$\beta$S | $G_{q/11}$ | h373 P47900 |
| $P2Y_2$ | UTP$\gamma$S | – | – | $G_{q/11}$ | h377 P41231 |
| $P2Y_4$ | UTP$\gamma$S | – | – | $G_{q/11}$ | h365 P51582 |
| $P2Y_6$ | UDP | – | – | $G_{q/11}$ | h328 Q15077 |
| $p2Y_{11}$ | ATP | – | – | $G_s/G_{q/11}$ | h371 AF030335 |
| $P2Y_{ADP}$ | ADP | ARL66096 | – | $G_{i/o}$ | – |
| Prostanoid receptors: *Annu. Rev. Pharmacol. Toxicol.*, 2001, **41**, 661 | | | | | |
| DP | L644698; BW245C | BWA868C | [[3]H]$PGD_2$ | $G_s$ | h359 Q13258 |
| FP | Fluprostenol | – | [[3]H]$PGF_{2\alpha}$ | $G_{q/11}$ | h359 P43088 |
| IP | cicaprost | – | [[3]H]Iloprost | $G_s$ | h386 P43119 |

(*continued overleaf*)

**Appendix 2: (*continued*)**

| | *Agonists* | *Antagonists* | *Radioligands* | *Coupling* | *Structure* |
|---|---|---|---|---|---|
| TP | U46619 | BMS180291; ONO3708 | [$^3$H]SQ29548; [$^{125}$I]SAP | $G_{q/11}$ | h369 P21731 |
| $EP_1$ | Iloprost | SC51089 | [$^3$H]$PGE_2$ | $G_{q/11}$ | h402 P34995 |
| $EP_2$ | Butaprost; AH13205 | – | [$^3$H]$PGE_2$ | $G_s$ | h358 P43116 |
| $EP_3$ | SC46275 | – | [$^3$H]$PGE_2$ | $G_{q/11}/G_{i/o}/G_s$ | h390 P43115 |
| $EP_4$ | – | AH23848 | [$^3$H]$PGE_2$ | $G_s$ | h488 P35408 |
| Protease-activated receptors: *Yakugaku Zasshi-J. Pharm. Soc. Jpn.*, 2001, **121**, 1 | | | | | |
| PAR1 | thrombin; trypsin; TFLLR-$NH_2$ | – | – | $G_{q/11}/G_{i/o}/G_{12/13}$ | h425 P25116 |
| PAR2 | trypsin; SLIGRL | – | – | – | h397 P55085 |
| PAR3 | thrombin, trypsin, factor Xa | – | – | – | h374 U92971 |
| PAR4 | thrombin, trypsin; GYPGQV | – | – | – | h385 AF80214 |
| Somatostatin receptors: *Annu. Rep. Med. Chem.*, 1999, **34**, 209 | | | | | |
| $sst_1$ | des-Ala$^{1,2,5}$-[DTrp$^8$, Iamp$^9$]SRIF | – | – | $G_{i/o}$ | h391 P30872 |
| $sst_2$ | Octreotide; BIM23027 | Cyanamid 154806 | [$^{125}$I][Tyr$^3$]octreotide | $G_{i/o}$ | h369 P30874 |
| $sst_3$ | – | – | – | $G_{i/o}$ | h418 P32745 |
| $sst_4$ | NNC269100 | – | – | $G_{i/o}$ | h388 P31391 |
| $sst_5$ | BIM23268 | BIM23056 | [$^{125}$I][Tyr$^3$]octreotide | $G_{i/o}$ | h363 P35346 |
| Steroid hormone receptors: *Clin. Sci.*, 2000, **90**, 1 | | | | | |
| MR | aldosterone | RU28318 | [$^3$H]aldosterone | – | h984 P08235 |

| | | | | | |
|---|---|---|---|---|---|
| GR | RU28362 | RU38486 | [$^3$H]dexamethasone | – | h777 P04150 |
| PR | ORG2058; progesterone | RU38486 | [$^3$H]ORG2058 | – | h993 P06401 |
| AR | DHT, R1881 | hydroxyflutamide | [$^3$H]DHT; [$^3$H]R1881 | – | h919 P10725 |
| Tachykinin receptors: *Neuropeptides*, 2000, **34**, 303 | | | | | |
| $NK_1$ | SP-OMe; [Pro$^9$]SP; septide | L742694; SR140333; LY303870 | [$^3$H]SP; [$^{125}$I]L703606 | $G_{q/11}$ | h407 P25103 |
| $NK_2$ | [$\beta$-Ala$^8$]$NK_{4-10}$; GR64349 | GR94800; MEN10627 SR48968 | [$^3$H]SR48968; [$^3$H]GR100627; [$^{125}$I]NKA | $G_{q/11}$ | h398 P21452 |
| $NK_3$ | senktide; [MePh$^7$]NKB | SR1428002; SB223412 | [$^3$H]senkide; [$^{125}$I] [MePh$^7$]NKB | $G_{q/11}$ | h468 P29371 |
| Thyrotropin-releasing hormone receptor: *Physiol Rev.*, 1996, **76**, 175 | | | | | |
| $TRH_1$ | Phe$^2$TRH | midazolam; diazepam | [$^3$H]MeTRH | $G_q$ | h398 P34981 |
| Vasoactive intestinal and pituitary cyclase activating peptide receptors: *Trends Pharmacol. Sci.*, 1999, **20**, 324 | | | | | |
| $VPAC_1$ | [Lys$^{15}$, Arg$^{16}$, Leu$^{27}$]$VIP_{1-7}$-$GRF_{8-27}$-$NH_2$ | | [$^{125}$I]VIP | $G_s$ | h457 P32241 |
| *Antagonist:* [Ac-His$^1$, D-Phe$^2$, Lys$^{15}$, Arg$^{16}$] $VIP_{3-7}$-$GRF_{8-27}$-$NH_2$ | | | | | |
| $VPAC_2$ | Ro251553 | – | [$^{125}$I]VIP | $G_s$ | h438 P41587 |
| $PAC_1$ | maxadilan | $PACAP_{6-38}$ | [$^{125}$I]PACAP | $G_s$ | h468 P41586 |
| Vasopressin and oxytosin receptors: *Physiol. Rev.*, 2001, **81**, 629 | | | | | |
| $V_{1a}$ | F180 | SR49059 | [$^3$H]SR49059 | $G_{q/11}$ | h418 P37288 |
| $V_{1b}$ | d[D-3-Pal$^2$]VP | – | [$^3$H]VP | $G_{q/11}$ | h424 P47901 |
| $V_2$ | OPC51803; VNA932 | SR121463; OPC31260 | [$^3$H]SR121463 | $G_s$ | h371 P30518 |
| OT | [Thr$^4$,Gly$^7$]OT | L372662 | | $G_s$ | h389 P30559 |
| *Radioligand*: [$^{125}$I]d$(CH_2)_5$[Tyr(Me)$^2$, Thr$^4$, Orn$^8$, Tyr-$NH_2$$^9$]OT | | | | | |

# Appendix 3

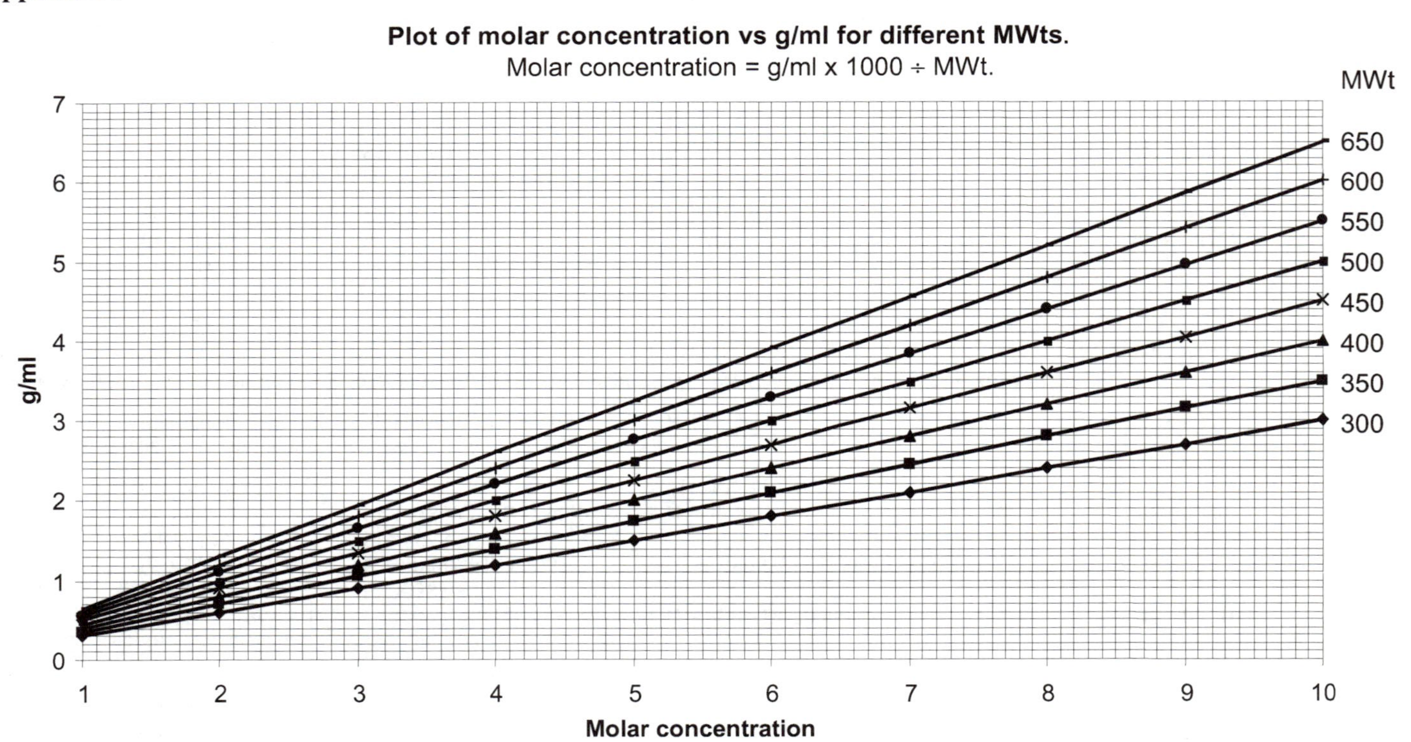
Plot of molar concentration vs g/ml for different MWts.
Molar concentration = g/ml x 1000 ÷ MWt.
MWt
650
600
550
500
450
400
350
300
g/ml
0
1
2
3
4
5
6
7
1
2
3
4
5
6
7
8
9
10
Molar concentration

## Appendix 4: Table of Molar Concentration *vs* g ml$^{-1}$ for Different MWs

| $M$ | g ml$^{-1}$ | g ml$^{-1}$ | g ml$^{-1}$ | g ml$^{-1}$ | g ml$^{-1}$ | g ml$^{-1}$ | g ml$^{-1}$ | g ml$^{-1}$ |
|---|---|---|---|---|---|---|---|---|
| 10 | 3 | 3.5 | 4 | 4.5 | 5 | 5.5 | 6 | 6.5 |
| 9 | 2.7 | 3.15 | 3.6 | 4.05 | 4.5 | 4.95 | 5.4 | 5.85 |
| 8 | 2.4 | 2.8 | 3.2 | 3.6 | 4 | 4.4 | 4.8 | 5.2 |
| 7 | 2.1 | 2.45 | 2.8 | 3.15 | 3.5 | 3.85 | 4.2 | 4.55 |
| 6 | 1.8 | 2.1 | 2.4 | 2.7 | 3 | 3.3 | 3.6 | 3.9 |
| 5 | 1.5 | 1.75 | 2 | 2.25 | 2.5 | 2.75 | 3 | 3.25 |
| 4 | 1.2 | 1.4 | 1.6 | 1.8 | 2 | 2.2 | 2.4 | 2.6 |
| 3 | 0.9 | 1.05 | 1.2 | 1.35 | 1.5 | 1.65 | 1.8 | 1.95 |
| 2 | 0.6 | 0.7 | 0.8 | 0.9 | 1 | 1.1 | 1.2 | 1.3 |
| 1 | 0.3 | 0.35 | 0.4 | 0.45 | 0.5 | 0.55 | 0.6 | 0.65 |
| MW | 300 | 350 | 400 | 450 | 500 | 550 | 600 | 650 |

## Appendix 5: Conversion Table for $IC_{50}$ ($K_i$) to $pIC_{50}$ ($pK_i$)

| $IC_{50}$ nM | 1 | 2 | 3 | 4 | 5 | 6 | 7 | 8 | 9 | 10 | 100 | 1 μM |
|---|---|---|---|---|---|---|---|---|---|---|---|---|
| $pIC_{50}$ | 9 | 8.7 | 8.52 | 8.4 | 8.3 | 8.22 | 8.15 | 8.1 | 8.05 | 8 | 7 | 6 |

| $pIC_{50}$ | 9 | 8.9 | 8.8 | 8.7 | 8.6 | 8.5 | 8.4 | 8.3 | 8.2 | 8.1 | 8 | 7 | 6 |
|---|---|---|---|---|---|---|---|---|---|---|---|---|---|
| $IC_{50}$ nM | 1 | 1.25 | 1.4 | 2 | 2.5 | 3.2 | 4 | 5 | 6.3 | 8.0 | 10 | 100 | 1 $\mu$M |

# Subject Index


Absorption, 120
ACE, 85
Acetylation, 151
Acyclovir, 348
ADME, 254
Administration, routes of, 121
Agonism, 5, 102
AIDS, 397
Aldehyde oxidase, 149
Alosetron, 32
Ames test, 170
Aminoglycosides, 45
Amlopidine, 35
Amoxycillin, 78
AMPA, 31
Amprenavir, 403
Aneugens, 169
Animal models, 113
Antagonism, 18, 102
Anti-emesis, 424
Anxiety, 116
Apparent inhibition constant, 70
Arachidonic acid pathway, 408
Artificial intelligence, 238
Artificial neural networks, 239
Aspartyl protease, 397
Aspirin, 408
Atracurium, 29
AUC, 123, 174
Automation, 192
  synthesis, 374
  purification, 376
  analysis, 376
Azithromycin, 43

Bacampicillin, 348
Bacteriorhodopsin, 252
Back-up, 335
Best mode, 271
Bioavailability, 124
Binding energy, 249
Bioinformatics, 322
Bioisosteres, 341, 387
$B_{max}$, 13, 98
$\beta$-lactamase, 79
Blood-brain barrier, 256
BMS-204352, 33
BMS-284756, 51
Brain penetration, 211
BRL 24682, 344
BRL 26175, 344

Caffeine, 147
Calcium channel blocker, 380
Captopril, 86
Carbamazepine, 36
Cathepsin D inhibitor, 379
CGP 38 560, 88
Chemical development, 182
Chemistry, 355
  multicomponent condensations, 371
  solid phase, 364
  Ugi reaction, 372
Chemotherapy-induced emesis, 422
Cheng–Prusoff equation, 208
Chromosomes, 303
Cimetidine, 7, 21
Ciprofloxacin, 49
Cisapride, 39
Clarithromycin, 43
Clark's occupation theory, 11, 15
Clastogens, 169
Clearance, 124, 125
Clebopride, 344
Clozapine, 147

Cluster analysis, 222
$C_{max}$, 123, 175
Collagenase inhibitor, 379
CoMFA, 236
Combinatorial chemistry, 359
  arrays, 360
  lead generation, 377
  lead optimisation, 378
  libraries, 360
  solution phase, 368
Computational chemistry, 243
Computational properties, 208
Conformational restriction, 342
COX-2 inhibitor, 380, 408
CP-96, 345, 416
CP-108, 671, 88
Craig plot, 220
Crestor®, 84
Cytochromes P450, 141
  CYP1A2, 147
  CYP2C8, 148
  CYP2C9, 145
  CYP2C19, 147
  CYP2D6, 145
  CYP3A4, 142
Cytotoxicity, 163, 166

ΔLogP, 202
Dalfopristin, 44
Data display, 230
  interpretation, 351
Deconvolution, 362
*De novo* design, 246, 250
Depression, 382, 426
DEREK, 160
Development candidate, 334
Diazepam, 31
Diclofenac, 408
Diltiazem, 35
Diazoxide, 38
Discriminant analysis, 233
Disposition, 120
Dissociation constant, 12, 70, 97
Dissolution, 196
Distribution, 120
DMPK, 111
DNA, 291
  gyrase, 47
  helicase, 295
  Okazaki fragments, 295
  recombinant technologies, 315
  replication, 294
  restriction mapping, 306
  sequencing, 319
  transcription, 296
Docking programs, 246
Dofetilide, 39
Drug development, 185
Drug efficacy, 16
Drug selectivity, 9
DuP, 697, 409

EGFR, 60
Electronic properties, 207, 219
Elimination, 120
Enalopril, 86
Encoded libraries, 366
Endothelin antagonists, 346
Enkephalins, 343
Enthalpy of binding, 81
Enzymes, drug metabolising, 140
Enzyme inhibitors, 64
  aldose reductase, 74
  assays, 69
  $\beta$-lactamase, 79
  competitive, 75
  hydroxyethylamine, 398
  irreversible, 77
  minimum inhibitory sequence, 399
  non-competitive, 74
  slow binding, 77
  substrate analogue, 67
  tight binding, 77
  transition state analogue, 67
  uncompetitive, 75
Erythromycin, 43
Excretion, 120
Extraction ratio, 127

Flavin monooxygenase, 148

FLIPR, 104
Fluorescence-based assays, 95
Fluoroquinolones, 47
Follow-up, 335
Free energy of binding, 209
FRET, 96
$Fu_B$, 131
Functional activity, 103
$Fu_T$, 131

Gastrointestinal absorption, 210
Gemafloxacin, 51
Gene expression, 296
Gene
  probes, 308
  labelling, 309
  libraries, 315
  libraries – screening, 317
Genetic algorithm, 238
Genetic code, 300
Genetic toxicology, 156
Genomics, 302
Gentamycin, 45
Glibendamide, 38
Glucoronidation, 150
Glutathione, 152
Granisetron, 32
Graphics programs, 243
Grepafloxacin, 51
GRID, 236
GR127935, 105, 108
GTP$\gamma$S, 104

Haematotoxicity, 172
Half-life, 124, 133, 174
Hammett equation, 217
Hansch equation, 219
Hepatotoxicity, 171
High throughput screening, 92, 330
HIV, 397
  protease, 397
HMGCoA, 83
Homology modelling, 246, 251
Human genome, 293, 324
Hydrogen bonding, 203, 248
Hydrophobicity, 200, 219, 225, 248

Ibuprofen, 146
$IC_{50}$, 70, 97
Indinavir, 403
Indomethacin, 407
Inosine monophosphate (IMP) dehydrogenase, 73
Intellectual property, 188, 262
Intracellular targets, 42
Intrinsic activity, 15
Inventions, 264
Inventive step, 266
Inventors, 271
Inverse agonism, 21, 102
Ion channels, 25
  GABA, 29
  glutamate, 29
  5-$HT_3$, 32, 339
  inward rectifiers, 28
  intracellular, 28
  KCNQ, 37
  L-type calcium, 35
  Ligand gated, 3, 28, 29
  nicotinic acetylcholine, 29
  N-type calcium, 35
  ryanodine, 33
  sodium, 27, 35
  voltage-gated, 28, 34, 37
Ionisation constants, 205
  estimation, 206
Isothermal titration calorimetry, 80

*k*-nearest-neighbour, 232
Kainic acid, 31
Kappa opioid antagonist, 380
$K_d$, 97
$K_i$, 97
Kinases
  ATP-competitive inhibitors, 59
  EGFR-tyrosine, 60, 74, 84
  mitogen-activated protein (MAP), 58
  p38 MAP, 61
  protein, 57

protein tyrosine, 57
serine-threonine, 57

L-741671, 422
Lamotrigine, 36
Lead identification, 330
Lead optimisation, 332
Lignocaine, 36
Linezolid, 45
Lineweaver–Burk plots, 68
Lipophilicity, 200, 389
Lisonopril, 86
Log D, 200
Log P, 200
estimation, 203
information content, 202
measurement, 202
Lopinavir, 403

Macrolides, 43
Manufacturing, 189
Maximum absorbable dose, 196
Maximum tolerated concentration, 122
Membrane proteins, 252
Mephobarbital, 148
Mepyramine, 7
Metabolism, 120, 138
Phase I, 139
Phase II, 139, 150
Me-too, 328
Me-better, 328
Methoxyphenamine, 146
Methylprednisolone, 348
Metoclopramide, 328
Mevinolin, 83
Mianserin, 146
Michaelis complex, 66
Michaelis–Menten equation, 11, 14, 67
Minimum effective concentration, 122
Mix and split, 361
MK-869, 421
MK-966 – see rofecoxib
Moclobemide, 148
Molar refractivity, 225
Molecular biology, 291
Molecular connectivity, 223
Molecular descriptors, 254
Molecular modelling, 385
Molecular size, 204
Monoamine oxidase, 149
Morphine, 343
Moxifloxacin, 51
MPTP, 190
MULTICASE, 160
Multiple linear regression, 219, 226
Multicomponent condensations, 371
Mutagens, 169
Mycophenolic acid, 73

Nabumetone, 348
Nalidixic acid, 48
Native tissues, 101
Naproxen, 146
Negative efficacy, 22
Nelfinavir, 403
Nephrotoxicity, 172
Neural networks, 239
Neurokinin receptors, 416
Nicotine, 29
Nifedipine, 35
NMDA, 29
NMR, 178, 247
No effect dose, 174
Nociception, 116
Non-linear mapping, 231
Norfloxacin, 48
Novelty, 265
NSAIDs, 407
NSL-95031, 345
Nuclear receptors, 52
Nucleic acids, 291
blotting, 307
electrophoresis, 305
extraction, 304
hybridisation, 307
Nucleotide sequencing, 319
automated fluorescent, 321

Ofloxacin, 49
Omeprazole, 148, 348
Ondansetron, 19, 32

$pA_2$, 20
Parkinson's disease, 115
Paroxetine, 114, 82
Partial agonists, 15
Partial least squares, 228
Partition coefficients, 200, 218
Patents, 261, 357
  litigation, 277
  internet sites, 285
PCR – see polymerase chain reaction
PCT application, 273
PD-153035, 59
Pharmaceutical development, 189
Pharmacodynamic assays, 111
Pharmacokinetics, 118
  estimation, 212
Pharmacophore 338, 419
Phenylbutazone, 409
Phenytoin, 36
Phosphodiesterase-4 inhibitor, 380
Physicochemical, properties, 195
Piperoxan, 7
Polymerase chain reaction, 309
Polymorphism, 186
PPAR, 55
PPAR$\gamma$ agonist, 380
Pregnane X receptor, 144
Primary screen, 92
Principle component analysis, 228
Priority application, 272
Procaine, 36
Prodrugs, 347
Promoter, 296
Propranolol, 147
Prosecution, 276
Proteases, 65
Proteins, 293
  synthesis, 297
  translation, 299
Protein kinases, 57
Protein modelling programs, 245
Protein sequence alignment, 251
Protein synthesis inhibitors, 42
Proteomics, 177
Prozac, 261

QSAR 216, 253
Quinolones, 47
Quinupristin, 44

Radiofrequency tagging, 367
Radioligand binding, 94
Raltitrexed, 82
Ranitidine, 7
Receptors, 2
  agonists, 5
  antagonists, 18
  aquaporins, 39
  $\beta$-adrenergic, 5
  connexins, 39
  G-protein coupled 3, 104
  hERG, 39
  histamine, 7
  5-hydroxytryptamine, 8
  5-$HT_{2C}$ antagonists, 382
  intracellular, 3, 52
  inverse agonists, 21
  muscarinic, 8
  nuclear, 52
  partial agonists, 15
  pregnane X, 144
  somatostatin-1, 380
  subtypes, 6
  terminal autoreceptor, 109
  Tyrosine kinase-linked, 3
Recombinant systems, 101
Regression coefficients, 226
Regulatory affairs, 189
Regulatory exclusivity, 262
Renin, 85
Reporter gene assays, 107
Retigabine, 38
RGD mimetics, 345
Rhodopsin, 252
Ritonavir, 146, 403
RNA, 291

polymerase, 297
processing, 297
Ro 31-8959 – see saquinavir
Rofecoxib, 408
Rosuvastatin, 84
Route of manufacture, 191
Routes of administration, 121
Roxithromycin, 43
Rule of five, 255

Safety assessment – see toxicology
Saquinavir, 397
SAR, 216
SAR by NMR, 247
SB-203580, 60
SB-206553, 384
SB-214857, 345
SB-221284, 390
SB-228357, 292
SB-234985, 391
SB-236057, 93, 100, 105, 112
SB-243213, 393
SB-271046, 331
SB-452466, 95
SB-477790, 95
SC-64762, 347
SC-67655, 347
Schild plot, 20
Screening cascade, 333
Selectivity, 345
Single nucleotide polymorphs – see SNPs
Sitafloxacin 51
SKF 107260, 345
SNPs, 303
SNX-111, 35
Soft drugs, 350
Solid phase synthesis, 364
on-bead monitoring, 366
Solubility, 196
calculation, 199
measurement, 198
Somatostatin-1 antagonist, 380
Sparfloxacin, 49
Split and mix – see mix and split
SSRI, 114, 382
Standard deviation, 352
Standard error of the difference, 352
Standard error of the mean, 352
Statistics, 226, 352
*F*, 227
Steric constants, 219
Structural database, 247
Streptogramins, 43
Structure-based design, 81
Substance P, 415
Sufficiency, 267
Sulfation, 151
Sulfaphenazole, 146
Supply route, 186
Support bound reagents, 370
Synercid, 44

Tacrine, 147
Tachykinins, 416
Target validation, 328, 331
Telomeres, 293
Terfenadine, 39
$T_{max}$, 123
Thrombin inhibitor, 379
Thymidylate synthase, 80
Tolbutamide, 146
Tomudex®, 80, 82
Topoisomerase (bacterial) inhibitors, 47
Tosufloxacin, 49
Toxicogenomics, 176
Toxicokinetics, 156, 174
Toxicology, 155, 189
aneugens, 169
clastogens, 169
cytotoxicity, 163, 166
genetic, 156
haematotoxicity, 172
hepatotoxicity, 171
mutagens, 169
nephrotoxicity, 172
high throughput, 165
*in silico* predictors, 159
*in vitro* systems, 161

Toxicology *(cont'd)*
*in vivo* studies, 173
target organ, 170
Transcription, 296
Translation, 299
Triclosan, 77
TRIPS agreement, 279
Trovafloxacin, 49

UCL-1848, 33
Ugi reaction, 372
Utility, 268

Vioxx – see rofecoxib
Virtual screening, 248
Volume of distribution, 124, 129

Warfarin, 145

X-ray crystallography, 246
Xanthine oxidase, 150

ZD1839, 60, 85
ZD4522, 84
ZD5522, 74